Chemical Modelling
Applications and Theory

Volume 3

Specialist Periodical Reports

RS•C

Systematic and detailed review coverage in major areas of chemical research

This series provides a unique service for the active research chemist, offering regular, critical in-depth accounts of progress in particular fields of chemistry

- Free site-wide access to the electronic version available for print purchasers of individual titles – including access to back volumes published from 1998 to the present day
- Individual chapters available online on a pay-to-view basis
- Contents pages can be viewed online free of charge
- Commissioned and overseen by an Editorial Board comprising renowned experts in each field

Editorial Board

Professor Bruce C Gilbert **(Chair)**, *University of York, UK*
Professor David W Allen, *Sheffield Hallam University, UK*
Dr Graham Barrett, *Oxford, UK*
Dr George Davidson, *University of Nottingham, UK*
Dr John Davies, *University of Wales, Swansea, UK*
Dr Michael Davies, *The Heart Research Institute, Sydney, Australia*
Dr Ian R Dunkin, *University of Strathclyde, UK*
Professor Michael Green, *University of Bristol, UK*
Dr Alan Hinchliffe, *UMIST, Manchester, UK*
Dr Damien M Murphy, *University of Cardiff, UK*
Professor J Jerry Spivey, *Louisiana State University, Baton Rouge, USA*
Professor John C Tebby, *Staffordshire University, Stafford, UK*
Professor Graham Webb, *formerly University of Surrey, UK*

Available titles:

Amino Acids, Peptides and Proteins
Carbohydrate Chemistry
Catalysis
Chemical Modelling
Electron Paramagnetic Resonance
Nuclear Magnetic Resonance
Organometallic Chemistry
Organophosphorus Chemistry
Photochemistry
Spectroscopic Properties of Inorganic and Organometallic Compounds

For further details, visit **www.rsc.org/spr**

Orders & further details Sales & Customer Care Dept · Royal Society of Chemistry
Thomas Graham House · Science Park · Milton Road · Cambridge · CB4 0WF · UK
T +44(0)1223 432360 · F +44(0)1223 426017 · E sales@rsc.org
Or visit our websites: www.rsc.org and www.chemsoc.org
Registered Charity No. 207890

advancing the chemical sciences

A Specialist Periodical Report

Chemical Modelling
Applications and Theory
Volume 3

A Review of the Literature Published between June 2001 and May 2003

Senior Reporter
A. Hinchliffe, *Department of Chemistry, UMIST, Manchester, UK*

Reporters
D. Babić, *Rudjer Bošković Institute, Zagreb, Croatia*
D.J. Klein, *Texas A&M University, Galveston, Texas, USA*
R.A. Lewis, *Lilly Research Centre, Surrey, UK*
D. Pugh, *University of Strathclyde, Glasgow*
T.E. Simos, *Democritus University of Thrace, Athens, Greece*
M. Springborg, *University of Saarland, Saarbrücken, Germany*
B. Sutcliffe, *Universite Libre de Bruxelles, Brussels, Belgium*
K.P. Travis, *University of Sheffield, Sheffield, UK*
N. Trinajstić, *Rudjer Bošković Institute, Zagreb, Croatia*
J. von Knop, *The Heinrich-Heine University, Düsseldorf, Germany*
S. Wilson, *Rutherford Appleton Laboratory, Didcot, Oxfordshire*

advancing the chemical sciences

> If you buy this title on standing order, you will be given FREE access to the chapters online. Please contact sales@rsc.org with proof of purchase to arrange access to be set up.
>
> Thank you

ISBN 0-85404-264-4
ISSN 1472-0965

A catalogue record for this book is available from the British Library

© The Royal Society of Chemistry 2004

All rights reserved

Apart from any fair dealing for the purpose of research or private study for non-commercial purposes, or criticism or review as permitted under the terms of the UK Copyright, Designs and Patents Act, 1988 and the Copyright and Related Rights Regulations 2003, this publication may not be reproduced, stored or transmitted, in any form or by any means, without the prior permission in writing of The Royal Society of Chemistry, or in the case of reprographic reproduction only in accordance with the terms of the licences issued by the Copyright Licensing Agency in the UK, or in accordance with the terms of the licences issued by the appropriate Reproduction Rights Organization outside the UK. Enquiries concerning reproduction outside the terms stated here should be sent to The Royal Society of Chemistry at the address printed on this page.

Published by The Royal Society of Chemistry,
Thomas Graham House, Science Park, Milton Road,
Cambridge CB4 0WF, UK

Registered Charity Number 207890

For further information see our web site at www.rsc.org

Typeset by Charlesworth, Huddersfield, West Yorkshire, UK
Printed by Athenaeum Press Ltd, Gateshead, Tyne and Wear, UK

Preface

Dear Reader,

Welcome to Volume 3 of this SPR. It follows the ground rules that I set out in Volume 1; colleagues reporting on 'new' topics are asked to give the rest of us an easily understandable historical perspective, together with their critical comments on recent developments. Colleagues reporting on continuing topics are simply asked to give a critical review of the literature for the period under consideration.

First of all, I have to thank the regular contributors, whose names and topics you will readily identify from the Contents. It is a pleasure to work with such a nice bunch of people.

I believe that the SPR is fast approaching equilibrium. It is a dynamic equilibrium, with colleagues leaving and joining, so first I should say that Jan van Knop has joined Nenad Trinajstić's team to review enumeration. Second, it is a real pleasure to welcome Karl Travis, who has taken over the liquid state mantle from David Heyes.

Finally I want to mention my old friend Brian Sutcliffe, who writes on the new topic of 'Calculations of the Vibration-Rotation Spectra of Small Molecules'.

The idea of cyclical time has arisen in many different civilizations. The ancient Chinese believed in a cyclical interplay between the opposing cosmic principles of yin and yang, and calculated a cycle of 23 639 years.* They might have had a point for Brian and I both contributed articles to the precursor SPR *Theoretical Chemistry*, which was edited by Richard Dixon and Colin Thomson (although not 23 639 years ago).

It only remains for me to add my usual comment, that there are still a few gaps in the coverage. I'm sure you will have your own ideas what is good and what is bad with this SPR. Rather than grumbling to your colleagues at coffee-time, why not volunteer your expertise? I am always willing to listen to constructive suggestions and can be reached at Alan.Hinchliffe@umist.ac.uk

Better still, I am always willing to consider new topics for inclusion in the SPR. It won't make you a (£ sterling) millionaire, but it gives you the opportunity to promote your subject.

Adam Hinchliffe

* 'Achilles in the Quantum Universe' by Richard Morris, Souvenir Press, London, 1997.

Contents

Chapter 1 Calculations of the Vibration–Rotation Spectra of Small Molecules   **1**
By B.T. Sutcliffe

1 Introduction   1
2 History   2
3 Symmetry   9
4 The Eckart-Watson Hamiltonian and its Context   12
    4.1 The Permutational Symmetry of the Eckart-Watson Hamiltonian   14
5 The General Form of a Tailor-made Hamiltonian   17
    5.1 The Permutational Symmetry in the General Form of the Hamiltonian   23
6 Computational Considerations   25
    6.1 Perturbational Computations   29
    6.2 Variational Computations   36
References   41

Chapter 2 Computer-aided Drug Design 2001–2003   **45**
By Richard A. Lewis

1 Introduction   45
2 ADME/Tox and Druggability   45
    2.1 Metabolism by Cytochrome P450   46
    2.2 Human Ether-a-go-go-related Gene K+ Channel   47
    2.3 What makes a Compound Drug-like?   47
    2.4 General Models   49
3 Docking and Scoring   49
    3.1 Validation Protocols   50
    3.2 Entropy   51
    3.3 Pharmacophore and Local Docking Schemes   51
    3.4 Protein Flexibility   52
    3.5 Docking and Virtual Screening   53

Chemical Modelling: Applications and Theory, Volume 3
© The Royal Society of Chemistry, 2004

4	De Novo	54
5	3D-QSAR	55
6	Pharmacophores	56
	6.1 Conformational Analysis	56
7	Library Design	57
8	Cheminformatics and Data Mining	59
	8.1 Data Mining	59
	8.2 Similarity and Descriptors	60
9	Inverse QSAR and Automated Iterative Design	61
10	Structure-based Drug Design	61
	10.1 G-Protein Coupled Receptors (GPCRs)	62
	10.2 A Case Study	63
11	Conclusions	63
	References	63

Chapter 3 Density Functional Theory   **69**
By Michael Springborg

1	Introduction	69
2	Theoretical Foundations	70
3	Structure and Energies	74
	3.1 Ionization Potentials of Nickel–benzene Clusters	74
	3.2 Brønsted Acidity of Some Zeolites	76
	3.3 Van der Waals Interactions	78
	3.4 Structure and Energetics for N-Acetyl-L-Glutamate-N-Methylamide	79
	3.5 Photodissociation of Triplet Acetaldehyde	80
	3.6 Vibrations of Benzimidazole	82
	3.7 Chemical Reactions Involving Hydrogen Bonds	84
	3.8 Metal–Sulphur bonds for Sulphur on the Au(111) Surface	85
	3.9 NO_x on MgO	88
	3.10 Rotational Stability of Substituted Acetanilides	89
	3.11 The Reaction of Methylformate with Ammonia	89
	3.12 The Rearrangement of Azulene to Naphthalene	92
	3.13 Dissociation of Azomethane	93
	3.14 Reactions Between Transition Metals and Ammonia	94
	3.15 Dynamics of Chemical Reactions	95
	3.16 Hydrogen Bonds	95
	3.17 DNA Base-stacking Interactions	97
	3.18 Vibrations of Fe_2CO	97
	3.19 Anharmonic Vibrational Modes in Adenine	98
	3.20 Cumulenes and Polyynes	99
4	Orbitals and Densities	100
	4.1 Conductivity in DNA	100

		4.2	Momentum-space Densities in Ethane	102

        4.2 Momentum-space Densities in Ethane   102
        4.3 Electron Density in [FeCo(CO)$_8$]$^-$ 103
    5 Excitations   104
        5.1 Bandgap in Molecular Crystals   104
        5.2 Binding Energies of Electrons   106
        5.3 Band Gap in Conjugated Oligomers   108
        5.4 (Hyper)polarizability of Si_4 109
        5.5 Electronic Absorption in Polycyclic Aromatic
             Hydrocarbons   110
        5.6 Nonadiabatic Processes   111
        5.7 NMR Chemical Shifts of Benzoxazine Oligomers   111
        5.8 NMR Shielding Constants   112
    6 Getting Further Information with Density-functional
       Calculations   113
        6.1 Classification of Reactions   113
        6.2 Hydrogen Bond Descriptors   115
    7 Limitations and Perspectives   116
        7.1 Polyenes and Current-density-functional Theory   116
        7.2 Anions and Exact-exchange Methods   119
        7.3 The Long-ranged Behaviour of Exchange
             Interactions   120
        7.4 Other Improvements   122
    8 Conclusions   122
       References   124

Chapter 4 Combinatorial Enumeration in Chemistry   126
By D. Babić, D.J. Klein, J. von Knop and N. Trinajstić

    1 Introduction   126
    2 Current Results   126
        2.1 Isomer Enumeration   126
        2.2 Kekulé Structure Count   135
        2.3 Counting Walks   138
        2.4 Combinatorial Measures of Molecular Complexity   144
        2.5 Other Enumerations   154
    3 Concluding Remarks   157
       References   159

Chapter 5 Photo-reduction and -oxidation   171
By Andrew Gilbert

    1 Introduction   171
    2 Electric Field Related Properties   172
        2.1 High Level Ab Initio and DFT Calculations   172
        2.2 Semi-empirical Calculations   190
        2.3 Solvent and Other Environmental Effects   195

		2.4	Polymers	197
		2.5	Dendritic Structures	199
		2.6	Octupolar Molecules	200
		2.7	Crystals	201
		2.8	Clusters	202
		2.9	Theoretical Developments	203
	3	Magnetic Field Response Functions in Diamagnetic Molecules		205
		3.1	Magnetizability and Nuclear Shielding	205
		3.2	Higher Order Response Functions – The Cotton-Mouton Effect	208
		References		210
Chapter 6	**Simulation of the Liquid State**			**217**
	By Karl P. Travis			
	1	Introduction		217
	2	Transport Properties		218
		2.1	Theories of Diffusion	218
		2.2	Computational Methodology	222
		2.3	Bulk Liquid Transport Properties	224
		2.4	NEMD	227
	3	Phase Equilibria		229
	4	Supercooled Liquids and Glasses		231
		4.1	Phenomenology	232
		4.2	Structural Models for Supercooled Liquids	235
	5	Confined Liquids		240
		5.1	Structure	240
		5.2	Phase Equilibria	241
		5.3	Glass Transition	243
		5.4	Diffusion	244
		5.5	Rheology	249
		5.6	Chemical Reactions	251
		5.7	Water Adsorption	252
		5.8	Applications of Density Functional Theory	252
	6	Water and its Solutions		253
	7	Mesoscale Simulations		255
		7.1	Dissipative Particle Dynamics	256
		7.2	Lattice Boltzmann	257
	8	Simulation Methodology		258
		8.1	Reverse Monte Carlo	258
		8.2	New Monte Carlo Algorithms	259
		8.3	Intermolecular Potentials	260
		8.4	Multiscale Methods	262
		8.5	Miscellaneous Developments	263
		References		263

Chapter 7 Numerical Methods in Chemistry — 271
By T.E. Simos

1 Introduction — 271
2 Multi Derivative Methods — 273
   2.1 Stability and Phase-lag Analysis of Multi-derivative Methods — 273
   2.2 A New Family of Multi-derivative Methods — 274
   2.3 A New Family of Multi-derivative Methods with Minimal Phase-lag — 275
   2.4 Computational Implementation — 276
   2.5 Numerical Illustrations — 280
   2.6 Remarks and Conclusions — 282
3 Symplectic Methods for the Numerical Solution of the Radical Schrödinger Equation — 283
   3.1 Introduction — 283
   3.2 Symplectic Integrators – Basic Theory — 283
   3.3 Construction of Symplectic Integrators — 284
   3.4 Development of the New Methods — 285
   3.5 Numerical Examples — 288
   3.6 Remarks and Conclusions — 291
4 Numerical Solution of the Two-dimensional Schrödinger Equation — 292
   4.1 Introduction — 292
   4.2 Partial Discretisation of the Two-dimensional Equation — 292
   4.3 Application of Symplectic Methods — 293
   4.4 Application of Numerov-type Methods — 299
   4.5 Remarks and Conclusions — 304
5 General Comments on the Bibliography of the Numerical Methods in Chemistry — 305
   Appendix A — 313
   Appendix B — 317
   Appendix C — 340
   Appendix D — 345
   Appendix E — 346
   Appendix F — 349
   Appendix G — 366
   References — 377

Chapter 8 Many-body Perturbation Theory and Its Application to the Molecular Structure Problem — 379
By S. Wilson

1 Introduction — 379

2 Diagrammatic Many-body Perturbation Theory of
   Molecular Structure Including Nuclear and Electronic
   Motion   380
   2.1 The Total Molecular Hamiltonian Operator   381
   2.2 The Hartree-Fock Theory of Nuclei and Electrons   383
   2.3 The Many-perturbation Theory of Nuclei and
        Electrons   384
   2.4 The Diagrammatic Perturbation Theory of Nuclei
        and Electrons   389
   2.5 Prospects   398
3 Diagrammatic Many-body Perturbation Theory of
   Molecular Electronic Structure: Low Order Approximants   399
   3.1 Summation Approximants   399
   3.2 Feenberg Scaling   401
   3.3 Digression: Scaled Many-body Perturbation Theory
        and Systematically Extended Basis Sets   404
   3.4 Padé Approximants   405
   3.5 Quadratic Padé Approximants   406
   3.6 Prospects   408
4 Diagrammatic Many-body Perturbation Theory of
   Molecular Electronic Structure for Larger Systems   409
   4.1 Local Correlation Methods   409
   4.2 Linear Scaling Correlation Methods   410
   4.3 Local "MP2" Methods   410
   4.4 Splitting of the Coulomb Operator   411
   4.5 Multipole Expansion of Long Range Integrals   412
   4.6 Density Fitting Approximations   414
   4.7 Prospects   416
5 Diagrammatic Many-body Perturbation Theory of
   Molecular Electronic Structure: A Review of
   Applications   416
   5.1 Incidence of the String "MP2" in Titles and/or
        Keywords   416
   5.2 Comparison with Other Methods   417
   5.3 Synopsis of Applications of Second Order
        Many-body Perturbation Theory   418
   5.4 Prospects   420
6 Summary and Prospects   420
   References   421

1
Calculations of the Vibration-Rotation Spectra of Small Molecules

BY B.T. SUTCLIFFE

1 Introduction

It would, I believe, be widely agreed that the modern theory of molecular spectra began with publication by Carl Eckart in 1935 of his paper *Some Studies Concerning Rotating Axes and Polyatomic Molecules*.[1] It would probably also be widely agreed that the apogee of this work occurred in 1968 when James K. G. Watson[2] published *Simplification of the molecular vibration-rotation hamiltonian* which put Eckart's classical mechanical form into a proper quantum mechanical one. This leads to the wave mechanical problem for molecular vibration-rotational motion specified by what we shall call the *Eckart-Watson* Hamiltonian.

This report begins with an account of the theories of molecular spectra that preceded the work of Eckart and the interpretation of spectra that followed his paper during the nineteen forties and fifties. This discussion will involve some consideration of diatomic molecules but they will not subsequently be discussed. So this article is concerned entirely with polyatomic molecules and, in particular those that become linear, only in somewhat excited states. The initial historical discussion, it is hoped, will put the computational work that began in the nineteen seventies into a proper context. During the nineteen eighties and nineties it will be seen that two strands develop in the computational study of molecular spectra. The first is an essentially perturbation theoretic approach, confined almost entirely to the Eckart formulation. The second is a variation theoretic approach which, although sometimes using the Eckart formulation, has found greatest use in formulations using Hamiltonians specifically constructed to describe particular molecules. In referring to such a class of Hamiltonians, they will be called *tailor-made*.

The aim of this this report is to provide an informative context in which relevant examples of computational work on the spectra of small molecules can be presented in a way that, it is hoped, is balanced, fair and comprehensible to the non-expert reader. It is not aimed to provide a comprehensive survey of the literature, since that can nowadays be done in an effective and timely fashion with the aid of facilities on the Internet. Rather it is aimed to provide representative examples of work so that the reader can gain some feeling for what has been done, what is being done and, perhaps, what might be done.

2 History

In December 1926 the US National Research Council published in its Bulletin[3] a Report of the Committee on Radiation in Gases entitled *Molecular Spectra in Gases*. The members of the Committee were Edwin C. Kemble, Raymond T. Birge, Walter F. Colby, F. Wheeler Loomis and Leigh Page. The coordinating editor seems to have been Kemble who, in his *Preface* thanks Professor R. S. Mulliken "whose suggestions and criticisms have been numerous and invaluable".

To put this report in context. Heisenberg's first paper on "the new quantum mechanics" had appeared late in 1925, as had Dirac's first paper, too. Schrödinger's first paper on wave mechanics appeared during February of 1926 and others followed throughout the year. The report was thus written a time of real flux in the underlying theory and its theoretical aspects give testimony to a somewhat uneasy co-existence between the old and new quantum theories, with a strong overlay of classical mechanics, in the theory of molecular spectra.

In his introductory chapter Kemble says that

> the foundation for the present theory of band spectra was laid in 1892 by the older Lord Rayleigh[4] when he pointed out that if an oscillator which at rest emits and absorbs light of frequency v_0 is caused to rotate with a frequency v_r about an axis perpendicular to the axis of vibration, then it should emit and absorb in about equal proportions the two frequencies $v_0 + v_r$ and $v_0 - v_r$.

and he goes on to say:

> As early as 1904 Drude[5] from the study of the dispersion of various crystals was led to the conclusion that the infrared absorption spectra and emission bands of most substances, including gases, must be due to the vibrations of electrically charged atoms and molecules rather than to the oscillations of electrons inside the atoms, and in 1912 Niels Bjerrum[6] called attention to the fact that the breadth of the as yet unresolved infrared absorption bands of gases was of the order of magnitude to be expected from the superposition of molecular rotations on molecular vibrations.

These considerations were, at the time being described here, effective only for diatomic molecules and in this case it follows from the Maxwell-Boltzmann law and the classical mechanics of a rigid rotor that a group of molecules in thermal equilibrium at the temperature T the number having rotational frequencies between v_r and $v_r + dv_r$ is

$$dn = \frac{4\pi^2 nI}{kT} e^{\frac{-2\pi^2 I v_r^2}{kT}} v_r dv_r$$

where I is the moment of inertia and k is Boltzmann's constant. The quantity dn/dv_r should be proportional to the absorption coefficient for either of the two frequencies $v_0 \pm v_r$, and so the band should be a doublet with a splitting

$$\Delta v = \frac{1}{\pi}\sqrt{\frac{kT}{I}}$$

By 1913, however, experimental techniques had advanced sufficiently for the predicted doublet to be observed and the calculated moments of inertia led to what Kemble justly observed, were "plausible values" for bond lengths. For example the bond length of CO was estimated to be 1.14Å while that of HCl was put at 1.34Å.[7] A portion of the near infrared spectrum of of HCl at this level of resolution is given as Figure 30 in Chapter II 2 of ref. 8. By 1914 Bjerrum had developed a theory for CO_2,[9] that treated the vibrations of the system in terms of atoms moving in a potential with a minimum at an isosceles triangle geometry. Among the potentials that he tried were a central field one and a valence field one, this last expressed in terms of a pair of bond oscillators coupled to a bond angle bending oscillator. This was an extremely important step for it introduced a molecular model into molecular spectroscopy. The model idea was that the infrared spectrum of a molecule could be understood if the molecule was looked upon as a vibrating-rotating entity whose vibrations could be interpreted in terms of a collection of point masses moving in a potential with a minimum at a particular geometry with the whole system undergoing free rotation. If a way could be found of attributing particular spectral features to the rotational motion, then it would be possible to establish the moments of inertia of the molecule as a nearly rigid body and from these moments of inertia to determine the geometry of the potential minimum.

Almost simultaneously with these developments however and at about the same time that Bohr's quantum theory came on the scene, further instrumental advances led to the discovery that the diatomic infrared bands were not really continuous but were resolvable into fine structure. A diagram of a portion of the near infrared spectrum of HCl showing this fine structure can be found as Figure 32 in Chapter II 2 of ref. 8. As Kemble remarks:

> It was immediately evident that the existence of this fine structure must be regarded as conclusive evidence for the of the quantization of the rotational motion of the molecules.

In fact the possibility of such quantization had been suggested in 1911 by Nernst and in 1912 by Lorentz but it was Ehrenfest in 1913 who suggested that diatomic quantization should be in multiples of $\frac{1}{2}hv_r$, with

$$v_r = \frac{mh}{4\pi^2 I} \qquad m = 0, 1, 2\ldots$$

and this gave rise to a re-interpretation of the diatomic spectra which, although it yielded a moments of inertia really quite close to those obtained from the older theory, was by 1916 providing the generally accepted way of interpreting diatomic spectra. Although the molecular model developed by Bjerrum was still central to the understanding of molecular spectroscopy, by 1920, the model had been incorporated into the old quantum theory and it is discussed, chiefly in relation to diatomics, in Page's contribution (Chapter II) to ref. 3 and somewhat more generally in Kemble's Chapter VII in the same volume. It is also discussed in the context of band spectra in a textbook by Baly[10] in 1927. In his chapter on emission band spectra, Baly pays particular tribute to the work of Kratzer in describing a rotating non-harmonic oscillator and also to his recognition, simultaneously with that of Loomis in 1920, that the infrared spectrum of HCl could be interpreted as due to the vibrations of two distinct molecular species, corresponding to the 35 and 37 isotopes of chlorine.

The new quantum theory proved immediately attractive to workers in the field and in August 1926, Dennison[11] published a paper called *The Rotation of Molecules* in which Heisenberg's form of quantum mechanics was used to describe the motions of a rigid rotor and of a symmetric top. In later developments however, it was Schrödinger's form of quantum mechanics, generally called wave mechanics, that was the preferred basis for theoretical descriptions. The molecular model in its wave mechanical form is described with great elegance and economy in a 1931 review by Dennison.[12] In this review the wave mechanical form of the molecular model is realised by taking the Hamiltonian for the whole system as a sum of one for the internal motions and one to describe the rotations. The internal motions are assumed to be those of s atomic nuclei which have a possible equilibrium position which is non-linear. There will be $n = 3s - 6$ independent displacement coordinates chosen so that they are infinitesimal when compared with the normal distances between nuclei. It is then possible to choose a linear combination of these coordinates as normal coordinates so that both the kinetic and potential energy operators are quadratic forms and so that the solution to the internal motion problem can be written as a product of simple harmonic oscillator functions, one for each normal coordinate. The rotational motion is described by treating the molecule as a rigid body and in terms of its principal moments of inertia. The rotational Hamiltonian consists therefore simply of a kinetic energy term of the form

$$\mathsf{H}_{rot} = \frac{\mathsf{L}_x^2}{2I_{xx}^0} + \frac{\mathsf{L}_y^2}{2I_{yy}^0} + \frac{\mathsf{L}_z^2}{2I_{zz}^0} \tag{1.1}$$

where the L_α are the components of the total angular momentum operator and $I_{\alpha\alpha}^0$ are the principal moments of inertia and are regarded as constants. If the moments of inertia are all different, the most general case, this is the Hamiltonian for an asymmetric top and its energy levels as functions of the principal moments of inertia were already well known by the time the review was written, as were the selection rules for electric dipole transitions between the levels. And spectroscopists were able to identify segments of spectra with rotational motion and from the observed lines were often able to determine moments of inertia. By using the model of a rigid framework of point masses, it was then possible to determine a molecular geometry yielding bond lengths and angles. The structures so determined made very good sense in classical chemical terms. It should be remembered that spectra that can be associated solely with rotational transitions did not really become accessible until after the development of microwave sources during World War II. And so what is being spoken of here actually arises from the analysis of rotational fine structure on vibrational bands.

The Hamiltonian for the molecular model described in Dennison's review is *sui generis* and is not derived from any more general model. We shall call it the *Bjerrum* model when a short designation is required. In 1934 however, Eckart in a paper entitled *The Kinetic Energy of Polyatomic Molecules*[13] attempted to describe the motion of a non-rigid assembly of particles in such a way that the rotational, vibrational and coupling terms could be distinguished. Eckart worked in classical mechanics. At nearly the same time, Hirschfelder and Wigner were trying to do the same in wave mechanics.[14] Eckart actually concerned himself only with the kinetic energy and Hirschfelder and Wigner did not explicitly specify their potential choice. However their arguments are valid for any potential that is invariant under uniform translations and rotation-reflections of all the particle coordinates. When a brief designation of such a potential is required, it will be called *geometrical* for it depends only on the geometry of the particles and not on their position or orientation.

What they both did was to choose a frame, fixed or embedded in the molecule whose orientation is described by the handedness of the frame and three Eulerian angles. They actually chose the Eulerian angles with which the rotational motion is described, to define an orthogonal matrix that diagonalises the molecular inertia tensor. This yields moments of inertia and puts the Lagrangian into principal axis form, just the form appropriate to describe a rigid rotator in classical mechanics. However doing this yields a rotational Hamiltonian of the form

$$H_{rot} = \frac{I_{xx}L_x^2}{2(I_{zz}-I_{yy})^2} + \frac{I_{yy}L_y^2}{2(I_{xx}-I_{zz})^2} + \frac{I_{zz}L_z^2}{2(I_{yy}-I_{xx})^2} \qquad (1.2)$$

Here no rigid body assumptions have been made and the moments of inertia are not constants but functions of the internal coordinates. The operator is obviously not at all like the rigid rotor operator given above. Here the operator is divergent whenever two moments of inertia are the same. It is thus quite impossible to describe a symmetric top molecule in this formulation. It seemed to pose such a severe problem that Eckart observed in the abstract of his paper[13] that:

> The ordinary moments of inertia appear in the Lagrangian kinetic energy but these are replaced by other functions of the radii of gyration in the Hamiltonian. This throws doubt upon all molecular configurations assigned on the basis of empirical values of moments of inertia.

Indeed it turns out more generally, inspite of some heroic efforts by Van Vleck,[15] that the Hamiltonian so derived is largely ineffective in describing molecules in terms of their traditional geometrical structures and so it found no use in the elucidation of molecular spectra.

In his second paper,[1] which has already been spoken of above, Eckart still continued to use classical mechanics but began his development by considering the potential in the problem. In order to choose a set of axes fixed in the molecule, he took his potential to be such that, when described in the chosen axis system (frame), it could be expressed in terms of the (dependent) cartesian coordinates z_i referred to the centre of mass of the system and which took constant values a_i at the minimum of the potential. The bold face is used to denote a column matrix of three cartesian components. The coordinates must satisfy the three constraints

$$\sum_{i=1}^{s} m_i \mathbf{z}_i = 0 \qquad (1.3)$$

to be displacement independent. By expanding the potential about the minimum and considering how the the rotation-reflection invariance requirements may be satisfied in the expanded form while maintaining the displacement invariance requirements, he showed that his frame choice could be achieved if the $\mathbf{z}_i$ satisfied the further three constraints, using an obvious vector notation,

$$\sum_{i=1}^{s} m_i \vec{\mathbf{a}}_i \times \vec{\mathbf{z}}_i = \vec{0} \qquad (1.4)$$

provided that the $\mathbf{a}_i$ do not define a straight line.

The two conditions (1.3) and (1.4) are usually called the Eckart conditions (though sometimes also the *Sayvetz* conditions). They can therefore, simply be regarded as specifying a reference geometry or *framework* for the system. They are also often referred to as the conditions for fixing or embedding a coordinate frame in the body and a set of six such conditions must always be chosen even for a tailor-made Hamiltonian.

Although the derivation of the second Eckart condition arises by considering a potential expandable up to quadratic terms about a minimum potential energy geometry specified by the $\mathbf{a}_i$, the condition can be imposed on any Hamiltonian with a geometrical potential. It can thus be chosen with a potential defined by the coulomb interaction between charged particles, for example. The minimum energy requirement on the $\mathbf{a}_i$ geometry is of no consequence in the expression of the kinetic energy operator and is important in the potential energy only if it is wished to expand it in a Taylor series about the minimum. Of course to get normal coordinates such an expansion is essential, so the *physical* relevance of the choice is vitally important.

If the Eckart conditions are imposed on the classical kinetic energy operator then, provided that the displacements

$$\boldsymbol{\rho}_i(\mathbf{z}_i) = \mathbf{z}_i - \mathbf{a}_i$$

are sufficiently small, the rotational part of the Hamiltonian becomes exactly what was hoped for, namely (1.1) with the moments of inertia given by those arising from the principal axis form given to the geometrical figure defined by $\mathbf{a}_i$ specifying the positions of the mass points m_i. With sufficiently small displacements the terms coupling the rotation and internal (vibrational) parts of the kinetic energy become negligible. The potential can also be expanded in Taylor series about the reference geometry. The first order terms vanish and the second order terms are sufficient. The expansion can be accomplished in $n = 3s - 6$ normal coordinates which are themselves invariant under uniform translations and of any orthogonal transformations of the original particle coordinates. The model obtained from the Eckart approach matches exactly, in this approximation, the model that Dennison put into wave mechanical form. There is one slight ambiguity however. In his discussion of the problem Eckart always refers to the particles as "atoms". This usage follows that of Bjerrum and Eckart's work can be regarded as fixing classical mechanics firmly to the Bjerrum model of the molecule. However Dennison in his treatment of the Bjerrum model[12] refers to the particles as "nuclei", a somewhat more natural usage from the wave mechanical point of view that he is taking. In what follows we shall try to indicate which convention is used in any particular work but when speaking generally we shall refer simply to electrons and to heavy particles.

To achieve an agreed quantum mechanical form for the Eckart Hamiltonian took some time because of puzzlement about how properly to incorporate the jacobian that arises during the transformation from the laboratory defined cartesian coordinates to the Eulerian angles and internal coordinates in the Eckart frame. A full account of what is involved here can be found in Section 35b of Kemble.[16] An initial attempt by Wilson and Howard in 1936 was subjected to critical scrutiny by Darling and Dennison in 1940 and the agreed form of the quantum mechanical Eckart Hamiltonian that emerged is given as Equation (10) in Chapter 11 of the 1955 textbook by Wilson, Decius and Cross.[17] An expression for this Hamiltonian including electronic motion is given in the review by Nielsen[18] but in this work the centre of nuclear mass is treated as if it is the centre of mass and thus not quite all aspects of electronic structure are included.

In Chapter 1 of their book, Wilson, Decius and Cross specify the Bjerrum model in its Eckart form concisely and clearly. They say

> The model which will be used in this book consists of particles held together by certain forces. The particles, which are to be endowed with mass and certain electrical properties, represent the atoms and are to be treated as if all the mass were concentrated at a point. It is assumed that the atoms may be electrically polarized by an external electric field, ... and that they may or may not be permanently polarized .... Finally the atoms may possess an internal degree of freedom or nuclear spin which introduces certain symmetry restrictions.

> The forces between the particles may be crudely though of as weightless springs which only approximately obey Hooke's law .... This picture of the forces as springs ... is not sufficiently general for all cases. For example, it does not cover cases of restricted rotation about single bonds such as may occur in ethane.......
>
> The statement that the model obeys the laws of quantum mechanics is a essential part of its specification. However, since atoms are fairly heavy particles (compared to electrons), it will sometimes be true that classical mechanics when properly used gives results which are good approximations to those of quantum mechanics.
>
> Since the atoms of this model have been regarded as point masses with certain electrical properties, there is an apparent disagreement with the fact that many experiments require that atoms be made up of electrons and nuclei. It is possible to reconcile these two points of view. If the wave equation for a molecule made up of electrons and nuclei is set up, a procedure[19] exists whereby this equation may be separated into two equations, one of which governs the electronic motions and yields the forces between the atoms, whereas the other is the equation for the rotation and vibration of the atoms and is identical with the equation for the model adopted here. In principle, therefore, the forces between the atoms can be calculated *a priori* from the electronic wave equation, but in practice this is not mathematically feasible (except for H_2) ......
>
> This separation of the electronic motion and the nuclear motions is only an approximation which may break down in certain cases ... If there were no interaction between the two types of motion, there would be no Raman effect of any importance. However, the coupling is small for the lowest electronic state.

In Chapter 8 they also say a little more about the potential

> [I]t has been assumed that the potential energy function could be expanded in a power series involving the displacement coordinates, and that only the quadratic terms need be considered. Moreover, it was implied that the coefficients in this expression were known constants and that the problem to be solved was the determination of the vibration frequencies as functions of these constants. Actually the force constants have been determined *a priori* for only a few diatomic and very simple polyatomic molecules (J H Van Vleck and P. C. Cross,[20] have made the necessary quantum mechanical calculation for the H_2O molecule, for example.) so that in practice it is usually the reverse problem which is most important, that is, the determination of the force constants from the known vibrational frequencies.

This exposition can be regarded as defining for the purposes of calculation and fitting, the Eckart model of the spectroscopic molecule.

As noted in the quotation above from Chapter 1 of Wilson, Decius and Cross, the Eckart model does not fit all molecules of spectroscopic interest because it is not designed to describe systems with large amplitude internal motions. It was Sayvetz, following a suggestion by Eckart, who first undertook in 1939 to modify the Eckart conditions to describe systems with one large amplitude internal motion. What Sayvetz actually did was well expressed by Ezra[21]

> the nuclear dynamics are ... described in terms of the intuitive picture, introduced by Sayvetz,[22] in which the nuclei execute rapid, small amplitude (vibrational) motion about an 'equilibrium' configuration that is itself performing some sort of slow large-amplitude (internal or contortional[23]) motion as well as undergoing overall rotation.

In future when it is necessary to consider the Eckart approach as modified by Sayvetz it will be explicitly noted but since it is essentially a development, it will not be considered in the context of tailor-made Hamiltonians.

In 1950 Curtiss, Hirschfelder and Adler published the first of a series papers[24] with the title *The Separation of the Rotational Coordinates from the N-Particle Schroedinger Equation.* They were not really looking for a molecular Hamiltonian but rather for one that would describe atom-molecule and molecule-molecule collisions. They showed, using the techniques developed earlier in ref. 14, that they could write down a set of coupled differential equations to describe the internal motions of the system but instead of choosing the earlier embedding, which had proved so unfruitful, they made an explicit choice of standard configuration in terms of a sub-set of particle positions. They actually chose the embedding by using two particle positions: one defined the z–axis and the other the positive $x - z$ plane. The resulting equations were too cumbersome, even for a triatomic, to find any use at that time. But their work should properly be regarded as the first attempt at tailor-made forms of the Hamiltonian, though little of the subsequent work in this area actually refers to them.

Of course an observable will have the same value when exactly calculated from wave functions arising from any pair of molecular Hamiltonians sharing the same potential and having a common solution domain. At the most fundamental level, therefore, the choice of a coordinate system fixed in the molecule is unimportant. However we cannot have exact solutions to any realistic model Hamiltonian and we need approximate solutions, preferably ones that give rise to helpful pictures. We choose the embedded system to aid both calculation and, if possible, visualization.

3 Symmetry

At about the same time that theoretical models to explain spectra began to be constructed, the mathematical theory of invariants and the theory of invariance groups reached very high stages of development. It was natural therefore that attempts should be made to categorise the models used in explaining spectra using these techniques. As has been implied in what has been said already, this led to the classification of the invariances of the Bjerrum model as being those of the rigid asymmetric rotor and of a system of harmonic oscillators. However in order to achieve these invariances in a system of particles obeying the laws of mechanics, as has been seen, involves making a very specific choice of geometrical potential. In considering the invariances initially however, an otherwise unspecified geometrical potential will be assumed.

A Hamiltonian with a geometrical potential when expressed in a frame fixed in the laboratory is invariant under all uniform translations and under all orthogonal transformations of the s particle coordinates and so can be studied quite generally from the viewpoint of the theory of invariants and the theory of groups.

Starting from a space $\mathbf{R}^{3s}$ in which the motion of the particles is described, the space which is invariant to translations is the quotient space $\mathbf{R}^{3s}/\mathbf{T}(3)$, where $\mathbf{T}(3)$ is the translation group in three dimensions. Because this group is Abelian, the translationally invariant quotient space is a vector space $\mathbf{R}^{3s-3}$ and so it is possible to choose a set of $s-1$ translationally invariant Cartesian coordinates expressed in terms of a set of interparticle coordinates and to separate off the translational motion completely in terms of the motion of the centre of particle mass. To form from the translationally invariant set a further set of internal coordinates that are invariant under orthogonal transformations of the original coordinates, it must be recognised that $\mathbf{R}^{3s-3}/\mathbf{O}(3)$, where $\mathbf{O}(3)$ is the orthogonal group in three dimensions, though still a quotient space, is not a vector space but simply a manifold so that it can be a vector space $\mathbf{R}^{3s-6}$ only locally. (Clearly when $s = 2$, a special case arises and it will be ignored in what follows.) For the purposes of the present discussion let us choose, following ref 25, the translationally invariant coordinates to be $s-1$ interparticle distances. Since every set of translationally invariant coordinates is linearly related to any other set, the important results will be independent of the precise choice. Each member of the set of *all* $s(s-1)/2$ scalar products of these coordinates will be invariant under translations and orthogonal transformations of the original cartesian coordinates. Each will also be an analytic function of the cartesian coordinates

Of the $s(s-1)/2$ scalar products only $3s-6$ will be algebraically independent invariants. This result, though perhaps intuitively obvious, may be established by the use of classical invariant theory, but establishing this result simultaneously establishes that any analytic function of the cartesian coordinates which is invariant under translations and orthogonal transformations, will need to be expressed in terms of all the $s(s-1)/2$ scalar products and that no single choice of $3s-6$ coordinates will do (see ref. 25). However, locally, any choice of $3s-6$ internal coordinates will do, for locally, any member of a set of internal coordinates is obviously an algebraic function of any other set. But it is not possible to construct a *global* internal coordinate system for the manifold, describing the geometrical potential. Only local systems are possible, each $\mathbf{R}^{3s-6}$ and constituting local charts. In general sufficient charts may be constructed to provide an atlas for the whole space. The topological ideas mentioned here are discussed in, for example, ref. 26.

These results mean that, locally, an eigenfunction of the Hamiltonian with a geometrical potential can be written as

$$\Psi^{J,M,r}(\mathbf{x}) \to T(\mathbf{X})\Psi^{J,M,r}(\phi,\mathbf{q})$$
$$\Psi^{J,M,r}(\phi,\mathbf{q}) = \sum_{k=-J}^{+J} \Phi_k^{J,r}(\mathbf{q})|JMkr\rangle$$
(1.5)

Here the $\mathbf{x}_i$ are the particle coordinates expressed in the frame fixed in the laboratory, $T(\mathbf{X})$ is a translational eigenfunction depending only on the centre of particle mass and the $\mathbf{q}_k$ are any set of internal coordinates. The $|JMkr\rangle$ are angular momentum eigenfunctions of definite parity, depending upon J and r, and are functions only of the Eulerian angles ϕ. The internal coordinate function on the right side cannot depend on M because, in the absence of a field, the energy of the system does not depend on M.

Because any set of internal coordinates can be written as a function of the scalar products of the inter-particle coordinates, a permutation of the particle coordinates will, at most, send a set of internal coordinates into a set of functions of the internal coordinates. However such a permutation will, in general, send an Eulerian angle into a function both of the Eulerian angles and the internal coordinates. Beyond this, however, the role played by permutational symmetry in classifying spectroscopic states depends upon the permutational invariants of the kinetic and potential energy specifications of the model. In future the kinetic energy operator will be taken to be invariant under any interchange of coordinates between particles of the same mass but the chosen geometrical potentials will have different invariances.

If the spectroscopic molecular model is taken as formed from electrons and nuclei, described by a geometrical potential that depends only on the particle charges, then the system is invariant under all permutations of coordinates representing particles which have the same charges and masses. If the particles are regarded as quantum particles, then the allowed irreducible representations (irreps) of the invariance group, which will label the allowed molecular states, are restricted to those satisfying the Pauli principle. For particles of spin I the spin-eigenfunctions form a basis for irreps corresponding to Young diagrams with $2I + 1$ rows. Thus for a set of identical particles with spin $\frac{1}{2}$ two-rowed Young diagrams specify the irrep carried by the spin-eigenfunctions and the allowed irreps for the space functions correspond to two-column Young diagrams conjugate to them. The combination of the space and spin parts yields the antisymmetric one-dimensional irrep satisfying the Pauli antisymmetry principle. All particles with odd half-integer spin will be similarly restricted to produce totally antisymmetric states. The allowed irreps for particles with integer spin will correspond $2I + 1$-rowed Young diagrams identical with those which specify the irrep carried by the spin-eigenfunctions. The combination of the space and spin parts yields the symmetric one dimensional irrep to satisfy the Pauli symmetric requirement. Labeling generated by the permutational invariances specifies the statistical weights of the relevant spectroscopic states.

If the spectroscopic molecular model is regarded as formed from particles of which some, at least, need not to obey the Pauli principle, then a potential may be invoked that is not invariant under some of the possible permutations of particles with the same masses and charges. Such a description is perhaps appropriate to the Bjerrum model as quantified by Dennison. The assumptions there are that the (implicit) electronic wave function is properly antisymmetric but that the nuclear permutational invariance is confined to those permutations that can be achieved by the point group operations on the molecular framework. Statistical weight calculations in such a description are confined to the allowed sub-set of permutations. This state of affairs is appropriate to the Eckart-Watson Hamiltonian which will be examined in more detail in the next section.

4 The Eckart-Watson Hamiltonian and its Context

The quantum mechanical form of the Eckart Hamiltonian agreed in ref. 17 is a bit clumsy and is difficult to use unless it is possible to treat the inertia tensor for the system as a constant, which is fortunately often an effective approximation. However in 1968 Watson,[2] in an algebraic *tour de force* succeeded in showing that, by incorporating the jacobian and by working the commutation and sum rules hard, a relatively simple expression could be obtained for the Hamiltonian describing rotation-vibration motion namely:

$$\mathsf{H} = \frac{1}{2}\sum_{\alpha\beta}\mu_{\alpha\beta}(\Pi-\pi)_\alpha(\Pi-\pi)_\beta + \frac{1}{2}\sum_k \mathsf{P}_k^2 + \mathsf{U} + \mathsf{V} \qquad (1.6)$$

with

$$\mathsf{U} = -\frac{1}{2}\hbar^2 \sum_\alpha \mu_{\alpha\alpha} \qquad (1.7)$$

Here α and so on denote the cartesian components x and so on. Π_α is the αth component of the total angular momentum and a function of the Eulerian angles only. The components obey the anomalous commutation conditions. π_α is the αth component of the Coriolis coupling operator (often inaccurately called the vibrational angular momentum) which is a function of the vibrational coordinates only. P_k is the momentum operator for the kth vibrational coordinate and $\mu_{\alpha\beta}$ are the elements of an effective reciprocal inertia tensor which can be written as:

$$\boldsymbol{\mu} = \mathbf{I}''^{-1}\mathbf{I}^0\mathbf{I}''^{-1} \qquad (1.8)$$

1: Calculations of the Vibration-Rotation Spectra of Small Molecules

where $\mathbf{I}^0$ is the inertia tensor for the molecule at the reference geometry

$$\mathbf{I}^0 = \sum_i m_i \hat{\mathbf{a}}_i^T \hat{\mathbf{a}}_i$$

and so is a constant matrix and

$$\mathbf{I}'' = \sum_j m_j \hat{\mathbf{z}}_j^T \hat{\mathbf{a}}_j \tag{1.9}$$

where the skew-symmetric matrix $\hat{\mathbf{a}}_i$

$$\hat{\mathbf{a}}_i = \begin{pmatrix} 0 & -a_{zi} & a_{yi} \\ a_{zi} & 0 & -a_{xi} \\ -a_{yi} & a_{xi} & 0 \end{pmatrix} \tag{1.10}$$

and similarly for z_i. The volume element of integration for the Hamiltonian (1.6) is

$$dV = d\phi \sin\theta d\theta d\gamma dQ_1 dQ_2 \ldots dQ_n$$

where ϕ, θ, and γ are the Eulerian angles with θ, in the range $[0, \pi)$ and the Q_k are the vibrational coordinates. The vibrational coordinates are defined in terms of the disolacement coordinates by

$$Q_k = \sum_i \sum_\alpha l_{\alpha i,k} m_i^{\frac{1}{2}} \rho_{\alpha i} \tag{1.11}$$

where the $l_{\alpha i,k}$ are constants chosen in such a way that the inverse transformation exists and, if the translational and rotational coordinates are set to zero, yields an inverse

$$\rho_{\alpha i} = m_i^{-\frac{1}{2}} \sum_k l_{\alpha i,k} Q_k$$

Treating the $l_{\alpha i,k}$ as the αth elements of column matrix l_{ik} the Coriolis coupling operator has components

$$\pi_\alpha = \frac{\hbar}{i} \sum_{k,l} \varsigma_{kl}^\alpha Q_k \frac{\partial}{\partial Q_l}$$

in which

$$\varsigma_{kl}^\alpha = \sum_i (\hat{l}_{ik} l_{il})_\alpha$$

The Eckart conditions are satisfied by choosing

$$\sum_i m_i^{\frac{1}{2}} \mathbf{l}_{ik} = 0, \quad \sum_i m_i^{\frac{1}{2}} \hat{\mathbf{a}}_i \mathbf{l}_{ik} = 0$$

and the kinetic energy is put in diagonal form by choosing

$$\sum_i \mathbf{l}_{ik}^T \mathbf{l}_{il} = \delta_{kl}$$

The vibrational coordinates are such as to put the vibrational kinetic energy term in diagonal form and though Watson sometimes refers to them as "normal" coordinates, they are not in fact required to diagonalise simultaneously a potential form bilinear in the displacement coordinates. Watson takes the particles to be atoms and ignores any electronic structure and he assumes a geometrical potential.

In 1976 Louck[27] derived the Eckart-Watson form of the Hamiltonian by direct coordinate transformation from the ordinary Schrödinger Hamiltonian with a geometrical potential and the coordinates regarded as being those of otherwise unspecified particles. Very shortly after this, Makushkin and Ulenikov[28] provided another account of the direct transformation approach. They assumed a geometrical potential and dealt explicitly with both electrons and nuclei as the particles involved in the kinetic energy operator. They also provided an account of how the "particles" defining the coordinates had been identified and treated in previous work, together with a clear and careful treatment of the jacobian for the transformation.

4.1 The Permutational Symmetry of the Eckart-Watson Hamiltonian. – The orthogonal matrix **C** that defines the Eckart frame relates the set of cartesian coordinates for the particles in the frame fixed in the body to those for the frame fixed in the laboratory by

$$\mathbf{x}_i - \mathbf{X} = \mathbf{C}\mathbf{z}_i \tag{1.12}$$

where **X** is the centre of particle mass coordinate.

Using the two Eckart conditions given by (1.3) and (1.4) it follows (see the discussion in Section 3 of Eckart[1]) that

$$\mathbf{C} = \mathbf{F}(\mathbf{F}^T \mathbf{F})^{-\frac{1}{2}} \tag{1.13}$$

where

$$\mathbf{F} = \sum_i m_i (\mathbf{x}_i - \mathbf{X})\mathbf{a}_i^T \tag{1.14}$$

The 3 by 3 matrix $\mathbf{F}^T\mathbf{F}$ is symmetric and therefore diagonalisable, so functions of it may be properly defined in terms of its eigenvalues. There are in principle, eight (2^3) possible distinct square root matrices, all such that their square yields the original matrix product. Consistency (see Ezra[21]) requires, however, that the positive square roots of each of the eigenvalues be chosen. No eigenvalue may be negative or zero or else the Eckart frame cannot be properly defined. Since the $\mathbf{a}_i$ are, by definition, constant vectors, they are unaffected by any coordinate transformation.

If we denote the 3 by s matrix composed of the columns $\mathbf{x}_i$ by $\mathbf{x}$ and denote the similarly constructed matrices of the $\mathbf{a}_i$ by $\mathbf{a}$ and those from $\mathbf{x}_i - \mathbf{X}$ by $\mathbf{w}$ then $\mathbf{F}$ can be written as

$$\mathbf{F} = \mathbf{wma}^T \qquad (1.15)$$

where $\mathbf{m}$ is a diagonal matrix with particle masses along the diagonal. It follows from the definitions (1.12), (1.13) and (1.14) that

$$\mathbf{z} = (\mathbf{amw}^T \mathbf{wma}^T)^{-\frac{1}{2}} \mathbf{amw}^T \mathbf{w} = (\mathbf{F}^T \mathbf{F})^{-\frac{1}{2}} \mathbf{F}^T \mathbf{w} \qquad (1.16)$$

In this formulation it is seen that $\mathbf{w}^T \mathbf{w}$ is a square matrix of scalar products, making obvious the orthogonal invariance of the $\mathbf{z}_i$ and thus it follows that

$$\mathbf{w}^T \mathbf{w} = \mathbf{z}^T \mathbf{z}$$

To see that the Eckart formulation is simply a local one consider a motion that causes the particles to form a straight line. It is then easy to see that $\mathbf{F}^T \mathbf{F}$ becomes singular and so the frame definition fails (the Eulerian angles cannot be uniquely defined) and the internal coordinate definition too, fails with this failure. Such a failure manifests itself directly in the Eckart-Watson Hamiltonian by the elements of μ (1.8) increasing without limit because $\mathbf{I}''$ (1.9) becomes singular. Thus the Hamiltonian definition fails. Of course this does not mean that a molecule cannot exhibit large amplitude internal motions, it simply means that, if it does, the Eckart-Watson Hamiltonian cannot be used to describe them.

To determine permutational invariances in the Eckart formulation it is first necessary to consider what changes in the $\mathbf{z}_i$ are induced by a permutation of the $\mathbf{x}_i$. If the permutation $\mathcal{P}$ is realised by the orthogonal permutation matrix $\mathbf{P}$ then under the permutation

$$\mathbf{x} \to \mathbf{xP}, \ \mathbf{w} \to \mathbf{wP}$$

The changes induced in $\mathbf{z}$ under the permutation $\mathcal{P}$ are obtained from (1.16) by the linear substitution

$$\mathbf{w} \to \mathbf{wP}^T$$

so that

$$\mathbf{F} \to \mathbf{wP}^T \mathbf{ma}^T$$

For all permutations that preserve the invariance of the kinetic energy operator, however,

$$\mathbf{mP} = \mathbf{Pm}$$

and for such permutations we may rewrite the above as

$$\mathbf{F} \to \mathbf{wmP}^T\mathbf{a}^T = \mathbf{CzmP}^T\mathbf{a}^T = \mathbf{CF}_P$$

Provided that $\mathbf{F}_P$ is non-singular then it follows that under the permutation $\mathcal{P}$

$$\mathbf{C} \to \mathbf{CF}_P(\mathbf{F}_P^T\mathbf{F}_P)^{-\frac{1}{2}} = \mathbf{CR}_P^T$$

where $\mathbf{R}_P$ is an orthogonal matrix which will, in general, be a function of the vibrational coordinates, thus illustrating the point that a permutation can induce a change in the Eulerian angles that depends on the internal coordinates. For the internal coordinates in this case the induced change is

$$\mathbf{z} \to \mathbf{R}_P \mathbf{z} \mathbf{P}^T$$

which can be expressed entirely in terms of the vibrational coordinates.

However, unless the permutation preserves as positive-definite, the matrix whose square root is to be taken, it will actually cause the frame definition to fail. It is hard to say, generally, when such a failure will occur, but since **w** may be replaced by **z** on the right hand sides of the equations above, the reference geometry case may be studied by replacing **z** with **a**. It is then easy to show that failure occurs with most of the permutations for a system comprising four or more identical particles. A detailed analysis in the case of ethene may be found in ref. 29. It is thus the case that only a limited sub-set of all the possible permutations of the coordinates of particles with the same mass, may properly be considered within the Eckart approach. The relevant sub-set is that for which

$$\mathbf{aP} = \mathbf{Ua}, \quad \mathbf{U}^T\mathbf{U} = \mathbf{E}_3$$

so that an allowed permutation acting on the reference geometry can be realised as an orthogonal transformation of the reference geometry, then

$$\mathbf{z} \to \mathbf{UzP}^T, \quad \mathbf{C} \to \mathbf{CU}^T$$

and

$$\mathbf{a} \to \mathbf{UaP}^T = \mathbf{a}$$

so that

$$\boldsymbol{\rho} = \mathbf{z} - \mathbf{a} \to \mathbf{UzP}^T - \mathbf{a} = \mathbf{U}\boldsymbol{\rho}\mathbf{P}^T$$

The operations are those for which the effect of an orthogonal transformation of the framework is cancelled out by a permutation of particles with the same masses. They are often called *perrotations* following Gilles and Philippot.[30] Clearly the operations comprising the standard molecular point group are perrotations. They seem to have been first studied by Wigner[31] in 1930 when he considered the symmetry properties of vibrations in a potential having a minimum at a reference geometry. (There is an English translation of the relevant parts of this paper in ref. 32.)

Under such a permutation the vibrational coordinates undergo the change

$$Q_k \to \sum_{ij}\sum_{\alpha\beta} U_{\beta\alpha} l_{\alpha i,k} m_i^{\frac{1}{2}} \rho_{\beta j} P_{ji}$$

The permutation will at most reorder displacements which correspond to the same masses and so the double sum over coordinates is simply a single sum giving

$$Q_k \to \sum_{i}\sum_{\alpha\beta} U_{\beta\alpha} l_{\alpha i,k} m_i^{\frac{1}{2}} \rho_{\beta i}$$

The work of Wigner cited above establishes that the vibrational coordinates can always be written as point group symmetry coordinates so that the l_{ik} can be chosen to carry the point group symmetry of the problem with an operation represented by the matrix U. Then the Q_k will at most map into orthogonal combinations of each other and so the Eckart kinetic energy operator will be invariant under these permutations. The geometrical potential must also be invariant under these permutations and it is easy to see that provided that the potential can be chosen as a quadratic form in the vibrational coordinates, this invariance is supplied.

The statistical weights of the various possible vibration-rotation states are determined in the Eckart-Watson approach entirely in terms of the permutations that correspond to perrotations.

If the geometrical potential is such as to require more permutational symmetry to be incorporated, then the Eckart formulation must be abandoned. The internal coordinates must be chosen in the light of the invariances in the potential and the permutations which it is wished to allow, so that the frame definition does not fail under a permutation of coordinates which is of interest. It is difficult to say anything general about these matters. Those forms arising from developments of the Eckart-Watson approach usually arise from Longuet-Higgins' idea of *feasible* operations[33] and the construction of more and more flexible molecular models, starting from the Eckart-Sayvetz one. The group relevant to the chosen model is usually called a *molecular symmetry group*. These matters are discussed by Ezra[21] and by Bunker and Jensen.[23] Some tailor-made approaches also use these ideas.

5 The General Form of a Tailor-made Hamiltonian

To remove the centre of mass motion from the full Hamiltonian all that is needed is a coordinate transformation symbolised by

$$(\mathbf{t}\mathbf{X}_T) = \mathbf{x}\mathbf{V} \tag{1.17}$$

In (1.17) $\mathbf{t}$ is a 3 by $s-1$ matrix and $\mathbf{X}_T$ is a 3 by 1 matrix, so that the combined (bracketed) matrix on the left of (1.17) is 3 by s. $\mathbf{V}$ is an s by s matrix which, from the structure of the left side of (1.17), has a special last column whose elements are

$$V_{is} = M_T^{-1} m_i, \quad M_T = \sum_{i=1}^{s} m_i. \tag{1.18}$$

Hence $\mathbf{X}_T$ is the standard centre of mass coordinate.

$$\mathbf{X}_T = M_T^{-1} \sum_{i=1}^{s} m_i \mathbf{x}_i \tag{1.19}$$

As the coordinates $\mathbf{t}_j, j = 1, 2, \ldots s-1$ are to be translationally invariant, we require on each remaining column of $\mathbf{V}$

$$\sum_{i=1}^{s} V_{ij} = 0, \quad j = 1, 2, \ldots s-1 \tag{1.20}$$

and it is easy to see that (1.20) forces $\mathbf{t}_j \to \mathbf{t}_j$ as $\mathbf{x}_i \to \mathbf{x}_i + \mathbf{a}$, all i. and the $\mathbf{t}_i$ are independent if the inverse transformation to (1.17) exists.

When we write the column matrix of the cartesian components of the partial derivative operator as $\partial/\partial \mathbf{x}_i$, the coordinate change (1.17) gives

$$\frac{\partial}{\partial \mathbf{x}_i} = \sum_{j=1}^{s-1} V_{ij} \frac{\partial}{\partial \mathbf{t}_j} + m_i M_T^{-1} \frac{\partial}{\partial \mathbf{X}_T} \tag{1.21}$$

and when it seems more convenient this column matrix of derivative operators will also be denoted as the vector grad operator $\vec{\nabla}(\mathbf{x}_i)$.

Assuming a geometrical potential, the Hamiltonian in the new coordinates becomes

$$\mathsf{H}(\mathbf{t},\mathbf{X}_T) = -\frac{\hbar^2}{2} \sum_{i,j=1}^{s-1} \frac{1}{\mu_{ij}} \vec{\nabla}(\mathbf{t}_i) \cdot \vec{\nabla}(\mathbf{t}_j) + \mathsf{V}(\mathbf{t}) - \frac{\hbar^2}{2M_T} \nabla^2(\mathbf{X}_T). \tag{1.22}$$

with the inverse reduced masses given by

$$1/\mu_{ij} = \sum_{k=1}^{s} m_k^{-1} V_{ki} V_{kj} \quad i,j = 1, 2, \ldots s-1 \tag{1.23}$$

The first two terms in (1.22) which will be denoted collectively by $\mathsf{H}(\mathbf{t})$ and referred to as the *translationally invariant* Hamiltonian.

For a system with more than two particles one can transform the coordinates $\mathbf{t}$ such that rotational motion can be expressed in terms of three orientation variables, with the remaining motions expressed in terms of variables (commonly called internal coordinates) which are invariant under all orthogonal transformations of the **t**. For $s = 2$ only two orientation variables are required and this case is rather special and is excluded from all subsequent discussion. To construct the frame fixed in the body it is supposed that the three orientation variables are specified by means of an orthogonal matrix **C**, the elements of which are expressed as functions of three Eulerian angles ϕ_m, $m = 1, 2, 3$ which are orientation variables. We require that the matrix **C** is specified in terms of the translationally invariant coordinates **t**. Thus the cartesian coordinates **t** are considered related to a set **z** by

$$\mathbf{t} = \mathbf{C}\mathbf{z} \tag{1.24}$$

so the matrix **C** may be thought of as a direction cosine matrix, relating the laboratory frame to the frame fixed in the body. Since **z** are fixed in the body, not all their $3s - 3$ components are independent, for there must be three relations between them. Hence components of $\mathbf{z}_i$ must be writable in terms of $3s - 6$ independent internal coordinates q_i, $i = 1, 2, \ldots, 3s - 6$. Some of the q_i may be components of $\mathbf{z}_i$ but generally q_i are expressible in terms of scalar products of the $\mathbf{t}_i$ (and equally of the $\mathbf{z}_i$) since scalar products are the most general constructions that are invariant under orthogonal transformations of their constituent vectors. Equation (1.24) *defines* the cartesian form of the variables in the frame fixed in the body by means of **C**.

The jacobian matrix elements for the transformation from the $(\phi, \mathbf{q})$ to the $(\mathbf{t})$ can be written

$$\frac{\partial \phi_m}{\partial t_{\alpha i}} = (\mathbf{C}\mathbf{\Omega}^i \mathbf{D})_{\alpha m} \tag{1.25}$$

$$\frac{\partial q_k}{\partial t_{\alpha i}} = (\mathbf{C}\mathbf{Q}^i)_{\alpha k} \tag{1.26}$$

In the above, the elements of $\mathbf{Q}^i$ and of $\mathbf{\Omega}^i$ are dependent on internal variables only, while the elements of **C** and of **D** are functions of the Eulerian angles only.

It is usually possible to express the internal coordinates of a problem as explicit functions of the scalar products of the translationally invariant coordinates. Thus it is usually possible, though always error-prone and tedious, to derive the Jacobian matrix elements given by (1.26). However it is seldom possible to express the Eulerian angles as explicit functions of the translationally invariant coordinates and usually the evaluation of the Jacobian matrix elements given by (1.25) involves great ingenuity and much good-luck and sometimes it is a labour undertaken without success. There are a couple of tricks which are sometimes used in this context. Although not all the $z_{\beta i}$ can be linearly independent nevertheless they all possess derivatives with respect to the $t_{\epsilon j}$ which have the form

$$\frac{\partial z_{\beta i}}{\partial t_{\epsilon j}} = (\mathbf{C}\mathbf{\Omega}^j \hat{\mathbf{z}}_i)_{\epsilon\beta} + C_{\epsilon\beta}\delta_{ij} \qquad (1.27)$$

It is also sometimes possible to express the constraint conditions on the $z_{\beta i}$ in the form $f_m(\mathbf{z}) = 0$, $m = 1,2,3$ and hence as $f_m(\mathbf{C}^T\mathbf{t}) \equiv g_m(\mathbf{t}) = 0$, $m = 1,2,3$. In that case

$$\frac{\partial g_m}{\partial t_{\epsilon j}} = (\mathbf{C}(-\mathbf{\Omega}^j \mathbf{T} + \mathbf{S}^j))_{\epsilon m} = 0$$

where

$$S^j_{\epsilon m} = \frac{\partial f_m}{\partial z_{\epsilon j}}, \quad \mathbf{T} = \sum_{i=1}^{s-1} \hat{\mathbf{z}}_i^T \mathbf{S}^i$$

The derivative with respect to $z_{\epsilon j}$ is perfectly well defined in the usual way even though the $z_{\epsilon j}$ are not all independent variables because $f_m(\mathbf{z})$ is an explicit function of all of them. If $\mathbf{T}$ is non-singular then one can write

$$\mathbf{\Omega}^i = \mathbf{S}^i \mathbf{T}^{-1} \qquad (1.28)$$

The derivatives of the translationally invariant coordinates in terms of the orientation and internal coordinates are

$$\frac{\partial}{\partial t_i} = \mathbf{C}\left(\mathbf{\Omega}^i \mathbf{D}\frac{\partial}{\partial \phi} + \mathbf{Q}^i \frac{\partial}{\partial \mathbf{q}}\right) \qquad (1.29)$$

where $\partial/\partial\phi$ and $\partial/\partial\mathbf{q}$ are column matrices of 3 and $3s - 6$ partial derivatives respectively and $\partial/\partial t_i$ column matrices of 3 partial derivatives.

In these coordinates the translationally invariant angular momentum operator becomes

$$\mathsf{L}(\mathbf{t}) = -\frac{\hbar}{i}|\mathbf{C}|\mathbf{C}\mathbf{D}\frac{\partial}{\partial \phi} = -|\mathbf{C}|\mathbf{C}\mathsf{L}(\phi) \qquad (1.30)$$

where $|\mathbf{C}|$ is either plus or minus one according to whether $\mathbf{C}$ corresponds to a proper rotation or to an improper rotation.

There is at this stage an element of choice for the definition of the angular momentum in the frame fixed in the body and in (1.30) it can be seen that we have chosen

$$\mathsf{L}(\phi) = \frac{\hbar}{i}\mathbf{D}\frac{\partial}{\partial \phi} \qquad (1.31)$$

Often, indeed perhaps more usually, the negative of this operator is chosen and the angular momentum operators so chosen, obey the anomalous commutation conditions. (This choice is made in the construction of the Eckart-Watson Hamiltonian already discussed.) However a little algebra shows that in either case $L^2(\phi) \equiv L^2(t)$ and that $\mathbf{L}_z(\phi)$ and $\mathbf{L}_z(t)$ commute with $\mathbf{L}^2$ so one can find a complete set of angular momentum eigenfunctions.

The kinetic energy operator describing rotation-vibration motion may be written as

$$K(\mathbf{q}) + K(\phi, \mathbf{q}) \tag{1.32}$$

The transformation of the translationally invariant kinetic energy operator from (1.22) into the coordinates ϕ and $\mathbf{q}$ is long and tedious, but the final result can be stated directly; as the derivation is mechanical, simply involving letting (1.29) operate on itself and summing over i and j, there is no need to go into details. The resulting operators are

$$k(\phi, \mathbf{q}) = \frac{1}{2}(\sum_{\alpha\beta} \kappa_{\alpha\beta} L_\alpha L_\beta + \hbar \sum_\alpha \lambda_\alpha L_\alpha) \tag{1.33}$$

and

$$K(\mathbf{q}) = -\frac{\hbar^2}{2}\left(\sum_{k,l=1}^{3s-6} g_{kl} \frac{\partial^2}{\partial q_k \partial q_l} + \sum_{k=1}^{3s-6} h_k \frac{\partial}{\partial q_k}\right) \tag{1.34}$$

κ is an inverse generalised inertia tensor defined as the 3 by 3 matrix

$$\kappa = \sum_{i,j=1}^{s-1} \frac{1}{\mu_{ij}} \mathbf{\Omega}^{i^T} \mathbf{\Omega}^j \tag{1.35}$$

and

$$\lambda_\alpha = \frac{1}{i}\left(v_\alpha + 2\sum_{k=1}^{3s-6} \tau_{k\alpha} \frac{\partial}{\partial q_k}\right) \tag{1.36}$$

with the $3s - 6$ by 3 matrix τ defined as

$$\tau = \sum_{i,j=1}^{s-1} \frac{1}{\mu_{ij}} \mathbf{Q}^{i^T} \mathbf{\Omega}^j \tag{1.37}$$

and

$$v_\alpha = \sum_{i,j=1}^{s-1} \frac{1}{\mu_{ij}}\left(\sum_\beta (\mathbf{\Omega}^{i^T} \mathbf{M}^\beta \mathbf{\Omega}^j)_{\beta\alpha} + \sum_{l=1}^{3s-6}\left(\mathbf{Q}^{i^T} \frac{\partial}{\partial q_l} \mathbf{\Omega}^j\right)_{l\alpha}\right) \tag{1.38}$$

The term (1.36) is associated with the Coriolis coupling and so no coordinate system can be found in which it will vanish.

The $3s-6$ by $3s-6$ matrix **g** is given by

$$\mathbf{g} = \sum_{i,j=1}^{s-1} \frac{1}{\mu_{ij}} \mathbf{Q}^{iT} \mathbf{Q}^{j} \tag{1.39}$$

and

$$h_k = \sum_{i,j=1}^{s-1} \frac{1}{\mu_{ij}} \left(\sum_{\beta} (\boldsymbol{\Omega}^{iT} \mathbf{M}^{\beta} \mathbf{Q}^{j})_{\beta k} + \sum_{l=1}^{3s-6} \left(\mathbf{Q}^{iT} \frac{\partial}{\partial q_l} \mathbf{Q}^{j} \right)_{lk} \right) \tag{1.40}$$

Although a specific embedding must be chosen to evaluate h_k it can be shown[29] that h_k is invariant under any orthogonal transformation of the $\mathbf{t}_i$ and so any convenient embedding may be used. Any geometrical potential will be expressible entirely in terms of the $\mathbf{q}_k$ so that states of the system for which $J = 0$ may be described entirely in terms of the sum of the kinetic energy operator (1.34) and the potential operator.

It is perfectly possible to use this more general Hamiltonian to obtain the Eckart-Watson form. How this is done is discussed in Section III G of ref. 29.

Since the angular momentum eigenfunctions are known it is possible to integrate out over the rotational coordinates and to write an effective Hamiltonian for each rotational state in terms of the internal coordinates alone. Using the functions (1.5), but dropping the parity index the effective matrix elements are, on integrating over the Eulerian angles:

$$< J'M'k' | K(\mathbf{q}) + V(\mathbf{q}) | JMk > = \delta_{J'J} \delta_{M'M} \delta_{k'k} (K(\mathbf{q}) + V(\mathbf{q})) \tag{1.41}$$

where $V(\mathbf{q})$ is the potential.

In what follows explicit allowance for the diagonal requirement on J and M will be assumed and the indices suppressed to save writing. Similarly the fact that the integration implied is over ϕ only will be left implicit.

To treat the angular term is much more complicated and best done by re-expressing the components of **L** in terms of $L_{\pm}(\phi)$ and $L_z(\phi)$. When this is done

$$< JM k' | K(\phi,\mathbf{q}) | JMk > =$$

$$\frac{\hbar^2}{4}(b_{+2} C^+_{Jk+1} C^+_{Jk} \delta_{k'k+2} + b_{-2} C^-_{Jk-1} C^-_{Jk} \delta_{k'k-2})$$

$$+ \frac{\hbar^2}{4}(C^+_{Jk}(b_{+1}(2k+1) + \lambda_+) \delta_{k'k+1} + C^-_{Jk}(b_{-1}(2k-1) + \lambda_-) \delta_{k'k-1})$$

$$+ \frac{\hbar^2}{2}((J(J+1) - k^2)b + b_0 k^2 + \lambda_0 k) \delta_{k'k} \tag{1.42}$$

In this expression

$$b_{\pm 2} = (\kappa_{xx} - \kappa_{yy})/2 \pm \kappa_{xy}/i$$
$$b_{\pm 1} = \kappa_{xz} \pm \kappa_{yz}/i$$
$$b = (\kappa_{xx} + \kappa_{yy})/2 \qquad b_0 = \kappa_{zz} \qquad (1.43)$$

and in terms of the λ_α in ([lambda]) λ_0 is λ_z and the $\lambda_\pm$ are

$$\lambda_\pm = (\lambda_x \pm \lambda_y/i) \qquad (1.44)$$

The apparently odd positioning of the complex unit as $1/i$ when i might have been expected, is because the standard commutation conditions have been chosen for the internal angular momentum components.

If the Eckart-Watson form of the Hamiltonian is chosen, since the angular momentum functions in that form obey the anomalous commutation conditions, the rotational matrix elements will not be quite the same as those implicit in the derivation of (1.42). The required matrix elements for these angular momentum functions can be found in equations (6) to (14) of ref. 34. However the modified form of (1.42) resulting from their use is not essentially different from that given (see (1.57) later).

A tailor-made quantum mechanical Hamiltonian may be constructed from the relevant classical Hamiltonian by transforming it. Such an approach is widely used and an early attempt in general form is found in ref. 35. The formalism, with examples, is reviewed by Meyer.[36] It was shown by Lukka,[37] and explained in ref. 36, that it is possible to avoid explicit choice of any specific rotational coordinates in formulating the Hamiltonian. Lukka's ideas have been developed by several workers, perhaps most generally by using Hestenes' geometrical algebra. The Hestenes approach is expounded and reviewed in an article by Pesonen and Halonen.[38]

5.1 The Permutational Symmetry in the General Form of the Hamiltonian. –
The effect of a permutation on the translationally invariant coordinates is:

$$\mathcal{P}(\mathbf{tX}_T) = (\mathbf{tX}_T) \begin{pmatrix} \mathbf{H} & 0 \\ 0 & 1 \end{pmatrix} \qquad (1.45)$$

where

$$(\mathbf{H})_{ij} = (\mathbf{V}^{-1}\mathbf{PV})_{ij} \quad i, j = 1, 2, \ldots N-1 \qquad (1.46)$$

The matrix $\mathbf{H}$ is not necessarily in standard permutational form neither is it orthogonal, even though it has determinant ± 1 according to the sign of $|\mathbf{P}|$. Using (1.46) it is seen that a permutation

$$\mathcal{P}\mathbf{t} = \mathbf{tH} = \mathbf{t'} \qquad (1.47)$$

and that

$$\mathbf{S'} = \mathbf{H}^T \mathbf{SH} \qquad (1.48)$$

where the matrices **S** and **S'** comprise the scalar products of the translationally invariant coordinate before and after permutation.

Recognising again that a permutation can induce changes in the Eulerian angles that involve the internal coordinates too, the permuted variable **t'** may be written equivalently as

$$\mathbf{t'} = \bar{\mathbf{C}}(\phi,\mathbf{q})\bar{\mathbf{z}}(\mathbf{q}) \tag{1.49}$$

where the bars denote the functional forms induced by the permutational change of variables. Equating (1.47) and (1.49) using (1.24) it follows that

$$\bar{\mathbf{z}} = \bar{\mathbf{C}}^T \mathbf{C} \mathbf{z} \mathbf{H} = \mathbf{U}^T \mathbf{z} \mathbf{H} \tag{1.50}$$

Naturally some care must be taken about the domain in which the orthogonal matrix **U** exists but since the expression above can be at most a function of the internal coordinates it follows that, where it exists, its elements are, at most, functions of the internal coordinates.

However there is no general form for the change induced in the q_k by a permutation and so the most that can be said is that a permutation of particles induces a general function change

$$\Phi_k^{J,r}(\mathbf{q}) \rightarrow \Phi_k'^{J,r}(\mathbf{q}) \tag{1.51}$$

where the precise nature of the function change depends on the permutation, the chosen form of the internal coordinates and on the chosen functional form. Thus the general change induced in (1.5) by $\mathcal{P}$ is

$$\Psi^{J,M,r}(\phi,\mathbf{q}) \rightarrow \sum_{k=-J}^{+J} \sum_{n=-J}^{+J} \mathcal{D}_{nk}^{rJ}(\mathbf{U}) \Phi_k'^{J,r}(\mathbf{q}) | JMnr \rangle$$

$$= \sum_{n=-J}^{+J} \bar{\Phi}_n^{J,r}(\mathbf{q}) | JMnr \rangle \tag{1.52}$$

where $\mathcal{D}_{nk}^{rJ}$ is a matrix element of a parity extended Wigner matrix whose precise form is given in Section IV of ref. 29. The function (1.5) should carry an additional label **s** to specify to which irreps of the various symmetric groups in the problem it belongs. If the function is such that it belongs to a one dimensional irrep of the symmetric group containing the permutation $\mathcal{P}$, then the resulting function (1.52) can differ from the original one by at most a sign change. In the case of a many dimensional representation, the resulting function (1.52) will be at most a linear combination of the set of degenerate functions providing a basis for the irrep. So in spite of possible coordinate mixing, there are no difficulties in principle. However in practice one must construct approximate wavefunctions which are not immediately adapted to the permutational symmetry of the problem and which must be explicitly adapted by, for example, the use of projections. In these circumstances coordinate mixing can cause tremendous complications. The expression (1.52) will clearly be very difficult to handle for not only will a U be difficult to determine, but one must be found for each distinct permutation of the identical particles and in a problem of any size there will be a very large number of such permutations.

It should be noted here that this coupling of rotations by the permutations can mean that certain rotational states do not occur because they cannot satisfy the correct symmetry requirements under the permutation of identical particles. Whether or not this is the case has to be determined in any particular occurrence by the changes induced according to (1.52) and this would be exceptionally tricky, in general. However Ezra has discussed the problem in detail in some special cases.[21] This possibility is relevant in assigning statistical weights to rotational states.

6 Computational Considerations

Contemporary calculations of spectra invariably start from potential energy surfaces which originate in clamped nuclei electronic structure calculations. The relevant Hamiltonian is:

$$H^{cn}(\mathbf{a},\mathbf{x}^e) = -\frac{\hbar^2}{2m}\sum_{i=1}^{N}\nabla^2(\mathbf{x}_i^e) - \frac{e^2}{4\pi\varepsilon_0}\sum_{i=1}^{s}\sum_{j=1}^{N}\frac{Z_i}{\left|\mathbf{x}_j^e - \mathbf{a}_i\right|} + \frac{e^2}{8\pi\varepsilon_0}\sum_{i,j=1}^{N}{'}\frac{1}{\left|\mathbf{x}_i^e - \mathbf{x}_j^e\right|} \quad (1.53)$$

where we have taken N to be the number of electrons with the electronic variables denoted by $\mathbf{x}_i^e$ and in future we shall think of s as specifying the number of heavy particles,* each of charge Z_i in electronic charge units. The mass of the heavy particles will generally be the nuclear mass, but in some circumstances slightly different masses are assumed.

It is not always the case that this Hamiltonian is used directly: sometimes the electronic repulsion terms are modeled to decrease the computational effort involved in approximate solutions. A widely used contemporary set of model approaches are based upon density functional theory and referred to collectively as DFT models. They vary in the form of the potential chosen.

The electronic Hamiltonian is put in a suitable computational context in the textbook by F. Jensen.[39] This book also contains definitions of the acronyms (such as HF, MP2 and so on) which are used in describing the kind of electronic structure calculations performed. We shall not redefine these acronyms here and it is perhaps sufficient to note for general comprehension, the longer the acronym, the more sophisticated the computational method. It also explains what is meant by expressions such as 3-21G* or cc-pVDZ, that are used to describe the basis sets of gaussian orbitals used in an electronic structure calculation. Again, for general comprehension, it is sufficient to note

---

* In the discussion of the Eckart formulation in Section 4 the s variables can only be heavy particles. In the discussion of the tailor-made formulation in Section 5, no such restriction need be effected.

that the longer the symbol, the more accurate will be the associated calculation.

The basic potential for any electronic eigenvalue $E_p^{cn}(\mathbf{a})$ at any nuclear disposition $\mathbf{a}$ is given by the sum of the eigenvalue and the classical nuclear repulsion at that disposition.

$$V_p^a(\mathbf{a}) = E_p^{cn}(\mathbf{a}) + \frac{e^2}{8\pi\epsilon_0} \sum_{i,j=1}^{s}{}' \frac{Z_i Z_j}{|\mathbf{a}_j - \mathbf{a}_i|} \qquad (1.54)$$

This equation is such that

$$V_p^a(\mathbf{a}) \to V_p(\mathbf{a}), \quad \text{as } \mathbf{a}_i \to \mathbf{a}_i + \mathbf{d}, \text{ and/or } \mathbf{a}_i \to U\mathbf{a}_i, \quad i = 1, 2 \ldots, H$$

if $\mathbf{d}$ is a constant column matrix and U is a constant orthogonal matrix. Thus if $V_p^a(\mathbf{a})$ is to be presented as a function of the $\mathbf{a}_i$, for single valuedness, the domain of the $\mathbf{a}_i$ must be confined to those geometries which are distinct under uniform translations and/or rigid rotation-reflections. This means that in a given domain for a given set of *internal* coordinates there is a function $V_p(\mathbf{q})$ which is single valued and continuous and which maps on to $V_p^a(\mathbf{a})$. This function is usually called the *potential surface* and is constructed, within the given domain, by fitting an analytic form in the variables $\mathbf{q}$ to the potential calculated at given values of $\mathbf{a}$. The potential $V_p^a(\mathbf{a})$ is also treated s if it is a particular value of the potential $V_p^x(\mathbf{x})$ which is defined for all values of $\mathbf{x}$ where $\mathbf{x}$ are the heavy particle coordinates. It should be noticed that because the electronic Hamiltonian and the nuclear repulsion term depend only upon the particle charges and the heavy particle geometry, the potential $V_p^x(\mathbf{x})$ is independent of the particle masses and will be the same for all the isotopomers[†] of any molecule. However it is often convenient to choose the internal coordinates q_k to be mass dependent so that they will differ between isotopomers. This is the case with the vibrational coordinates as defined in (1.11). Thus a single set of electronic structure calculations will map onto a different potential for each isotopomer. The potentials can usually be chosen to have the same functional form for different isotopomers, it is just the definitions of the internal coordinates that differ between them.

A bound system is anticipated for every minimum in $V_p(\mathbf{q})$ and it is these minima that dominate spectral interpretation. It is assumed as a first approximation that it is sufficient to expand the potential in a Taylor series about a minimum thus:

$$V(\mathbf{q}) = V(0) + \frac{1}{2}\sum_{kl=1}^{3s-6} f_{kl} q_k q_l + \cdots$$

[†] IUPAC recommends that what are here, and commonly, called isotopomers, should be called isotopologues. The term isotopomer should be reserved to describe isomers due to the positions of nuclear isotopes.

where, for brevity, the electronic state designator p has been dropped. The origin of the internal coordinates has been taken to be the equilibrium geometry and designated as zero. The linear term is absent because the first derivatives of the potential vanish at the equilibrium geometry, by definition. The matrix **f** is composed of the second derivatives of the potential with respect to the q_k evaluated at the equilibrium geometry. Higher order terms, if necessary will involve higher order derivatives.

Chapter 2 (by Császár, Allen, Yamaguchi and Schaefer) in the book edited by P. Jensen and P. R. Bunker and entitled *Computational Molecular Spectroscopy*[40] provides a detailed account of how potentials are actually calculated. It is appropriate too, to draw attention to two other compilations which, like this one, contain articles relevant not only to the computation of potentials but also to aspects of computational spectroscopy. They are *Encyclopedia of Computational Chemistry*[41] edited by P. von Ragué Schleyer and *Handbook of Molecular Physics and Quantum Chemistry*[42] edited by S. Wilson. In Chapter 20 of Part 3 of this last, an account by C. Bissonette of how diatomic molecules are dealt with, a topic not otherwise considered here, can be found. Although there are no articles in the first that deal exclusively with diatomics, they do feature in some of them.

Modern electronic structure packages often include facilities that make possible the evaluation of the first and second derivatives of the potential energy, as calculated at a particular level of approximation, with respect to the cartesian coordinates that specify these heavy particle positions. If the clamped nuclei electronic problem is associated with the full problem in such a way that the heavy particle cartesian coordinates in it may be written in terms of three cartesian coordinates **X** describing the translational motion and of three rotational coordinates **R** describing the orientation of the system together with $3s - 6$ internal coordinates q_k describing the heavy particle geometry of the system, then we may write the cartesian heavy particle derivatives as

$$\frac{\partial}{\partial x_{\alpha i}} = \frac{\partial X_\alpha}{\partial x_{\alpha i}} \frac{\partial}{\partial X_\alpha} + \sum_{m=1}^{3} \frac{\partial R_m}{\partial x_{\alpha i}} \frac{\partial}{\partial R_m} + \sum_{k=1}^{3s-6} \frac{\partial q_k}{\partial x_{\alpha i}} \frac{\partial}{\partial q_k}$$

If a heavy particle configuration can be found at which all these cartesian first derivatives of the potential vanish, then the equivalent derivatives with respect to translation, rotation and internal motion should also vanish. So such a point should be a genuine stationary point on the potential surface $V_p(\mathbf{q})$. If all the cartesian second derivatives at his point are known, then the matrix of these second derivatives, the cartesian *Hessian* at the stationary point, should have six zero eigenvalues. If the stationary point is a minimum, then the remaining $3s - 6$ eigenvalues should be positive.

If the first and second derivatives with respect to the internal coordinates are required elsewhere than at the minimum, then it is necessary to perform the appropriate transformations from heavy particle cartesians to internal coordinates explicitly and, as might be anticipated from the earlier discussion of the jacobian matrix elements in the construction of tailor-made Hamiltonians, is generally, no easy matter. Alternatively, the surface may be fitted directly to the internal coordinates and the relevant derivatives determined from the fitted surface. This is generally a computationally intensive process and, with certain choices of internal coordinates, must be repeated for each isotopomer of the chosen system. The papers[43,44] provide an account of the basis of a widely-used scheme for transforming between derivatives, which is realised in the program INTDER developed by Allen and co-workers at the Center for Computational Chemistry, University of Georgia. (See also ref. 45.)

This sort of transformation is more important than might at first be supposed because not always is the potential arising from a particular electronic structure calculation, developed in a power series about the equilibrium geometry specified by the vanishing of the first derivatives in that calculation: rather the equilibrium geometry is assigned from another, superior, calculation or perhaps on the basis of experimental evidence.

More accurate attributions of spectra require more extensive knowledge of the potential surface and if this knowledge is to be expressed in terms of higher order derivatives of the internal coordinates, then it is necessary to obtain them explicitly. If one starts from an electronic structure program one will often have analytic first and second derivatives in terms of heavy particle cartesians and it is perfectly possible, though computationally intensive, to obtain higher-order derivatives in terms of internal coordinates from them by finite difference methods. There is a widely used package SPECTRO,[46] which computes the cubic and quartic force constants of a normal coordinate expansion to accomplish the relevant calculations about the equilibrium geometry. A context for this kind of work is provided by K. Sarka and J. Demaison in Chapter 8 of ref. 40. Alternatively it is possible, as suggested above, to construct an analytic fit of the computed potential points to produce a suitable form for $V(\mathbf{q})$ that can be manipulated directly.

Highly accurate calculations need to consider how the clamped nuclei electronic problem fits into the full problem as specified by the Schrödinger Hamiltonian for the electrons *and* the nuclei. Corrections arising from these considerations are usually called *Born-Oppenheimer* corrections. A brief account of the simplest of such corrections, the so-called *diagonal* one, can be found in Chapter 3 of ref. 39 and also in section 2.3.2 of ref. 40. A more general discussion of the breakdown of the Born-Oppenheimer approach forms Part 4 of ref. 40.

Corrections to account for relativistic effects, not considered in the Schrödinger formulation, also need to be considered. An introductory account of what is involved here can be found in Chapter 8 of ref. 39 and in section 2.3.3 of ref. 40. A more developed account can be found in B. Hess and C. Marian's article that forms Chapter 7 of ref. 40 and a pretty full account, including a consideration of quantum electrodynamics, constitutes Part 4 of Volume 2, by H. Quiney, of ref. 42.

A single basic approach is possible to computing solutions to the Eckart-Watson Hamiltonian. The form of the Hamiltonian is a given and can be simplified in a first approximation using an effective quadratic field. The solutions of the simplified problem can then be used as a basis for approximate solutions to the full problem. It can perhaps be considered as a problem analogous to the atomic problem in electronic structure calculations, where the Hamiltonian is of definite form and can be usefully simplified using an effective central field to yield orbitals and orbital energies as a start for less approximate solutions.

Tailor-made Hamiltonians, however, have many forms. They are constructed afresh for each problem and are expressed in coordinates that seem appropriate for the solution of the chosen problem. There cannot in such cases be a single solution strategy. Although the situation here is, in some aspects, analogous to the molecular electronic structure problem, it is not the case that we can develop a general effective field for the problem as is possible for the electronic problem in terms of the LCAO-MO-SCF approach. Of course a solution strategy here may well be appropriate to a whole class of problems. Thus a similar strategy might be usable for all three-nuclei systems expressed in internal coordinates coordinates involving two lengths and an angle and so on.

6.1 Perturbational Computations. – Although it is perfectly possible to perform perturbational computations based on any specified Hamiltonian, for reasons that have, it is hoped, become clear in the discussion of earlier sections, a perturbation approach is a very natural way to approximate solutions of the Eckart-Watson Hamiltonian. Since this is the traditional Hamiltonian with which to describe molecular spectra, the overwhelming number of perturbational calculations make use of it and in this section attention will be confined to it.

If the displacements are small then it is sufficient simply to consider the quadratic approximation to the potential, then it is possible to construct a special set of internal coordinates which diagonalise the force constant matrix and maintain the kinetic energy operator forms. Let the normal coordinates be denoted as Q_k. If the displacements are small then it is plausible as well, to treat the coordinates Q_k as small quantities and to neglect the Coriolis coupling operator and to treat the inverse generalised inertia tensor as $\mu^0 = I^{0-1}$, the inverse of the equilibrium inertia tensor. In these circumstances one can choose the frame fixed in the body so that the equilibrium inertia tensor is diagonal (the principal axis choice) with $\mu^0_{\alpha\beta} = \delta_{\alpha\beta}/I^0_{\alpha\alpha}$.

If this is the case, then the equations for the kinetic energy simplify to yield, together with the quadratic potential, a Hamiltonian

$$H^0(\phi, \mathbf{Q}) = K^0(\phi) + H^0(\mathbf{Q})$$

in which

$$K^0(\phi) = \frac{1}{2}\sum_\alpha \mu^0_{\alpha\alpha} \Pi^2_\alpha \equiv \frac{1}{2}\sum_\alpha \frac{\Pi^2_\alpha}{I^0_{\alpha\alpha}} \tag{1.55}$$

$$H^0(\mathbf{Q}) = -\frac{\hbar^2}{2}\sum_l \frac{\partial^2}{\partial Q_l^2} + \frac{1}{2}\sum_l \lambda_l Q_l^2 \tag{1.56}$$

where the zero of the potential energy in (1.56) has been chosen to incorporate the constant terms. It has also been assumed that the coordinates are normal ones, chosen to diagonalise the quadratic approximation to the potential. Thus the λ_l are the eigenvalues of the quadratic form **f**. All the eigenvalues must be positive for a stable molecule because the matrix of second derivatives of the internal coordinates evaluated at a minimum (the Hessian at the minimum) must be positive definite. If for any reason the expansion is made not about the equilibrium geometry, but rather at some other geometry, then it is possible to get negative values of λ_l and hence imaginary vibration frequencies. The two above equations represent exactly the Bjerrum model for molecular motion as realised by David Dennison and discussed earlier.

Because the elements of $\boldsymbol{\mu}^0$ are constants then (1.55) is just the Hamiltonian for an asymmetric top. The rotational matrix element (dropping the parity designator cf (1.42)) simplifies to

$$<JMk' | k^0(\phi) | JMk> =$$

$$\frac{\hbar^2}{4}\left[\frac{(\mu_{xx}^0 - \mu_{yy}^0)}{2}(C_{Jk+1}^+ C_{Jk}^+ \delta_{k'k+2} + C_{Jk-1}^- C_{Jk}^- \delta_{k'k-2})\right]$$

$$+\frac{\hbar^2}{2}\left[\frac{(\mu_{xx}^0 + \mu_{yy}^0)}{2}(J(J+1) - k^2) + \mu_{zz}^0 k^2\right]\delta_{k'k} \quad (1.57)$$

The $(2J+1)$ dimensional secular problem composed of these matrix elements cannot generally be solved to give an energy expression in closed form but the rotational wave functions solutions are of the form

$$^M x_\tau^J(\phi) = \sum_{k=-J}^{k=J} c_{\tau k}^J |JMk>, \quad \tau = -J, -J+1, \ldots, J$$

The $c_{\tau k}^J$ are constant coefficients and each rotational wave function is associated with an energy $E_{J\tau}$.

If two of the equilibrium moments of inertia are the same (the symmetric top case) then these may be designated as x and y and the first term in (1.57) vanishes. The energy is then given by the last term in (1.57) and the $|JMk>$ are individually angular eigenfunctions. Thus for the symmetric top, k is a good quantum number.

The Hamiltonian (1.56) simply represents a sum of non-interacting Harmonic oscillators, each with a wavefunction of the form

$$\psi_{n_l}(Q_l) = N_l e^{-\frac{\alpha_l Q_l^2}{2}} H_{n_l}(\sqrt{\alpha_l} Q_l)$$

$$\alpha_l = \frac{\sqrt{\lambda_l}}{\hbar} \equiv \frac{\omega_l}{\hbar}$$

and the energy of the oscillator is

$$\epsilon_{n_l} = (n_l + \frac{1}{2})\hbar\omega_l \equiv (n_l + \frac{1}{2})h\nu_l$$

The full vibrational wave function is then usually written

$$\Psi_{\mathbf{n}}(\mathbf{Q}) = \prod_l \psi_{n_i}(Q_i) \tag{1.58}$$

and the total vibrational energy of the system is just

$$E_v = \sum_l \epsilon_{n_l}$$

The assumption here is that it is not necessary to consider explicitly changes induced in the normal coordinates by permutations of identical particles. Further, that it is not necessary to consider nuclear spin statistics. Thus the normal coordinates are regarded as specifying identifiable entities and a product form for the wave function is acceptable.

The total wavefunction for the internal motion is the product

$$\phi_p(\mathbf{c},\mathbf{z})^M x_\tau^J(\phi)\Psi_{\mathbf{n}}(\mathbf{Q})$$

in which $\phi_p(\mathbf{c}, \mathbf{z})$ is the electronic wavefunction taken at the equilibrium nuclear geometry, $\mathbf{c}$. The wavefunction for the full problem in the single product approximation is

$$T(\mathbf{X}_T)^M x_\tau^J(\phi)\Psi_{\mathbf{n}}(\mathbf{Q})\phi_p(\mathbf{c},\mathbf{z}) \equiv |T>|J,\tau;\mathbf{n}>|p> \tag{1.59}$$

The total energy of the molecule in this approximation is

$$E = E_T + E_{J\tau} + E_{\mathbf{n}} + V(0)$$

The translational energy E_T is usually ignored as is the translational wavefunction and the fact that in this approximation the energy is the sum of an electronic and a rotational and a vibrational part is often said to specify the Born Oppenheimer approximation.

In practice the coordinates $\mathbf{Q}$ are not known in advance so that the Taylor expansion in them is not directly possible. Traditionally the Wilson **B** matrix method[#] (see also Chapter 21 by Bissonette in Part 3 of ref. 42) is used to

---

[#] It would perhaps be more just to call this the Wilson-El'yashevich **B** matrix method, for it was certainly developed by this Russian worker at the same time as it was developed by Wilson. It would also be fair to cite the book by El'yashevich et al. alongside that of Wilson, Decius and Cross,[17] as covering much the same ground. A first version of this book was published in 1949 and it was revised and shortened by Gribov for a second edition ref. 47. I am grateful to Prof. Császár for this information.

construct internal coordinates relative to an assumed equilibrium geometry, which satisfy the Eckart constraints and in terms of which, force constants can be assigned to yield a bilinear form for the potential energy. Diagonalising this form while simultaneously diagonalising the kinetic energy operator yields a set of trial normal coordinates. In order to specify the coordinates for use here it is necessary to specify the equilibrium geometry, the particle masses and the elements of the **B** matrix that specify internal coordinates that describe bond stretching, valence angle bending, bending in an angle between a bond and a plane defined by two bonds, torsional motion about a bond and so on. They are specified so as to enable force constants to be chosen according to a physical picture of the molecular vibrations. In order to satisfy the Eckart conditions the elements of **B** are chosen[‡] so that

$$\sum_i B_{k,\alpha i} = 0, \quad \sum_i \sum_\alpha (\hat{\mathbf{a}}_i)_{\beta\alpha} B_{k,\alpha i} = 0$$

When expressed in matrix form this scheme produces internal coordinates according to

$$q_k = \sum_i \sum_\alpha B_{k,\alpha i} \rho_{\alpha i} \qquad (1.60)$$

If we define three translational and three rotational coordinates as

$$\mathbf{T} = M^{-1} \sum_i m_i \rho_i, \quad \mathbf{R} = \mathbf{I}^{0-1} \sum_i \hat{\mathbf{a}}_i \rho_i$$

where M is the total heavy particle mass. They vanish when the Eckart conditions are satisfied. We may define a complete inverse transformation as

$$\rho_i = \mathbf{T} + \hat{\mathbf{a}}_i^T \mathbf{R} + (\mathbf{m}^{-1} \mathbf{B}^T \mathbf{G})_i$$

where **m** is the diagonal matrix of heavy particle masses and

$$\mathbf{G}^{-1} = \mathbf{B}\mathbf{m}^{-1}\mathbf{B}^T$$

If it were desired simply to obtain a set of coordinates that allowed the kinetic energy operator for internal motion to be written in diagonal form, it would be sufficient to determine an orthogonal matrix that diagonalised **G**. It is however usual to adopt a rather less direct route. A set of force constants is assigned for the problem and these are made up into a square matrix **f**, of dimension $3s - 6$. The matrix is then transformed to one of dimension $3s$ according to

$$\mathbf{f}^x = \mathbf{B}^T \mathbf{f} \mathbf{B}$$

[‡] Here the conventional ordering is used for the subscripts on the elements of **B**. It is the transpose of the ordering used on the subscripts of $\mathbf{I}_{il}$.

which is thought of as the force constant matrix expressed in the laboratory coordinates. It is then further transformed to a matrix $\bar{\mathbf{f}}^x$ with elements

$$\bar{f}^x_{\alpha i,\beta j} = m_i^{-\frac{1}{2}} f^x_{\alpha i,\beta j} m_j^{-\frac{1}{2}}$$

and this matrix is then diagonalised by means of an orthogonal matrix **C**. The resulting diagonal matrix λ has six zero entries, corresponding to the three translational and three rotational modes and $3s-6$ non-zero entries corresponding to the vibrations. If we label the columns of **C** corresponding to these non-zero entries by k, $k = 1, 2, \ldots 3s-6$ then it follows that the coordinates given by

$$Q_k = \sum_i \sum_\alpha C_{\alpha i,k} m_i^{\frac{1}{2}} \rho_{\alpha i} \equiv \sum_i \sum_\alpha l_{\alpha i,k} m_i^{\frac{1}{2}} \rho_{\alpha i} \qquad (1.61)$$

not only diagonalise the kinetic energy operator but also the potential energy operator as specified by the force constant matrix. They are, therefore, normal coordinates and special cases of the general vibrational coordinates represented in (1.11).

Computer programs to perform such calculations have been available since the 1960s. An early and influential one was developed by J.H. Schachtschneider and forms the basis of many later programs, for example the program UMAT by McIntosh and Peterson, made available through the Quantum Chemistry Program exchange (QCPE) at the University of Indiana as QCMP067. A later one is is the system ASYM: this arises from the work of Hedberg and Mills[48,49] and can be accessed on the web-site.[50] The output from such a program can be compared with experiment and the input successively modified until a satisfactory output is achieved. Such programs follow spectroscopic practice by using heavy particle masses derived from atomic masses rather than from nuclear masses. Although perhaps a little dated, an extensive and detailed account of the assignment of force constants by this sort of fitting, which also contains many references and a critical discussion of **B** matrix construction, may be found in the Specialist Periodical Report[51] in 1975. A more recent discussion of force constants and fitting can be found in the article by Sarka and Demaison mentioned above, forming Chapter 8 in ref. 40. This article also considers force constants arising from electronic structure calculations.

As was noted above, modern non-relativistic electronic structure packages often include facilities that make possible the evaluation of the first and second derivatives of the potential energy, as calculated at a particular level of approximation, with respect to the cartesian coordinates that specify the heavy particle positions. From this information the equilibrium geometry may be obtained while the matrix of second derivatives evaluated at that geometry (the cartesian Hessian at the minimum) is a directly calculated form that is related to the matrix $\mathbf{f}^x$ above. The sort of thing that is involved here is illustrated in Chapter 11 of ref. 39 where a systematic study of computations on the water molecule is presented. This chapter also considers force constants as calculated in some density-functional (DFT) approaches.

Were the minimum in the cartesian form to have been located sufficiently accurately then the first derivatives of the potential with respect to translational, rotational and internal motion would all vanish there too and the Hessian at the minimum would be exactly of the form f^x. In that case six of the eigenvalues of the Hessian would vanish and the remaining eigenvalues would correspond directly to harmonic vibrational frequencies and the associated eigenvectors would define normal coordinates in terms of the heavy particle cartesian coordinates. In practice, however, the cartesian minimum is not always located sufficiently accurately and the Hessian often has three very small eigenvalues and three somewhat larger ones, usually well separated from a set of much larger ones, but this will depend on whether the calculation is sufficiently converged and whether there are any genuine low-frequency modes too. Thus the output for water from *Gaussian98* in an HF/3-21G* calculation recorded in the manual "Vibrational Analysis in *Gaussian* (1999)" looks like this for the nine lowest frequencies expressed in cm^{-1}.

Low frequencies   −0.0008   0.0003   0.0013   40.6275   59.3808   66.4408
Low frequencies   1799.1892   3809.4604   3943.3536

Many packages contain routines to transform or project the "impure" Hessian in some way so as to make sure that the internal modes are uncontaminated and, if they are to be treated as if they were normal coordinates in the Eckart-Watson scheme, that they actually satisfy the Eckart conditions. There are routines in, for example, the *Gaussian*,[52] or GAMESS,[53] packages that may be invoked to perform such projections or transformations to yield normal coordinates in Eckart form. An elementary account of how a projection is performed is given in section 13.1 of ref. 39. An equivalent account is provided in the vibrational analysis section of the *Gaussian* manual.

It should be remembered that while the potential for the internal motion in the heavy particle problem must be invariant under translations and rotation-reflections, as explained above in discussing the Eckart conditions (1.3) and (1.4), it is not required that the potential be expanded about any particular configuration for a proper formulation of the internal motion problem. For example, in a quite early paper on the assignment of vibrational spectra of diatomics from potentials calculated *ab initio*, Schwendeman[54] argued very persuasively that the potential should be expanded about the experimentally assigned equilibrium. Of course this means that there will be linear terms (forces) in the expansion but if these are properly considered in solving the internal motion problem, nothing will go wrong. In fact general expansions are widely and effectively used in all kinds of calculations. However it should also be remembered that potentials containing odd powers of a variable with the range $(-\infty, +\infty)$ are not bounded below and must therefore be treated in practice as having only a finite range.

In general the heavy particle masses used in molecular structure packages are derived from the atomic masses and not from the nuclear masses and so yield results consistent with standard spectroscopic usage. This usage is generally justified by saying that it might be expected that the electrons associated with a particular nucleus when in atomic form, are so light that they will follow the nuclear motion so closely as to contribute to the overall mass of the moving particle.

Although not of immediate spectroscopic significance it should be noted that there have been developments in the calculation of molecular properties such as dipole moments, bond lengths and bond angles to allow for heavy particle vibrations. The theoretical foundations for such developments were laid some time ago, chiefly by Bishop and his co-workers[55] (see also S. Sauer and M. Packer in Chapter 7 of ref. 40) but the program system DALTON may be used to perform such property calculations[56] within the context of the Eckart-Watson Hamiltonian for the heavy particle motion. This program system has the interesting feature that generally in it, the potential is expanded not about the geometry at an electronic minimum, but about a geometry which minimises the sum of the electronic and zero-point vibrational energy.

In broad general terms it can be said that if one can perform an electronic structure calculation on a molecule whose spectrum it is wished to interpret, and if that calculation yields an isolated fairly deep minimum in the potential surface, then one can readily perform perturbation calculations to obtain estimates of most if not all, of the relevant spectroscopic parameters. How well the computed parameters compare with those experimentally assigned, when these are known, naturally depends on the quality of the electronic structure calculation and the extent to which the perturbation calculations are carried.

If the determination of equilibrium geometry is considered, and this is the central element in the calculation of moments of inertia and hence of rotational constants, then it seems that very good basis sets are required to get decent equilibrium geometries. At the HF level (and indeed at all higher levels of calculation) the better the basis set, the shorter the bond length and extrapolation leads one to believe that at the basis set limit, bond lengths will be too short perhaps by 2 or 3%. It is much more difficult to say anything about bond angles, which can be too large or too small but in triatomics they are often too large and in water, for example, the bond angle is about 2% too large.

Commonly accepted values for the "experimental" harmonic frequencies for water are (in cm^{-1}) 3943, 3832 and 1649 and these compare quite well with the calculated figures quoted above. The "folklore" in these matters seems to be that at the HF level with a reasonable basis the harmonic frequencies calculated at the computed optimum geometry are usually about 7% too high and it is usual to scale computed values accordingly. Anharmonic *corrections* at this level naturally depend upon the accuracy with which the harmonic frequencies are calculated but they can sometimes be 10–13% too high. At the MP2 level the harmonic frequencies seem to be about 5% too high.

A more complete discussion of such matters can be found in Chapter 11 of ref. 39 where there is also a discussion of what can be done with density functional methods and for this last, see also ref. 57. DFT methods sometimes give pretty good equilibrium geometries but it seems rather difficult to determine any systematic features, however they generally seem to give harmonic frequencies that are about 1% too high.

It would perhaps not be too unfair to say that really quite low level calculations are capable of giving ball-park figures for most spectral parameters which will enable tolerable guesses to be made about the identity of molecules known only through their spectral signatures.

However, it should not be thought that the use of modern electronic structure packages limits one to pretty approximate results. Using them it is also possible to pass beyond the quadratic force field to calculate anharmonic corrections and the like and to incorporate these effects by using perturbation theory. The results of the perturbation theory are achieved by means of a contact transformation first introduced in this context by Van Vleck but an accessible account the theory and examples of results can be found in ref. 58 and also in the contribution of Sarka and Demaison that forms Chapter 8 of ref. 40 and in the article by Császár on *Anharmonic Molecular Force Fields* in ref. 41.

Examples of what sort of things are possible here can be found in two special issues of *Spectrochimica Acta Part A*. The first,[59] in 1997, is entirely devoted to molecular spectra and in the second,[60] in 1999, the last seven papers are relevant. Some feeling for the best that can be done on a small molecule can be gained looking at Table 2.2 in Chapter 2 of ref. 40. This table presents the force fields up to fourth order for the electronic ground state of water using a aug-cc-pVQZ basis at the HF, MP2, CCSD and CCSD(T) levels of computation and compares them with the force field that is derived empirically from the water spectrum itself. The computed force fields result from expansions about an empirical estimate of the equilibrium geometry and so contain, albeit small, linear terms (forces). And this sort of achievement is not confined to triatomics only. It is possible to deal with quite large systems such as *cis*-1-chloro-2-fluoroethene (*cis*-CHCl=CHF) at the CCSD(T) level of electronic computation to yield very satisfactory agreement with observation.[61] A review of interest in this context is ref. 62 and much of the perturbation theory literature is reviewed, *inter alia*, in ref. 63. Here a perturbational approach in terms of generalised Rayleigh-Schrödinger theory rather than the more usual contact transformation formulation, is presented, together with a numerical attempt on the spectrum of methane.

When there is not a fairly deep isolated minimum in the potential, a case often described colloquially as the molecule being *floppy*, then the perturbation approach outlined above cannot be used effectively. As will be seen, the variational approach can be and that will be described in the next section, but following the work of Sayvetz mentioned above, perturbation approaches have been developed that allow for the large amplitude motions (LAMs) that are characteristic of such floppy molecules. These approaches are based upon allowing for a LAM in the form of the Hamiltonian, a typical LAM being rotation about a single bond, treating that LAM specially and applying perturbation theory to all the other motions. The theory here, based as it is on the Eckart Hamiltonian, has been briefly treated above by reference to the work of

Ezra[21] and of Bunker and Jensen.[23] This last reference contains an account of the perhaps the first effective computational approach to a LAM using the Sayvetz-Eckart approach in describing the work of Hougen, Bunker and Johns.[64] A further development along these lines, the MORBID (Morse oscillator rigid bender internal dynamics) method of Jensen, is also described. This interesting method can usefully be thought of as transitional between perturbational and variational approaches. An account of some of the computational work done in this area can be found in Chapter 12 by J. Makarewicz of ref. 40. This sort of approach was taken, one might say, to extremes, in work by Miller, Handy and Adams[65] in 1980 to describe a reaction-path Hamiltonian with the LAM along the reaction-path. Such an approach was later incorporated into the variational program 'MULTIMODE' to be described below.

6.2 Variational Computations. – Variational computations are based upon the calculation of matrix elements between functions of all the internal coordinates. Such many variable functions are exemplified by (1.58), which is the direct product form for normal coordinates constructed from a basis of single variable functions. A collection of such many variable functions is called an *internal coordinate* basis. The single variable functions can be thought of as a bit like molecular orbitals in electronic structure calculations. The symmetry of the functional forms reflects the symmetry of the potential and permutational symmetry in relation to the requirements of the Pauli principle is not usually considered. It is sometimes the case that the single variable functions are coupled as, for instance, happens in the case of atomic orbitals in spherical polar coordinates, where the coupled form is $P_m^l(\theta)\Phi_m(\phi)$ with m limited to values that depend on l. Thus although variational trial functions are usually of product form, they are not always of *direct* product form.

As has been shown in (1.41) and (1.42), an effective Hamiltonian may be constructed that depends only upon the internal variables **q** but which contains the rotational state quantum numbers J, k and k' as parameters. The effective Hamiltonian thus forms a square matrix of side $2J+1$. If an internal coordinate basis is chosen to be of extent S and independent of k then the secular problem arising by applying the linear variation theorem will be of dimension $(2J+1)S$.

The history of the variational approach to the computation of molecular spectra from its inception in the early seventies up to 1978 is summarised incisively and informatively in the review by Carney, Sprandel and Kern in 1978.[66] The authors themselves were extremely active in the development of the field and they describe in that review the early attempts on triatomic molecules using both the Eckart-Watson Hamiltonian and tailor-made Hamiltonians. They also give a critical account of the uses of the perturbation theory approach.

Clearly one can write an individual internal coordinate function as a linear combination of known, fixed functions much as molecular orbitals are written as linear combinations of atomic orbitals. Thus if one were attempting to describe a triatomic in bond-length, bond-angle coordinates one might well attempt to realise the functions corresponding to the bond-stretching as linear combinations of spherical oscillators in the bond-length coordinate. Assuming that all the integrals were do-able then one could set up a sort of SCF scheme, a bit like the standard electronic LCAO scheme, to determine the optimum form of such a linear combination. It would, in general, have to be a matrix, SCF approach except when $J = 0$ but the idea is clear enough and it is discussed by Jung and Gerber in Chapter 11 of ref. 40. The SCF process would generate a set of such functions and these could be used to form an internal coordinate basis for use in a linear variational computations, a bit like electronic CI calculations.

Although the variation approach offers a way of treating systems that cannot be treated by perturbation theory, it does so only at a price. If the potential is not separable, matrix elements involving it must be evaluated numerically and if there are $3s - 6$ variables and if one chose an m point integration scheme in each variable, the computational effort scales as $(3s - 6)^m$ and is thus an exponentially hard computing problem. This can be tolerated when $s = 3$ but for tetratomic systems and beyond, special steps have to be taken to render variational approaches feasible. When normal coordinates are used as in, for example, the MULTIMODE program, then reduced piecewise fitting of the potential is sometimes attempted.[67] In this way the potential can be effectively presented as a sum of products of one-dimensional cuts, much easing the computational effort involved. A review by T. Carrington of what has been attempted here for $J = 0$ cases can be found as the Chapter *Vibrational Energy Level Calculations* in ref. 41. It also contains a measured and careful assessment of those theoretical and numerical factors that need to be kept in mind in order to ensure that any calculation undertaken, actually yields meaningful results.

Even for triatomic systems, calculations of higher J states rapidly become intractable unless it is possible to separate, at least partially, the rotational and vibrational motions of the problem. What is usually done here is that calculations are performed for a given J value in which only the last term in (1.42) is retained, so that the calculations remain diagonal in k. The resulting solutions for given J and each k are then "contracted", to yield a new basis as a fixed linear combination of functions in the internal coordinate basis and this contracted basis is then used to solve the rotational problem. It is rather analogous to the use of a contracted gaussian basis in electronic structure calculations. An account of how this is done for a triatomic can be found in ref. 68. Of course contraction is, quite generally, an attractive strategy, even for $J = 0$ cases, enabling the construction of effective trial functions that lead to decent results from lower dimension secular problems than would otherwise be possible. This is particularly important if one does not use a full eigensolver, such as the Householder, but rather uses an iterative direct method, such as the Lanczos. The article[69] considers contracted methods rather generally but in the light of a rather specialised approach to using an iterative eigensolver.

In the foregoing discussion it has been assumed that the single variable functions from which the product functions are formed, are given explicitly and matrix elements are to be evaluated directly in terms of them. Such an approach is said to be in a *Finite Basis Representation* or an FBR. However this is not the only way in which functions can be expressed. It is, for example, perfectly possible to represent an orthogonal polynomial in terms of its weights and positions in a gaussian quadrature scheme and such representations have come to be widely used in the context of variational calculations. In this context they are usually said to provide a *Discrete Variable Representation* or a DVR. There is a review of such methods in an article by Light and Carrington.[70] If the DVR is chosen to represent exactly the functions which are the imagined single variable functions, then the energy upper bound properties, implicit in the variational formulation implied above, are maintained. (See also Section 9.4.1 in Tennyson's contribution, *Variational Calculations of Rotation-Vibration Spectra*, to ref. 40.) However DVR methods as used in practice are not strictly variational although, in the hands of experts, they usually exhibit only very small departures from variational behaviour.

There are two related difficulties for any successful attempt on a problem by variational means: one is the choice of embedding scheme and the associated choice of internal coordinates and the other is the construction of an effective internal coordinate basis.

In the case of a triatomic system there may seem to be no real choice about the embedding other than to choose the molecular plane. However that still leaves the orientation of the embedded axis system open. In the Eckart approach to the triatomic the z-axis fixed in the body will be chosen perpendicular to the plane but in many tailor-made approaches the plane is chosen to be the $x - z$ plane with the y-axis perpendicular to the plane. Of course, were exact solutions possible, this choice would be a matter of indifference, but in any approximate scheme a wise choice can do much to simplify the calculation. The choice made in the Eckart approach ensures that the coupling of the angular to the internal motions is minimised and thus the internal coordinate basis may be chosen to be largely independent of J and k. However, with the standard choice of internal coordinate basis, it is not easy to deal with large amplitude motions. The placing of the molecule in the $x - z$ plane allows the elements of the effective vibration-rotation operator to be chosen real and thus simplifies their calculation. If bond-length and bond-angle internal coordinates are chosen, the single variable function of bond-angle coordinate must be chosen to depend on k to be effective, yielding internal coordinate basis functions which are coupled to the rotational motion. But such a choice makes possible an internal coordinate basis that avoids the unpleasantnesses that arise from large amplitude motions. It might be supposed that the angular coupling in the triatomic functions could be avoided by choosing three bond-length internal coordinates, but if this choice were made, the integration ranges of these variables would be coupled and would render the calculation of matrix elements between the internal coordinate basis functions deeply nasty. Bond-length bond-angle coordinates are not orthogonal, so that the kinetic energy operator contains cross-terms between the radial and the angular coordinates and between the two radial coordinates. This means that the internal motions are coupled by the kinetic energy operator and it also makes it difficult to apply the DVR method to such a form. Orthogonal coordinates in the Radau form can be chosen, which closely approximate the bond-length bond-angle coordinates if one of the atoms is much heavier than the other two. But it is not, of course, always possible to choose such an effective approximative set. These and related problems in variational calculations on triatomic molecules are discussed fully and effectively in Chapter 9 of ref. 40 by Tennyson.

The general thrust of the argument should now be clear. Individual but entangled choices have to be made about the embedding, the internal coordinate specification and the choice of basis for any problem to be faced. Within a chosen embedding, any set of internal coordinates may be used. This is because, as noted above at the beginning of Section 3, that locally, any set of internal coordinates is an algebraic function of any other set, so that any given set of internal coordinates may, in principle, be transformed into any other set. Of course in practice the relations between two sets are usually implicit and so a transformation in closed form is not possible. (It is possible, see ref. 29 B.1, to re-express in closed form, the bond-length bond-angle coordinates for a triatomic molecule in terms of the Eckart normal coordinates. It is not however possible to invert this transformation explicitly. It is possible to produce closed forms for transformations between all the internal coordinate systems mentioned above for the tailor-made triatomic Hamiltonian.) It would seem, that the choices of embedding and of internal coordinates will often, in practice, remain entangled. And once one passes beyond triatomics, this entanglement becomes more. Thus both NH_3 and H_2O_2 are tetratomic systems but intuitively it would seem inappropriate to embed in a similar manner in both nor would it seem at all helpful to try for the same internal coordinates. We are guided towards our views about the embedding by our intuitions about the potential for heavy particle motion in each system and to our views on the internal coordinates by a view of what internal motion functions might be appropriate to the description of the expected internal motion. Thus, since we anticipate "umbrella" motion in NH_3 we would choose an angular internal coordinate to reflect that with an appropriate angular function in mind. Similarly we might well go for an inter-OH group torsion coordinate and an associated function for the peroxide. In the end one must construct an approach in response to the details of the potential for the heavy particle motion as revealed by electronic structure calculation.

It is, therefore, not the case in the variational calculation of spectra that "one size fits all" and it is unfortunately not possible to say, that anyone with access to a electronic structure package can make a decent job of accounting for a spectrum using this approach. At the level of obtaining the potential it requires much more extensive electronic structure calculation than does a perturbation approach for it is essential to be able to represent the potential well everywhere that the internal basis functions are significant. At present this probably means that systems with no more than 5 or 6 nuclei and less than 20 electrons are the most that can be tackled. Up to this number of heavy particles there are a number of forms available of the internal motion Hamiltonian. Thus variational calculations remain what electronic structure calculations were in the 1960s and 70s, an endeavour that requires specialist care and attention. What can be done with such care and attention is exemplified in another special issue of *Spectrochimica Acta*[71] that appeared in 2002.

However that may be, it is perhaps useful to provide an account of two variational approaches to the vibration-rotation problem for any who might think of performing variational calculations to elucidate vibration-rotation spectra.

MULTIMODE Based on the Eckart-Watson Hamiltonian. – The ultimate development of the variational approach to the Eckart-Watson formulation is probably the MULTIMODE program of Bowman and Carter. It can, in one version, allow for a large amplitude internal motion. This program is not yet in the public domain but its closeness to the Eckart formulation makes it very accessible and there is a web site that presents it as if one were to be a user so that it may be studied. The theoretical under-pinning of the program in its basic form is described in ref. 34 and information about its use can be found on the web site.[72] A form of the program which allows for one large amplitude motion to be considered is described using glyoxal as an example in ref. 73 and in ref. 74. How the potential energy surface for the chosen problem might be represented is presented in ref. 67.

The program can be used to perform vibrational self-consistent field (VSCF) calculations. It can also be used to perform extended basis calculation utilising a given single variable basis. These are called V-CI calculations. Alternatively the single variable functions may be chosen from the SCF results and used in VSCF-CI calculations. It is possible to deal with states for which $J > 0$ by either exact or approximate means. The vibrational basis in the VSCF-CI approach is non-orthogonal and so a generalised eigenvalue problem results. If the basis is sufficiently small, less than 5000 functions or so, a full eigensolver may be used, otherwise a Davidson-Lanczos process can be used to provide a few of the lower roots.

TRIATOM Based on a Tailor-made Hamiltonian for Triatomics. – A general description of the system TRIATOM, developed by Tennyson and his co-workers, may be found in ref. 75. An account of the theory behind and the construction of the tailor-made Hamiltonian can be found in ref. 76. The programs are in the public domain, and are freely accessible. They can deal with any triatomic system in which large amplitude bending motions are possible. TRIATOM in the FBR is presented in the article[77] and may be obtained from ref. 78. A DVR form is presented most recently in the article[79] and may be found at ref. 80. Associated with these programs are programs to compute dipole moment surfaces so that transition intensities between states can be calculated.

Among the earliest of systems considered by using the variational approach was H_3^+. The work was carried out using a potential based on *ab initio* electronic structure calculations carried out by Carney and Porter in 1974 and subsequently used by them in 1976 to perform vibration-rotation calculations[81] and this molecule became, and to some extent remains, the test-bed for *ab initio* variational calculations. Because it is only a two-electron system it is possible to perform very high quality electronic structure calculations on it and because it has only three internal coordinates it is possible, without excessive computation, to fit a set of computed points to produce a potential that is extensive and effective at reproducing experimental results and anticipating unobserved results. Among the more recent vibration-rotation calculations on this system is the one by Polyansky and Tennyson using TRIATOM[82] which seems to yield results of outstanding accuracy. However, to date, the *tour de force* in triatomic calculations in this scheme is a calculation made with the DVR form of the program, DVR3D, on the rotation-vibration transitions in water,[83] that starts from an exceptionally accurate and extensive potential surface and includes corrections not only for the breakdown of the Born-Oppenheimer approximation and relativistic corrections but also quantum electrodynamic corrections.

Acknowledgments

I am very grateful to the authors who replied to my often somewhat impertinent queries about their work with such good humour. I hope that those who spoke to me but are not cited here will continue to believe in my goodwill and attribute their absence simply to imperfections in my judgment.

I am grateful too, to Prof. P. Cassam-Chemaï, Prof. A. G. Császár, Prof A. Kalemos and Prof. J. Tennyson for reading and commenting upon drafts of this article.

References

1. C. Eckart, *Phys. Rev.*, 1935, **47**, 552.
2. J.K.G. Watson, *Mol. Phys.*, 1968, **15**, 479.
3. Bulletin of the National Research Council, **11** Pt 3, No. 27, 1926.
4. Lord Rayleigh, *Phil. Mag.* 1892, **24**, 410.
5. P. Drude, *Ann. d. Phys.*, 1904, **14**, 677.
6. N. Bjerrum, Nernst Festschrift, Halle, 1912.
7. N. Bjerrum, *Verh. d. D. Phys. Ges.*, 1914, **16**, 640.
8. G. Herzberg, Molecular Spectra and Molecular Structure, I Spectra of Diatomic Molecules. Van Nostrand, New York, 1950.
9. N. Bjerrum, *Verh. d. D. Phys. Ges.*, 1914, **16**, 737.
10. E.C.C. Baly, Spectroscopy, Vol III, 3rd Edn. Longmans, Green, London, 1927.
11. D. Dennison, *Phys. Rev.*, 1926, **28**, 318.
12. D. Dennison, *Rev. Mod. Phys.*, 1931, **3**, 280.
13. C. Eckart, *Phys. Rev.*, 1934, **46**, 384.
14. J.O. Hirschfelder and E. Wigner, *Proc. Nat. Acad. Sci.*, 1935, **21**, 11.
15. J.H. Van Vleck, *Phys. Rev.*, 1935, **47**, 487.
16. E.C. Kemble, The fundamental principles of Quantum Mechanics, McGraw-Hill, New York, 1937.
17. E.B. Wilson, J.C. Decius and P.C. Cross, Molecular Vibrations, McGraw-Hill, New York, 1955.

18. H.H. Nielsen, *Rev. Mod. Phys.*, 1951, **23**, 90.
19. M. Born and J. R. Oppenheimer, *Ann. der Phys.*, 1927, **84**, 457.
20. P.C. Cross and J.H. Van Vleck, *J. Chem. phys.*, 1933,**1**, 357.
21. G. Ezra, Symmetry properties of molecules, Lecture Notes in Chemistry 28, Springer-Verlag, Berlin, 1982.
22. A. Sayvetz, *J. Chem. Phys.*, 1939, **7**, 383.
23. P.R. Bunker and P. Jensen, Molecular Symmetry and Spectroscopy, 2nd Edition, National Research Council (Canada), 1998.
24. C.F. Curtiss, J.O. Hirschfelder and F.T. Adler, *J. Chem. Phys.*, 1950, **18**, 1638.
25. M.A. Collins and D.F. Parsons, *J. Chem. Phys.*, 1993, **99**, 6756.
26. B. Schutz, Geometrical methods of mathematical physics, Cambridge University Press, Cambridge, 1980.
27. J.C. Louck, *J. Mol. Spec.*, 1976, **61**, 107.
28. Yu. S. Makushkin and O.N. Ulenikov, *J. Mol. Spec.*, 1977, **68**, 1.
29. B.T. Sutcliffe, *Advances in Chemical Physics*, 2000, **114**, 1.
30. J.M.F. Gilles and J. Philippot, *Int. J. Quantum Chem.*, 1972, **6**, 225.
31. E. Wigner, *Nachrict. Ges. d. Wiss. Göttingen*, 1930, 133.
32. R.S. Knox and A. Gold, Symmetry in the Solid State, Benjamin, New York, 1964.
33. H.C. Longuet-Higgins, *Mol. Phys.*, 1963, **6**, 445.
34. S. Carter, J. Bowman and N. Handy, *Theor. Chem. Acc*, 1998, **100**, 191.
35. H.M. Pickett, *J. Chem. Phys.*, 1972, **56**, 1715.
36. H. Meyer, *Ann. Rev. Phys. Chem.*, 2002, **53**, 141.
37. T.J. Lukka, *J. Chem. Phys*, 1995, **102**, 3945.
38. J. Pesonen and L. Halonen, *Advances in Chemical Physics*, 2003, **125**, 269.
39. F. Jensen, Introduction to Computational Chemistry, Wiley,New York, 1999.
40. P. Jensen and P.R. Bunker, Ed. Computational Molecular Spectroscopy, Wiley, New York, 2000.
41. P. von Ragué Schleyer, N.L. Allinger, T. Clark, J. Gasteiger, P.A. Kollmann, H. F Schaefer III, P.R. Schreiner, Eds. Encyclopedia of Computational Chemistry, Wiley, New York, 1998
42. S. Wilson, P. Bernath and R. McWeeny Eds. Handbook of Molecular Physics and Quantum Chemistry, Wiley, New York, 2003.
43. W.D. Allen and A.G. Császár, *J. Chem. Phys*, 1993, **98**, 2983.
44. W.D. Allen, A.G. Császár, V. Szalay and I.M. Mills, *Mol. Phys.*, 1996, **89**, 1213.
45. W. Thiel, *Mol. Phys*, 1989, **68**, 427.
46. A. Willets, J.F. Gaw, W.H. Green Jr. and N.C. Handy, SPECTRO, as second order rovibrational perturbation theory program, Version 3.0, University Chemical Laboratory, Cambridge UK, 1994.
47. M.V. Volkenstein, L.A. Gribov, M.A. El'yashevich and B.I. Stepanov, Kolebaniya Molekul, Izd. Akad. Nauka SSSR, Moscow, 1972.
48. L. Hedberg and I.M. Mills, *J. Mol. Spectrosc.*, 1993, **160**, 117.
49. L. Hedberg. and I.M. Mills, *J. Mol. Spectrosc*, 2000, **203**, 82.
50. www.uni-ulm.de/ typke/progbe/asym.html
51. J.L. Duncan, *Specialist Periodical Reports*, 1975, **3**, 104.
52. Gaussian 98 (Revision A.11.3), M.J. Frisch, G.W. Trucks, H.B. Schlegel, G.E. Scuseria, M.A. Robb, J.R. Cheeseman, V.G. Zakrzewski, J.A. Montgomery, Jr., R.E. Stratmann, J.C. Burant, S. Dapprich, J.M. Millam, A.D. Daniels, K.N. Kudin, M.C. Strain, O. Farkas, J. Tomasi, V. Barone, M. Cossi, R. Cammi, B. Mennucci, C. Pomelli, C. Adamo, S. Clifford, J. Ochterski, G.A. Petersson, P.Y. Ayala, Q. Cui, K. Morokuma, P. Salvador, J.J. Dannenberg, D.K. Malick, A.D. Rabuck, K. Raghavachari, J.B. Foresman, J. Cioslowski, J.V. Ortiz, A.G. Baboul,

B.B. Stefanov, G. Liu, A. Liashenko, P. Piskorz, I. Komaromi, R. Gomperts, R.L. Martin, D.J. Fox, T. Keith, M.A. Al-Laham, C.Y. Peng, A. Nanayakkara, M. Challacombe, P.M.W. Gill, B. Johnson, W. Chen, M.W. Wong, J.L. Andres, C. Gonzalez, M. Head-Gordon, E.S. Replogle, and J.A. Pople, Gaussian, Inc., Pittsburgh PA, 2001.
53. M.W. Schmidt, K.K. Baldridge, J.A. Boatz, S.T. Elbert, M.S. Gordon, J.H. Jensen, S. Koseki, N. Matsunaga, K.A. Nguyen, S.J. Su, T.L. Windus, M. Dupuis, J.A. Montgomery, *J. Comput Chem.*, 1993, **14**, 1347.
54. R.H. Schwendeman, *J. Chem. Phys.*, 1966, **44**, 2115.
55. D.M. Bishop and B. Kirtman, *J. Chem. Phys*, 1991, **95**, 2646.
56. T. Helgaker, H. Jensen, P. Jorgensen, J. Olsen, K. Ruud, H. Ågren, A Auer, K. Bak, V. Bakken, O. Christiansen, S. Coriani, P. Dahle, E. Dalskov, T. Enevoldsen, B. Fernandez, C. Hättig, K. Hald, A. Halkier, H. Heiberg, H. Hettema, D, Jonsson, S. Kirkepar, R. Kobayashi, H. Koch, K. Mikkelsen, P. Norman, M. Packer, T. Pedersen, T. Ruden, A. Sanchez, T. Saue, S. Sauer, B. Schimmelpfennig, K. Sylvester-Hvid, P.R. Taylor and O. Vahtras, Dalton, an ab initio electronic structure program, Release 1.2. http://www.kjemi.uio.no/software/dalton/dalton.html
57. C. Bauschlicher and D. Partridge, *J. Chem. Phys.*, 1995, **103**, 1788.
58. D. Papousek and M.R. Aliev, Molecular Vibrational-Rotational Spectra, Elsevier, Amsterdam, 1982.
59. T.J. Lee, Ed. *Spectrochimica Acta Part A*, 1997, **53** Issue 8.
60. T.J. Lee and M. Head-Gordon, Eds. *Spectrochimica Acta Part A*, 1999, **55** Issue 3.
61. A. Gambi, C. Puzzarini, G. Cazzoli, L. Dre and P. Palmieri, *Mol. Phys.*, 2002, **100**, 3535.
62. M. Herman, J. Liéven, J. Vander Auwera and A. Campargue, *Advances in Chemical Physics*, 1999, **108**, 1.
63. P. Cassam-Chenaï and J. Liéven, *Int. J. Quantum Chem.*, 2003, **93**, 246.
64. J.T. Tougen, P.R. Bunker and J.W.C. Johns, *J. Mol. Spectrosc.*, 1970, **34**, 395.
65. W.H. Miller, N.C. Handy and J.E. Adams, *J. Chem. Phys.*, 1980, **72**, 99.
66. G.D. Carney, L.L. Sprandel and C.W. Kern, *Advances in Chemical Physics*, 1978, **37**, 305.
67. S. Carter and N. Handy, *Chem. Phys. Letts*, 2002, **352**, 1.
68. J. Tennyson and B.T. Sutcliffe, *Molec. Phys.*, 1986, **58**, 1067.
69. X-G. Wang and T. Carrington, *J. Chem. Phys.*, 2002, **117**, 6923.
70. J.C. Light and T. Carrington, *Advances in Chemical Physics*, 2000, **114**, 263.
71. A.G. Császár, Ed. *Spectrochimica Acta Part A*, 2002, **58** Issue 4.
72. http://www.emory.edu/CHEMISTRY/faculty/bowman/multimode/
73. D. Tew, N. Handy and S. Carter, *Molec. Phys.*, 2001, **99**, 393.
74. D. Tew, N. Handy and S. Carter, *Phys. Chem. Chem. Phys*, 2001, **3**, 1958.
75. http://www.tampa.phys.ucl.ac.uk/pub/vr/
76. B.T. Sutcliffe and J Tennyson, *Int. J. Quant Chem.*, 1991, **39**, 183.
77. J. Tennyson, S. Miller and R. Le Sueur, *Comput. Phys. Commun.*, 1993, **75**, 339.
78. http://www.tampa.phys.ucl.ac.uk/pub/vr/cpc92
79. J. Tennyson, M.A. Kostin, P. Barletta, G.J. Harris, O.L. Polyansky and N.F. Zobov, *Comput. Phys. Commun.*, in press, 2003.
80. http://www.tampa.phys.ucl.ac.uk/pub/vr/cpc03
81. G.D. Carney and R.N. Porter, *J. Chem. Phys*, 1976, **65**, 3547.
82. O. Polyansky and J. Tennyson, *J. Chem. Phys*, 1999, **110**, 5056.
83. O. Polyansky, A. Császár, S. Shirin, N. Zobov, P. Barletta, J. Tennyson, D. Schwenke and P. Knowles, *Science*, 2003, **299**, 539.

2
Computer-Aided Drug Design 2001–2003

BY RICHARD A. LEWIS

1 Introduction

Research in computer-aided drug design over the last two years has focussed on understanding and developing models for ADMET, docking and scoring, and library design, reflecting the truth that he who pays the piper, calls the tune. I have tried to visit some of the other quieter areas, in the belief that some of them will become tomorrow's stars. In consequence, this review will be at best a subjective success, at worst, a skim through the rich variety of modelling papers that have appeared this year. I acknowledge that some key areas are missing in fields where the work is still too young to assess, for example chemogenomics. As a premium has been placed on new work, rather than minor improvements to existing solutions, it is inevitable that some key papers will be missing. As in my previous reviews, the emphasis has been placed on refereed, widely available journals, rather than on books or patents. This will cut down the number of papers reviewed, but it is the author's hope that this approach may give a better insight into the directions in which modelling is moving. I have tried to abstract out some of the key points from the papers reviewed. Inevitably, this will lead to gross simplifications; the goal is to stimulate interest, and encourage reading of the primary literature. The main themes of this review are: ADME/Tox, docking and scoring, 3D-QSAR, pharmacophores, library design, cheminformatics and structure-based drug design.

2 ADME/Tox and Druggability

The introduction of drug-like properties is a hot topic in the pharmaceutical industry. The current physical hurdles are reviewed by Di and Kerns[1] and van de Waterbeemd,[2] and include solubility, permeability, metabolism and stability; fortunately, progress in modelling has been be made in all of these areas.[3] Probably one of the hottest areas over the review period has been the modelling of ADMET (absorption, metabolism, toxicity) phenomena. There have been many advances in our understanding of cytochrome p450s (cyp450s), the most important metabolising enzymes in the liver. One of the key challenges in taking a drug from concept to clinic is to get the right profile of metabolism. If a drug is metabolised too quickly, one has to have frequent and/or large doses. The ideal dose is once a day and less than 10 mg/kg. One also wants to avoid drugs that are metabolised too slowly (leading to build-up and possibly toxic side-effects), that inhibit the cyp450s (leading to drug-drug interactions) or that have an unbalanced profile (that is, being metabolised exclusively by only one cyp450). The last characteristic can lead to the drug having different clearance rates depending on the age and genetic profile of the patient. These considerations have generated intense effort geared towards understanding the structure and function of the cyp450's, crowned by the publication of the structure of the human 2C9 isozyme[4] and announcements that the human 2D6 and 3A4 structures have also been solved (but not published). More structures with bound ligands and mechanistic studies are still required before the cyp450's are understood, but this new experimental data is a welcome addition to the field.

2.1 Metabolism by Cytochrome P450. – The cyp450 family of enzymes oxidise their substrates using molecular oxygen and a heme cofactor. Medicinal chemists would like to predict the exact site of metabolism in the substrate, so that the rate of metabolism can be modulated or blocked. Understanding the structure of the active site allows one to predict the possible binding modes of the substrates; understanding the mechanism allows one to predict the rate of reaction or lability of the site of metabolism. In the absence of the key experimental data, many workers have focussed on surrogates for this process. Due to its dissimilarity to known structures (17% homology to rabbit 2C9) and its importance (50% of marketed drugs are substrates), cyp3A4 has been the subject of several papers. Singh et al.[5] propose a simple model of hydrogen radical formation, the energy of which can be computed easily using AM1, as the rate-limiting step and solvent accessible surface area. To speed the calculation up, they derive a QSAR from the AM1 radical formation energy based on local environment, and derive an upper limit of 27 kcal/mol for protons that might be susceptible. A limit for accessible surface area, which should be > 8.0 $Å^2$ to qualify as a potential site of metabolism, is also proposed. These two rules catch 78% of known metabolic sites in the test set. Schneider et al.[6] developed a PLS model based on data from 581 compounds, using a classification scheme and high-throughput screening data. They were able to classify potential inhibitors at 90% success rate, which makes the procedure very suitable as a filter for flagging problematic molecules. A protocol that uses more of the information contained in the active site (even though it might be from a homology model) is described by Zamora, Afzelius and Cruciani.[7] The active site is abstracted using the flexible mode of GRID, to allow for protein side-chain motion, and the fields are transformed into pharmacophore-like descriptors. Any query ligand is treated in an equivalent manner, so that ligands can be matched into the active site using a similarity-driven docking. The protons that are predicted to be close to the heme are candidates for metabolism. The predictions agree very well with experimentally observed sites of metabolism, with the experimental site of metabolism corresponding to the top candidate 50% of the time, the second 25%, and the third 15%, giving a cumulative score of 90%. Later unpublished work from this group considers a generalised correction for reactivity (e.g. hydrogen abstraction from methylene is more favoured than from methyl) to improve discrimination.

2.2 Human Ether-a-go-go-Related Gene K+ Channel. – The Human Ether-a-go-go-Related Gene K+ channel (HERG) has achieved notoriety over the past decade, as interaction with this channel can cause the development of fatal cardiac arrhythmias; several drugs have had to be withdrawn (e.g. terfernadine, sertindole) or given black-box labels as a result of this side effect. Great efforts have been made to understand the SAR around HERG. The area has been reviewed by Pearlstein *et al.*:[8] Of the three recent models of the SAR, the CoMFA model of Recanatini *et al.*[9] seems to be the best validated (the other models were generated using CoMSIA[10] and Neural Networks[11]). In this study, they took 31 diverse compounds for which inhibition values had been measured in mammalian cells; the affinities varied from 1 to 130000 nM. A pharmacophore was built up, starting with the most potent inhibitor. As not all compounds matched the pharmacophore fully, alignments were fixed manually for some molecules. The accuracy of prediction on the test set was better than 1 log unit. Some of the key issues in the area are consistency of experimental data, and diversity of structure. Progress is also being made in mapping out the binding site experimentally, so proper experimental validation of the QSAR hypotheses should be available soon. Finally the relationship between in vitro measurements, and the in vivo models (where plasma protein binding comes into play) is not clear-cut.

2.3 What Makes a Compound Drug-like? – Further studies have been carried out to mine the historical information contained in databases of orally available drugs either in development or on the market (for a review see Rishton[13]). Wenlock *et al.*[14] examined the statistical differences in overall distribution of molecular weight, lipophilicity (logP and log$D_{7.4}$), hydrogen-bonding and flexibility. If one makes the assumption that the attrition rate is mainly due to physicochemical issues (rather than toxicity or efficacy or even cost of synthesis), then one can deduce the 'desirable characteristics' that drug candidates should have. This is very much in the same vein as Lipinski's work.[15] The trends observed were that marketed orally available drugs have a marked lower molecular weight than the drugs in development. The same trend towards simpler molecules is observed for the other descriptors too (the descriptors are not independent, so this behaviour is expected). Their new rules are (with Lipinski's values in brackets): Weight – 473 (500); ClogP – 5 (5); H-bond acceptors – 7 (10); H-bond donors – 4 (5). Workers from GSK[16] looked at their in-house database of 1100 compounds for which permeability and clearance had been determined to derive their own set of rules: compounds with fewer than 10 rotatable bonds and a polar surface area <140 Å^2 have a high probability of having good oral bioavailability in the rat. It is noted that molecular weight, rotatable bond and ClogP are all strongly correlated, so the results are consistent with other work in this area.

Bergstrom et al.[17,18] have tackled the issue of druggability from a different angle. Observing that inter-laboratory variation in measured solubility is as much as for CaCO2 permeability, they re-measured data for 23 diverse compounds, and correlated it with computed surface area variables using the PLS technique. Solubility was predicted as a function of non-polar surface area and hydrogen-bonding, and permeability of polar surface area and size. In both cases, the relationships are inverse, and the variables are contributing negatively to the overall prediction. The errors were about 1 order of magnitude when applied to an external test set. When the models were used to classify the test set into a Biopharmaceutical Classification Scheme (BCS) class, the accuracy was 87%. The authors' conclusion was that the method showed much promise, but a much larger data set was needed. The same group has also recently reviewed the area of theoretical predictions of drug absorption.[19] The ChemGPS system of Oprea et al.[20] also tries to classify compounds into the BCS framework. The principle is the same as GPS: a set of 423 satellite compounds is used to define the coordinates of a chemical space, and the position of a new compound can be determined with reference to these satellites and a set of orally available drugs (avoiding extrapolation). The descriptors used are derived from VolSurf and other standard connectivity indices. The data is analysed by PCA. It is shown that the GPS scores (derived from the principal components) correlate well with external biological data like BBB penetration or CaCO2 permeability, even though the data was not used to derive the scores. It is also noted that the components for solubility and permeability are not orthogonal, and it is easier to improve permeability than it is to improve solubility. ChemGPS can be used to rank large libraries in terms of permeability and solubility, thereby giving a BCS-like global mapping device. Clark et al.[21] used descriptors derived from AM1 calculations, and concluded that factors such as size and shape do not discriminate between drugs and non-drugs, but surface electrostatics do. This result was translated into a Kohonen map based on 3 QM-derived descriptors; the map showed a good spatial separation of the two categories, and could even be extended to pick out the steroid subclass.

Another twist to the 'drugs vs non-drugs' debate has been provided by the work of McGovern and Shoichet[22,23] who have demonstrated the role that aggregation plays in the identification of false positive screening hits. Not only did they show that several published leads were in fact only active because they formed aggregates that fooled the screening protocol, they have also found the same phenomenon in some widely-used non-specific kinase inhibitors. These compounds also appear to inhibit such diverse enzymes as beta-lactamases and chymotrypsin with micromolar affinities, but again the inhibition is due to the formation of aggregates. The formation of aggregates can be detected by light scattering, so in the future more information about which compounds are prone to aggregation should be available, allowing the effect to be modelled.

2.4 General Models. – More generalised models for metabolism have been reported, based on data obtained from human S9 homogenate, and using the kNN method.[12] The reported accuracy of true positive prediction was 83% and true negative was 75%. There were two interesting techniques used in the paper: (i) the use of exponential weighting based on the distance to the nearest neighbour; (ii) a threshold for decided whether a prediction is an unreasonable extrapolation. In trying to model ADMET effects, when the data set is often chemically diverse, several groups have looked at 3D alignment-free methods. The latest approach has been to describe diverse molecules using the smallest moments of inertia and the dipole to set up a common, arbitrary frame of reference for alignment.[24] CoMFA and CoMSIA fields are then generated for each molecule and are used as input in SIMCA (rather than PLS, as the observations are better deal with as classes, due to the low experimental accuracy). Only one conformation is used (the results do not seem to affected by this), and the models offer 3–6 fold improvement over random.

Finally, two interesting QSAR models have been published and are worth citing (even though classical QSAR is outside the scope of this review). A new model for predicting human intestinal permeability has been published by Winiwarter et al.,[25] using more sophisticated descriptions of hydrogen-bonding behaviour and strength. Several equivalent models were found which used a general hydrogen-bonding descriptor, an H-bond donor descriptor and lipophilicity.

$$\log P_{eff} = -3.1 - 0.115*(\text{number of h-bonding atoms}) - 0.993*(\text{sum of partial charges of H's attached to O,N,S}) + 0.140*\text{ClogP}$$

There is also a model for binding to Human Serum Albumin[26], which predicts affinity to be mainly determined by hydrophobic forces, with some shape factors as secondary determinants.

$$\log K_{hsa} = 0.020 + 0.055*\text{dipole} - 1.22*\text{JursRPSA} - 0.02*\{E_{HOMO} + 7.4\}^2 + 0.15*\text{ClogP} - 3.48*\{0.185 - {}^6\chi_{ring}\}$$

3 Docking and Scoring

Docking is an integral part of the drug designer's toolkit. There are several methods for performing docking: for recent reviews, see Taylor, Jewsbury and Essex,[27] or Nussinov et al.[28] The topic of scoring has been thoroughly reviewed by Gohlke and Klebe.[29] A more general overview of virtual screening has been provided by Lyne.[30] The debate over docking and scoring continues to attract attention, because of the growing realisation that knowledge-based drug design (e.g. virtual screening) is given much better value than the numbers paradigm (e.g. HTS and diverse combinatorial chemistry). Although the two will never be independent, the ability to plug different scoring and optimisation functions into docking routines has allowed workers in the field to dissect out some key issues, and make general recommendation about which approaches are best suited to which targets. Schultz-Gasch & Stahl[31] have classified three strategies for docking (Monte Carlo, Rigid based on multiconformer libraries e.g. FRED or Glide, incremental build-up e.g. FlexX) and six strategies for scoring (empirical, force-field and knowledge-based) with hard and soft variants of each. Aside from the general conclusions, they identified the conformational sampling in Glide as a possible source of error. For lipophilic sites where shape is more important than h-bonding (e.g. nuclear hormone receptors), the best method is multiconformer docking using a soft docking function and a hard scoring function. Sites that have key polar interactions but are otherwise lipophilic respond well to incremental approaches. Sites with a network of polar interactions can be dealt with by either incremental or multiconformer methods combined with harder scoring functions. This work highlights the value of performing extensive preliminary studies before embarking on a cpu-intensive virtual screen. Wang, Lu and Wang[32] performed another validation study of different scoring metrics, with the twist of using multiple docking configurations generated by Autodock, rather than just the highest scoring. This detaches the scoring from the docking to a greater extent. The scoring functions are tested to see if they can select the configuration closest to the x-ray from the diverse decoys; there is still a bias in that the configurations may not be minima as far as the scoring functions are concerned so that the fitness may be reduced even for a 'correct' configuration. This is examined by looking at near neighbours of the x-ray configuration, and also by plotting the score against rmsd for all of the available configurations to see if there is a smooth trend. As others have found,[33] they demonstrate that consensus scoring improves the success rate. No one scoring function outperformed all the rest in all the tests but the empirical functions like PLP or DrugScore had a slight edge. Another study confirming the strong biasing effect of molecular weight in docking score has been published by Pan et al.;[34] they propose a correction based on $N^{1/3}$ to bring the MW profile on the hits more into line with databases like MDDR.

3.1 Validation Protocols. – Although many researchers use a wide array of complexes to aid in validating docking and scoring protocols, sometimes a more simple system is better. Shoichet et al.[35] have used T4 lysozyme to provide fine control of the character of the binding sites. A small apolar site can be created by a Leu->Ala mutation, and a polar site by a second mutation of a Met->Asn. A large database was seeded with some known binders. It was found that by using a better charge model for the ligands, and by including a desolvation term (both derived from AMSOL), that better enrichment rates could be obtained. Some of the hits were assayed experimentally to confirm the predictions, both for affinity and for the x-ray pose. It was found that the new protocol tied in well with experiment, both in terms of binding and the docking rmsd, which was <0.4Å. These model systems look to be good test beds for future scoring function development, due their simplicity and the relative ease with which predictions can be verified.

The development of a clean data set for training and validating docking and scoring protocols has also been addressed by Nissink et al.,[36] who have examined all the available complexes in the PDB, to check for factors that would make the complex unsuitable for inclusion in a training set. They looked for ligand-protein clashes, the interactions between the ligand and other proteins in the unit cell, and structural errors in ligand placement. 61 complexes were identified as being problematic, and the paper discusses in detail the reasons behind the assignments. The remaining 305 complexes comprise the clean data set, which can be further refined using resolution as a parameter. GOLD was used to redock the ligands for the clean data set and it is shown that GOLD performs much better than against the 61 complexes with problems, underlying the need for careful construction of the training and test sets used in this field.

3.2 Entropy. – The influence of entropy on affinity is very hard to estimate, and probably is the main reason why it is hard to estimate binding energy of a ligand-receptor complex. Murray and Verdonk[37] have presented a detailed analysis of the translational and rotational entropy lost on binding, and estimate it to be 15–20 kJ/mol at 298K. The estimate is based on the experimental measurements of fragments binding at independent sites within the same protein. They carry on to argue that if one could connect low-affinity fragments in a way that preserves their optimal binding modes and the linker is not strained, one will obtain a large jump in affinity for the combined molecule, as the entropy penalty only has to be paid once. Although other workers have cited a dependence of entropy on molecular weight, it is assumed that this dependence is negligible. It follows that optimisation of fragments is tractable, and that HTS screening and combinatorial chemistry will give poorer hits, simply due to the interplay between optimal binding of the key fragments in the pockets and the constraints imposed by sub-optimal linkers found in most screened compounds. Their analysis also explains why some regions of a lead molecule are very sensitive to change (disregarding changes that introduce steric bumps). It will be interesting to see how these findings can be translated into the next generation of empirical scoring functions. Cozzini et al.[38] have also looked at bringing in entropy by using the HINT program which is based on experimental logP values. It is argued that this brings in entropy involved in solvation, and to demonstrate, the authors look at the HINT scores against experimentally measured free energies of binding for 15argets. The errors they get are reasonable (2.6 kcal/mol) but it would be nice to see a larger training and test set employed. The method is also very sensitive to changes in conformation and configuration of the ligand and active site.

3.3 Pharmacophore and Local Docking Schemes. – Glick, Grant and Richards[39] have examined the docking problem from a more general standpoint, considering the case of docking to a target whose active site has not been specified. To do this, they use a pharmacophore-like representation of the ligand, and perform a quick search against the target using look-up grids, before performing a slower optimisation. The ligand conformation space is reduced using energy, rmsd and dihedral filters to cluster the conformers. While the authors can successfully redock ligands into the holo-targets, they acknowledge that when the ligand is flexible, or the active site is not in the right conformational state, their method will have problems.

Scoring can also be handled using local models. This was first described by Vieth & Cummins,[40] and now by Golhke & Klebe.[41] The active site of the protein is mapped onto a grid using all the atom types in the DrugScore potential developed for general scoring.[42] The potentials on the grid are then multiplied (in a distance-dependent manner) by the occurrence of those atom types in ligands derived by experimental x-ray data. The fields can be correlated by PLS with the experimental affinities of the ligands to give a CoMFA-like (the authors use the term AFMoC in homage) local scoring function. In the two test cases examined, thermolysin and glycogen phosphorylase, the quality of the affinity predictions compared to the DrugScore figures were much improved. This would seem to offer significant advantages if iterative structure-based drug design is being performed.

A standard trick to improve docking is to use pharmacophore points to constrain the poses to be more sensible. Kuhn et al.[43] have taken this a step further in their SLIDE program by using a knowledge base to locate the constraining regions more precisely. Flexibility is handled at a later stage through a relaxation of the poses. Templates for hydrogen-bonding and hydrophobic regions are used, and points from different side chains are clustered to reduce the number of constraints; the old method used a fixed grid, which can cause placement errors if the resolution of the grid is inappropriate. In test, they could improve the accuracy of docking against apo structures of thrombin and glutathione-S-transferase by 15–25%. Similarly, the use of pharmacophore constraints in FlexX[44] improved the accuracy of the dockings, particularly in the cases where the native FlexX docking was poor. An important caveat noted by the authors that over-constraining the search could lead to poorer performance in virtual screening, as potential hits could be missed.

3.4 Protein Flexibility. – The issue of protein flexibility has been a thorn in the side of docking research. Does flexibility matter, and to what extent, and how does one allow for it within the docking protocol? Protein flexibility is key to understanding ligand-protein complexes, and we should change our design paradigms accordingly.[45] Several methods for performing docking using flexible proteins have been reviewed by Carlson,[46] and overall usefulness of docking as a discovery tool by Shoichet.[47] To try to understand the extent of the problem, McGovern and Shoichet[48] docked a large number of small molecules against 10 targets using the apo-, holo- and modelled forms of the binding site. Using enrichment rates as their measure, they found that the holo form gave the best enrichment (7/10 cases) followed by the apo (2/10) and then the modelled (1/10) form. As observed by Bissantz et al.,[49] the holo-form can be over influenced by the ligand in the holo complex, due to site collapse, leading to lower retrieval of larger but highly similar ligands; apo forms can be markedly different from holo forms. The take-home message is that any target to dock against will give better than random hit retrieval, but the holo-forms routinely given an enrichment factor to make the exercise worthwhile. A similar study of the effect of induced fit on docking accuracy has also been performed by Birch et al.[50] using GOLD and neuraminidase as the test case. All the targets were from complexes; the active sites that were most 'open' (as opposed to closed or collapsed due to induced fit) gave the best results in the docking experiments. Scoring functions that use a soft clash term will perform better than those with harder van der Waals penalties.

3.5 Docking and Virtual Screening (VS). – There are now studies into virtual screening comparing the results to those of an HTS run in parallel to really assess the activity of the decoys used in the VS[51]. It is claimed from this study that VS has the great utility when used to clean up the HTS hits and to select which need to be profiled most thoroughly (6-fold enrichment), in addition to the more traditional use of preselecting compound to submit to the screen (21-fold enrichment). In a study of PTB-1B, Doman et al.[52] found that compared to an HTS run against the same set of 40000 compounds, docking enriched the hit rate by 1700 over random, and that the hits found by the two techniques were complementary. Grueneberg et al.[53] have also looked at the effect of filtering on hit rates in virtual screening. They used a hierarchy of 2D (druglike), 3D (pharmacophores derived from hot-spots in the binding site) and similarity filters (superimposing on 2 known potent inhibitors) before docking the remaining compounds. This reduces their database down from 90000 to 100, using the test case of human Carbonic Anhydrase II. The approach successfully retrieved 13 new hits (including 3 that were subnanmolar) and correctly predicted the binding mode of two compounds whose complexes with hCA-II were subsequently solved by x-ray. A case study comparing docking and library design for lead generation has been performed, using DHFR as the target.[54] A virtual library of 9448 compounds was reduced into three smaller libraries using docking (library 1 – compounds with good dock scores; library 2 – compounds with poor scores or no dockings) and diversity selection based on 3D similarity (library 3). The three libraries were made and tested; it was found that library 1 gave more and better hits (21%) than library 3 (3%), whereas library 2 contained a few very weakly active compounds. The authors conclude that the power of structure-based virtual screening generally lies in its ability to filter out undesirable compounds rather than picking out specific actives. The hit rate allows the chemists to focus on the most promising leads, reducing the discovery cycle time. Docking has also been used successfully for lead generation for CDK2.[55]

A different slant on scoring, exploiting the interface with 3D-QSAR, is provided by the work of Zamora et al.,[56] in which they use VolSurf descriptor to correlate with binding affinity for two datasets, with reasonable success. The effect of hydration and protonation states in the actives sites were investigated and were found not to be crucial. Polar and water interaction volumes contribute positively to affinity, as do larger regions of hydrophobic interaction. The intriguing message is that VolSurf descriptors, which have already been shown to be valuable in the modelling of druggability, might also be useful in predicting affinity, so that the two phenomena might be optimised at the same time.

4 De Novo

De novo design methods have been used to generate potent antagonists of ERα based on a novel chemical scaffold.[57] Although the details of the methodology are sketchy, the process is to generate a virtual receptor which can reliably reproduced observed binding activities, then to apply an evolutionary algorithm to build the new structures. It seems that the quality of the structures still has to be assessed manually before synthesis. To deal with this issue, Vinkers et al.[58] have developed a program, SYNOPSIS, that includes a synthesis route for each generated compound. This is achieved by starting from a database of known compounds and applying a small number (70) of known feasible reactions. This is combined with a fitness function to guide the de novo design. In a test case against HIV reverse transcriptase, 18 of the 28 proposals were made and tested, and 10 were found to be active. This is a much higher rate of feasibility and enrichment than has been seen with other programs, so that SYNOPSIS represents a significant step forward in the area. Honma et al.[59] used the program LEGEND to design novel potent inhibitors of CDK4; when these scaffolds were exploded using combinatorial libraries, a more potent inhibitor (42 nM) was discovered. An x-ray study validated the predicted binding mode proposed by the de novo protocol.

One of the difficulties in the field of de novo design is the validation of the methods, usually because the chemotypes suggested may be unknown and the effort required to make them prohibitive. Stahl et al.[60] address this by comparing the generated chemotypes with known inhibitors for 4 targets; the comparison is necessarily subjective but even so it was found that most of the major classes of known inhibitor were regenerated.

5 3D-QSAR

A continuing issue with 3D-QSAR is the requirement for a protocol for aligning the ligands in a consistent way to allow both the creation and utilisation of a meaningful model. Considerable effort has been devoted to the study to alignment independent descriptors, the most recent of which are the MaP descriptors of Stiefl and Baumann.[61] The descriptors are prepared by creating an evenly sampled dot surface around a structure (one surface per conformation), assigning property values to the surface points, then computing a frequency fingerprint based on the distance between all pairs of surface points and a binning function. The advantage of using surface maps over classical pharmacophore keys is the extra weighting given to a carbonyl group over an ether. Three QSARs were constructed as a test, and the key variables found could be interpreted in terms of a pharmacophore model, another beneficial effect of the methodology.

A new approach to the alignment problem is provided by the program FIGO[62] which uses the molecular interaction fields (MIFs) derived from the GRID program as the basis for determining optimal superposition (of the MIFs) and hence alignment of the underlying molecules. However, the presence of a rigid reference molecule still seems to be required for the approach to work successfully.

The alignment issue in 3D-QSAR can also be dealt with by increasing the 'dimensionality' of the model. 4D-QSAR includes conformational ensembles of the molecules, and 5D-QSAR[63] allows for induced fit by incorporating multiple models of the active site. Using NK-1 as a test case, the results were not significantly better than for 4D-QSAR but the models were more diverse so presumably will be more predictive.

A standard statistic used to assess the quality of regression models is q^2. However, q^2 may be misleading, as Golbraikh and Tropsha argue[81]. They demonstrate that there is no correlation between q^2 and predictive ability for a test set, leading to the conclusion that a good q^2 is necessary but not sufficient for a model to have high predictive power. Fortunately, they also give a series of other criteria that can be used to evaluate predictive power, for example the correlation between observed and predicted values for a completely external test set should be very close to 1.

Techniques from 3D-QSAR have been used to classify kinases on the basis of their binding sites. The kinases were built by homology with, or were aligned to PKA as a template. GRID probes were used to construct field maps, which were analysed using PCA. The first two principal components were sufficient to divide the kinases up into subfamilies (PKA, MAP, CDK

Table 1 *Some representative high-quality 3D-QSAR models*

Target	Method	Alignment	Q^2
Melatonin[64]	CoMFA	Reference ligand	0.77
EGFR Inhibitors[65]	FLARM	Reference ligand	0.92
Nicotinic a4b2[66]	CoMFA	Reference ligand/pharmacophore	0.7
Paclitaxel[67]	CoMFA	Reference ligand	0.7
NMDA[68]	CoMFA	Reference ligand	0.8
CYP2C9[82]	ALMOND	N/A	0.64
Herg[9]	CoMFA	pharmacophore	0.57
VLA-4[69]	CoMFA	X-ray template	0.7
HLA-A2[70]	CoMSIA	Reference template	0.68
5HT4[71]	CoMFA	Reference template	0.79
1,4-dihydropyridines[72]	CoMFA, COMSIA, GRID/GOLPE	Pharmacophore	0.6
HIV1 integrase[73]	CoMFA/CoMSIA MCS, docking		0.6, 0.8
Napthyliosquinolines[74]	ComSIA	FLEXs	0.82
5-HT, NE reuptake[75]	CoMFA	GASP	0.67
GABAa[76]	4D-QSAR	N/A	0.8
D4[77]	CoMFA/CoMSIA	Pharmacophore	
A1 adrenoceptor[78]	GRID/GOLPE	Pharmacophore	0.65
5HT7[79]	CoMFA	Reference ligand	0.7
CDK1[80]	CoMFA/CoMSIA	Docking	0.63

and SRC), and to identify specific interaction sites for each subfamily, some which contributed mainly to affinity, some to selectivity. From this, the authors were able to design specific potent inhibitors of CDK1. A similar analysis of the Cyp450 2C family has been performed:[82] from the results of the variable loadings of the various probe maps, they were able to identify the features responsible for selectivity within the family of isozymes, and the pharmacophore for 2C9.

6 Pharmacophores

The three commonly used programs for generating pharmacophore models, Catalyst, DISCO and GASP, have been compared based on their ability to generate pharmacophores for known complexes.[83] The ligands were used as training sets, and a visual examination of the complexes was used to deduce the common pharmacophore for the protein target. 5 different datasets were used. GASP and Catalyst were found to perform better than DISCO, and both the target pharmacophores were, if not top, ranked very highly. The advantage of Catalyst is the ability to customise the pharmacophore definitions; that of GASP is the ability to consider the steric overlap of the ligands when deriving the model.

6.1 Conformational Analysis. – There are several popular programs for performing conformational analyses that can then be fed into docking or pharmacophore generation programs. Bostrom[84,85] has investigated how well these programs can regenerate the observed bioactive conformations. 30 ligands from the high-resolution, well-resolved complexes from the PDB were taken as targets; the criterion for success was an rmsd < 0.5Å (symmetry and mirror images were corrected for). Ab initio calculations were used to determine whether the bioactive conformation was a minimum, as this will throw methods that perform local optimisation. Macromodel found only 22/30 of the bioactive conformations, but still performed the best. Catalyst did not perform well in either BEST (15/30) of FAST (16/30) modes; Flo99did well (21/30); ConFORT poorly (11/30), and Omega was in between (17/30). However, Omega is by far the fastest to run. For comparison, using a single conformer from Concord or Corina was worst (8/30). In the second paper, the various parameters that control the analysis affect the number the parameters controlling one program, Omega, were varied using an experimental design protocol. It was found that eliminating duplicate conformers (rmsd < 0.6) as well as an energy cut-off of 5 kcal/mol (this finding will not be transferable to other force fields) gave the best results. Diller and Merz[86] have also looked at the problem of identifying the bioactive conformer, and they find that the bioactive conformers are generally more extended, with higher polar surface area, fewer internal interactions than other low-energy conformers. These descriptors could be used as a secondary filter after the force field energy.

Table 2 *Some high quality pharmacophore models published during the review period*

Target	Method	Features	Tolerances
Angiotensin II[87]	Catalyst	7	N/A
GABA[88]	Manual	4 + sterics	N/A/
5HT-1a[89]	Catalyst	6	N/A
Influenza endonuclease[90]	Manual	3	1.0
EDG3[91]	Catalyst		N/A
CYP17[92]	Catalyst	4–5	
UDP-Glucuronosyltransferase 1A4[93]	Catalyst	3	N/A
p-glycoprotein[94–96]	Manual, GASP, in-house	4,6,4	0.8–1.5, 0.3–1.0
Squalene synthase[97]			
Imidazole glycerol phosphate dehydratase[98]			
NK2 antagonists[99]	Manual	4	N/A
5HT-4[100]	Catalyst	5	N/A
AMPA[101]	Catalyst	4	
VLA-2[102]	Catalyst	5	N/A

The fFLASH program[103] approaches the issue of flexibility in similarity scoring of databases, by pre-sampling conformational space using a built-up/ rules-based sampling. Only data on the fragments and the features within the fragments are stored. This pharmacophore-like fingerprint is the primary search object before final reconstruction and overlay of all the retrieved molecules. The size of the fragmentation is important, and the authors use fragments of size 20–40 atoms. In validation, fFLASH seems to perform as well as FLEXs,[104] and is sufficiently fast to be used for database searching of larger sets of molecules.

7 Library Design

Some of the heat seems to have gone out of the area of library design, partly due to a realisation that the bottleneck is not access to design tools, but deciding on the design criteria and then performing the iterative cycles of design and synthetic assessment of feasibility. A new method for library design based on information theory has been published.[105] The concept is to include the maximum amount of information within the library, so that the results of screening a library will enable the maximum number of conclusions to be drawn as quickly as possible (similar to experimental design strategies). The compounds should only contain sufficient redundant information to reinforce any observations. To allow for druggability considerations, the fitness score also includes terms for property distributions. The main score is based on pharmacophore fingerprints. No concrete examples are given, leaving the reader wondering if the compounds might not contain too much information, as discussed by Hann *et al.*[106]

The interface between 3D-QSAR and library design has been investigated by Cramer[107] in his work on topomer CoMFA. The topomer methodology can be used to align fragments to a common library core. The fragment is generated in a single (CONCORD) conformer. The 3D-QSAR is then regenerated. If the method works, the alignment issue is largely finessed and the QSAR can be used as a virtual screen. To validate these assumptions, 14 literature CoMFA studies were repeated using the new protocol. Although the topomer models are weaker in terms of q^2, it is not by much, and the prediction errors on an independent test set were very comparable. Database searches yielded alternative substituents to the core with predicted improvements in activity of 10–20 fold. The author admits that it is surprising how well topomer COMFAs perform, but the use of a template structure with fully extended substituents may not be so far from the true bioactive conformation. This approach is well suited towards rapid lead optimisation.

The OptiSim program has been extended to perform library design as well[108] through the introduction of pivoting between the reagent sets that together make the final product. Further refinement can be introduced by assigning the reagents to classes (cheap, hard to obtain etc) and by using a roulette wheel approach to pick preferentially from the classes with better properties. To maintain efficiency, one can limit the number of reagents of a particular functionality, for example if $Cij = Ai + Bj$, the upper limits of i and j can be fixed to force better plate layouts. Designs that are found to contain reagents subsequently discarded by the chemist can be used to seed future runs, so that redesign is quick.

A practical application (p38 inhibitors) of library design using a Monte Carlo algorithm is given by McKenna *et al.*[109] The objective function included a term for druggability and efficiency. When the library was assayed, it was found that the designed compounds had more drug-like profiles, while maintaining activity. The authors note that the SAR in the series is not obvious, so it is unlikely that similar results could have been obtained by the common incremental approach to medicinal chemistry.

Two papers from Gillet *et al.*,[110,111] describe the use of multiobjective genetic algorithms in library design. The basic concept is that libraries ought to be designed to optimise several factors, for example, druggability, diversity, affinity, at the same time. This can be achieved by combining all the factors into a single objective function, but then the issue is the relative weightings that should be used. The approach that the authors use is based on Pareto optimality, that is, solutions that are better or equivalent to all other solutions in one or more dimensions have higher fitness values, so propagate. The library size and configuration is also allowed to vary. Examples of libraries selected by this approach are described, and the various designs are shown to be compromises between the competing demands of diversity, library size and so on; the user can then make an informed choice as to which of the equivalent designs to taken forward. Pre-assigning weights to the different factors would obscure the interplay between the different factors.

8 Cheminformatics and Data Mining

8.1 Data Mining. – The first approach that is usually tried to generate hits or SAR is to search chemical structure databases for analogues. There are many different data-mining protocols that can be used. Sheridan and Kearsley[112] have reviewed many of these methods, and present their experiences based on practical examples. The take-home message is that there is no single best method, and the more approaches that are tried, the better. One method for identifying commonality among structures is to look for the maximal common substructure; developments in this area have been reviewed by Willett.[113] The similarity principle (similar molecules will have similar biological profiles) has underpinned many researchers thinking about diversity and QSAR. However, Martin et al.[114] demonstrate that this principle is not as strong as its proponents claim. They used Daylight fingerprints, with a Tanimoto threshold of 0.85 as their definition of similarity, and a complete database of screening results. Although the proportion of similars that have similar profiles is 0.3, it does mean that the definition of similarity is still too coarse to allow one to only screen, say, one representation from a family of similar analogues: the chance of missing activity is 70%. These results imply that a strategy of focussed diversity might be better, in which several samples of each diverse cluster are included in any screening set.

Further study has been done on looking for privileged or promiscuous structures,[115] by identifying (not necessarily connected) common substructures in compounds that have similar structures but either very similar or dissimilar biological profiles. As more compounds are examined, so the most significant substructures are discovered. The usual issues around the accuracy of the biological profile in public databases are dealt with via an inspection phase. Skeletons like steroids and diphenylamines are found, as well as ergot and benzodiazepine motifs. By judicious use of the results it may be possible to design targeted libraries of avoid promiscuous motifs in following up hits from HTS.

A similar approach by Xu[116] classifies compounds by their (connected) chemical scaffold. The scaffolds are dues to define dimensions of complexity and cyclicity (number of rings). Complexity is defined in terms of rings, atoms, bonds and elemental diversity. In an analysis of the CMC, NCI and ACD databases, the drug-like molecules are less complex and less cyclic. It is also clear that some regions of the complexity/cyclicity space are never sampled, and probably ought not to be. Average electronegativity is also a good discriminator between drug and non-druglike. It is claimed that classification based on scaffolds is more intuitive to the medicinal chemist, and the results do seem to make sense. One of the difficulties in modelling complexity as perceived in chemical structures, is that few chemists can agree what it is. In a fascinating study, Takaoka et al.[117] asked five chemists to score 3980 compounds according to drug-likeness and ease of synthesis. The variance of the scores was very high, and the only agreement was around the really ugly compounds, which were 'obvious'. For the other compounds, the score was much more subjective. In spite of this, the authors managed to build binary classification models that could detect molecules that were hard to make and that were not drug-like, which can be used for filtering screening runs or compound acquisition catalogues.

Sheridan[118] has looked at the 1:1 replacements of chemical groups in molecules with similar biological activities, to look for possible bioisosteres or critical points in the SAR. The common parts of the molecules are identified by a maximal-common substructure algorithm. The most common replacements seem to correspond with medicinal chemistry intuition (or maybe reflect historic prejudice), for example aromatic carbon <-> aromatic nitrogen, O <-> S. Tables of other common replacements are given, forming a useful source of ideas for future lead optimisation efforts. Lewell et al.[119] have developed a database of ring structures, with the same objective of identifying similar replacements. The rings are classified by the core structure and the substitution points. Various descriptors for feature counts, complexity, geometry and frequency of occurrence are computed and stored. They illustrate the search capabilities by searching for replacements of 1,3,5-substituted indoles. The results and their context within the parent molecules are valuable idea-generators for medicinal chemistry. Holliday et al.[120] have presented some new work about bioisosteres. Although bioisosterism can only be determined in the context of an SAR, several groups have attempted to devise descriptors and similarity metrics that have a more absolute property. Here the R-groups are described in terms of standardised vectors for charge, hydrogen-bonding, lipophilicity and so on, the similarity metric was Euclidean, and the measure of success was whether the groupings of R-groups described in the BIOSTER database could be reproduced. All the descriptors do seem to give a statistically significant separation, based on this validation dataset.

8.2 Similarity and Descriptors. – Raymond et al.[121] compared 4 clustering methods based on graph or fingerprint metrics. They found that most methods do well when the dataset is characterised by large and/or unique ring templates, whereas diverse collections are poorly differentiated by any method. Jarvis-Patrick clustering seems to be slightly better but the performance of this and the other algorithms was dependent on the parameter settings. Fingerprint metrics did as well as the more sophisticated graph-based similarity measures, but neither method was significantly better. Further research needs to be performed to correlate the optimal settings for clustering and the relative diversity of the data set.

The study of physicochemical properties in QSAR has a long history, but the current emphasis on druggability is giving it new impetus. Although several method exist for computing reliable estimates of logP, much less work has been published on estimating pKa. The computation of pKa and of hydrogen-bonding basicity and acidity are becoming commonplace, and there are the first signs that tautomerisation is being addressed. Hennemann and Clark[122] have used their AM1-derived descriptors to produce a QSPR for the basicity of nitrogen heterocycles. Xing and Glenn[123] have used an atom path descriptor rooted at any ionisable atom or group and PLS to derive a regression model that performs well ($q^2 = 0.83$). The descriptor encodes the atoms solely in terms of the sybyl atom types. Other descriptors such as partial charge did not help the model. The approach can also rank ionising centres in correct order of pKa.

One of the technical difficulties around the use of fingerprints based around pharmacophores was the fingerprints rapidly grew in size with the number of pharmacophoric features and the granularity of the distance bins. Abrahamian et al.[124] have used run-length compression methods to reduce the bit string fingerprints into bitmaps, which can still be logically manipulated in the same way as their much larger brothers. The method for creating the fingerprints follows previous methods, but attention is paid to the order of bits to make the encoding and decoding of individual bits a unique operation, and to maximise the benefits of compression. 4 Bitmaps are stored for each molecule: concord, 2 conformational samples, and a full conformational analysis. This allows one to look at the effect of flexibility on bitmap similarity. The usefulness of the bitmaps approaches that of full 3D searching with discrete queries for hit retrieval. This work is a significant advance in an area that was seemingly stalled.

9 Inverse QSAR and Automated Iterative Design

There has been renewed interest in this area, as advances in graph theory and measures of chemical sensibility improve. The key driver is to understand how to utilise the large number of high-quality regression models to suggest new molecules for synthesis. In a pair of papers, Faulon et al.[125] propose a new set of descriptors called molecular signatures that seem to perform just as well as other descriptors like MolConn-Z on standard data-sets. The equivalence is then shown mathematically. In the second paper,[126] they show how these signatures can be deconvoluted into molecules. They argue that all molecules that have the same signature as an active reference should also be active. This does not allow for extrapolation into molecules that have similar signatures, as this is an exact solution. Visual examination of the signature isomers reveal that large pendant groups are being shuffled around the core molecule, which probably will not result in retention of activity. Other compounds possess unusual groups, underscoring the need for subsequent filtering by some means.

10 Structure-based Drug Design

With the advent of high-throughput crystallography,[127] structure-based drug design is starting to fulfil much of its early promise. The next big advances will come when we can relate protein sequence to the structures of preferred ligands and vice versa; the term being used for this chemogenomics, but the seminal papers have yet to be published. One of the foundations for this approach will be tools like ReliBASE. ReliBASE is a well-known method for data-mining protein-ligand complexes but the key papers describing the program[128] and some of the potential applications[129] have only just been published. In essence, ReliBASE is a relational database of proteins and their associated waters, ligands and cofactors. The data can be queried using many different techniques, including by 3D query. The hits from the queries are superposed, making the analysis of the output simpler. Examples of its use are given, including the geometry of interaction between positively charged amines and aromatic rings, the hydrogen-bonding ability of nitro groups and so on. This allows a modeller involved in structure-based design to see how a pocket is filled in homologous proteins, the sidechain flexibility of a residue across different complexes and the preferred geometry of interaction between a ligand group and a sidechain. The same group[130] has also analysed active site similarity based on pharmacophoric properties (hydrogen-bonding, hydrophobic/ aromatic), to see if molecular recognition patterns are conserved across different proteins. The regions of the site surface that display a feature are abstracted into a distance matrix graph. These graphs can be compared by standard clique detection algorithms. The solutions are superposed and scored according to the overlap of the surface points. Using this methodology, it was possible to detect binding sites for the same ligand despite low sequence homology, and co-factor motifs in non-homologous proteins. The authors speculate how the information could be mined for use in de novo drug design, and cite an example of a pocket in HIV-1 protease showing similarity to a pocket in PKA, where there are known ligands. These results have been integrated via the module CavBase. Further development of ReliBASE is eagerly awaited.

10.1 G-protein Coupled Receptors (GPCRs). – GPCRs are a common target for the pharmaceutical industry, but as they are transmembrane proteins they pose a tough set of challenges; for an overview of the area, see Klabunde & Hessler.[131] Bissantz *et al.* continue their pioneering work in docking and scoring by examining the question of whether homology models built on the low resolution (2.8 Å) structure of bovine rhodopsin are useful for virtual screening. Models of D3, m1 and VIP1a were constructed and used as targets against a database of non-binders seeded with known antagonists and agonists to the receptors. Docking and scoring was carried out using the consensus protocol previously published by the same authors. Homology models were built either with a known antagonist or agonist to mimic the ground and activated state of the GPCRs. Reasonable hits rates could be obtained (2–40 fold higher than random), but agonists were not found using the ground-state model, and antagonists with the activated model, pointing out the need for the right model for each situation. The activated state models showed bias towards the ligands used to derive the complex, so protein flexibility in agonist models, and the original rhodopsin structure (which was in its ground state), is important.

10.2 A Case Study. – A nice story about how X-ray and SAR methods were combined to create druggable inhibitors of SH2 is told by Lesuisse et al.[132] The main issue with SH2 domains is to find surrogates for the Tyr phosphate group in the endogenous ligands. The group carried out a screening program to look for simple aromatic fragments that showed some affinity for the Tyr-phosphate binding pocket. At the same time, they performed soaking experiments to determine that the fragments did bind in the same mode at the Tyr phosphate; the best was benzoic acid ortho-substituted with an aldehyde. The final optimised grouping replaced the acid with malonate, and the aldehyde with an ester. The full inhibitor was stable over 24 hours in plasma. This approach is a nice complement to SAR by NMR. In a similar vein is the work of the Protherics team,[133] using a virtual screening protocol to drive the design and synthesis of libraries against Factor Xa, starting from a template structure that can be elaborated using simple chemical reactions. For each point of substitution, reagent lists are obtained and scored within the context of the active site. The best were redocked and scored, before synthesis. After a few iterations, several lead molecules with potent affinities were found.

11 Conclusions

There have been many significant advances in computer-aided drug design during the period of this review. Perhaps the most impressive progress has come in the field of ADME/Tox; only a few years ago, the models were limited and very crude. Increased amounts of high-quality data have altered that state to one where we can confidently predict not just which molecules will be subject to extensive metabolism, but even the site of metabolism. The development of better scoring functions for docking is continuing, again fuelled by access to better and more varied data sets. There is also a realisation that perhaps in the field of Cheminformatics, we will have to step back and take stock of how we think about high-throughput screening and combinatorial chemistry, as those two tools have not delivered as much as was hoped for. The discipline of computer-aided drug design is a healthy state, and that the field will continue to grow and advance, consolidating its position as an invaluable aid to drug discovery.

References

1. L. Di and E.H. Kerns, *Curr. Opin. Chem. Bio.*, 2003, **7**, 402.
2. H. van de Water beemd, D.A. Smith, K. Beaumont and D.K. Walker, *J. Med. Chem.*, 2001, **44**, 1313.
3. D. Butina, M.D. Segall and K. Frankcombe, *Drug Disc. Today*, 2002, **7**, S83.
4. P.A. Williams, J. Cosme, A. Ward, H.C. Angove, D.A. Vinkovi and H. Jhoti, *Nature*, 2003, **424**, 464.
5. S.B. Singh, L.Q. Shen, M.J. Walker and R.P. Sheridan, *J. Med. Chem.*, 2003, **46**, 1330.

6. J. Zuegge, U. Fechner, O. Roche, N.J. Parrott, O. Engkvist and G. Schneider, *QSAR*, 2002, **21**, 249.
7. I. Zamora, L. Afzelius and G. Cruciani, *J. Med. Chem.*, 2003, **46**, 2313.
8. R. Pearlstein, R. Vaz and D. Rampe, *J. Med. Chem.*, 2003, **46**, 2017.
9. A. Cavalli, E. Poluzzi, F. De Ponti and M. Recanatini, *J. Med. Chem.*, 2002, **45**, 3844.
10. R.A. Pearlstein, R.J. Vaz, J.S. Kang, X.L. Chen, M. Preobrazhenskaya, A.E. Shchekotikhin, A.M. Korolev, L.N. Lysenkova, O.V. Miroshnikova, J. Hendrix and D. Rampe, *Bioorg. Med. Chem. Letts.*, 2003, **13**, 1829.
11. O. Roche, G. Trube, J. Zuegge, P. Pflimlin, A. Alanine and G. Schneider, *Chembiochem*, 2002, **3**, 455.
12. M. Shen, Y.D. Xiao, A. Golbraikh, V.K. Gombar and A. Tropsha, *J. Med. Chem.*, 2003, **46**, 3013.
13. G.M. Rishton *Drug Disc. Today*, 2003, **8**, 86.
14. M.C. Wenlock, R.P. Austin, P. Barton, A.M. Davis and P.D. Leeson, *J. Med. Chem.*, 2003, **46**, 1250.
15. C.A. Lipinski, F. Lombardo, B.W. Dominy and P.J. Feeney, *Advanced Drug Delivery Reviews*, 1997, **23**, 3.
16. D.F. Veber, S.R. Johnson, H.Y. Cheng, B.R. Smith, K.W. Ward and K.D. Kopple, *J. Med. Chem.*, 2002, **45**, 2615.
17. C.A.S. Bergstrom, U. Norinder, K. Luthman and P. Artursson, *Pharm. Res.*, 2002, **19**, 182.
18. C.A.S. Bergstrom, M. Strafford, L. Lazorova, A. Avdeef, K. Luthman and P. Artursson, *J. Med. Chem.*, 2003, **46**, 558.
19. P. Stenberg, C.A.S. Bergstrom, K. Luthman and P. Artursson, *Clin. Pharm.*, 2002, **41**, 877.
20. T.I. Oprea, I. Zamora and A.L. Ungell, *J. Comb. Chem.*, 2002, **4**, 258.
21. M. Brustle, B. Beck, T. Schindler, W. King, T. Mitchell and T. Clark, *J. Med. Chem.*, 2002, **45**, 3345.
22. S.L. McGovern, E. Caselli, N. Grigorieff and B.K. Shoichet, *J. Med. Chem.*, 2002, **45**, 1712.
23. S.L. McGovern and B.K. Shoichet, *J. Med. Chem.*, 2003, **46**, 1478.
24. P.R.N. Wolohan and R.D. Clark, *J. Comp.-Aided Mol. Des.*, 2003, **17**, 65.
25. S. Winiwarter, F. Ax, H. Lennernas, A. Hallberg, C. Pettersson and A. Karlen, *J. Mol. Graph. Mod.*, 2003, **21**, 273.
26. G. Colmenarejo, A. Alvarez-Pedraglio and J.L. Lavandera, *J. Med. Chem.*, 2001, **44**, 4370.
27. R.D. Taylor, P.J. Jewsbury and J.W. Essex, *J. Comp.-Aided Mol. Des.*, 2002, **16**, 151.
28. I. Halperin, B.Y. Ma, H. Wolfson and R. Nussinov, *Proteins: Struct., Fun. Gen.*, 2002, **47**, 409.
29. H. Gohlke and G. Klebe, *Ang. Chem. Int. Ed.*, 2002, **41**, 2645.
30. P.D. Lyne, *Drug Disc. Today*, 2002, **7**, 1047.
31. T. Schulz-Gasch and M. Stahl, *J. Mol. Mod.*, 2003, **9**, 47.
32. R.X. Wang, Y.P. Lu and S.M. Wang, *J. Med. Chem.*, 2003, **46**, 2287.
33. R.D. Clark, A. Strizhev, J.M. Leonard, J.F. Blake and J.B. Matthew, *J. Mol. Graph. Mod.*, 2002, **20**, 281.
34. Y. Pan, N. Huang, S. Cho and A.D. MacKerell, Jr., *J. Chem. Inf. Comp. Sci.*, 2003, **43**, 267.
35. B.Q.Q. Wei, W.A. Baase, L.H. Weaver, B.W. Matthews and B.K. Shoichet, *J. Mol. Biol.*, 2002, **322**, 339.

36. J.W.M. Nissink, C. Murray, M. Hartshorn, M.L. Verdonk, J.C. Cole and R. Taylor, *Proteins – Struct. Fun. Gen.*, 2002, **49**, 457.
37. C.W. Murray and M.L. Verdonk, *J. Comp.-Aided Mol. Des.*, 2002, **16**, 741.
38. P. Cozzini, M. Fornabaio, A. Marabotti, D.J. Abraham, G.E. Kellogg and A. Mozzarelli, *J. Med. Chem.*, 2002, **45**, 2469.
39. M. Glick, G.H. Grant and W.G. Richards, *J. Med. Chem.*, 2002, **45**, 4639.
40. M. Vieth and D.J. Cummins, *J. Med. Chem.*, 2000, **43**, 10.
41. H. Gohlke and G. Klebe, *J. Med. Chem.*, 2002, **45**, 4153.
42. H. Gohlke, M. Hendlich and G. Klebe, *J. Mol. Biol.*, 2000, **295**, 337.
43. M.I. Zavodszky, P.C. Sanschagrin, R.S. Korde and L.A. Kuhn, *J. Comp.-Aided Mol. Des.*, 2002, **16**, 883.
44. S.A. Hindle, M. Rarey, C. Buning and T. Lengauer, *J. Comp.-Aided Mol. Des.*, 2002, **16**, 129.
45. S.J. Teague, *Nature Reviews (Drug Discovery)*, 2003, **2**, 527.
46. H.A. Carlson, *Curr. Opin. Chem. Bio.*, **6**, 447.
47. B.K. Shoichet, S.L. McGovern, B. Wei and J.J. Irwin, *Curr. Opin. Chem. Bio.*, **6**, 449.
48. S.L. McGovern and B.K. Shoichet, *J. Med. Chem.*, 2003, **46**, 2895.
49. C. Bissantz, P. Bernard, M. Hibert and D. Rognan, *Proteins – Struct. Fun. Gen.*, 2003, **50**, 5.
50. L. Birch, C.W. Murray, M.J. Hartshorn, I.J. Tickle and M.L. Verdonk, *J. Comp.-Aided Mol. Des.*, 2002, **16**, 855.
51. J.L. Jenkins, R.Y.T. Kao and R. Shapiro, *Proteins – Struct. Fun. Gen.*, 2003, **50**, 81.
52. T.N. Doman, S.L. McGovern, B.J. Witherbee, T.P. Kasten, R. Kurumbail, W.C. Stallings, D.T. Connolly and B.K. Shoichet, *J. Med. Chem.*, 2002, **45**, 2213.
53. S. Gruneberg, M.T. Stubbs and G. Klebe, *J. Med. Chem.*, 2002, **45**, 3588.
54. P.C. Wyss, P. Gerber, P.G. Hartman, C. Hubschwerlen, H. Locher, H.P. Marty and M. Stahl, *J. Med. Chem.*, 2003, **46**, 2304.
55. E. Vangrevelinghe, K. Zimmermann, J. Schoepfer, R. Portmann, D. Fabbro and P. Furet, *J. Med. Chem.*, 2003, **46**, 2656.
56. I. Zamora, T. Oprea, G. Cruciani, M. Pastor and A.L. Ungell, *J. Med. Chem.*, 2003, **46**, 25.
57. J.M. Schmidt, J. Mercure, G.B. Tremblay, M. Page, A. Kalbakji, M. Feher, R. Dunn-Dufault, M.G. Peter and P.R. Redden, *J. Med. Chem.*, 2003, **46**, 1408.
58. H.M. Vinkers, M.R. de Jonge, F.F.D. Daeyaert, J. Heeres, L.M.H. Koymans, J.H. van Lenthe, P.J. Lewi, H. Timmerman, K. Van Aken and P.A.J. Janssen, *J. Med. Chem.*, 2003, **46**, 2765.
59. T. Honma, K. Hayashi, T. Aoyama, N. Hashimoto, T. Machida, K. Fukasawa, T. Iwama, C. Ikeura, M. Ikuta, I. Suzuki-Takahashi, Y. Iwasawa, T. Hayama, S. Nishimura and H. Morishima, *J. Med. Chem.*, 2001, **44**, 4615.
60. M. Stahl, N.P. Todorov, T. James, H. Mauser, H.J. Boehm and P.M. Dean, *J. Comp.-Aided Mol. Des.*, 2002, **16**, 459.
61. N. Stiefl and K. Baumann, *J. Med. Chem.*, 2003, **46**, 1390.
62. F. Melani, P. Gratteri, M. Adamo and C. Bonaccini, *J. Med. Chem.*, 2003, **46**, 1359.
63. A. Vedani and M. Dobler, *J. Med. Chem.*, 2002, **45**, 2139.
64. S. Rivara, M. Mor, C. Silva, V. Zuliani, F. Vacondio, G. Spadoni, A. Bedini, G. Tarzia, V. Lucini, M. Pannacci, F. Fraschini and P.V. Plazzi, *J. Med. Chem.*, 2003, **46**, 1429.

65. T. Peng, J. Pei and J. Zhou, *J. Chem. Inf. Comp. Sci.*, 2003, **43**, 298.
66. H. Gohlke, S. Schwarz, D. Gundisch, M.C. Tilotta, A. Weber, T. Wegge and G. Seitz, *J. Med. Chem.*, 2003, **46**, 2031.
67. M.N. Islam, Y. Song and M.N. Iskander, *J. Mol. Graph. Mod.*, 2003, **21**, 263.
68. I.G. Tikhonova, I.I. Baskin, V.A. Palyulin and N.S. Zefirov, *J. Med. Chem.*, 2003, **46**, 1609.
69. J. Singh, H. van Vlijmen, W.C. Lee, Y. Liao, K.C. Lin, H. Ateeq, J. Cuervo, C. Zimmerman, C. Hammond, M. Karpusas, R. Palmer, T. Chattopadhyay and S.P. Adams, *J. Comp.-Aided Mol. Des.*, 2002, **16**, 201.
70. I.A. Doytchinova and D.R. Flower, *Proteins – Struct. Fun. Gen*, 2002, **48**, 505.
71. M.L. Lopez-Rodriguez, M. Murcia, B. Benhamu, A. Viso, M. Campillo and L. Pardo, *J. Med. Chem.*, 2002, **45**, 4806.
72. K.J. Schleifer and E. Tot, *QSAR*, 2002, **21**, 239.
73. J.K. Buolamwini and H. Assefa, *J. Med. Chem.*, 2002, **45**, 841.
74. G. Bringmann and C. Rummey, *J. Chem. Inf. Comp. Sci.*, 2003, **43**, 304.
75. J. Wellsow, H.J. Machulla and K.A. Kovar, *QSAR*, 2002, **21**, 577.
76. X. Hong and A.J. Hopfinger, *J. Chem. Inf. Comp. Sci.*, 2003, **43**, 324.
77. J. Bostrom, M. Bohm, K. Gundertofte and G. Klebe, *J. Chem. Inf. Comp. Sci.*, 2003, **43**, 1020.
78. T. Balle, J. Perregaard, M.T. Ramirez, A.K. Larsen, K.K. Soby, T. Liljefors and K. Andersen, *J. Med. Chem.*, 2003, **46**, 265.
79. R.E. Wilcox, J.E. Ragan, R.S. Pearlman, M.Y.K. Brusniak, R.M. Eglen, D.W. Bonhaus, T.E. Tenner and J.D. Miller, *J. Comp.-Aided Mol. Des.*, 2001, **15**, 883.
80. T. Naumann and H. Matter, *J. Med. Chem.*, 2002, **45**, 2366.
81. A. Golbraikh and A. Tropsha, *J. Mol. Graph. Mod.*, 2002, **20**, 269.
82. M. Ridderstrom, I. Zamora, O. Fjellstrom and T.B. Andersson, *J. Med. Chem.*, 2001, **44**, 4072.
83. Y. Patel, V.J. Gillet, G. Bravi and A.R. Leach, *J. Comp.-Aided Mol. Des.*, 2002, **16**, 653.
84. J. Bostrom, *J. Comp.-Aided Mol. Des.*, 2001, **15**, 1137.
85. J. Bostrom, J.R. Greenwood and J. Gottfries, *J. Mol. Graph. Mod.*, 2003, **21**, 449.
86. D.J. Diller and K.M. Merz, *J. Comp.-Aided Mol. Des.*, 2002, **16**, 105.
87. E.M. Krovat and T. Langer, *J. Med. Chem.*, 2003, **46**, 716.
88. P. Kahnberg, E. Lager, C. Rosenberg, J. Schougaard, L. Camet, O. Sterner, E.O. Nielsen, M. Nielsen and T. Liljefors, *J. Med. Chem.*, 2002, **45**, 4188.
89. L. Orus, S. Perez-Silanes, A.M. Oficialdegui, J. Martinez-Esparza, J.C. Del Castillo, M. Mourelle, T. Langer, S. Guccione, G. Donzella, E.M. Krovat, K. Poptodorov, B. Lasheras, S. Ballaz, I. Hervias, R. Tordera, J. Del Rio and A. Monge, *J. Med. Chem.*, 2002, **45**, 4128.
90. K.E.B. Parkes, P. Ermert, J. Fassler, J. Ives, J.A. Martin, J.H. Merrett, D. Obrecht, G. Williams and K. Klumpp, *J. Med. Chem.*, 2003, **46**, 1153.
91. Y. Koide, T. Hasegawa, A. Takahashi, A. Endo, N. Mochizuki, M. Nakagawa and A. Nishida, *J. Med. Chem.*, 2002, **45**, 4629.
92. O.O. Clement, C.M. Freeman, R.W. Hartmann, V.D. Handratta, T.S. Vasaitis, A.M.H. Brodie and V.C.O. Njar, *J. Med. Chem.*, 2003, **46**, 2345.
93. P.A. Smith, M.J. Sorich, R.A. McKinnon and J.O. Miners, *J. Med. Chem.*, 2003, **46**, 1617.
94. A. Garrigues, N. Loiseau, M. Delaforge, J. Ferte, M. Garrigos, F. Andre and S. Orlowski, *Mol. Pharm.*, **62**, 1288.
95. I.K. Pajeva and M. Wiese, *J. Med. Chem.*, 2002, **45**, 5671.

96. J.E. Penzotti, M.L. Lamb, E. Evensen and P.D.J. Grootenhuis, *J. Med. Chem.*, 2002, **45**, 1737.
97. I.J.S. Fairlamb, J.M. Dickinson, R. O'Connor, S. Higson, L. Grieveson and V. Marin, *Bioorg. Med. Chem.*, 2002, **10**, 2641.
98. B.A. Schweitzer, P.J. Loida, C.A. CaJacob, R.C. Chott, E.M. Collantes, S.G. Hegde, P.D. Mosier and S. Profeta, *Bioorg. Med. Chem. Letts.*, **12**, 8.
99. A. Poulsen, T. Liljefors, K. Gundertofte and B. Bjornholm, *J. Comp.-Aided Mol. Des.*, 2002, **16**, 273.
100. R. Bureau, C. Daveu, S. Lemaitre, F. Dauphin, H. Landelle, J.C. Lancelot and S. Rault, *J. Chem. Inf. Comp. Sci.*, 2002, **42**, 962.
101. M.L. Barreca, R. Gitto, S. Quartarone, L. De Luca, G. De Sarro and A. Chimirri, *J. Chem. Inf. Comp. Sci.*, 2003, **43**, 651.
102. J. Singh, H. van Vlijmen, Y.S. Liao, W.C. Lee, M. Cornebise, M. Harris, I.H. Shu, A. Gill, J.H. Cuervo, W.M. Abraham and S.P. Adams, *J. Med. Chem.*, 2002, **45**, 2988.
103. A. Kramer, H.W. Horn and J.E. Rice, *J. Comp.-Aided Mol. Des.*, 2003, **17**, 13.
104. C. Lemmen, T. Lengauer and G. Klebe, *J. Med. Chem.*, 1998, **41**, 4502.
105. J.L. Miller, E.K. Bradley and S.L. Teig, *J. Chem. Inf. Comp. Sci.*, 2003, **43**, 47.
106. M.M. Hann, A.R. Leach and G. Harper, *J. Chem. Inf. Comp. Sci.*, 2001, **41**, 856.
107. R.D. Cramer, *J. Med. Chem.*, 2003, **46**, 374.
108. R.D. Clark, J. Kar, L. Akella and F. Soltanshahi, *J. Chem. Inf. Comp. Sci.*, 2003, **43**, 829.
109. J.M. McKenna, F. Halley, J.E. Souness, I.M. McLay, S.D. Pickett, A.J. Collis, K. Page and I. Ahmed, *J. Med. Chem.*, 2002, **45**, 2173.
110. V.J. Gillet, P. Willett, P.J. Fleming and D.V. Green, *J. Mol. Graph. Mod.*, 2002, **20**, 491.
111. T. Wright, V.J. Gillet, D.V.S. Green and S.D. Pickett, *J. Chem. Inf. Comp. Sci.*, 2003, **43**, 381.
112. R.P. Sheridan and S.K. Kearsley, *Drug Disc. Today*, 2002, **7**, 903.
113. J.W. Raymond and P. Willett, *J. Comp.-Aided Mol. Des.*, 2002, **16**, 521.
114. Y.C. Martin, J.L. Kofron and L.M. Traphagen, *J. Med. Chem.*, 2002, **45**, 4350.
115. R.P. Sheridan, *J. Chem. Inf. Comp. Sci.*, 2003, **43**, 1037.
116. J. Xu, *J. Med. Chem.*, 2002, **45**, 5311.
117. Y. Takaoka, Y. Endo, S. Yamanobe, H. Kakinuma, T. Okubo, Y. Shimazaki, T. Ota, S. Sumiya and K. Yoshikawa, *J. Chem. Inf. Comp. Sci.*, 2003, **43**, 1269.
118. R.P. Sheridan, *J. Chem. Inf. Comp. Sci.*, 2002, **42**, 103.
119. X.Q. Lewell, A.C. Jones, C.L. Bruce, G. Harper, M.M. Jones, I.M. Mclay and J. Bradshaw, *J. Med. Chem.*, 2003, **46**, 3257.
120. J.D. Holliday, S.P. Jelfs, P. Willett and P. Gedeck, *J. Chem. Inf. Comp. Sci.*, 2003, **43**, 406.
121. J.W. Raymond, C.J. Blankley and P. Willett, *J. Mol. Graph. Mod.*, 2003, **21**, 421.
122. M. Hennemann and T. Clark, *J. Mol. Mod.*, 2002, **8**, 95.
123. L. Xing and R.C. Glen, *J. Chem. Inf. Comp. Sci.*, 2002, **42**, 796.
124. E. Abrahamian, P.C. Fox, L. Naerum, I.T. Christensen, H. Thogersen and R.D. Clark, *J. Chem. Inf. Comp. Sci.*, 2003, **43**, 458.
125. J.L. Faulon, D.P. Visco and R.S. Pophale, *J. Chem. Inf. Comp. Sci.*, 2003, **43**, 707.
126. J.L. Faulon, C.J. Churchwell and D.P. Visco, *J. Chem. Inf. Comp. Sci.*, 2003, **43**, 721.
127. A. Sharff and H. Jhoti, *Curr. Opin. Chem. Bio.*, 2003, **7**,
128. M. Hendlich, A. Bergner, J. Gunther and G. Klebe, *J. Mol. Biol.*, 2003, **326**, 607.

129. J. Gunther, A. Bergner, M. Hendlich and G. Klebe, *J. Mol. Biol.*, 2003, **326**, 621.
130. S. Schmitt, D. Kuhn and G. Klebe, *J. Mol. Biol.*, 2002, **323**, 387.
131. T. Klabunde and G. Hessler, *Chembiochem*, 2002, **3**, 929.
132. D. Lesuisse, G. Lange, P. Deprez, D. Benard, B. Schoot, G. Delettre, J.P. Marquette, P. Broto, V. Jean-Baptiste, P. Bichet, E. Sarubbi and E. Mandine, *J. Med. Chem.*, 2002, **45**, 2379.
133. J.W. Liebeschuetz, S.D. Jones, P.J. Morgan, C.W. Murray, A.D. Rimmer, J.M.E. Roscoe, B. Waszkowycz, P.M. Welsh, W.A. Wylie, S.C. Young, H. Martin, J. Mahler, L. Brady and K. Wilkinson, *J. Med. Chem.*, 2002, **45**, 1221.

3
Density Functional Theory

BY MICHAEL SPRINGBORG

1 Introduction

Four years have passed since the first report on density-functional theory in this series,[1] and two since the second report.[2] Compared to the almost 40 years that the density-functional theory of Kohn et al.[3,4] has been applied in physics and the almost 20 years it has been applied in chemistry, the four years seem to be a fairly short period of time, and one could have expected that the field has reached a level of maturity that would justify reports on its progress at only larger time intervals. That this is not the case may be ascribed to two facts. First, although the currently applied density-functional theory most often is accurate, it is not exact and, accordingly, can be improved and, second, the 'technical equipment' (i.e., computers, programs, and methods) have improved. Thus, at the moment, the research within the concept of density-functional theory of Kohn et al. can be split into two lines of research, partly exemplified through our two earlier reports.

First, density-functional theory is a very useful tool in extending, supporting, and explaining experimental studies of specific systems although often only idealized systems can be treated. The computations give information that often not is directly available with experimental methods, but due to the above-mentioned lack of 'exactness', simultaneously such studies give information that can be useful in improving the density functionals. In our first report[1] we concentrated on showing – through a number of examples – which information can be calculated with density-functional methods and, simultaneously, discussed the obtained accuracy. In the second report[2] we, instead, focused on the limitations of the currently applied density functionals as well as on suggestions for improvements. In this third report it seems, therefore, natural to return to the applications and see where we are today. Thus, we shall in the next section give a brief introduction to the fundamentals of density-functional theory and, subsequently, turn to various selected examples of applications. It shall be strongly stressed that the examples we will discuss are chosen partly according to the author's subjective choice and partly according to their diversity. Very, very many other examples of equal scientific quality could, but were not, have been chosen, simply because of the space restrictions that the present author put on this report.

2 Theoretical Foundations

Almost exclusively, theoretical studies of the structural and electronic properties of materials invoke the Born-Oppenheimer approximation, implying that the nuclei are treated as point charges at fixed positions. They contribute with one term to the total energy of the system of interest (i.e., the Coulomb interaction energy for those point charges, E_n), and the remaining part of the total energy, the electronic energy E_e, is found from the electronic Schrödinger equation

$$\hat{H}_e \Psi_e = E_e \Psi_e. \tag{1}$$

$\hat{H}_e$ contains the kinetic energy of the electrons, the Coulomb energy from their mutual interactions, and the Coulomb energy from the interactions between the electrons and the nuclei.

With M and N being the number nuclei and electrons, respectively, $\vec{R}_1, \vec{R}_2, \ldots, \vec{R}_M$ and $\vec{r}_1, \vec{r}_2, \ldots, \vec{r}_N$ their positions, $Z_k e$, $k = 1, \ldots, M$ and $-e$ their charges, and M_k, $k = 1, \ldots, M$ and m_e their masses, respectively, we have

$$\hat{H}_e = \hat{H}_{ke} + \hat{H}_{ee} + \hat{H}_{en}, \tag{2}$$

with

$$\hat{H} = -\frac{\hbar^2}{2m_e} \sum_{i=1}^{N} \frac{1}{2} \nabla_{\vec{r}_i}^2$$

$$\hat{H}_{ee} = \frac{1}{2} \sum_{i \neq j=1}^{N} \frac{e^2}{4\pi\epsilon_0 |\vec{r}_i - \vec{r}_j|}$$

$$\hat{H}_{en} = -\sum_{k=1}^{M} \sum_{i=1}^{N} \frac{Z_k e^2}{4\pi\epsilon_0 |\vec{R}_k - \vec{r}_i|}, \tag{3}$$

as well as

$$E_{\text{tot}} = E_n + E_e \tag{4}$$

with

$$E_n = \frac{1}{2} \sum_{k \neq l=1}^{M} \frac{Z_k Z_l e^2}{4\pi\epsilon_0 |\vec{R}_k - \vec{R}_l|}. \tag{5}$$

3: Density Functional Theory

There exists two fundamentally different approaches for calculating E_e for a given structure, i.e., for a given $\{\vec{R}_k = 1, \ldots M\}$. In practice, with only very few exceptions for almost all systems of interest, within both approaches one or more approximations have to be invoked and, therefore, no method is guaranteed to give accurate results for a given system and it is rather a question of experience, belief, and background which of the two methods is applied. The two approaches are the so-called wavefunction-based methods and the density-functional methods. They are described in detail in many textbooks (see, e.g., [5]), but since some of the current developments within density-functional theory bring this approach into close contact with the wavefunction-based methods (see later), we shall briefly outline the foundations of both.

Within the wavefunction-based methods, Eq. (1) is most often solved by first approximating Ψ_e as a single Slater determinant. Thereby correlation effects are per definition ignored. (Parts of) these may, however, be added subsequently either directly or via perturbation theory. The N single-particle functions $\phi_1, \phi_2, \ldots, \phi_N$ of the Slater determinant are calculated by solving the Hartree-Fock single-particle equations

$$\hat{F}\phi_k(\vec{x}) = \epsilon_k \phi_k(\vec{x}) \tag{6}$$

with

$$\hat{F} = \hat{h}_1 + \sum_{i=1}^{N}(\hat{J}_i - \hat{K}_i). \tag{7}$$

Here,

$$\hat{h}_1 = -\frac{\hbar^2}{2m_e}\nabla^2 - \sum_{k=1}^{M}\frac{Z_k e^2}{4\pi\epsilon_0|\vec{r} - \vec{R}_k|}, \tag{8}$$

and

$$\sum_{i=1}^{N}\hat{J}_i\,\phi_k(\vec{x}) = \sum_{i=1}^{N}\int\frac{e^2|\phi_i(\vec{x}_2)|^2}{4\pi\epsilon_0|\vec{r}_2 - \vec{r}|}d\vec{x}_2\,\phi_k(\vec{x})$$

$$\sum_{i=1}^{N}\hat{K}_i\,\phi_k(\vec{x}) = \sum_{i=1}^{N}\int\frac{e^2\phi_i^*(\vec{x}_2)\phi_k(\vec{x}_2)}{4\pi\epsilon_0|\vec{r}_2 - \vec{r}|}d\vec{x}_2\,\phi_i(\vec{x}). \tag{9}$$

Then,

$$E_e = \sum_{i=1}^{N}\langle\phi_i|\hat{h}_1|\phi_i\rangle + \sum_{i,j=1}^{N}\left[\langle\phi_i\phi_j|\frac{e^2}{4\pi\epsilon_0|\vec{r}_1 - \vec{r}_2|}|\phi_i\phi_j\rangle - \langle\phi_i\phi_j|\frac{e^2}{4\pi\epsilon_0|\vec{r}_1 - \vec{r}_2|}|\phi_j\phi_i\rangle\right]. \tag{10}$$

The first term on the right-hand side contains the largest part of the kinetic energy of the electrons and the Coulomb energy from the electrons-nuclei interactions. The second term contains the classical Coulomb energy (first term in the bracket) and the non-classical exchange energy (second term) for the inter-electronic interactions.

Going beyond the Hartree-Fock approximation, the assumption that the electronic wavefunction can be approximated through a single Slater determinant is abandoned, and an expansion in terms of more (many) determinants is applied. Thereby, per definition correlation effects are included, but, in addition, the computations become considerably more time consuming.

The density-functional theory of Hohenberg and Kohn[3] (see also [1, 2, 6]) provides a useful alternative. According to the original work of Hohenberg and Kohn[3] any ground-state property, including E_e, can be calculated once the electron density $\rho(\vec{r})$ is known. Thus, E_e is a functional of $\rho(\vec{r})$,

$$E_e = E_e[\rho(\vec{r})] \tag{11}$$

but, unfortunately, the precise form of this functional is unknown. The starting point for practical calculations within density-functional theory was provided by Kohn and Sham[4] who showed that the problem of calculating $E_e[\rho(\vec{r})]$ of Eq. (11) can be formulated as that of solving a set of single-particle equations,

$$\hat{h}_{\text{eff}}\psi_i(\vec{r}) = \epsilon_i\psi_i(\vec{r}) \tag{12}$$

with

$$\hat{h}_{\text{eff}} = -\frac{\hbar^2}{2m_e}\nabla^2 + V_{\text{eff}}(\vec{r}). \tag{13}$$

The potential V_{eff} of Eq. (13) contains the external Coulomb potential from the nuclei, the Coulomb potential from the electrons, as well as a potential describing the exchange interactions and correlation effects,

$$V_{\text{eff}}(\vec{r}) = V_{\text{ext}}(\vec{r}) + V_{\text{C}}(\vec{r}) + V_{\text{xc}}(\vec{r}). \tag{14}$$

In this case, the electronic energy becomes

$$E_e = \sum_{i=1}^{N}\langle\psi_i|\hat{h}_1|\psi_i\rangle + \frac{1}{2}\int V_{\text{C}}\rho(\vec{r})d\vec{r} + E_{\text{xc}}. \tag{15}$$

E_{xc} and $V_{\text{xc}}(\vec{r})$ are related through

$$V_{\text{xc}}(\vec{r}) = \frac{\delta E_{\text{xc}}}{\delta\rho(\vec{r})}, \tag{16}$$

and each can be split into an exchange and a correlation part,

$$E_{\text{xc}} = E_{\text{x}} + E_{\text{c}}$$
$$V_{\text{xc}} = V_{\text{x}} + V_{\text{c}}. \tag{17}$$

3: Density Functional Theory

Usually, the exchange-correlation energy is written as

$$E_{xc} = \int \epsilon_{xc}(\vec{r})\rho(\vec{r})d\vec{r} = \int \epsilon_{x}(\vec{r})\rho(\vec{r})d\vec{r} + \int \epsilon_{c}(\vec{r})\rho(\vec{r})d\vec{r}. \tag{18}$$

Since the precise functional dependencies of ϵ_{xc}, ϵ_x, and ϵ_c on ρ are unknown, one has to restore to approximations. The most common ones are the local-density approximations (LDAs) according to which $\epsilon_{xc}(\vec{r})$ is a function (and not functional) of $\rho(\vec{r})$. With the generalized-gradient approximations (GGAs) also dependencies on $|\nabla\rho(\vec{r})|$ and on $\nabla^2\rho(\vec{r})$ are included. Finally, with the hybrid approximations (of which the B3LYP is the most widespread one) E_x is written as a combination of the exchange energy expression obtained when inserting the Kohn-Sham orbitals into the Hartree-Fock expression, Eq. (10), and of a GGA expression, whereas the correlation energy is treated according to the GGA expression.

Both in the Hartree-Fock approximation and in the Kohn-Sham approach the total electron density is given as a sum over the N energetically lowest orbitals, i.e., in the Kohn-Sham case as

$$\rho(\vec{r}) = \sum_{i=1}^{N} |\psi_i(\vec{r})|^2 \simeq \sum_{i=1}^{N} |\phi_i(\vec{r})|^2, \tag{19}$$

but, whereas the Hartree-Fock approximation per construction gives only an approximate electron density, the Kohn-Sham approach gives, in principle, the exact density, so that the two may differ when the Hartree-Fock approximation is less good.

There is a number of differences between wavefunction- and density-based methods that makes it non-trivial to choose one or the other as superior, and that also has to be taken into account when accessing the accuracy of the results of a calculation. Thus, the density-functional methods include in principle all correlation effects, whereas with wavefunction-based methods these are included first on top of a Hartree-Fock calculation. On the other hand, the wavefunction-based methods are based on the exact (Schrödinger) equation making systematic improvements of the results **in principle** possible. In contrast thereto, for the density-functional methods, the approaches of Kohn and coworkers do not directly provide some calculational framework but only a proof of existence of the relation between density and ground-state properties. Thus, in any practical calculation one has to restore to approximate descriptions, first of all of exchange and correlation effects, which means that the starting point is approximate equations so that it is in principle unknown how to improve a calculation systematically. However, assuming that the approximate functionals are accurate, density-functional methods are computationally much more efficient than wavefunction-based methods, making, in particular, it possible to study large systems or to perform detailed structure optimizations.

Of practical importance is also the fact that most of the presently applied density-functional methods are based on calculating the electronic ground-state energy E_e using the fictitious Kohn-Sham orbitals $\{\psi_i\}$. Thus, although in principle any other ground-state property also can be obtained through the electron density, the precise procedure for doing so is in most cases largely unknown and, in particular, the role of the Kohn-Sham orbitals is unknown. Luckily, experience (e.g., through comparison with results of wavefunction-based calculations or with experimental results) has shown that in most cases accurate results are obtained when simply treating the Kohn-Sham orbitals as were they electronic orbitals. But one should remember that this is an approximation that needs careful control whenever applied.

A further aspect deserves to be mentioned here (also because we will study it below). Thus, as discussed above, the original density-functional theory of Hohenberg and Kohn was explicitly a ground-state theory, meaning that properties of excited states are not included in the theory. Since this will exclude the theoretical study of the results of a very large number of experimentally accessible quantities (e.g., spectroscopical data), this would reduce the practical use of the theory enormously. Once again, experience has shown that one often obtains accurate results when ignoring this fundamental aspect and treating the Kohn-Sham orbitals as electronic orbitals. In our last two reports we have seen and discussed some examples of this approach. Alternatively, as also discussed in those reports, during the last years another approach has become increasingly popular. This approach, the so-called time-dependent density-functional theory[7] (see also [8]), makes excitation energies accessible within the framework of density-functional theory. As the original density-functional theory of Hohenberg and Kohn, the time-dependent density-functional theory of Runge and Gross[7] provides at first only a proof of existence and, accordingly, practical applications of this theory are based on approximate functionals. Since this theory only during the last few years has become more widely applied, the experience that so far has been obtained regarding the accuracy of the currently applied approximate functionals is considerably more limited as that regarding the approximations within the original ground-state density-functional theory. Nevertheless, below we shall discuss some few examples of applications of this approach.

3 Structure and Energies

3.1 Ionization Potentials of Nickel-benzene Clusters. – As mentioned above, the theorems of Hohenberg and Kohn give only a proof of existence of functionals of the electron density for any ground-state property, but do not give their precise form for most observables. The approximate functionals that have been derived are first of all functionals for the electronic energy for a given structure, whereas approximate functionals for most other observables are extremely scarce. Due to this concentration on energetics one would first of all expect that the currently applied density-functional methods give accurate energies and corresponding structures. This shall be considered in this section.

As a first example we discuss the study of Rao and Jena[9] on $Ni_n(benzene)_m$ complexes. The goal of the study was to analyse some experimental results on the singly negatively charged anions with n up to around 10 and m up to around 5. In the theoretical study, Rao and Jena considered somewhat smaller systems, i.e., $n = 1-3$, $m = 1, 2$. Moreover, they optimized the structure completely, both for the neutral molecule and for the negatively charged anion and considered different spin states. The calculations used a generalized-gradient approximation (GGA) for exchange and correlation effects. The difference in the total energies of the neutral systems with the geometry of the anion yields vertical electron detachment energies,

$$\text{VDE} = E_0(\{\vec{R}_{X^-}\}, X^-) - E_0(\{\vec{R}_{X^-}\}, X). \tag{20}$$

Equivalently, a similar comparison but for the neutral systems with relaxed geometries gives the adiabatic electron affinities.

$$\text{AEA} = E_0(\{\vec{R}_{X^-}\}, X^-) - E_0(\{\vec{R}_X\}, X). \tag{21}$$

Here, $E_0(\{\vec{R}_Y\}, Z)$ is the ground-state energy of the system Z in the ground-state structure of system Y. The vertical detachment gives first of all two prominent transitions when the spin multiplicities of the two systems differ by ± 1. Finally, also vertical and adiabatic ionization potentials were calculated,

$$\text{VIP} = E_0(\{\vec{R}_X\}, X) - E_0(\{\vec{R}_X\}, X^+) \tag{22}$$

and

$$\text{AIP} = E_0(\{\vec{R}_X\}, X) - E_0(\{\vec{R}_{X^+}\}, X^+), \tag{23}$$

respectively.

The resulting ionization potentials and electron affinities are shown in Table I together with the available experimental information. Although only

Table I *Vertical and adiabatic ionization potentials (VIP and AIP) as well as vertical and adiabatic electron affinities (VDE and AEA) for $Ni_n(benzene)_m$ clusters as functions of n and m. The numbers in parenthesis are experimental values, and for the clusters with $n = 1$ the VDE are non-existing (the clusters cannot attach an extra electron). From ref. 9*

n	m	VIP(eV)	AIP(eV)	VDE(eV)	AEA(eV)
1	1	6.36 (5.99–6.42)	6.17	—	−0.35
2	1	5.81	5.70	0.72	0.48
3	1	5.70	5.62	0.81	0.78
1	2	6.12 (5.86)	5.68	—	−0.39
2	2	6.37	5.76	0.44	0.17
3	2	6.34	6.12	0.61	0.40

little experimental information is available, it is clear that the agreement between experiment and theory is good. This result deserves some further comments, because the result is less trivial as it may appear at first sight. Often a Koopmans-like theorem is assumed valid, meaning that ionization potentials are the energies of the occupied orbitals and electron affinities those of the unoccupied orbitals. However, current density-functional methods suffer from two short-comings: the occupied orbitals appear at too high energies, whereby the ionization potentials become too small, and the energy gap is too small, which in combination with the first problem means that accurate electron affinities may be merely a result of a lucky cancellation of errors. On the other hand, in the approach of Rao and Jena, the ionization potentials and electron affinities are calculated as differences of total energies. These total energies are themselves many orders of magnitudes larger than the differences themselves, so that it is a highly non-trivial matter to obtain the differences accurately.

Subsequently, Rao and Jena identified the spin multiplicities of the ground state of the neutral and charged clusters. These are reproduced in Table II. Due to the even number of electrons for the neutral clusters most of those have singlet ground states with, however, some exceptions. Finally, Rao and Jena identified the vertical transition energies from the ground state of the anion to the neutral system with a spin multiplicity differing by ± 1, as well as of similar transitions from the neutral system to the cation. These are reproduced in Table III. Here it shall be emphasized that spin effects usually have an only weak effect of the total energies so that an accurate description of the energies of the different multiplets is a non-trivial task. Therefore, a theoretical prediction like that of Table III can be obtained only through careful calculations.

3.2 Brønsted Acidity of Some Zeolites. – Another important issue of theoretical studies of chemically interesting questions is that of calculating energies involved in chemical reactions. In this subsection we shall discuss the reaction energies for adding or removing hydrogen atoms or ammonia molecules to a zeolite. Zeolites are crystalline materials with large pores. Therefore, these materials can be used as catalysts, and by varying their composition one may hope to vary the properties in a controlled way. For a theoretician, the fact

Table II *The spin multiplicities of the neutral, anionic, and cationic $Ni_n(benzene)_m$ clusters as functions of n and m. From ref. 9*

n	m	neutral	anion	cation
1	1	1	2	2
2	1	3	2	2
3	1	3	2	4
1	2	1	2	2
2	2	1	2	2
3	2	1	2	2

Table III Vertical transition energies for $Ni_n(benzene)_m$ clusters as functions of n and m for transitions where the spin multiplicities differ by ± 1. M_a, M_n, and M_c denotes the spin multiplicities of the anionic, neutral, and cationic system, respectively, and E the excitation energy. The left part shows anion $\rightarrow$ neutral transitions and the right part neutral $\rightarrow$ cation transitions. In two cases ($m = 1$ and $n = 2, 3$), the ground state of the neutral system is a triplet, whereas the lowest singlet lies only slightly higher in energy (0.19 and 0.42 eV, respectively), so that also the neutral $\rightarrow$ cation transitions involving the singlets have been included. The results are from ref. 9

n	m	M_a	M_n	E(eV)	M_n	M_c	E(eV)
1	1				1	2	6.36
2	1	2	1	0.72	3	2	5.81
2	1	2	3	1.38	3	4	6.78
2	1				1	2	5.57
3	1	2	1	1.30	3	2	5.88
3	1	2	3	0.81	3	4	5.70
3	1				1	2	5.54
1	2				1	2	6.12
2	2	2	1	0.44	1	2	6.37
2	2	2	3	1.92			
3	2	2	1	0.61	1	2	6.34
3	2	2	3	1.95			

that the materials are extended (roughly infinite) but also open (i.e., not closely packed) poses serious complications on the calculations. Here, the most important problem is that the open structure allows for significant structural relaxations when the material is reacting with some molecule, and for the calculation of the energy changes related to such reactions it is extremely important to identify the correct structures. Moreover, the open structure makes it difficult to apply methods based on plane waves as basis functions.

In order to study the catalytic properties of some zeolites as function of the composition, Yuan et al.[10] considered a fragment of the zeolite crystal and how the structure and the total energy changed when this fragment was allowed to interact with either hydrogen or ammonia. Considering finite segments allowed for using 'standard' quantum-chemical programs. Moreover, since they were interested in total energies for structures where both the composition and the number and types of chemical bonds were changed, they could not apply a local-density approximation but used the so-called B3LYP functional that has been devised to accurately reproduce total energies and reaction energies for compounds involving lighter elements. The functional combines density-functional and Hartree-Fock descriptions of exchange interactions.

They studied two different segments of the zeolite, i.e., $(OH)_3Si-O(H)-M(OH)_3$ and $((HO)_3SIO)_3Si-OH-M(OSi(OH)_3)_3$. Here, M is a metal atom which in their study was chosen as B, Al, Ga, or Fe, and the two clusters are

Table IV *Proton affinities (PA) and NH_3 adsorption energies (ΔE_{ads}) for different zeolites. The zeolites have been modeled through finite segments (defined through 'segment'), and different basis sets have been used (defined through 'basis set' – their precise definition is not important). M describes the metal of the zeolite. For further details, see the text. The results are from ref. 10*

segment	basis set	M	PA(kcal/mol)	ΔE_{ads}(kcal/mol)
M-2TH	a	Al	341.2	
M-2TH	b	Al	341.0	
M-2TH	c	Al	338.6	
M-2TH	d	Al	344.6	
M-8TH	d	Al	337.9	
M-2TH	a	Ga	342.1	
M-2TH	b	Ga	342.5	
M-2TH	c	Ga	340.7	
M-2TH	d	Ga	345.5	
M-8TH	d	Ga	339.5	
M-2TH	a	Fe	348.6	
M-2TH	b	Fe	344.4	
M-2TH	c	Fe	342.2	
M-2TH	d	Fe	352.2	
M-8TH	d	Fe	344.5	
M-2TH	a	B	410.7	
M-2TH	b	B	408.6	
M-2TH	c	B	403.9	
M-2TH	d	B	390.1	
M-8TH	d	B	357.3	
M-8TH		Al		25.6
M-8TH		Ga		24.0
M-8TH		Fe		21.5
M-8TH		B		19.8

denoted M-2TH and M-8TH, respectively. They studied two chemical reactions, i.e., the energy for removing the acidic proton (this energy is the proton affinity) and the energy for adsorbing a NH_3 molecule on the cluster.

Some of their results are summarized in Table IV. The authors studied the dependence of the results on the basis set as well as on the size of the segment. The table shows that the properties of their interest are well converged with respect to both quantities. It shall be stressed that only through careful choice of both such a convergence can be achieved. Moreover, the results show that the Al-, Ga-, and Fe-containing zeolites behave fairly similarly, whereas the B-containing one is different. In total they find an acid strength increasing from B, over Fe and Ga, to, finally, Al. This sequence is consistent with experimental observations.

3.3 Van der Waals Interactions. – For many years, density-functional methods were used mainly in physics, and the systems of interest were most often infinite, periodic crystals for which the atoms of different structures often had similar coordinations and where, in addition, the electrons were more or less delocalized. Then, the local-density approximation for exchange and correlation effects often was sufficiently accurate. This was, however, not the case when attempting to study chemical systems where the electrons are localized and where, when considering chemical reactions, the bonding situations may change. Then, the observation that the local-density approximation led to too strong chemical bonds (where, moreover, the overestimate largely depends only on the types of the two atoms forming a chemical bond) led to unacceptably inaccurate results. Often the calculations would predict the structure with the most chemical bonds as the most stable one (in some cases, this could even be the transition state in a chemical reaction!). This situation was, of course, not acceptable, but with the introduction of the generalized-gradient approximation (whereby not only the density but also its gradients are used in modeling the exchange and correlation effects) the major parts of these inaccuracies were removed so that the so-called chemical accuracy (reaction energies with an accuracy of about 1 kcal/mol) was achieved. Even further improvement has been obtained with functionals that include both Hartree-Fock and density-functional treatments of exchange effects. Here, the B3LYP functional is the most well-known functional.

However, the thereby obtained accuracies are still not so high that one can be confident that weak bonds are accurately described. For hydrogen bonds the accuracy is most often acceptable, but for the even weaker van der Waals bonds, the situation is complex, and highly inaccurate results may result. There has been some attempt to modify the existing density functionals so that the bond lengths and the binding energies of van der Waals complexes are accurately described, and some of those were studied recently by Kamiya *et al.*[11] Table V summarizes their findings.

From the table it is first of all clear that the energies that are been sought are much below the above-mentioned chemical accuracy. Moreover, even the specially designed functionals for such weak bonds show some scatter in the accuracy and, accordingly, one has to conclude that these systems belong to a class where current density-functional methods should be applied with extreme care, if at all.

3.4 Structure and Energetics for N-Acetyl-L-Glutamate-N-Methylamide. – As the computers become larger and faster and, parallel thereto, the computer programs become more efficient, the complexity of the systems that are been treated with theoretical methods increases. The increased complexity has, however, as a further consequence that the structure may not be uniquely defined, i.e., that there may be many meta-stable structures, so that the identification of the stabler one may be difficult.

Masman *et al.*[12] studied the relative total energies of different structures of N-acetyl-L-glutamate-N-methylamide, CO_2–CH_2–CH_2–CH(NH–$COCH_3$) (CO–$NHCH_3$), using both a density-functional method (with the B3LYP functional and an extended basis set) and Hartree-Fock methods with both a

Table V *Calculated bond lengths (R_e) and dissociation energies (D_e) for various van der Waals complexes in comparison with experimental results (denoted 'Exp'). DFT1, DFT2, DFT3, and DFT4 distinguish different density functionals that were explicitly constructed for studying such systems, and MP2 denotes results with Hartree-Fock-based methods for which correlation effects were added. For more details, see the original literature, i.e., [11]*

System	Method	$R_e(\text{Å})$	D_e(kcal/mol)
He$_2$	DFT1	3.04	0.017
He$_2$	DFT2	3.09	0.042
He$_2$	DFT3	3.16	0.052
He$_2$	DFT4	—	—
He$_2$	MP2	3.22	0.011
He$_2$	Exp	2.97	0.022
Ne$_2$	DFT1	3.20	0.080
Ne$_2$	DFT2	3.34	0.077
Ne$_2$	DFT3	3.40	0.101
Ne$_2$	DFT4	3.05	0.055
Ne$_2$	MP2	3.42	0.027
Ne$_2$	Exp	3.09	0.084
Ar$_2$	DFT1	3.85	0.336
Ar$_2$	DFT2	4.62	0.049
Ar$_2$	DFT3	4.69	0.056
Ar$_2$	DFT4	3.91	0.072
Ar$_2$	MP2	3.92	0.194
Ar$_2$	Exp	3.76	0.284

smaller and the extended basis set. The authors identified 21 meta-stable structures with the density-functional method, whereas 27 and 32 were located with the Hartree-Fock calculations with the smaller and larger basis set, respectively. Thus, the calculations do show some differences depending on the method. Surprisingly, however, is their finding that the relative total energies and, in particular, the relative ordering of the structures were significantly less dependent on the method. Even the Hartree-Fock calculations with the small basis set were able to give an accurate description of the trends. Moreover, although differences in the structural parameters could be obtained, these were fairly small.

All these results suggest that the currently applied methods can be of relevance for also large, e.g., biological, systems. Moreover, the results may show some differences in the details, but the overall trends seem to be observed with all the different types of methods.

3.5 Photodissociation of Triplet Acetaldehyde. – In our first report[1] we gave some examples of the accuracy that could be obtained with density-functional methods when applying them to chemical reactions. Thereby, structural changes, in particular the structure of the transition state, has to be obtained

Table VI *The calculated energy barrier for the reaction $CH_3CHO \to CH_3 +$ HCO (denoted direct barrier) and the barrier for the reverse reaction $CH_3 + HCO \to CH_3CHO$ (denoted reverse barrier) on the triplet surface. HF marks Hartree-Fock results, Corr. such ones extended with correlation effects, DFT density-functional results, and Exp. marks experimental results. The results are from ref. 13*

Method	Direct barrier (kcal/mol)	Reverse barrier (kcal/mol)
HF	34.4	10.8
HF	30.2	
Corr.	13.9	
Corr.	14.8	17.9
DFT	14.9	5.4
DFT	13.3	
DFT	14.3	6.3
Corr.	12.3	13.1
Corr.	12.4	
Corr.	13.3	8.8
Corr.	8.0	
Exp.	12.6	
Exp.		5.7
Exp.		6.0, 6.8
Exp.		7.5

with great accuracy, and also the energies of the different structures need to be accurate.

In a recent study, Cordeiro et al.[13] have given an example of how even more information can be extracted from such studies. They studied the photoinduced reaction $CH_3CHO \to CH_3 + HCO$, whereby the acetaldehyde is initially not in a singlet but in a triplet state. Using spin-restricted density-functional calculations they identified the initial and final states as well as the transition state, including the structures. Table VI summarizes their findings together with other available information as also reported by Cordeiro et al.[13]

The table shows clearly that Hartree-Fock calculations give inaccurate results, whereas the density-functional results are accurate (notice, that this has been obtained by using the B3LYP functional that combines Hartree-Fock and density-functional descriptions of exchange effects), and that the calculations that include correlation effects beyond a Hartree-Fock calculations are the most accurate ones. Except for being a slightly more complicated case than what we discussed earlier (i.e., the excited triplet state and not the ground-state singlet configuration), the conclusion is essentially a confirmation of our earlier results.

However, Cordeiro et al. showed that it was possible to do a complete dynamical calculation on top of the structure determinations. Starting from the transition state with an excess energy of 2.34 kcal/mol (this is the energy difference between the experimental photolysis energy and the barrier height) a series of dynamical calculations were performed which ultimately showed how

Table VII *Average partitioning of the energy for the dissociation $CH_3CHO \rightarrow CH_3 + HCO$ on the triplet surface. The results are from ref. 13*

Product	Mode	Calculated energy (kcal/mol)	Experimental energy (kcal/mol)
HCO	Translation	1.9	2.6 ± 0.5
HCO	Rotation	0.9	1.15, 1.70
HCO	Vibration	6.6	
CH_3	Translation	3.7	5.0 ± 1.0
CH_3	Rotation	0.7	
CH_3	Vibration	20.7	

the energy was distributed among the various translational, rotational, and vibrational modes of the system. As shown in Table VII, thereby information of direct experimental relevance is obtained. Moreover, it is seen that the results are accurate. Thus, this study is an example of how density-functional calculations nowadays also can be used in giving precise information on the dynamical properties of the systems of interest, at least for small systems.

3.6 Vibrations of Benzimidazole. – The calculation of vibrational properties of molecules is a field where density-functional methods can be very useful. Such calculations require, first, that the structure of the lowest total energy is accurately determined and, second, that also the second-order derivatives of the total energy with respect to structural parameters are accurate. In addition, experimental information from, e.g., infrared or Raman spectroscopy can be used in assigning the accuracy of the results. Moreover, if, in addition, also the relevant matrix elements are calculated then not only the positions but also the strengths of the experimentally observed peaks can be calculated. Thus, through comparison between experimental and theoretical studies a very detailed examination of the accuracy of the theoretical approach can be carried through.

Morsy *et al.*[14] calculated the vibrational properties of benzimidazole using both density-functional and Hartree-Fock methods. For the density-functional calculations they used both a generalized-gradient approximation and the B3LYP functional that combines density-functional and Hartree-Fock descriptions of exchange effects. Moreover, they considered the effects of the quality of the basis set on the results.

Fig. 1 shows the experimentally observed and theoretically calculated results for the optimized structure of the benzimidazole molecule. It is seen that the density-functional results are highly accurate, which confirms earlier findings for covalently bonded molecules containing lighter atoms.

Subsequently, experimental Raman and infrared spectra were reported both on the crystalline and on the gas phase of the benzimidazole molecule, and the results were compared with theoretical ones. It was found that the Hartree-Fock results were the least accurate ones (as often is the case – usually the Hartree-Fock results are scaled with some constant between 0.8 and 0.9 in

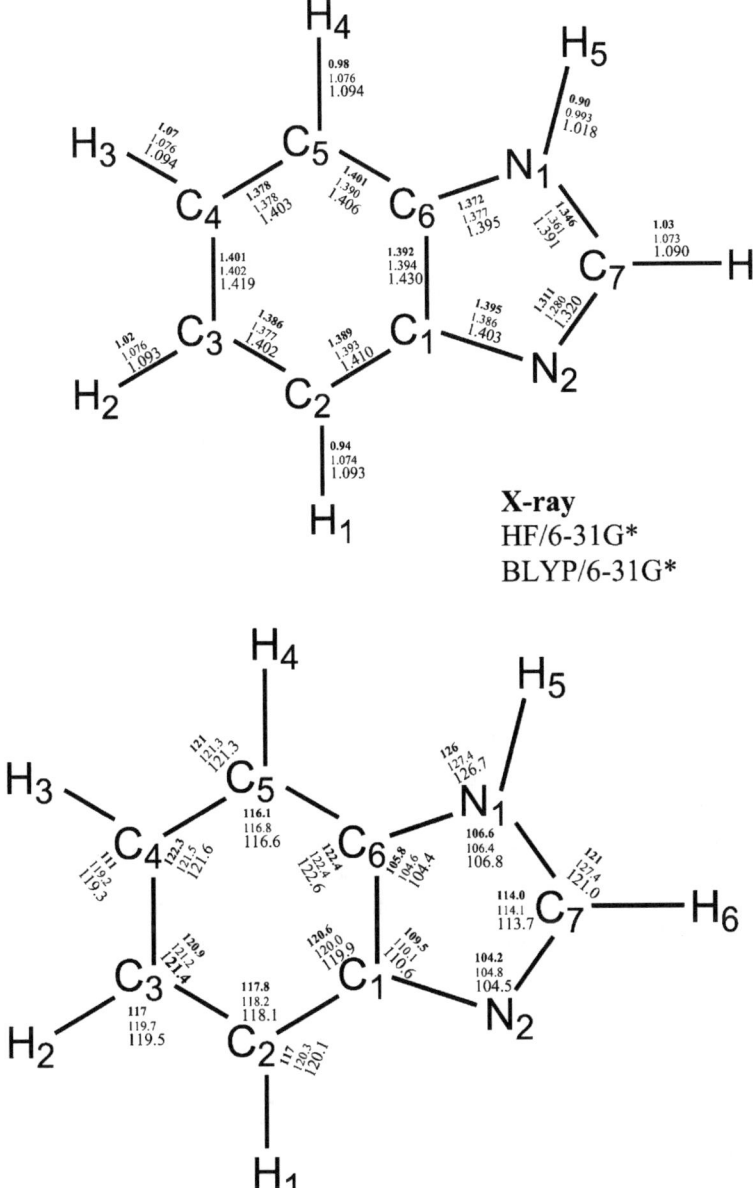

Figure 1 *Experimentally observed (denoted X-ray) and theoretically calculated structural parameters for benzimidazole. The calculations were performed using the extended 6-31G* basis set with either the Hartree-Fock (HF) or a generalized-gradient approximation within density-functional theory (BLYP). Reproduced with permission of The American Chemical Society from ref. 14*

order to obtain accurate results), whereas almost all vibrational modes, including their intensities, were accurate with the different density-functional methods, with the results of the BLYP calculations (i.e., of the generalized-gradient-approximation ones) being the most accurate ones. This finding is interesting since the B3LYP has been designed explicitly for calculating energetics accurately, but, obviously, structural parameters may thereby become accurate whereas the second-order derivatives may suffer from some inaccuracies.

3.7 Chemical Reactions Involving Hydrogen Bonds. – As discussed in our previous reports,[1,2] the relatively small strength of hydrogen bonds (up to at most 0.5 eV per bond) together with the tendency of the local-density approximations to overestimate the strength of any bond, has made theoretical studies within the density-functional formalism of systems involving hydrogen bonds difficult. In particular, chemical reactions where hydrogen bonds are formed and/or broken are often not correctly described within local-density approximations. On the other hand, generalized-gradient approximations as well as the methods, like B3LYP, that combine Hartree-Fock and density-functional treatments of exchange effects are often accurate.

As a slightly more complicated situation, Rauhut[15] studied a number of reactions involving the simultaneous transfer of two hydrogen atoms, i.e., double-proton-transfer reactions. Some of the systems of this study are shown in Fig. 2, i.e., the 1H-1,2,3-triazole and the 2H-1,2,3-triazole together with the complexes of two 1,2,3-triazole molecules, involving two hydrogen bonds. As may be interfered from the figure, for the complexes one may consider a process in which both hydrogen atoms simultaneously are transferred along the hydrogen bonds.

In order to access the accuracy of B3LYP calculations Rauhut performed also calculations based on the Hartree-Fock approximation with addition of correlation effects (these are often considered of paramount importance in describing hydrogen bonds). In Table VIII we show some of the resulting total

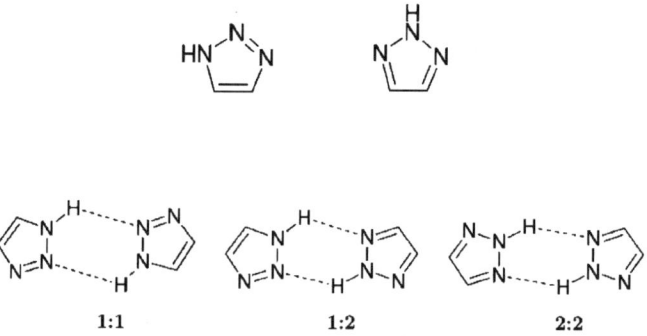

Figure 2 *Upper row: structure of 1H-1,2,3-triazole and 2H-1,2,3-triazole, and lower row: complexes of 1,2,3-triazole involving two hydrogen bonds. We shall refer to the five structures as 1H, 2H, 1H-1H, 1H-2H, and 2H-2H, respectively. Reproduced with permission of the PCCP owner societies from ref. 15*

Table VIII *Relative energies (in kcal/mol) for the structures of Fig. 2 as well as of transition states (denoted TS). The notation of the structures are from Fig. 2. Corr. denotes results based on Hartree-Fock calculations augmented with the inclusion of correlated effects, whereas B3LYP are density-functional results with the functional that combines Hartree-Fock and density-functional treatment of exchange effects. The results are from ref. 15*

System	Corr.	B3LYP
1H	4.3	4.7
2H	0.0	0.0
TS(1H→2H)	52.6	52.8
1H-1H	5.8	5.9
1H-2H	3.3	3.4
2H-2H	0.0	0.0
TS(1H-1H→2H-2H)	23.5	20.7
TS(1H-2H→2H-1H)	24.0	21.1

energies from this work. We notice that Rauhut reports only few structural results so that these can hardly be compared.

The table shows clearly that the energetics for these transitions are accurately described with the chosen density functional. It should be added that the B3LYP functional is a partly parameterized functional, so that the agreement may be considered less surprising. However, the learning set for the parameterizing of the functional does hardly contain any hydrogen-bonded systems. Moreover, for larger systems with multiple hydrogen bonds, their hydrogen-bond energies are often considered non-additive and instead cooperative effects are assumed to be important for the energies (see later). Thus, the double-hydrogen-bonded systems of Rauhut do indeed represent a set of systems for which the performance of theoretical approaches can be critically explored. On the other hand, unfortunately the study of Rauhut does not contain any studies using pure density functionals, e.g., of the generalized-gradient type.

3.8 Metal-sulphur Bonds for Sulphur on the Au(111) Surface. – Chemical reactions are continuously a subject of these reports. Chemistry are to a large extent concerned with producing systems (molecules, solids, etc.) with predefined properties, and in the process of synthesizing these systems one has to create, break, and modify chemical bonds through chemical reactions. Such processes can often be assisted through the application of catalysts. In a homogeneous catalysis, the reactants and the catalyst exist in the same phase, which makes it possible to fine-tune the catalyst but, on the other hand, often makes a separation of products from the catalyst difficult. On the other hand, for heterogeneous catalysis, the catalyst and the reactant exist in different phases, so that most often the catalyst is a solid, whereas the reactants are found in gas or liquid phases. Studying the details of these latter reactions requires accordingly studying how some molecules interact with a solid.

In the simplest case one would assume that the catalyst is a crystalline material, so that the molecules interact with a surface of such a crystal. Whereas theoretical studies of infinite, periodic, crystalline materials are often fairly simple, those of a catalyst plus some molecules are highly complicated. The existence of a surface breaks the symmetry in at least one direction, and the existence of the molecules may break the symmetry in the other two. In total, one ends up with a system that is very (if not roughly infinitely) large with a very low symmetry. Therefore, such theoretical studies have to consider simplified and idealized systems.

One example of a such is the recent one of Rodriguez et al.[16] These authors considered the (111) surface of an Au crystal on which sulphur atoms were deposited. They considered different coverages, i.e., 0.25, 0.5, and 1 monolayer of sulphur on gold. Here, x monolayer of sulphur means that for every $1/x$ gold atoms there is one sulphur atom. As a further complication, they also considered different lower-symmetry structures. E.g., for the 0.5 monolayer of sulphur one can imagine a situation where four gold atoms on the surface have two sulphur atoms adsorbed at two positions that are not equivalent. In order to describe this situation, one may use, e.g., a unit cell of 2×2 atoms from the gold surface and adsorb 2 sulphur atoms on those. Such situations were also considered by Rodriguez et al.[16]

The authors performed density-functional calculations on the system of interest using three different density functional, all of the generalized-gradient type. The calculations were actually performed on a so-called periodic-slab geometry, i.e., the surface was represented by a slab of a certain thickness that subsequently was periodically repeated in the direction perpendicular to the slab and with a certain vacuum region in between.

The results are summarized in Table IX. In the study, different positions of the sulphur atoms on the gold surface were considered, whose precise definitions for the present purpose are less important (although they are listed in the table). The table gives several aspects that are important for this report. First, it is satisfying that the different density functionals give roughly the same structural parameters once the overall structure has been chosen. Also, for a given coverage all density functional give the same relative ordering of the adsorption energies for the different positions of the sulphur atoms. On the other hand, the fact that the adsorption energies, after all, show some scatter (according to Rodriguez et al.,[16] the RPBE functional is supposed to give more accurate adsorption energies than the other two) is less satisfactory, and may be considered an example of one of the current problems of density-functional theory: there exist very many functionals that all have advantages and disadvantages, but which, unfortunately, do not give mutually consistent results, most notably regarding binding energies.

A further aspect that can be extracted from Table IX is important. Thus, neither structural parameters nor adsorption energies are independent of the

Table IX Au–S distances (d, in Å) and adsorption energies (E_a, in eV/(sulphur atom)) for 0.25, 0.5, and 1 monolayer (denoted by ML) of sulphur on Au(111). 'Cell' denotes the repeated unit of the surface, 'position' denotes where the sulphur atoms are adsorbed, and 'DFT' marks the different density-functionals. The results are from ref. 16 where further information can be found

ML	cell	position	DFT	d	E_a
0.25	2 × 2	a-top	PW91	2.26	2.23
			PBE	2.26	2.17
			RPBE	2.27	1.92
		Bridge	PW91	2.37	3.40
			PBE	2.36	3.31
			RPBE	2.38	3.03
		hollow hcp	PW91	2.41	3.66
			PBE	2.40	3.57
			RPBE	2.42	3.12
		hollow fcc	PW91	2.39	3.83
			PBE	2.39	3.72
			RPBE	2.39	3.31
0.50	2 × 1	a-top	PW91	2.31	2.36
			PBE	2.32	2.26
			RPBE	2.33	2.01
		Bridge	PW91	2.39	2.75
			PBE	2.40	2.69
			RPBE	2.41	2.38
		hollow hcp	PW91	2.47	2.87
			PBE	2.46	2.75
			RPBE	2.47	2.46
		hollow fcc	PW91	2.45	2.96
			PBE	2.44	2.83
			RPBE	2.45	2.54
0.50	2 × 2	hollow fcc, S	PW91	2.45	2.94
			PBE	2.43	2.82
			RPBE	2.44	2.53
		bridged + a-top	PW91	2.46, 2.45	3.00
			PBE	2.47, 2.46	2.88
			RPBE	2.48, 2.46	2.60
1.00	1 × 1	a-top	PW91	2.39	2.56
			PBE	2.37	2.48
			RPBE	2.40	2.09
		bridge	PW91	2.62	2.34
			PBE	2.61	2.26
			RPBE	2.65	1.85
		hollow hcp	PW91	2.72	2.32
			PBE	2.73	2.21
			RPBE	2.74	1.82
		hollow fcc	PW91	2.70	2.31
			PBE	2.71	2.22
			RPBE	2.73	1.81
	2 × 1	a-top, S	PW91	2.38	2.54
			PBE	2.39	2.47
			RPBE	2.40	2.07
		a-top + a-top	PW91	2.47	3.23
			PBE	2.46	3.03
			RPBE	2.48	2.69

coverage. This means that when studying theoretically the catalytic properties of solids it is important to study a coverage that is not unrealistically large. Moreover, it is also important to identify exactly the adsorption sites of the adsorbants.

3.9 NO_x on MgO. – Once it is known how different molecules adsorb on a given surface as well as the adsorption energies, one can start studying how the crystal surface may act as a catalysts, i.e., promote different reactions. A study in this direction was undertaken by Miletic et al.[17] They used density-functional calculations with a generalized-gradient approximation for exchange and correlation effects in studying the adsorption of neutral and charged NO_x molecules on the (100) surface of crystalline MgO. In contrast to the previous work, they modeled the crystal through a finite cluster containing either 32 or 50 atoms (two layers of each 4×4 or 5×5 ions) and added the adsorbants on the central region of the uppermost layer.

Miletic et al.[17] checked first their calculations by comparing the calculated structural and vibrational parameters of the isolated adsorbants with experimental information. Their results are reproduced in Table X, where it clearly is seen that accurate results are obtained.

Subsequently, the molecules were placed at a number at various positions on top of the cluster, and the structure was allowed to relax to its closest total-energy minimum. The binding energy was calculated as the energy of the total system minus that of the cluster and that of the isolated adsorbate (charged or neutral), and from this the authors found support for the so-called cooperative chemisorption model. I.e., their results suggest that a crystalline metal oxide, [MO], with M being the metal, will adsorb pairs of NO_x molecules, that are charged, like $[MO] + 2NO_2 \rightarrow [MO](NO_2^+)(NO_2^-)$, $[MO] + NO + NO_2 \rightarrow [MO](NO^+)(NO_2^-)$, and $[MO] + NO_2 + NO_3 \rightarrow [MO](NO_2^+)(NO_3^-)$. This information is extremely useful when attempting to understand and improve catalytic after-treatment of NO_x.

One may suggest that the finite cluster is too small to allow for accurate results, first of all since the bonding of MgO has a strong ionic character,

Table X *N–O bond length (d, in Å), O–N–O bond angle (α, in deg.), the stretch frequency (v, in cm^{-1}) from calculations (denoted 'calc') and experiment (denoted 'exp') for different neutral and charged NO_x molecules. The results are from ref. 17*

Compound	d(calc)	d(exp)	α(calc)	α(exp)	v(calc)	v(exp)
NO	1.162	1.15			1680	1870
NO_2	1.208	1.197	133.3	134.3	1646	1621
NO_3	1.246	1.236	120	120	1791	
NO^+	1.074	1.063			2375	2376
NO_2^+	1.132	1.15	180	180	2427	2375
NO_2^-	1.283	1.236	115.1	115.4	1292	1243
NO_3^-	1.272		120		1354	1356

which could lead to long-ranged electrostatic interactions. Whether this is the case or not, is beyond the purpose of this discussion. Instead we emphasize that, as always, the theoretical calculations have to be designed as a compromise between two opposing wishes: On the one side, the calculations shall be as accurate as possible, which, however, often means that they easily become so involved that only very simple, and in some case unrealistic, situations can be studied. On the other side, the calculations shall be able to address issues of relevance to basic or applied research, which implies that approximations have to be introduced that make the calculations possible. The study of Miletic et al.[17] is an example of how such a compromise can be reached and highly relevant information thereby be obtained without making so severe approximations that the results become unreliable.

3.10 Rotational Stability of Substituted Acetanilides. – The combination of experiment and theory is very often a very constructive way of obtaining detailed information about the system of interest. The two approaches can give supplementary information that, together, can make the conclusions of the study very reliable. There exists, accordingly, a very large number of such studies, and here we shall just consider some few examples.

Ilieva et al.[18] performed density-functional calculations with the B3LYP functional and performed furthermore vibrational IR spectroscopy experiment on some substituted acetanilides, Fig. 3. In particular, they were interested in the rotational properties of the amide group, $(NH)(CO)CH_3$, i.e., rotations about the N–C bond of this group. They substituted the hydrogen atom in the phenylene ring opposite of the amide group (i.e., *p*-substitution) by various groups, H, CH_3, OH, OCH_3, OC_2H_5, NH_2, Cl, COOH, NO_2, SO_2NH_2, and studied the above-mentioned rotation. Fig. 3 shows their results for the unsubstituted system, and in Table XI the energy barriers for the various compounds are listed.

As one intuitively would expect, the energies of the rotation depend only weakly on the substituent. The authors calculated and measured also the N–H and C–O vibrational frequencies of the amide group as a function of the substituent, and they observed that there is a clear correlation between the energy barriers for rotation, ΔE_1 and ΔE_2, and the C–O stretch frequency. Thus, this study shows in a very convincing way how relations between quantities, that at first hand may not be assumed strongly related, can be identified through careful studies.

3.11 The Reaction of Methylformate With Ammonia. – That part of chemistry where density-functional calculations have become most widely used is, most likely, that of studying chemical reactions between molecules. Already in our first report[1] we reviewed some examples of such applications as well as mentioned different descriptors that are been used in trying to identify the properties of a single molecule that would describe whether and how it would react with a given other molecule. Whereas the latter part still is in its infancy and not yet has found a widespread applicability (maybe partly because these

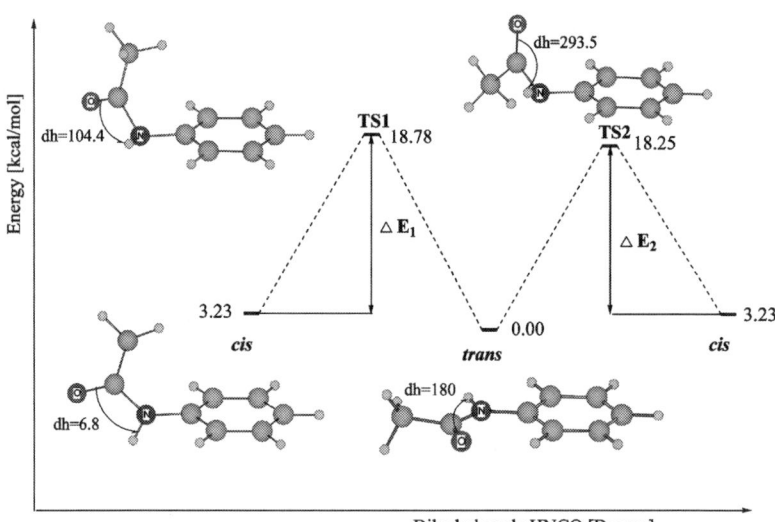

Figure 3 *Energy diagram for the rotation around the C–N bond in acetanilides. Reproduced with permission of The American Chemical Society from ref. 18*

Table XI *Energy barriers for rotation around the C–N bond in substituted acetanilides. The energies are given in kcal/mol and shown schematically in Fig. 3. The results are from ref. 18*

p-substituents	E_{TS1}	E_{TS2}	ΔE_1	ΔE_2
H	18.8	18.2	15.5	15.0
CH_3	18.9	18.4	15.8	15.3
OCH_3	19.3	18.8	16.5	16.1
OCH_2CH_3	19.4	19.1	16.4	16.1
Cl	18.9	18.3	15.2	14.6
COOH	18.0	17.3	14.4	13.8
OH	19.3	18.8	16.5	16.0
NH_2	19.6	19.2	17.0	16.5
NO_2	17.8	17.1	13.7	12.9
SO_2NH_2	18.3	17.5	14.4	13.6

descriptors are not giving measurable numbers), the study of chemical reactions is still a very important part of applied density-functional theory within chemistry. Here, the systems that are been studied have become larger.

As an example of such applications we discuss another work of Ilieva et al.[19] They studied the reaction between ammonia and methylformate, cf. Fig. 4, whereby two reaction paths were explored, i.e., the concerted and the stepwise, cf. Fig. 4. Using both density-functional calculations and Hartree-Fock calculations extended with correlation effects they found the relative energies shown in Table XII.

3: Density Functional Theory

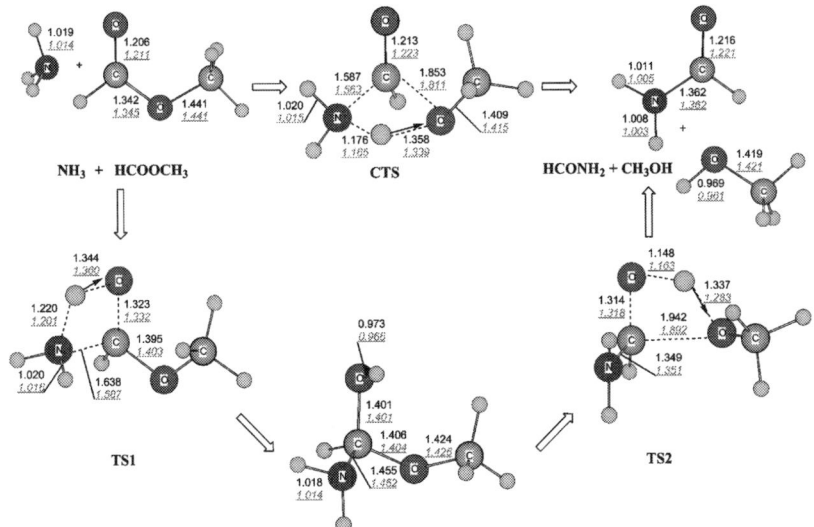

Figure 4 The reaction of methylformate with ammonia. Both the concerted (upper part) and the stepwise (lower part) reaction path is shown, and the upper (lower) structural data give the values calculated with density-functional (Hartree-Fock + correlation) calculations. Reproduced with permission of The American Chemical Society from ref. 19

Table XII Relative energies (relative to the reactants, and in kcal/mol) for the reaction of Fig. 4. 'B3LYP' and 'corr' represent results from density-functional and Hartree-Fock+correlation calculations, respectively. The results are from ref. 19

Structure	B3LYP	corr
CTS	40.1	46.7
TS1	31.1	44.8
I1	10.5	3.9
TS2	38.1	43.2
Products	4.6	5.0

The table shows a very good agreement between the density-functional calculations (that employed the B3LYP functional) and the correlated Hartree-Fock calculations, both in respect to structure (cf. Fig. 4) and in respect to the variation in the total energies (cf. Table XII). Most important is it that the energy barrier for the reaction was found to be large, i.e. about 40 kcal/mol. Therefore, the authors considered a possible catalyst-assisted reaction. First, it was observed that experiments involving the reactions between ammonia and methylformate take place in solution. Accordingly, effects of a solvent may be important. Therefore, they studied whether a second NH_3 molecule could have some catalytic effects. They found indeed that the energy barrier could be reduced somewhat.

The latter approach represents one way of including the effects of a solvent. Ilieva et al.[19] found that the second ammonia molecule does form a chemical bond to the intermediate complexes. Therefore, the more common approach where the effects of a solvent are included as those of a polarizable continuous medium surrounding the molecules of interest may not be so useful in this case, since in the present study there are bonds between the solvent and the reacting molecules.

3.12 The Rearrangement of Azulene to Naphthalene. – With the study of Ilieva et al.[19] that we discussed in the previous subsection, we demonstrated how, compared to earlier, density-functional studies on chemical reactions have been extended to consider more complex reactions. We stress that the study of chemical reactions is a far from trivial endeavour, since not only the stable ground states of the reactants and of the products have to be identified but also, what is considerably more complicated, the transition states have to be found. Transition states are characterized by being saddle points on the total-energy-vs.-structure hypersurface, but it is complicated to identify which such saddle points are relevant for a given reaction.

To some extent, the recent study by Alder et al.[20] on the rearrangement of azulene to naphthalene is even more impressive, as these authors considered very many different reaction paths. Whereas naphthalene and azulene both have the stoichiometry $C_{12}H_{10}$ and are planar, naphthalene consists of two fused hexagons sharing a side but azulene consists of a fused pentagon and heptagon also sharing a side. Naphthalene is considerably more stable than azulene (by 143.3 kJ/mol according to Alder et al.[20]), so starting from azulene, one may first expect that a smaller restructuring of the carbon backbone, involving only smaller energy changes can explain how azulene changes into naphthalene.

It turns out, however, that the simplest model involves a large energy barrier (almost 400 kJ/mol, cf. [20]). Therefore, other mechanisms have to be considered. As demonstrated very nicely by Alder et al.,[20] suggestions from experimental investigations can with advantage be incorporated into the theoretical studies – without such an input it would be almost impossible to guess and study all the possible reaction paths one can think of! They considered very many reactions, including various ones involving bond breakings as well as internal hydrogen- and radical-promoted mechanisms.

Here, we shall not try to discuss all their results, nor reproduce the different reaction paths including the total-energy variations. Instead we stress the importance of the combination of theory and experiment: From various experimental studies, different information on the various products and intermediates could be obtained, and with this information it is possible to

3: Density Functional Theory

guess a reaction paths. The theoretical calculations could then be used in verifying the proposals as well as in obtaining more precise information on the reaction paths and on the energetics.

3.13 Dissociation of Azomethane. – Another combined experimental and theoretical study of a chemical reaction is the work of Diau and Zewail,[21] who studied the dissociation of azomethane that can take place either through a concerted reaction, $(H_3C)-N=N-(CH_3) \rightarrow N_2+2CH_3$, or through a stepwise reaction, $(H_3C)-N=N-(CH_3) \rightarrow CH_3N_2+CH_3 \rightarrow N_2+2CH_3$. Here, the experimental work was done using femtosecond spectroscopy, which is so fast that information during a chemical reaction can be extracted. In parallel to the experimental studies, density-functional calculations, e.g., with the B3LYP functional, were carried through, and the combined information was used in identifying the reaction path.

As a first step in the theoretical study, the optimized structures of the reactants and the products as well as of various transition states were determined. Table XIII contains the calculated structural and energy parameters. For some of the energies, also experimental values are listed, and for these the theoretical ones are seen to be very close, which indicates that such studies are reliable when studying chemical reactions.

But the ultimate goal was to compare the theoretical calculations with the results of the real-time observations using femtosecond spectroscopy, which means that information on excitations has to be available. In order to obtain this information, Diau and Zewail used time-dependent density-functional theory, i.e., the extension of density-functional theory to describe excitations (this approach will be discussed further in Sec. 5). Using the experimental

Table XIII *Optimized structural parameters and energies from density-functional calculations for different systems that are of relevance for the dissociation of azomethane (AZM). d_{NN} and d_{CN} are the lengths of the nearest-neighbour N–N and C–N bonds, respectively, in Å, and α the bond angle in deg. Finally, ΔE is the energy relative to the trans isomer in kcal/mol, either from the density-functional calculations (DFT) or from experiment (exp). The results are from ref. 21*

Species	d_{NN}	d_{CN}	α_{CNN}	ΔE(DFT)	ΔE(exp)
I. trans-AZM	1.236	1.466	113.0	0.0	0.0
II. cis-AZM	1.237	1.483	120.2	10.1	9.1
III. $CH_3N_2 + CH_3$	1.170	1.514	122.7	46.3	50.2–55.4
IV. $N_2 + 2CH_3$	1.095			28.4	31.6
V. $2NCH_3$		1.414		102.0	
TS (I→II), rotamer	1.287	1.467	117.8	43.6	
TS (I→II), invertomer	1.222	1.512, 1.395	115.3, 180.0	49.2	
TS (I→IV)	1.133	2.126	112.3	50.0	
TS (III→IV)	1.135	1.820	119.6	47.9	

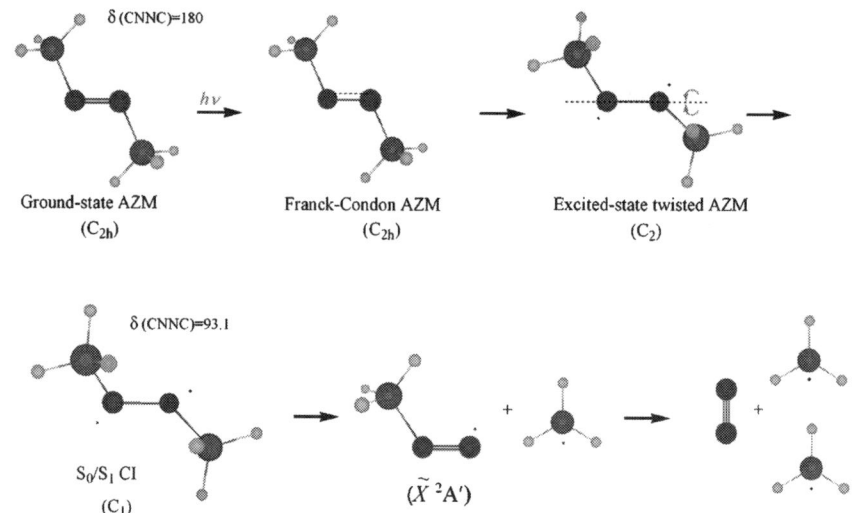

Figure 5 *The dissociation mechanism for azomethane. Reproduced with permission of Wiley-VCH Verlag from ref. 21*

information in combination with theoretical calculations on the ground, excited, and ion potential-energy surfaces, the authors arrived at the dissociation mechanism summarized in Fig. 5.

What makes this study different from the previous one, is two aspects. First, with femtosecond spectroscopy the timescale of the measurement is comparable to typical vibrations of molecules, which means that information at different times during a chemical reaction can be extracted. Thus, a chemical reaction can be followed in real time. Second, the work of Diau and Zewail is one example of calculations where time-dependent density-functional theory has been applied to obtain new information, and not merely as a testing ground. Such studies constitute one of those areas of applied density-functional theory that is growing most rapidly at the moment, and the study of Diau and Zewail[21] shows elegantly how two advances techniques nowadays can be exploited in getting very detailed information.

3.14 Reactions Between Transition Metals and Ammonia. – Transition metals form a class of materials, where Hartree-Fock-based methods often have serious problems giving accurate results. On the other hand, density-functional methods do most often not suffer from these shortcomings. As a single example of chemical reactions involving transition metals we discuss the work of Chen et al.[22] who used both experimental (infrared vibrational spectroscopy) and theoretical (B3LYP calculations) methods in studying the reactions between the early transition-metal atoms Sc, Ti, and V and a single NH_3 molecule.

Although at first this reaction may look very simple, the fact that there are many possible reaction channels make it significantly more complicated. We mention here $M + NH_3 \to MNH_3$, $M + NH_3 \to HMNH_2$, $M + NH_3 \to H_2MNH$, and $M + NH_3 \to MNH + H_2$. Here, the calculations in combination with the experimental information on the vibrational properties can help in identifying the products.

The results showed that the reaction $M + NH_3 \to MNH_3$ for all three early transition metals, Sc, Ti, and V took place at first. This reaction is exothermic ($\Delta E = -15.4$, -20.6, and -19.7 kcal/mol for Sc, Ti, and V, respectively), and the energy barrier is almost negligible. Subsequent irradiation with near ultraviolet or visible light changed the MNH_3 molecule into a planar $HMNH_2$ molecule. Also this reaction is exothermic ($\Delta E = -26.7$, -25.4, and -21.1 kcal/mol for Sc, Ti, and V, respectively), but there is a considerably energy barrier ($\Delta E = 17.4$, 17.5, and 20.9 kcal/mol for Sc, Ti, and V, respectively). For scandium, finally, a further endothermic reaction was found in which an H_2 molecule was eliminated.

3.15 Dynamics of Chemical Reactions. – In all the examples of chemical reactions we have discussed so far, the calculations are performed by identifying ground states of reactants and products as well as transition states. Subsequently, some paths in the configurational space is assumed but for the discussion this path is only of secondary importance.

More closely related to the experimental situation would, of course, be to perform a purely dynamical calculation in which the time development of the structure as the reaction proceeds is followed. This requires that both nuclear and electronic degrees of freedom are treated in a dynamical way, which is a computational demanding endeavour. In such studies not only the single structures of the ground states and of the transition states are important but also the surroundings of them. E.g., it will most likely play an important role whether the energy hypersurface around the transition states is very flat or very steep – this kind of information does not enter the above-mentioned approach.

As recently discussed by Frank,[23] such calculations are on their way, although so far only very few have been reported. However, they should be able to give a much more precise picture of how and why chemical reactions take place. Also here, density-functional methods are at the heart of the studies.

3.16 Hydrogen Bonds. – Hydrogen bonds are weak bonds with enormous importance for biological systems. Due to the low energies of these bonds, their theoretical description is particularly sensitive to inaccuracies, e.g., in the approximate density functionals. The B3LYP approach that combines density-functional and Hartree-Fock descriptions of exchange effects may be sufficiently accurate but it suffers from being computational complicated for large systems (since the exact exchange energy has to be calculated, and exchange interactions are only slowly converging as function of inter-electronic distance). On the other hand, the local-density approximation fails miserably for hydrogen bonds and many of the commonly used generalized-gradient approximations are also not sufficiently accurate. However, some more

Table XIV *Properties of hydrogen-bonded complexes. The complexes are denoted by x–y, with x and y being one of the following symbols: fm: formamide, nma: N-methylacetamide, w: water. In some cases more different complexes have been considered; they are distinguished through different numbers. n gives the number of hydrogen bonds, MP2 and PBE results of Hartree-Fock+correlation and density-functional calculations, respectively, d_{NH-O} and d_{CH-O} hydrogen bond lengths (in Å), and ΔE is the association energy upon forming the complex (in kcal/mol). The results are from ref. 25*

Complex	n	ΔE (MP2)	ΔE (PBE)	d_{NH-O} (MP2)	d_{NH-O} (PBE)	d_{CH-O} (MP2)	d_{CH-O} (PBE)
fm-fm (1)	2	−14.35	−14.34	1.825	1.81		
fm-fm (2)	2	−9.70	−9.11	1.825	1.84	2.234	2.28
fm-fm (3)	1	−7.34	−6.78	1.935	1.92		
fm-fm (4)	1	−6.76	−6.28	1.904	1.90		
nma-nma (1)	2	−17.18	−15.27	1.799	1.78		
nma-nma (2)	2	−10.76	−8.57	1.867	1.87	2.249	2.25
nma-w	1	−7.2	−7.51	1.979	1.84		

recently derived functionals (e.g., the so-called PBE functional[24]) are expected to be sufficiently accurate.

Ireta et al.[25] studied some different hydrogen-bonded systems. First, they accessed the accuracy of the PBE functional by calculating structure and bonding energies of some hydrogen-bonded complexes and compared the results with those of Hartree-Fock+correlation calculations. The results are summarized in Table XIV. From the table it is seen that in almost all cases that were studied by the authors, a very good agreement in the stabilization energies between Hartree-Fock+correlation and the density-functional calculations is found (i.e., of chemical accuracy, meaning below 1 kcal/mol per bond). Also the structural parameters are very accurate, except for one case, where a strong deviation is found. The exception is the only complex formed by two different molecules, and it is therefore uncertain whether the deviations are related to this fact.

The ultimate goal of the authors was to study the cooperativity of the hydrogen bonds in an α-helix. Thus, the fact that hydrogen bonds have a non-negligible electrostatic component suggests that different hydrogen bonds may interact electrostatically and, accordingly, that the hydrogen bonding energy of a large system with many (say p) hydrogen bonds, is not simply p times the hydrogen bonding energy of the system containing only one hydrogen bond. They studied oligomers of alinine and estimated, from an extrapolation, that for an infinite polyalinine α-helix, the hydrogen-bond energy per hydrogen bond was more than twice that of the system with just one hydrogen bond.

This study demonstrates two things: First, it is possible to explore the properties of the relatively weak hydrogen bonds using the present density

3: Density Functional Theory

functionals. However, it also shows that one has to be very careful choosing the appropriate density functional, and, furthermore, first check carefully that the functional is able to describe the properties of interest for related systems.

3.17 DNA Base-stacking Interactions. – Above, we studied van der Waals interactions between rare-gas atoms and found that the binding energies were so small that they hardly could be described through the present approximate density functionals, so that specifically designed functionals for those had to be developed.

A similar problem is encountered in many biologically relevant problems where the binding energies are comparable with room temperature. Kurita et al.[26] examined whether it was possible to develop density functionals that could describe such systems. They started out with one of the common and popular density functionals, the so-called PW91 approximation, i.e., a generalized-gradient approximation (GGA). Thereby they used (cf. the discussion above) that most other functionals, including the B3LYP functional, are completely incapable to describe the interactions between rare-gas atoms.

Starting from the PW91 functional they modified it so that the binding energies and bond lengths of the homoatomic diatomics He_2, Ne_2, Ar_2, and Kr_2 are well described (actually, it turned out that their improved functional, involving only a single adjustable parameter, could not describe the properties of all four diatomics simultaneously). Subsequently, they explored whether the thereby improved functional was capable of describing the binding-energy curve of two cytosine molecules, which were chosen as representing a part of DNA. It turned out that the results were not promising. Thus, the conclusion of this study must be that there still is some way to go before such weakly bonded systems are accurately described within density-functional theory.

3.18 Vibrations of Fe_2CO. – As mentioned above, particularly for transition-metal-containing molecules density-functional methods are useful in giving accurate structural and energetic properties. The combined experimental and theoretical study of Tremblay et al.[27] on the vibrational properties of the Fe_2CO demonstrates that reasonable accuracy also is obtained when discussing vibrational properties.

The authors calculated first the energetically lowest isomers of the neutral and the singly positively or negatively charged molecules. In its original formulation, density-functional theory applies only to the ground state, but this has later been extended to be valid for the energetically lowest structure of each symmetry representation of the system of interest, separately. Therefore, Tremblay et al. could study five different isomers with different symmetries. The results gave spin multiplicities of 7 for the ground state of the neutral molecule and 6 for that of the charged molecule. Moreover, using the total-energy differences between these ground states, an adiabatic electron affinity of 1.31 eV and an adiabatic ionization potential of 6.74 eV were found, which are similar to the experimental values for Fe_2 and FeCO, giving credibility to the calculations.

Table XV *Experimental and theoretical vibrational frequencies (ν, in cm^{-1}) and intensities (I, relative intensities for the experimental values, and in km/mol for the theoretical ones) for Fe$_2$CO. Two modes, no. 4 and 6, were not observed in the experiments (labeled 'exp'), and BPW91 and BLYP mark two different density functionals. The results are from ref. 27*

Mode	ν(exp)	I(exp)	ν(BPW91)	I(BPW91)	ν(BLYP)	I(BLYP)
1	1898.0	1000	1929	1229	1920	1214
2	483.2	9	468	26	429	27
3	371.6	4	337	7	314	1
4			306	0.2	284	0
5	291.5	6	250	21	253	19
6			51	1.5	44	1

Subsequently, a detailed analysis of the vibrational properties was presented. Here, we only show (in Table XV) the calculated and experimental frequencies and intensities for Fe$_2$CO. The authors used two different density functionals and argued, through the results of the table, that one functional is superior to the other. Nevertheless, both functionals give results that are in reasonable, but not perfect, agreement with experiment.

The study of Tremblay *et al.*[27] shows that vibrational properties of such systems, where spin multiplicities play a role, can be calculated with current density functionals. However, it once more emphasizes the problem related to having very many different available functionals, making it very difficult to choose the 'best' one for a given application.

3.19 Anharmonic Vibrational Modes in Adenine. – In a recent paper, Lappi *et al.*[28] showed that the combination of experimental and theoretical studies of the vibrational properties of a given compound (in their case, adenine) can be used in giving new information on the vibrations. The authors used experimental vibrational frequencies from matrix-isolation spectra and compared those with results from density-functional calculations. For the latter, they used both a generalized-gradient approximation and the B3LYP functional.

For some modes they observed a discrepancy between theory and experiment, and instead of ascribing these to inaccuracies in the calculations, they suggested that anharmonic effects could be responsible. Thus, the calculations assume that the total-energy hypersurface around the total-energy minimum is well approximated by a harmonic potential, but if this approximation is not sufficiently accurate, the calculated and measured vibrational frequencies will differ.

In order to explore this proposal in more detail, Lappi *et al.* calculated the total energy as function of the normal coordinate for such vibrations. As one example we show in Fig. 6 the total-energy variations as well as the calculated vibrational wavefunctions. The figure clearly shows anharmonicities.

3: Density Functional Theory

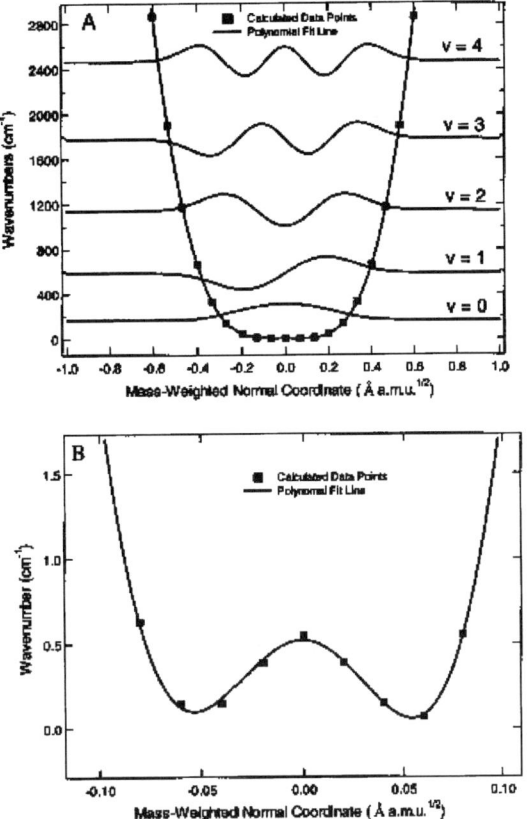

Figure 6 *The total-energy variation for a mode of adenine possessing anharmonicities. The upper part shows the total energy and the calculated vibrational modes, whereas the lower part is an expansion of the central region of the upper part. Reproduced with permission of The American Chemical Society from ref. 28*

This study demonstrates how the combination of theory and experiment leads to information that from each part separately would not have been obtained.

3.20 Cumulenes and Polyynes. – In this section we have discussed chemical reactions where some bonds have been broken and others been formed, and where the total-energy changes often were relatively large. We have also studied reactions involving the much weaker hydrogen bonds as well as the very weak van der Waals interactions. We found that the accuracy of the density-functional calculations often was sufficient to give precise descriptions of the chemical bonds as well as of the hydrogen bonds, whereas the van der Waals interactions were inaccurately described.

In the last subsection of this section we shall discuss a somewhat different scenario, i.e., the total-energy changes accompanying changes in the bond orders. Starting out with a linear chain containing an odd number of carbon atoms and including, furthermore, four hydrogen atoms, two relevant structures can be constructed: With two hydrogen atoms at each end, a structure with double bonds between the carbon atoms is obtained (i.e., a cumulene), whereas having three hydrogen atoms at one end and one at the other, a structure with alternating single and triple bonds between the carbon atoms is obtained (i.e., a polyyne). With the reasonable, but not exact, assumption that the C–H bond energies are the same for the two structures, the total-energy difference is that of the bond ordering.

Woodcock et al.[29] studied such chains using both density-functional and Hartree-Fock+correlation calculations. They employed both generalized-gradient approximations and the B3LYP functional in the density-functional calculations and they used a number of different basis sets in order to assure that their conclusions were not biased by the choice of the basis set.

For the smallest possible system of the above-mentioned type, i.e., the one containing three carbon atoms, the Hartree-Fock+correlation calculations as well as experiment gave that the polyyne with alternating bond lengths is the stabler one (by 1–5 kcal/mol). On the other hand, all density-functional calculations predicted the cumulene with essentially non-alternating bond lengths to be stabler (by around 4–6 kcal/mol). For the two larger systems of the study of Woodcock et al., i.e., containing five and seven carbon atoms, respectively, all types of calculations found the polyyne to be stabler than the cumulene, but once again the stability of cumulene relative to that of polyyne was overestimated by the density-functional calculations.

This study extends the class of test systems that can be used in exploiting the accuracy of approximate density functionals. It has been known for very many years that many density functionals are inaccurate when describing the promotion of electrons from, say, s orbitals to p orbitals, which suggests that comparing, for carbon, similar compounds but with different numbers of sp, sp^2, and sp^3 hybrids will give useful information on the accuracy of a given density functional. But as pointed out by Woodcock et al.,[29] also their study can be used in accessing the accuracy of density-functionals and, accordingly, in improving the functionals.

4 Orbitals and Densities

4.1 Conductivity in DNA. – Per construction, the currently applied density-functional methods are methods for calculating total energies. Accordingly, most applications are devoted to exactly this issue, as indirectly evidenced by the many examples of the previous section. In addition to the total energy, the density-functional methods give also the total electron density.

However, in the course of calculating the total energy for a given system and structure, one almost exclusively solves the Kohn-Sham single-particle equations, thereby obtaining a set of orbitals. Although, in the strict sense, these orbitals are those of a fictitious system, it can be argued (see, e.g., [1]) that they constitute a set of good approximations to the electronic orbitals and,

3: Density Functional Theory

accordingly, they are often treated as such. But from this discussion it should be clear that this is an approximation that needs to be justified, e.g., through comparisons with results of Hartree-Fock-based calculations or of experiments whenever possible.

As an example of how these orbitals are used in extracting further information from a calculation than what the total energy as a function of structure would give, we discuss the study of Gervasio et al.[30] on wet DNA. Simultaneously, this study represents an example of the complexity of the systems that nowadays can be treated with density-functional methods.

The authors studied the smallest nucleotide crystal structure that can be made in the laboratory, i.e., a structure containing a guanine-cytosine dodecamer in the unit cell. Their ultimate goal was to study charge transport for a realistic system for this system and, therefore, they added both Na atoms (thereby allowing a charging of the guanine-cytosine system to take place) and water molecules (representing a realistic medium for the system). In total this led to a unit cell containing no less than 1194 atoms, whose positions were optimized using the Car-Parrinello method with a generalized-gradient approximation for exchange and correlation effects within the density-functional formalism.

The calculations led to a small value of the energy gap separating occupied and unoccupied orbitals, 1.28 eV. As mentioned above, at first this gap has no relation to the energetically lowest electronic excitation energy (we shall return to this issue in the next section) but experience has shown that it is often a good, although underestimated, approximation to it. Thus, the results indicate a value of the lowest excitation energy that is well below the typical values for biological systems. The latter most often are as for insulators, whereas here the value resembles those of semiconductors. Removing the water molecules the authors found even that the value was reduced, so although the water molecules are not directly participating in the bonding, their polarity influences the electronic properties.

With such a low value for the lowest excitation energy, one may suggest that charge transport can take place. They used the Kubo-Greenwood expression

$$\sigma_{\alpha\alpha}(\omega) = \frac{2\pi e^2}{3m_e^2 \Omega \omega} \sum_{v,c} |\langle \psi_v | \hat{p}_\alpha | \psi_c \rangle|^2 \delta(\epsilon_c - \epsilon_v - \hbar\omega) \tag{24}$$

to calculate the optical conductivity tensor. In this expression, v and c denote occupied and unoccupied orbitals, respectively, with ψ_v and ψ_c being the single-particle orbitals and ϵ_v and ϵ_c their energies. Moreover, α equals x, y, or z, ω is the frequency, and Ω the volume of the unit cell. Finally, $\hat{p}_\alpha$ is the momentum operator.

This formula is, in its original formulation, the expression for the conductivity calculated from the real orbitals. Nevertheless, since the Kohn-Sham orbitals are known to provide good approximations to the former, these have been inserted, instead. Plotting the conductivity for different directions as function of frequency, Gervasio et al.[30] found for the lowest frequencies that the electrical conductivity along the strand is mainly due to excitations from guanine to the sodium ions, whereas the excitations perpendicular thereto have significant inner-strand contributions.

This study illustrates several issues. First, very large systems can be treated, although it is not routinely done. Second, the Kohn-Sham orbitals can often, and are often, be manipulated as were they electronic orbitals, whereby information of direct relevance to experiment is obtained. Third, when performing theoretical calculations on systems that are experimentally relevant, one most often has to make certain simplifying assumptions on the structure in order to make the calculations possible. The importance of the sodium atoms and water molecules in the present work shows that one has to be very careful removing parts that at first may seem irrelevant for the issue of interest.

4.2 Momentum-space Densities in Ethane. – In the preceding subsection the orbitals entered through matrix-element effects, cf. Eq. (24). In momentum spectroscopy one can, under certain assumptions, directly measure the orbital densities, thereby providing a unique tool for studying the accuracy of theoretical methods and, in our case, for exploring whether the Kohn-Sham orbitals constitute good approximations to the true orbitals.

In electron-momentum spectroscopy a high-energy photon is used in ionizing the target molecule. Assuming that the ionized and neutral system have identical orbitals (except for the one that is been emptied – this is the so-called target approximation), one measures thereby the momentum density of the orbital from which the electron has been removed.

Deng et al.[31] performed such experiments on ethane and compared the results with those of theoretical calculations using both Hartree-Fock and density-functional methods as well as with different basis sets. In the density-functional calculations they considered only functionals that use a combination of density-functional and Hartree-Fock descriptions of exchange effects (like the B3LYP functional), whereas correlation effects are described with a density-functional approximation.

Although the total momentum-space density suffers from being rather structure-less (all nuclei are placed at the origin, and the densities tend to resemble Gaussians more or less independent of system), they do show differences that are significant when discriminating between different theoretical approaches. Deng et al.[31] found in their work clearly, that the density-functional calculations led to better agreement with experiment than the Hartree-Fock calculations, thus giving strong support for using Kohn-Sham orbitals in interpreting chemical bonding as were they electronic orbitals.

The work of Deng et al.[31] is interesting of a further reason. According to density-functional theory any ground-state observable should be a functional of the ground-state electron density. This is also the case for the momentum density. However, whether the functional is simply the density obtained through the Fourier-transform of the Kohn-Sham orbitals, is most likely not the case. In fact, Lam and Platzman[32] have given explicit expressions for the momentum densities within density-functional theory and thereby shown that a correction term occurs, the so-called Lam-Platzman correction. Deng *et al.* chose to ignore this correction which, obviously, was a justified approximation. This may make one hope that similar corrections to other experimental observables (for instance to the conductivity as studied in the previous subsection) also can be ignored. But here, much work and experience is needed.

4.3 Electron Density in [FeCo(CO)$_8$]$^-$. – Besides the total energy, the total electron density is the quantity that per construction should be correctly reproduced by density-functional calculations, as given through Eq. (19). Since the electron density is experimentally accessible via X-ray diffraction experiments, a comparison between theory and experiment is possible, whereby the approximations that are made seemingly are controllable. However, differences may occur through differences in the systems that are being studied by the two approaches: the experimental studies require crystalline samples, whereas theoretical studies on molecular crystals most often are performed for isolated molecules.

Nevertheless, there exists a large number of combined theoretical and experimental studies on the electron density of different materials, and in this subsection we shall just discuss a single one, whereby we simultaneously will show how the theoretical studies can be used in extending the experimental ones. Macchi *et al.*[33] studied the metal-containing carbonyl [FeCo(CO)$_8$]$^-$. This molecule has two metal atoms and is one example of a series of related carbonyls containing two metal atoms. The interesting issue is that the two metal atoms are at the center of the molecule but not so close that a chemical bond between them can be unambiguously identified. Instead, it may be suggested that a CO group acts as a bridge between the metal atoms.

First, Macchi *et al.*[33] determined experimentally the electron density of the compound of interest and, subsequently, they studied theoretically different structures in order to identify the bonding properties. As a central result of their study we show in Fig. 7 the calculated total electron densities for different structures of the [FeCo(CO)$_8$]$^-$ anion. The different structures differ first of all in how the bridging CO group is placed relative to the Fe–Co 'bond'.

The electron densities illustrate a common problem with such studies: it is very large at the sites of the nuclei and in the regions, where one may expect chemical bonds, it changes only little from that of a superposition of atomic electron densities. Thus, the identification of individual atoms or of chemical bonds is very difficult (see, e.g., [5]). A useful approach is provided by the atoms-in-molecules approach of Bader.[34] In contrast to many other approaches, it is largely independent of the definition of the basis set used in the calculations, and it has been shown to be helpful in identifying chemical bonds. The middle row of Fig. 7 shows how this approach is used in separating the total electron density into contributions from the individual atoms and in identifying chemical bonds. It is seen that, as also schematically shown in the upper row, a chemical bond is found between the two metal atoms when the bridging CO group is placed highly asymmetric, i.e., closer to one of the two metal

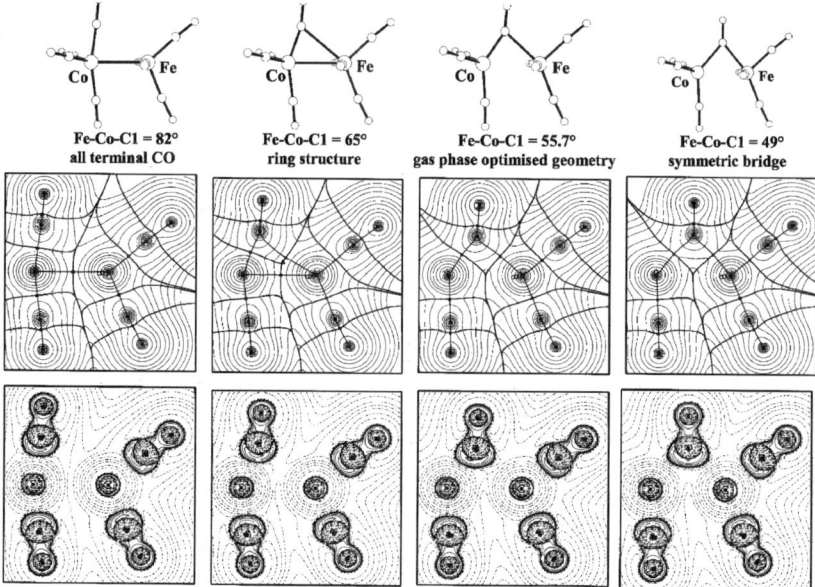

Figure 7 *Schematic representation of the bonding properties of the $[FeCo(CO)_8]^-$ anion (upper row), as well as the electron density (including both its separation into atomic contributions and the chemical bonds) (middle row) and the Laplacian of the electron density (lower row) for different geometrical structures of the anion. Reproduced with permission of The American Chemical Society from ref. 33*

atoms. This is a situation that occurs for some similar compounds containing two metal atoms, but not for others. On the other hand, when it is approximately symmetrically placed between the two metal atoms, the bond between the two metal atoms is absent, and, instead, the CO group does act as a bridge.

The authors noticed also that an analysis of the Laplacian of the electron density (also shown in Fig. 7) would give additional useful information. This is, however, beyond the scope of this presentation.

This study has shown how a careful control of the theoretical approach, here through a comparison with experiment, can allow for extending the theoretical studies into areas that only with difficulties can be reached with experiment, thus giving a much more detailed – and reliable – information, here on the bonding properties, that theory or experiment individually could have given.

5 Excitations

5.1 Bandgap in Molecular Crystals. – In Sec. III we discussed the structure and total energies of a number of different systems, and in many cases these were used in extracting information that could be compared with experimental studies. Often, in an experiment, the system of interest is being excited, so the theoretical calculations need to study such excitations. However, in many cases the fundamental problems related to the fact that the original density-functional theory applies only to the energetically lowest state of a given symmetry could be circumvented. Thus, vibrations could be studied by calculating the variations of the total energy around the structure of the lowest energy, and some electronic excitations and ionization processes could be studied by considering systems of different symmetries and/or charges.

In this section we shall go beyond that approach and discuss different, more or less exact and theoretically well-founded, approaches for studying the response of the systems to perturbations that are relevant in experiment. As a first example we consider the excitation of an electron from the highest occupied orbital to the lowest unoccupied orbital. For an infinite crystal, this energy is often referred to as the band gap or optical gap, and for these it has been known for years (see, e.g., [5]) that Hartree-Fock calculations grossly overestimate them (often by a factor much larger than 2), whereas density-functional calculations with a local-density or a generalized-gradient approximation underestimates them (often by a factor somewhat less than 2).

The B3LYP functional and related functionals (e.g., the B3PW one) that combine Hartree-Fock and generalized-gradient treatments of exchange effects were originally devised as an approximation to total energies, and the parameters entering the functionals were determined by optimizing the accuracy of total energies for various systems. However, due to the different behaviours in calculating band gaps of the Hartree-Fock and the density-functional methods one may hope that these functionals will give a band gap in better agreement with experiment than either Hartree-Fock or density-functional calculations would give. Such a finding would be very useful, but we stress that it is lacking mathematical rigor: Koopmans' theorem gives arguments for using the Hartree-Fock orbital energies in approximating ionization potentials and electron affinities, but a similar theorem for the Kohn-Sham orbital energies exists only for the highest occupied orbital, as well as in a differential form. I.e., with ϵ_i^{HF} and ϵ_i^{KS} being the Hartree-Fock and Kohn-Sham orbital energies, respectively, for the ith orbital, we have

$$-\epsilon_i^{HF} \simeq IP_i \quad \text{for} \quad i \leq \text{HOMO}$$

$$\epsilon_i^{HF} \simeq EA_i \quad \text{for} \quad i \geq \text{LUMO}$$

$$-\epsilon_i^{KS} = IP_i \quad \text{for} \quad i = \text{HOMO}$$

$$\epsilon_i^{KS} = \frac{\partial E_{\text{tot}}}{\partial n_i}, \tag{25}$$

with n_i being the number of electrons in the ith orbital.

In order to explore the proposal that hybrid functionals may give good band gaps, Perger[35] studied different molecular crystals, i.e., anthracene, pentaerythritol, pentaerythritol tetra-nitrate, and cyclotrimethylene trinitramine. In Table XVI we summarize the findings of Perger.

The table confirms the statement above about the errors of Hartree-Fock and of Kohn-Sham gaps. Moreover, not surprisingly, the gaps of the infinite

Table XVI *Calculated band gap for crystalline anthracene, pentaerythritol, pentaerythritol tetranitrate, and cyclotrimethylene trinitramine using different approaches together with the HOMO-LUMO energy differences for the isolated molecules. The table gives Hartree-Fock (denoted HF), local-density (labeled LDA), and two types of hybrid (B3LYP and B3PW) results in comparison with experiment (exp), and all quantities are given in eV. The results are from ref. 35*

System	Method	Gap for crystal	HOMO-LUMO gap for molecule
Anthracene	HF	8.18	8.42
	LDA	2.36	2.50
	B3LYP	3.45	3.64
	B3PW	3.56	3.65
	Exp	4.4	
Pentaerythritol	HF	15.6	17.5
	LDA	5.21	6.55
	B3LYP	7.54	8.80
	B3PW	7.87	9.19
pentaerythritol tetranitrate	HF	14.3	14.8
	LDA	3.89	4.22
	B3LYP	5.96	6.47
	B3PW	5.99	6.48
cyclotrimethylene trinitramine	HF	13.8	14.7
	LDA	3.59	3.69
	B3LYP	5.54	5.92
	B3PW	5.53	5.90
	Exp	3.4	

crystals are slightly smaller than the gaps of the isolated molecules. Finally, the table shows that the hybrid gaps do lie between those of the Hartree-Fock and of the density-functional calculations, what may not surprise, but also that the calculated gaps then may be (depending on system) accurate. However, the hybrid functionals are not as accurate for band gaps as they are for total energies, and, density-functional methods shall therefore only with extreme care be applied in calculating band gaps.

5.2 Binding Energies of Electrons. – The last identity of Eq. (25) relates the Kohn-Sham orbital energies to the energies for removal of infinitesimal fractions of an electron. However, ionization potentials are energies for removing whole electrons and one would, therefore, need to integrate the equation,

$$\Delta E_i = E_i(N-1) - E_i(N) = -\int_0^1 \frac{\partial E_{\text{tot}}}{\partial n_i} dn_i = -\int_0^1 \epsilon_i \, dn_i. \qquad (26)$$

Here, we have indirectly assumed that both the total energy E_{tot} and the orbital energies ϵ_i depend explicitly on the number of electrons of the system, that changes from $N-1$ to N.

3: Density Functional Theory

When letting i denote the HOMO, ΔE_i becomes the binding energy of the HOMO electron. Then, Eq. (26) shows that the negative of the HOMO Kohn-Sham orbital energy is not the binding energy, but a correction needs to be added,

$$\Delta E_{\text{HOMO}} = -\epsilon_{\text{HOMO}} + \Delta_{\text{HOMO}} \qquad (27)$$

with the first term being a finite-step-width approximation to the integral of Eq. (26).

This idea has been used by Jellinek and Acioli[36] to convert the HOMO orbital energies from density-functional calculations into the electron binding energies. The integrals in Eq. (26) were approximated by finite summations over a certain grid, and by gradually building the system of interest up by adding electron by electron (in fractions), the different ionization potentials (i.e., binding energies) could be extracted. Here, the corrections Δ_{HOMO} could be estimated from the results of the calculations with different number of electrons.

First, they made a careful check of their approach by analyzing results for a number of atoms, but here we shall turn directly to the application of their method to the calculation of ionization potentials for some molecules. The results are summarized in Table XVII.

Table XVII *The single-particle orbital energies from local-density (LDA) and generalized-gradient (GGA) calculations together with the binding energies (BE) for those orbitals in comparison with experimental values (exp) for different molecules and orbitals. i gives the orbital and its population. All quantities are given in eV. The results are from ref. 36*

Molecule	i	$-\epsilon_i$ (GGA)	$-\epsilon_i$ (LDA)	BE (GGA)	BE (LDA)	BE (exp)
F_2	$1\pi_g^4$	9.59	10.01	15.56	16.25	15.83
	$1\pi_u^4$	12.81	13.61	18.88	19.95	18.8
	$3\sigma_g^2$	15.38	16.28	21.54	22.65	21.1
	$2\sigma_u^2$	27.78	27.77	34.37	34.50	
	$2\sigma_g^2$	33.78	34.67	40.54	41.66	
C_2H_4	$1b_{2u}^2$	6.57	7.15	10.59	11.36	10.51
	$1b_{2g}^2$	8.54	8.85	12.62	13.07	12.85
	$3a_g^2$	10.15	10.59	14.27	14.85	14.66
	$1b_{3u}^2$	11.38	11.84	15.54	16.10	15.87
	$2b_{1u}^2$	14.24	14.50	18.52	18.82	19.10
	$2a_g^2$	18.70	19.03	23.08	23.51	
C_6H_6	$1e_{1g}^4$	6.18	6.79	9.22	9.94	9.25
	$3e_{2g}^4$	8.27	8.67	11.33	11.85	11.53
	$1a_{2u}^2$	8.83	9.50	11.89	12.69	12.38
	$3e_{1u}^4$	10.23	10.69	13.31	13.90	13.98
	$2b_{2u}^2$	10.78	11.41	13.87	14.63	14.86
	$1b_{1u}^2$	11.28	11.50	14.36	14.72	15.46
	$3a_{1g}^2$	12.79	13.35	15.92	16.60	16.84

The table shows first of all that the Kohn-Sham energies in general are much too high (less negative) than the experimental ionization potentials. This is a well-known fact (see, e.g., [5]) and, furthermore, we add that Hartree-Fock energies suffer from a related, but not so significant problem: they are too negative. Second, the table shows that the results of the local-density calculations, in general, not are less accurate than those of generalized-gradient calculations in contrast to what is found for the energies of chemical bonds. Finally, the table confirms the validity of the approach of Jellinek and Acioli,[36] i.e., orbital energies can be transformed into electron binding energies upon a proper treatment.

5.3 Band Gap in Conjugated Oligomers. – The preceding two subsections have shown that the orbital energies from Hartree-Fock or density-functional calculations provide fairly poor approximations to excitation energies. In particular the band gap between the HOMO and the LUMO is not well described. One may, however, attempt to calculate this excitation energy directly, e.g., using perturbation theory. Whereas this is, in principle, straightforward with wavefunction-based approaches (using the true Schrödinger equation), it was for a long time less clear how to carry such calculations through with density-functional methods. First, with the development of time-dependent density-functional theory[7,8] a theoretical basis for such calculations has been established, but due to numerical complexities it has not been before the most recent years that such calculations have been more generally carried through.

Here, we shall discuss the band gap of a large set of conjugated oligomers (in total, 60 molecules), which was studied by Hutchison et al.[37] Conjugated polymers have a relatively small band gap, making them semiconducting and, accordingly, interesting for various technological applications. Hutchison et al.[37] optimized the structures of the oligomers using both the semiempirical AM1 method and density-functional calculations with a generalized-gradient approximation. Second, they used three different approaches for calculating the excitation energies (i.e., time-dependent density-functional theory, the analogue for Hartree-Fock calculations, and the semiempirical ZINDO method), and for each they considered two approximations for including excitations (the RPA and the CIS, respectively – for the present purpose the difference is less important). Table XVIII contains the statistics of the accuracy of the calculations. Notice that the typical band gaps of the systems lie in the range from 2 to 6 eV.

The table shows that there is some dependence of the results on the geometry, although not too strong. Moreover, the errors in the absorption energies compared to experiment, i.e., of the order of 0.5 eV, are considerably smaller than those one encounters when simply using the HOMO-LUMO gap; these would for the present systems be some to several eV. On the other hand, the parameter-free methods, i.e., the Hartree-Fock and the density-functional calculations, are not yet able to yield more accurate results than are semiempirical methods (here, the ZINDO method) that have been parameterized specifically for the problem at hand. For the density-functional calculations the reason

Table XVIII *Summary of the accuracy of the calculated band gaps for 60 different conjugated oligomers. The methods are described in the text, and the structure denotes whether that optimized with the AM1 or with the density-functional calculations is used. Third, fourth, fifth, and sixth column gives the root-mean-squares error, the averaged absolute error, the maximal absolute error, and the average error, respectively. All results are in eV and are from ref. 37*

Method	Structure	RMS error	avg. abs. error	max. abs. error	avg. error
ZINDO/CIS	AM1	0.36	0.29	0.95	−0.09
	DFT	0.31	0.22	1.06	0.01
ZINDO/RPA	AM1	0.89	0.83	1.65	−0.65
	DFT	0.81	0.74	1.39	−0.59
HF/CIS	AM1	0.83	0.74	1.68	0.74
	DFT	0.91	0.81	2.03	0.81
HF/RPA	AM1	0.59	0.49	1.43	0.48
	DFT	0.73	0.58	1.87	0.58
TDDFT/CIS	AM1	0.49	0.40	1.36	−0.23
	DFT	0.47	0.38	1.25	−0.21
TDDFT/RPA	AM1	0.56	0.48	1.45	−0.42
	DFT	0.52	0.47	1.28	−0.37

may be that the approximate density functionals that over the years have been developed in order to describe the static structural properties of a system accurately, not yet have arrived a similar accuracy when studying dynamical properties or excitations.

5.4 (Hyper-)Polarizability of Si_4. – When a molecule, an atom, or a solid is exposed to an external electric (AC or DC) field, it will respond to this perturbation. The linear response of the total energy (or, equivalently, of the dipole moment) is the polarizability, and non-linear responses are described through the hyperpolarizabilities. The calculations of these can, accordingly, for static DC fields be done in two ways: either the field, of different strengths and directions, is directly included in the calculation, and the response is calculated and fitted to a polynomial in the field components or, alternatively, perturbation theory is employed. For AC fields, the latter approach has to be used.

Maroulis and Pouchan[38] performed calculations of the first type for Si_4 using different theoretical methods and obtained thereby the polarizability $\bar{\alpha}$ and the hyperpolarizability $\bar{\gamma}$ (the quantities depend on the direction the field and of the response, so, indicated through the bar, spherically averaged quantities are obtained; moreover, γ is the third-order response to the field). These quantities were compared with those obtained by others using time-dependent perturbation theory.

Table XIX shows a comparison of the calculated results from different types of calculations, i.e., both finite-field calculations and time-dependent perturbation-theory calculations. It is obvious that for this system the

Table XIX *Calculated spherical averages of the polarizability ($\bar{\alpha}$) and of the second hyperpolarizability ($\bar{\gamma}$) in Hartree atomic units for Si_4. TDDFT denotes time-dependent density-functional results, CCSD(T) results with Hartree-Fock + correlation calculations, and B3LYP density-functional results with the B3LYP functional. The two last ones have been obtained with finite-field methods. The results are from ref. 38*

Method	$\bar{\alpha}$	$\bar{\gamma}$
TDDFT	134.97	
TDDFT	136.86	
CCSD(T)	140.35	106.3×10^3
B3LYP	140.45	113.2×10^3

agreement between the different approaches is excellent, also for the hyperpolarizability which is considerably more difficult to calculate than the polarizability. However, as we shall see below, Si_4 is an example of a class of systems for which such an agreement is achieved, whereas for others, most notably chain compounds that are long in one and short in the other two directions, the results are considerably more scattered.

5.5 Electronic Absorption in Polycyclic Aromatic Hydrocarbons. – Halasinski et al.[39] presented results of a combined experimental and theoretical study of the absorption energies of perylene ($C_{20}H_{12}$), terrylene ($C_{30}H_{16}$), and quaterrylene ($C_{40}H_{20}$), i.e., of planar, polycyclic aromatic compounds containing 5, 8, and 11 hexagons, respectively. They considered both neutral and positively and negatively charged molecules, and the experiments were carried through in a Ne matrix in order to eliminate the interactions between the molecules.

In parallel to the experiments, time-dependent density-functional calculations with a generalized-gradient approximation were carried through. In Table XX we show those of their results for which both experimental and theoretical information was available (i.e., for the theoretical results we show the energies for the transitions with the largest intensities). The table shows that the agreement between theory and experiment is reasonable, but not overwhelming. The systems of this study are not chain compounds but nevertheless more extended in some directions than in others, whereby they represent intermediates between the very compact Si_4 molecule of the preceding section, for which the similar calculations were successful, and the chain compounds to be discussed further below where the calculations fail completely. Thus, what may not surprise, there seems to be a smooth decrease in the accuracy of the calculations the more quasi-one-dimensional the system becomes.

Table XX *Calculated (TDDFT) and experimental (exp) transition energies (in eV) for different molecules. The results are from ref. 39*

System	Transition	TDDFT	exp
Perylene, neutral	$^1A_g \to {}^1B_{3u}$	2.64	2.96
Perylene, cation	$^2A_u \to {}^2B_{3g}$	2.44	2.36
Perylene, anion	$^2B_{3g} \to {}^2A_u$	2.26	2.23
Terrylene, neutral	$^1A_g \to {}^1B_{3u}$	2.02	2.35
Terrylene, cation	$^2A_u \to {}^2B_{3g}$	1.93	1.80
Terrylene, anion	$^2B_{3g} \to {}^2A_u$	1.76	1.67
Quaterrylene, neutral	$^1A_g \to {}^1B_{3u}$	1.67	2.04
Quaterrylene, cation	$^2A_u \to {}^2B_{3g}$	1.62	1.48
Quaterrylene, anion	$^2B_{3g} \to {}^2A_u$	1.50	1.41

5.6 Nonadiabatic Processes. – As the last example in this section of electronic excitations of a system we mention the recent work of Doltsinis and Marx.[40] Usually, an electronic-structure calculation is performed by applying the Born-Oppenheimer approximation, i.e., the nuclei are held fixed and the electronic ground state for this structure is sought. With the Car-Parrinello method it is possible to optimize the structural and electronic degrees of freedom simultaneously, but the nuclei are supposed to move on the Born-Oppenheimer surface. However, in some cases electronic excitations may lead to structural changes, and such nonadiabatic processes are not accessible with the conventional Car-Parrinello method.

Doltsinis and Marx[40] showed how it is possible to modify the Car-Parrinello method so that also such processes can be treated. Subsequently they studied formaldimine, $HNCH_2$. This molecule possesses two equivalent ground-state structures. But, in order to change the system from one structure to another, an electronic excitation is needed, and, therefore, theoretical studies of this process need theoretical developments like that of Doltsinis and Marx.[40] We shall, however, not discuss the approach further here, since it still represents a very new development.

5.7 NMR Chemical Shifts of Benzoxazine Oligomers. – In the previous subsections of this section we have discussed the calculation of the response of a system to electronic excitations. There are, however, other experimental techniques that are of extreme value in characterizing a compound and where theoretical support is important. One such example is nuclear-magnetic-resonance (NMR) experiments, which will be the subject of the last two subsections of this section.

The calculation of NMR chemical shifts requires calculating the response of the system to an externally applied magnetic field, in particular of how the electrons in the vicinity of the nuclei response to this field. A non-trivial problem is the gauge-problem, i.e., the applied magnetic field can be described within different gauges, which all are physically equivalent but mathematically different. The results of the calculations shall, of course, also be independent of the gauge, but to achieve this has been non-trivial until the work of Ditchfield[41,42] who introduced the concept of gauge-including atomic orbitals (GIAOs) in order to circumvent this problem.

NMR chemical shifts are very sensitive to the structure of a molecule and as such very useful in determining the precise structure. Goward et al.[43] studied the structure of benzoxazine oligomers, using both ^{1}H NMR experiments and density-functional calculations. The precise structure of the molecule is not important for the present discussion. The calculations were performed by optimizing the structure with a generalized-gradient approximation to exchange and correlation effects within the density-functional theory. Subsequently, the chemical shifts were calculated for the optimized structure. As mentioned, these are very sensitive to the precise structure, so only if an accurate structure is calculated and if the calculation of these shifts is accurate, a good agreement between theory and experiment can be obtained.

Table XXI shows the results for some of the chemical shifts. The agreement between theory and experiment is indeed remarkable and does not require much discussion. It shall only be added that ultimately the authors used the results in proposing that the larger benzoxazine oligomers form helices that are stabilized by hydrogen bonds.

5.8 NMR Shielding Constants. – Magyarfalvi and Pulay[44] performed recently a detailed analysis of the performance of density-functional calculations in

Table XXI *Calculated (calc) and experimental (exp) ^{1}H chemical shifts for various benzoxazine oligomers. 'Resonance' describes the chemical surroundings of the proton, and in some cases there are more similar protons, whose shifts then are distinguished through the capital letters. The results are from ref. 43*

System	Resonance	exp	calc
Dimer	N–H (A)	11.7	12.2
	O–H	6.8	7.0
	Aromatic (C)	6.5–5.5	6.0–5.0
	Aliphatic (E)	3.0–0.0	3.0–0.0
Trimer	N–H (A')	12.3	12.2
	N–H (A)	10.7	10.7
	O–H	7.4	7.4
	Aromatic (C)	7.0–5.0	6.2–5.5
	Aromatic (D)	6.0–5.5	4.8–4.2
	Aliphatic (E)	3.0–1.5	3.0–1.5
	Aliphatic (F)	1.5–1.0	1.5–1.0
Tetramer	N–H (A")	12.9	13.1
	N–H (A')	11.7	11.8
	N–H (A)	10.9	11.1
	O–H	7.9	7.0
	Aromatic (C)	7.2–5.5	6.5–5.8
	Aliphatic (E)	3.5–1.0	4.2–2.0
	Aliphatic (F)	2.0–0.0	2.0–0.5

calculating nuclear-magnetic-resonance shieldings for a large set of smaller molecules. They compared Hartree-Fock results with those of density-functional calculations using either 'standard' generalized-gradient approximations, the B3LYP functional that combines the density-functional and Hartree-Fock treatments of exchange interactions, or more special functionals that have been derived explicitly for giving good performance when calculating these quantities. When realizing that the NMR responses are responses to a weak perturbation and, accordingly, accessible with perturbation theory, one of the sources for inaccuracies of density-functional calculations could be the too small energy gap between occupied and unoccupied orbitals. Thus, some of the more special functionals have been derived by explicitly correcting this problem. Alternatively, NMR experiments involve a magnetic field that in turn can create a current in the molecule. Therefore, other proposals have focused on deriving functionals that explicitly depend on the current (these will be discussed below). The study of Magyarfalvi and Pulay[44] was partly aimed at addressing which type of change in the density functional would lead to the best agreement with experimental results.

They found that Hartree-Fock calculations were superior to generalized-gradient and B3LYP calculations. Although this result may appear surprising, one has to remember that these approximate functionals have been derived in order to obtain an optimal agreement with experimental total energies and not with orbitals at the sites of the nuclei. However, in some cases, when correlation effects are important and, accordingly, the Hartree-Fock approximation is poor, the density-functional results are more accurate and here the generalized-gradient and the B3LYP calculations are of comparable accuracy.

Moreover, they found that the functionals that were explicitly devised for the calculation of NMR parameters performed better. Here, those that corrected the energy gap seemed to be better than those that included a dependence on the current thus indicating that the inaccuracies in density-functional NMR shielding are related to inaccuracies in the energy gap. On the other hand, it shall not be forgotten that the density-functional calculations are based on Kohn-Sham orbitals that, in the strict sense, not are electronic orbitals, so that it is not clear how to use those orbitals in calculating NMR parameters. This may, but need not, be another source for inaccuracies.

6 Getting Further Information With Density-functional Calculations

6.1 Classification of Reactions. – In our previous reports[1,2] we have discussed various quantities that have been introduced in order to help understanding, describing, and predicting why and how two molecules interact (or why not) by looking at properties of the individual molecules. Hardness and softness, as well as their position-space-resolved analogues, are two such quantities, as is the Fukui function. They are useful but have not (yet) become common tools in theoretical and/or experimental studies of specific chemical reactions. One reason may be that these quantities are not directly experimentally accessible and, therefore, their study is restricted to theoretical investigations.

Whereas the above-mentioned quantities focus on properties of the individual reactants, also properties of the combined system consisting of all (say, two) reactants can be used in predicting general properties of the reaction. One such example is the Hammond postulate,[45] stating that if the transition state is close in energy to an adjacent stable complex, then the structure of the transition state is similar to that of the stable complex. Notice, that since the structure of the transition state corresponds to a saddle point on the total-energy hypersurface whereas that of the stable complex corresponds to a local minimum, the postulate of Hammond is non-trivial.

The postulate of Hammond focuses on structure, i.e., on the positions of the nuclei, and does not directly include any information on the electronic structure. Amat et al.[46] explored whether the electronic distribution could be used in analysing an analogue of Hammond's postulate but for the electrons. They studied the four rearrangement reactions $F_2S_2 \rightarrow FSSF$, $HNC \rightarrow HCN$, $H_2SO \rightarrow HSOH$, and $H_2SCH_2 \rightarrow HSCH_3$ using both Hartree-Fock and density-functional (with the B3LYP functional) calculations. Table XXII shows the calculated total energies for the reactions together with the exothermicity factor

$$\gamma = \frac{E_{tot}(\text{products}) - E_{tot}(\text{reactants})}{2E_{tot}(\text{transition state}) - E_{tot}(\text{products}) - E_{tot}(\text{reactants})}$$
$$\equiv \frac{E_p - E_r}{2E_{ts} - E_p - E_r}. \tag{28}$$

Of the four reactions, the first two are supposed to follow Hammond's postulate, whereas the last two should not. The table shows that the exothermicity factor does not really make this distinction, independently of whether Hartree-Fock or density-functional results are used as the basis for the discussion (upon passing we remind that the total-energy variations as given in table XXII are most accurate for the density-functional calculations).

Table XXII *Hartree-Fock (denoted HF) and density-functional (labeled DFT) results for some rearrangement reactions. The quantities are defined in Eq. (28), and the energies are given in kcal/mol. The results are from ref. 46*

Reaction	Method	$E_{ts} - E_r$	$E_p - E_r$	γ
$F_2S_2 \rightarrow FSSF$	HF	64.5	−9.6	−0.069
	DFT	42.8	−3.6	−0.04
$HNC \rightarrow HCN$	HF	40.0	−9.5	−0.106
	DFT	33.9	−13.4	−0.165
$H_2SO \rightarrow HSOH$	HF	50.3	−32.9	−0.246
	DFT	38.2	−26.2	−0.255
$H_2SCH_2 \rightarrow HSCH_3$	HF	17.9	−80.2	−0.691
	DFT	25.9	−70.0	−0.575

3: Density Functional Theory

Amat et al.[46] studied whether one may alternatively use the momentum- or position-space densities in identifying Hammond-like behaviours. Although, from a strictly fundamental point of view, the Kohn-Sham orbitals not are electronic orbitals, and, accordingly, the electronic momentum density not necessarily is the density of the Fourier-transformed Kohn-Sham orbitals, Amat et al. chose to ignore this formal problem – experience has shown that the Kohn-Sham and the Hartree-Fock orbitals in position space in most cases are very similar, giving support for their approach.

From the momentum-space densities $\rho(\vec{p})$ they calculated their moments,

$$\langle p^n \rangle = \int p^n \rho(\vec{p}) d\vec{p}, \tag{29}$$

and a similar approach was used for the position-space densities $\rho(\vec{r})$, giving

$$\langle r^n \rangle = \int r^n \rho(\vec{r}) d\vec{r}. \tag{30}$$

Finally, the authors studied the changes in these quantities when passing from the reactants via the transition state to the products.

They found, indeed, that from these quantities, either in momentum or in position space, it was possible to distinguish clearly between reactions following Hammond's postulate and those that did not. Thus, this study demonstrates a close relation between structural and electronic properties, and that calculations, e.g., those based on density-functional formalism, can be used in characterizing chemical reactions. We should, however, add that the study we have discussed in this subsection is not the only one attempting to extract single parameters to be used in describing chemical reactions, nor is the success of this study a guarantee that the approach is generally applicable.

6.2 Hydrogen Bond Descriptors. – The fact that the energy of a hydrogen bond is comparable to room temperature makes hydrogen bonds active and important for very many systems in biology, chemistry, and materials science. Therefore, simple ways of describing the properties of hydrogen bonds also for complicated systems are extremely useful and important. Lamarche and Platts[47] studied whether it was possible to use basicity and acidity scales together with a polarity/polarizability scale in describing hydrogen bonds. The three parameters, denoted α, β, and π, are not relevant for our discussion, but rather the approach the authors took. They studied, with density-functional calculations using the B3LYP functional, more than 120 hydrogen-bonded complexes involving fairly small molecules (thereby making the calculations possible) and calculated the parameters of interest.

They found that exactly those three parameters were able to characterize the hydrogen bonds. The fact that they consider so many systems and could identify just three parameters in characterizing the hydrogen bonds gives strong support for suggesting that these three parameters can be used in describing any hydrogen-bonded systems, i.e., also such that are beyond simple computations. Thus, this example shows how a study of simpler systems can be used in quantitatively characterizing the bonding properties of much more complex systems.

7 Limitations and Perspectives

7.1 Polyenes and Current-density-functional Theory. – The original approximate density functionals, i.e., the local-density approximations, were constructed by considering the homogeneous electron gas. This system may to some extent be found in crystalline materials, first of all in metals. With the generalized-gradient approximations most often small, finite molecular systems were considered in order to determine parameter values entering the approximate functionals. In addition, the latter approximations were supposed to remain good for the crystalline systems for which the local-density approximation was good.

There are, however, systems that are significantly different from metallic crystals and from small, finite molecules and where, accordingly, it is not obvious whether the introduced approximations remain good. To these belong chain compounds and polymers that are roughly infinite in one dimension and finite in the other two. And a prominent example of these materials is that of the conjugated polymers that we discussed already in our previous report.[2] These are polymeric materials that are semiconducting and, accordingly, for which the electronic properties are important. We showed in the previous report that the presently employed approximate density functionals in some cases were not capable in giving accurate results. Here, we shall continue this discussion and discuss some further aspects of those problems.

Cai et al.[48] studied the excitation energies of finite $C_{2n}H_{2n+2}$ oligomers (polyenes) as function of n using different theoretical methods. Fig. 8 shows the resulting excitation energies. The time-dependent density-functional results, denoted 'TD-B3LYP', indicates that the excitation energy goes to 0 for a finite value of $2n$, in clear contrast to experimental information, giving an excitation energy somewhere between 1 and 2 eV for $n \to \infty$. Instead, using the standard B3LYP method in calculating the energy difference (denoted 'B3LYP' in the figure) gives a non-zero vertical excitation energy, but when the excited state is allowed to relax, also these calculations predict a vanishing excitation energy. On the other hand, the Hartree-Fock-based results do not suffer from these problems.

Subsequently, the authors used various density functionals in calculating the excitation energy from the ^1A_g structure with alternating bond lengths to the ^3B_u excited structure containing a soliton/antisoliton pair (i.e., the bond-length alternation is at two places in the chain interrupted and reversed) as well as the energy it costs to change the first structure to the latter. The results are summarized in Table XXIII. The table shows that all the approximations lead, for sufficiently large $2n$, to the (wrong) prediction that the ^3B_u state is the state of the lowest total energy, with the B3LYP functional performing best. More serious is that the generalized-gradient approximations predict that the soliton/

3: Density Functional Theory

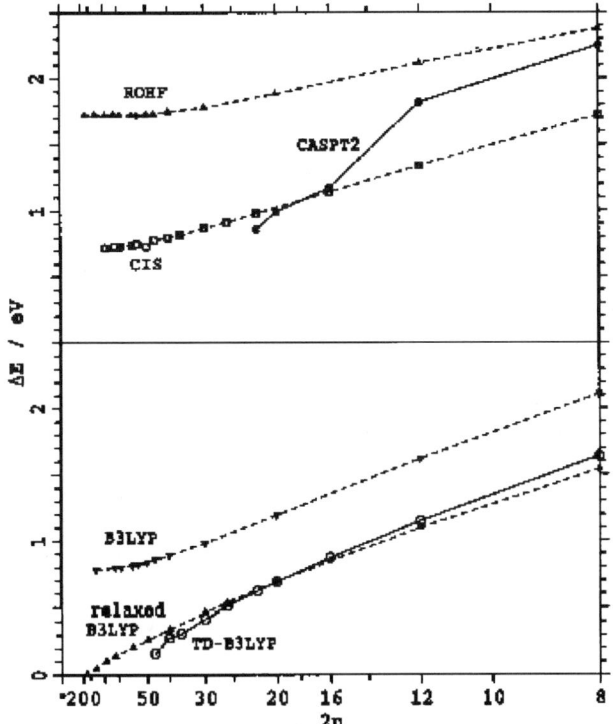

Figure 8 The calculated $^1A_g \to {}^3B_u$ excitation energies for $C_{2n}H_{2n+2}$ as function of $2n$. All energies are for the vertical excitation with the structure optimized with the B3LYP calculations, except for the curve denoted 'relaxed' where also the excited state has been relaxed. The upper part shows results with Hartree-Fock-based methods, whereas the lower part shows results from density-functional methods. Reproduced with permission of The American Institute of Physics from ref. 48

Table XXIII Calculated excitation energies $E_{triplet}$ from the ^1A_g state with alternating bond lengths to the ^3B_u structure with a soliton/antisoliton pair for $C_{2n}H_{2n+2}$ as well as the reorganization energy E_{reorg} for creating the soliton/antisolition pair for different values of $2n$. PB86 and PW91 are two different generalized-gradient approximations, and all quantities are given in eV. The results are from ref. 48

Property	$2n$	B3LYP	PB86	PW91
$E_{triplet}$	80	0.142	0.076	0.076
	124	0.048	−0.101	−0.095
	172	0.005	−0.231	−0.259
E_{reorg}	80	0.593	0.044	0.049
	124	0.614	−0.093	−0.131
	172	0.805	−0.193	−0.197

antisoliton pair has a lower total energy than the structure with alternating bond lengths. Thus, for these systems the presently applied density functionals have problems.

A solution to the problems may have been identified by van Faassen et al.[49,50] According to the original work of Hohenberg and Kohn, the electronic ground-state energy is a functional of the electron density. This means that we take the electron density and then do 'something'. This 'something' could, e.g., be to solve equations like the ones of Kohn and Sham (or some other equations), thereby obtaining some new quantities (for instance the Kohn-Sham orbitals) that, consequently, also are functionals of the electron density. From these new quantities we may perform yet new mathematical manipulations, and ultimately arrive at the ground-state energy. Vignale and Kohn[51] have argued that one should calculate the quantum-mechanical current from the Kohn-Sham orbitals and use a functional that depends on this current. Thereby, one arrives at the so-called current-density-functionals.

Similar arguments may be carried through for the time-dependent density-functional theory of Runge and Gross. van Faassen et al.[49,50] implemented such a current-density-functional in combination with time-dependent density-functional theory and applied it subsequently in studying the polarizability of various polymers, including polyenes, as function of chain length. Fig. 9 shows

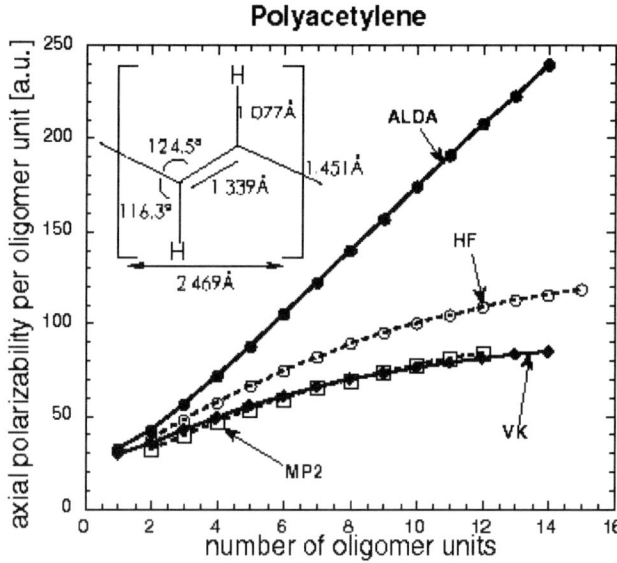

Figure 9 *Calculated static polarizability per unit along the chain axis for polyacetylene oligomers (polyenes) with either Hartree-Fock (HF), Hartree-Fock+ correlation (MP2), or time-dependent density-functional theory with either a local-density approximation (ALDA) or a current-density-functional (VK). The structure is given in the insert. Reproduced with permission of The American Physical Society from ref. 49*

their results for polyenes. The MP2 results are believed to be accurate, but also the Hartree-Fock results (HF) shows a convergence to a constant value per unit. This is contrasted by the results marked ALDA that are the time-dependent density-functional results with a local-density approximation, whereas the curve marked VK, using the functional of Vignale and Kohn, gives a close agreement with the MP2 results. In total, this suggests that current-density-functionals may become more important in the future, at least for certain systems and properties, like the excitations in polymeric systems.

7.2 Anions and Exact-exchange Methods. – As mentioned in the previous subsection, the Hohenberg-Kohn theorems state nothing about **how** to calculate, e.g., the total electronic energy from the electron density so that also the construction of intermediate objects like the Kohn-Sham orbitals is within the concept of the density-functional theory. A subsequent construction of a current-density-functional is one approach, but another one is to construct a functional that contains an exact expression for exchange interactions, but based on the Kohn-Sham and not the Hartree-Fock orbitals. This approach was discussed in our previous report[2] and here we shall give just a single recent example of its applications.

However, first we discuss in some details the exact-exchange functionals. No matter what functional we construct, the final Kohn-Sham equations will have the form of Eq. (13), i.e., also the exchange potential will enter the equations as a multiplicative potential acting on the Kohn-Sham orbitals. This is in marked contrast to the Hartree-Fock equations where the exchange interactions are described through a purely non-local operator and not through a local potential. Therefore, even if the total-energy expressions within Hartree-Fock and density-functional theory may look very similar, when using an exact-exchange density-functional theory, the resulting single-particle equations are fundamentally different and the orbitals will also be different.

One of the fundamental advantages of exact-exchange density-functional theory is the fact that the so-called self-interaction has been removed, i.e., the fact that there should be no electrostatic interactions between an electron and itself. Within Hartree-Fock theory, this interaction is canceled exactly by exchange interactions (cf. Eq. (10)), but within approximate density functionals for exchange interactions, the cancellation is not perfect. As a consequence of this imperfect cancellation, most of the presently employed approximate density-functionals suffer from the problem that many smaller or medium-sized anions are unstable. This result may be related to the finding of positive orbital energies for the highest occupied orbitals.

Weimer et al.[52] showed recently that with an exact-exchange density functional such anions could be made stable. Table XXIV shows some of their results. From this table it is seen that the methods where the self-interaction has been removed (i.e., HF and EXX) give the lowest orbital energies of the highest occupied orbital of the anions, and for the B3LYP where, through the partial inclusion of Hartree-Fock treatment of exchange effects, some self-interaction has been removed, these energies are lower as those of the

Table XXIV *Orbital eigenvalues of the highest occupied orbital of different anions as calculated with different theoretical methods in comparison with experimental electron affinities. HF, EXX, LDA, GGA, and B3LYP denote results with Hartree-Fock, exact-exchange density-functional, local-density, generalized-gradient, and B3LYP calculations. The results are in eV and from ref. 52*

System	Orbital	HF	EXX	LDA	GGA	B3LYP	Exp
H⁻	$1s$	−1.26	−1.26	1.28	1.26	1.05	−0.70
F⁻	$2p$	−4.92	−4.91	1.18	1.26	0.04	−3.40
Cl⁻	$3p$	−4.09	−4.08	0.10	0.13	−0.71	−3.62
Br⁻	$4p$	−3.79	−3.76	−0.06	−0.02	−0.75	−3.45
N_3^-	$1\pi_g$	−2,86	−2.76	0.71	0.82	0.04	
OH⁻	1π	−2.99	−2.99	1.36	1.51	0.87	−1.83
SH⁻	2π	−2.54	−2.57	0.79	0.82	0.17	−2.36
CN⁻	$5\sigma^+$	−5.25	−5.12	−0.15	−0.13	−1.18	−3.86
NCO⁻	2π	−4.34	−4.21	0.03	0.15	−0.78	−3.61
CNO⁻	2π	−4.10	−3.96	0.13	0.25	−0.63	
NCS⁻	3π	−3.85	−3.76	−0.38	−0.31	−1.01	−3.54
CNS⁻	3π	−3.45	−3.36	−0.18	−0.12	−0.76	
HCOO⁻	$6a_1$	−5.06	−4.94	0.09	0.16	−0.92	
CH_3COO^-	$13a'$	−5.15	−5.14	0.10	0.21	−0.97	

conventional density-functional calculations, independent of whether a local-density or a generalized-gradient approximation has been used. Moreover, as shown by the authors the anions do become stable.

Subsequently the authors studied all energies of the occupied orbitals of two systems and found (cf. Table XXV) that these appeared generally at lower energies than those of the normal density-functional calculations but not as low as those of Hartree-Fock calculations. Then, the orbital energies may actually become closer to experimental ionization potentials than those from Hartree-Fock or LDA/GGA calculations.

In total, the exact-exchange methods seem to show a way of improving some of the deficiencies of the more common density-functional methods, in particular the stability of negatively charged, smaller systems. In addition, these methods may give, as a product, orbital energies that are in better agreement with experiment also for the energetically deeper ones. Here, normal density-functional calculations tend to give too high energies, whereas Hartree-Fock calculations give too deep energies. But, some work needs still to be done: the present commonly used density functionals that approximate both exchange and correlation effects benefit from a cancellation of errors in the approximate treatments of the two quantities. When, then, introducing an exact treatment of the one quantity, this cancellation is removed and the overall accuracy may suffer.

7.3 The Long-ranged Behaviour of Exchange Interactions. – Above, we have studied two classes of systems where the conventional approximate density

Table XXV *Orbital eigenvalues of the different occupied orbitals of azide as calculated with different theoretical methods. HF, EXX, LDA, GGA, and B3LYP denote results with Hartree-Fock, exact-exchange density-functional, local-density, generalized-gradient, and B3LYP calculations. The results are in eV and from ref. 52*

Orbital	HF	EXX	LDA	GGA	B3LYP
$1\sigma_g^+$	−422.18	−380.93	−374.23	−377.90	−386.86
$2\sigma_g^+$	−417.33	−377.44	−370.84	−374.50	−383.18
$1\sigma_u^-$	−417.33	−377.44	−370.84	−374.50	−383.18
$3\sigma_g^+$	−32.99	−24.94	−20.94	−21.06	−23.43
$2\sigma_u^-$	−26.99	−20.72	−16.74	−16.75	−18.81
$1\pi_u$	−11.45	−8.83	−5.45	−5.29	−6.60
$4\sigma_g^+$	−11.56	−8.01	−4.34	−4.47	−5.90
$3\sigma_u^-$	−9.66	−6.76	−3.16	−3.16	−4.51
$1\pi_g$	−2.86	−2.76	0.71	0.82	0.04

functionals are too inaccurate to give reliable results. In both cases, new functionals directly based on the Kohn-Sham orbitals and not only on the electron density were seen to yield promising improvements. There are, however, other attempts of improving the density functionals, based only on the density. Here, we shall discuss one class of such studies.

An electron, that is far away from some finite systems, feels the exchange interactions from the remaining $N-1$ electrons essentially as an $\frac{1}{r}$ potential. To describe this potential with a functional that depends solely on the density and its derivatives in the point of interest (i.e., in the outer limits of the system) is very difficult: the density falls off exponentially and so will any derivative. Some studies are therefore focusing on this problem which should be highly relevant in time-dependent density-functional theory (where excitations/ionizations are studied). Although there are promising results, no conclusions have yet been arrived at and, therefore, these studies shall not be considered further here.

We saw above that the properties of long conjugated systems were erroneously described with the present density functionals. It has been argued hat this problem is, again, related to the long-ranged behaviour of the exchange interactions, which, per construction, are correctly described by Hartree-Fock theory. Therefore, approaches like B3LYP that combine Hartree-Fock and density-functional descriptions point to a possible improvements. This idea was, e.g., followed by Iikura *et al.*[53] who studied whether a space-separation of Hartree-Fock and density-functional treatments of exchange interactions could lead to improved results. They used a smooth cut-off function, erf(μr_{12}), with r_{12} the distance between two electrons and μ a chosen parameter, to gradually turn off the density-functional and turn on the Hartree-Fock exchange interactions.

Subsequently they applied the approach to two problems. First, they studied the energy for transferring a 4s electron to a 3d orbital for transition-metal atoms and, second, they studied the polarizability of long polyenes. They did find a significant improvement, supporting the consensus that the long-ranged behaviour of the exchange interactions is a source for inaccuracies in presently used approximate density functionals.

7.4 Other Improvements. – There exists many other studies devoted to improve density functionals in one way or another. One class of improvements are based on the Kohn-Sham orbitals through a kinetic-energy density

$$\tau = \frac{\hbar^2}{2m_e}\sum|\psi(\vec{r})|^2. \tag{31}$$

These improvements lead to the so-called meta-GGAs of which some functionals have been proposed during the latest years. These may actually be the ones that will survive.

Whenever new density functionals are been proposed, they may include some arbitrariness through some parameters (e.g., the parameter μ in the previous subsection). The more of these parameters that can be determined unambiguously through fundamental physical principles, the better. Alternatively, the concept of training sets is used. I.e., the parameters are optimized so that the properties (energies and structures, most often) of a certain set of systems are as accurately as possible reproduced. This makes the methods appear partly empirical and, moreover, it shall then be remembered that a similar accuracy as for the training set not shall be expected when systems that are markedly different from those of this set are studied (e.g., containing heavier atoms – the training sets most often contain molecules with light atoms, or large systems like the chain compounds).

Since the presently employed density functionals have their roots first in infinite, periodic crystalline compounds with delocalized electrons (where the local-density approximation over many years was successful) and later in smaller molecules with lighter atoms (for which the parameters of the generalized-gradient approximations were optimized), one shall always be aware of the fact that first of all for such systems the calculations may be accurate. However, even for such there are problem cases. A list (the so-called 'sick-list') of such cases has been collected by the National Institute of Standards, USA.[54]

8 Conclusions

Upon closing this report the natural question arises: how has the application of density-functional theory within chemistry developed compared with our previous reports. In the first report we tried to illustrate the diversity of the chemical problems that could be addressed with density-functional theory, although not even that could be completely covered. In the second report we addressed some of the problems of the applications of density-functional theory. The present report, dealing with only some few dozens of the papers appearing during the last two years where such applications have been reported (and, accordingly, discussing only less than a 1000th of all the papers that could have been included) has tried to show how some of the previous problems are approaching a solution as well as to illustrate how the application of density-functional theory has entered new areas. Thereby, neither the complete width of the applications, nor all the relevant questions related to the improvements could be discussed.

One thing is, however, clear. As in any other field in science, density-functional theory has prospered from the technical developments, so more complicated systems can be treated and more accurate results obtained. In the present case, 'technical developments' means three things: the computers have become more powerful and have larger memory, the computer programs have become more efficient, and new and more efficient algorithms and methods have been developed. As a consequence, larger and more complicated systems are now been treated compared with just some few years ago. Also in this overview we saw some examples of such systems that could not have been treated earlier. Certainly, this is a positive development.

On the other hand, it may also have become clear that density-functional suffers from one problem: the fact that the exact functional is unknown forces the development of approximations. However, the precise form of the approximations may depend on the systems and/or properties that shall be studied. Therefore, many different functionals have been proposed and, unfortunately, the choice of the functional can often, for the non-expert, seem rather arbitrary. This arbitrariness is counterproductive for density-functional theory and the development of parameter-free density-functionals is urgently needed. Here, current-density-functional theory and the exact-exchange density-functional theory are, to the present author's opinion, two promising proposals.

In the present report we saw repeatedly that the conjugated oligomers form a class of systems that so far have been included only marginally in the training set for testing and improving the density functionals, but for which serious problems occur with the currently applied approximate functionals.

The ultimate goal of electronic-structure calculations, for instance within density-functional theory, is to extend, supplement, support, and augment experimental studies. Therefore, the quantities that shall be calculated are to be as closely related to the experimentally measured ones as possible. This means first of all that the response of the system of interest to external perturbations shall be calculated. The development of time-dependent density-functional theory is an important step in this direction, and, as we have seen here, during the last years this method has become more widely used. However, also here further development is needed: the standard density-functional calculations deal with static, time-independent (or, equivalently, frequency-independent) properties for which the approximations have been developed. For time-dependent density-functional theory, frequency-dependent approximations have to be developed, which only partly has been done so far.

The calculation of structural properties and energy variations as functions of structure (either due to isomerisms, vibrations, or chemical reactions) remains the most important field of applied density-functional theory within chemistry. Here, the calculations are becoming highly accurate, as long as no weak bonds (first of all, no van der Waals bonds, but in some cases also no hydrogen bonds) are involved. Developments that yield accurate results also in those cases remain to be presented.

Although the Kohn-Sham energies and orbitals only are model objects, the examples we have presented show, hopefully, that useful and accurate information can be extracted when treating them as being the energies and orbitals of the electrons. Some of the more recent developments have thereby led to closer agreement with experiment.

Finally, we shall emphasize that the chosen examples do not cover the complete width of applied density-functional theory. Neither are they claimed to be of higher scientific quality than the very many ones that were not mentioned!

Acknowledgments

The author is very grateful to Fonds der Chemischen Industrie for very generous support.

References

1. M. Springborg, in *Chemical Modeling: Applications and Theory, Vol. 1*, ed. A. Hinchliffe, The Royal Society of Chemistry, Cambridge, UK, 2000, p. 306.
2. M. Springborg, in *Chemical Modeling: Applications and Theory, Vol. 2*, ed. A. Hinchliffe, The Royal Society of Chemistry, Cambridge, UK, 2002, p. 96.
3. P. Hohenberg and W. Kohn, *Phys. Rev.*, 1964, **136**, B864.
4. W. Kohn and L.J. Sham, *Phys. Rev.*, 1965, **140**, A1133.
5. M. Springborg, *Methods of Electronic-Structure Calculations*, John Wiley, Chichester, UK, 2000.
6. R.G. Parr and W. Yang, *Density-Functional Theory of Atoms and Molecules*, Oxford University Press, New York, USA, 1989.
7. E. Runge and E.K.U. Gross, *Phys. Rev. Lett.*, 1984, **52**, 997.
8. E.K.U. Gross, J.F. Dobson and M. Petersilka, in *Density Functional Theory*, ed. R.F. Nalewajski, Springer Verlag, Berlin, Germany, 1996, p. 81.
9. B.K. Rao and P. Jena, *J. Chem. Phys.*, 2002, **117**, 5234.
10. S.P. Yuan, J.G. Wang, Y.W. Li and H. Jiao, *J. Phys. Chem. A*, 2002, **106**, 8167.
11. M. Kamiya, T. Tsuneda and K. Hirao, *J. Chem. Phys.*, 2002, **117**, 6010.
12. M.F. Masman, M.A. Zamora, A.M. Rodríguez, N.G. Fidanza, N.M. Peruchena, R.D. Enriz and I.G. Csizmadia, *Eur. Phys. J. D*, 2002, **20**, 531.
13. M.N.D.S. Cordeiro, E. Martínez-Núñez, A. Fernández-Ramos and S.A. Vázquez, *Chem. Phys. Lett.*, 2003, **375**, 591.
14. M.A. Morsy, M.A. Al-Khaldi and A. Suwaiyan, *J. Phys. Chem. A*, 2002, **106**, 9196.
15. G. Rauhut, *Phys. Chem. Chem. Phys.*, 2003, **5**, 791.
16. J.A. Rodriguez, J. Dvorak, T. Jirsak, G. Liu, J. Hrbek, Y. Aray and C. González, *J. Am. Chem. Soc.*, 2003, **125**, 276.

17. M. Miletic, J.L. Gland, K.C. Hass and W.F. Schneider, *J. Phys. Chem. B*, 2003, **107**, 157.
18. S. Ilieva, B. Hadjieva and B. Galabov, *J. Org. Chem.*, 2002, **67**, 6210.
19. S. Ilieva, B. Galabov, D.G. Musaev, K. Morokuma and H.F. Schaefer III, *J. Org. Chem.*, 2003, **68**, 1496.
20. R.W. Alder, S.P. East, J.N. Harvey and M.T. Oakley, *J. Am. Chem. Soc.*, 2003, **125**, 5375.
21. E.W.-G. Diau and A.H. Zewail, *Chem. Phys. Chem.*, 2003, **4**, 445.
22. M. Chen, H. Lu, J. Dong, L. Miao and M. Zhou, *J. Phys. Chem. A*, 2002, **106**, 11456.
23. I. Frank, *Angew. Chem.*, 2003, **115**, 1607.
24. J.P. Perdew, K. Burke and M. Ernzerhof, *Phys. Rev. Lett.*, 1996, **77**, 3865.
25. J. Ireta, J. Neugebauer, M. Scheffler, A. Rojo and M. Galván, *J. Phys. Chem. B*, 2003, **107**, 1432.
26. N. Kurita, H. Inoue and H. Sekino, *Chem. Phys. Lett.*, 2003, **370**, 161.
27. B. Tremblay, G. Gutsev, L. Manceron and L. Andrews, *J. Phys. Chem. A*, 2002, **106**, 10525.
28. S.E. Lappi, W. Collier and S. Franzen, *J. Phys. Chem. A*, 2002, **106**, 11446.
29. H.L. Woodcock, H.F. Schaefer III and P.R. Schreiner, *J. Phys. Chem. A*, 2002, **106**, 11923.
30. F.L. Gervasio, P. Carloni and M. Parrinello, *Phys. Rev. Lett.*, 2002, **89**, 108102.
31. J.K. Deng, G.Q. Li, X.D. Wang, J.D. Huang, H. Deng, C.G. Ning, Y. Wang and Y. Zheng, *J. Chem. Phys.*, 2002, **117**, 4839.
32. L. Lam and P.M. Platzman, *Phys. Rev. B*, 1974, **9**, 5122.
33. P. Macchi, L. Garlaschelli and A. Sironi, *J. Am. Chem. Soc.*, 2002, **124**, 14173.
34. R.F.W. Bader, *Atoms in Molecules: a Quantum Theory*, Clarendon, Oxford, UK, 1990.
35. W.F. Perger, *Chem. Phys. Lett.*, 2003, **368**, 319.
36. J. Jellinek and P.H. Acioli, *J. Chem. Phys.*, 2003, **118**, 7783.
37. G.R. Hutchison, M.A. Ratner and T.J. Marks, *J. Phys. Chem. A*, 2002, **106**, 10596.
38. G. Maroulis and C. Pouchan, *Phys. Chem. Chem. Phys.*, 2003, **5**, 1992.
39. T.M. Halasinski, J.L. Weisman, R. Ruiterkamp, T.J. Lee, F. Salama and M. Head-Gordon, *J. Phys. Chem. A*, 2003, **107**, 3660.
40. N.L. Doltsinis and D. Marx, *Phys. Rev. Lett.*, 2002, **88**, 166402.
41. R. Ditchfield, *J. Chem. Phys.*, 1972, **56**, 5688.
42. R. Ditchfield, *Mol. Phys.*, 1974, **27**, 789.
43. G.R. Goward, D. Sebastiani, I. Schnell, H.W. Spiess, H.-D. Kim and H. Ishida, *J. Am. Chem. Soc.*, 2003, **125**, 5792.
44. G. Magyarfalvi and P. Pulay, *J. Chem. Phys.*, 2003, **119**, 1350.
45. G.S. Hammond, *J. Am. Chem. Soc.*, 1955, **77**, 334.
46. L. Amat, R. Carbó-Dorca, D.L. Cooper and N.L. Allan, *Chem. Phys. Lett.*, 2003, **367**, 207.
47. O. Lamarche and J.A. Platts, *Phys. Chem. Chem. Phys.*, 2003, **5**, 677.
48. Z.-L. Cai, K. Sendt and J.R. Reimers, *J. Chem. Phys.*, 2002, **117**, 5543.
49. M. van Faassen, P.L. de Boeij, R. van Leeuwen, J.A. Berger and J.G. Snijders, *Phys. Rev. Lett.*, 2002, **88**, 186401.
50. M. van Faassen, P.L. de Boeij, R. van Leeuwen, J.A. Berger and J.G. Snijders, *J. Chem. Phys.*, 2003, **118**, 1044.
51. G. Vignale and W. Kohn, *Phys. Rev. Lett.*, 1996, **77**, 2037.
52. M. Weimer, F. Della Sala and A. Görling, *Chem. Phys. Lett.*, 2003, **372**, 538.
53. H. Iikura, T. Tsuneda, T. Yanai and K. Hirao, *J. Chem. Phys.*, 2001, **115**, 3540.
54. http: //srdata.nist.gov/sicklist/edita.asp.

4
Combinatorial Enumeration in Chemistry

BY D. BABIĆ, D.J. KLEIN, J. VON KNOP AND N. TRINAJSTIĆ

1 Introduction

In this report we review the literature in the area of combinatorial enumeration in chemistry published in the last two years – from June 2001 to May 2003. In our earlier report[1] we presented historically important isomer enumerations, enumeration methods and then we reviewed the literature published between June 1999 and May 2001. Here we will focus only on the recent enumerative work since our previous report gives much background information on the history and methodology of counting objects in chemistry. Nevertheless, we will also mention some of the past achievements closely related to the discussed results most of which were not previously reviewed.

However, we wish to stress that the most fundamental problem underlying a part of modern chemistry called combinatorial chemistry[2,3] is that a very large number of different molecules is possible. This fact raises the pertinent question *How many different molecules are possible?* It is a rather simple question, but it does not have a simple answer. Combinatorial enumerations can provide an answer.

2 Current Results

2.1 Isomer Enumeration. – In this section we will review the current results regarding enumerations of acyclic structures, usually represented by trees, and benzenoid hydrocarbons whose enumeration represents a subject considered by many past and present authors. We will also mention the related past results in order that the reader can see how the current achievement fits into the development of a particular area of isomer enumeration.

2.1.1 Trees. – The enumeration of all possible graph-theoretical trees[4] and chemical trees[5] still attracts attention. Trees are connected acyclic graphs.[4] Chemical trees are trees in which no vertex has a degree greater than 4, whilst graph-theoretical trees do not have such a constraint. Chemical trees provide, among many other things, the natural graph-theoretical representation of acyclic carbon structures. There has also been introduced by Kroto et al.[6] in 1987 the concept of physical tree. Based on this concept a proposal was made for the simple mechanism by which acyclic molecules could be formed in interstellar space and circumstellar shells. Physical trees with no vertex-degree greater than 4 were generated and enumerated using the N-tuple code.[7,8]

Physical trees carry 'memory' of their origin, implemented by a special labeling of their vertices, unlike chemical trees, that is, trees without 'memory'. Thus, a physical tree is obtained by assigning labels to the vertices of a tree consecutively, and each vertex to be labeled must be adjacent to an already labeled vertex. Therefore, each vertex, except the vertex labeled 1, has exactly one neighbor with a lower label. This labeling results in the vertex-adjacency matrix[5] A of a physical tree that contains only one non-zero element in each column of its upper (or in each row of lower) triangle.[9] A consequence of the above is that the total number of different physical trees with N vertices and with no restriction upon the vertex-degree is simply $(N-1)!$. Since there are $(i-1)$ ways to pick the non-zero element in the i-th column of the upper (or in the i-th row of lower) triangle of the vertex-adjacency matrix, and since these choices are independent, the total number of ways equals $\prod_{i=2}^{N}(i-1)=(N-1)!$.[10] The number of linear (chain-like) physical trees is also given by a simple formula:[6] 2^{N-2}. For example, the number of physical trees with 13 vertices and the vertex-degree restriction to 4 or less is 311216626,[8] whilst this number for all physical trees with 13 vertices is $12! = 479001600$. The number of linear physical trees with $N = 13$ is $2^{11} = 2048$. Physical trees represent a subset of labeled trees. The number of different labeled trees is N^{N-2}, a result given by Cayley already in 1889.[11] Hence, the number of all possible labeled trees with 13 vertices is $13^{11} = 1792159794037$.

Lukovits and Gutman[12] have shown that Morgan-trees represent a subclass of physical trees. The formal definition and nomenclature of a Morgan-tree was introduced by Lukovits[13] in 1999 and has been named in honor of H.L. Morgan, who proposed the underlying concept in 1965, when working on the development of a computer-based chemical information system at the Chemical Abstracts Service in Columbus (Ohio), an algorithm for unique labeling of chemical structures.[14]

A Morgan-tree (MT) is a physical tree in which the numbering must obey an additional (with respect to physical trees) restriction: Each new label must be attached to a vertex which is adjacent to a vertex labeled with the lowest available ordinal. Example: Number 1 can be attached to any vertex. Number 2 must be attached to one of the vertices being adjacent to vertex 1, since an MT is also a physical tree. Number 3 must be attached to a vertex being adjacent to vertex 1, if there is such a vertex. If not, it must be attached to a vertex being adjacent to vertex 2, etc. The essence of the enumeration method based on MTs is that the number of MTs is a small fraction of physical trees. Any tree that can be labelled in such a way it is an MT. The codes representing MTs can be easily generated, and the redundant structures can be eliminated by using several rules. Note that there is not any step which involves a comparison of different structures.

Lukovits and Gutman[12] derived a formula for ennumerating all Morgan-trees (MT) with N vertices:

$$MT(N) = (2N-2)!/N! (N-1)! \qquad (1)$$

They used Morgan-trees for the exhaustive generation and enumeration of chemical trees (alkane isomers and alkyl derivatives). In a previous work, Lukovits encountered a problem of redundant structures[13,15] that appears now to be solved.[12,16]

Došlić[10] has shown that there is a simple bijection between the set of all Morgan-trees with a given number of vertices N and Dyck paths on $2(N-2)$ steps. A Dyck path on $2N$ steps is a lattice path in the coordinate plane (x,y) from $(0,0)$ to $(2N,0)$ with steps $(1,1)$ *up* and $(1,-1)$ *down*, never falling below the x-axis.[17] Dyck paths are one of many combinatorial families enumerated by Catalan numbers (C_N) $N \geq 0$.[18] These Dyck paths are also recognized as path diagrams (e.g. in ref. 19) corresponding to Yamanouchi symbols[20] which characterize spin-singlet coupling patterns for the symmetric group, as of use in many-particle quantum mechanics. A consequence of the said bijection is that the exact enumerative results for Morgan-trees with a given number of vertices can be obtained in terms of Catalan numbers. Došlić[10] proved that the number of all Morgan-trees with N vertices is C_{N-2}. Lukovits-Gutman equation (1) is really equal to C_{N-1}. The seeming discrepancy between Lukovits-Gutman and Došlić formulas is due to the different starting points.

For a given $N \geq 0$, the quantity:[18,21]

$$C_N = 1/(N+1)\binom{2N}{N} \qquad (2)$$

is called the N-th Catalan number. Formula (2) generates Catalan numbers: 1, 1, 2, 5, 14, 42, 132, 429, 1430, 4862, 16796, 58786, 208012, etc. In Lukovits and Gutman's paper[12] these first thirteen Catalan numbers appear as matrix-elements next to the zero-diagonal elements of the (14 × 14) accessibility matrix S (see Table 1 in their paper) as the numbers of Morgan-trees with up to 13 vertices, but have not been identified by these authors as Catalan numbers.

The N-tuple code deserves a few more words. Most of the literature on isomer counting has been focused on the enumeration of specific structural classes. Very little attention has been given to the sequence of generation and its possible relationship to nomenclature. An exception to this trend is the N-tuple code developed for trees by von Knop et al.[7] The N-tuple code represents a set of non-negative integers smaller than N (the number of vertices in a tree; the number of carbon atoms in an alkane), each representing the degree of a vertex in a tree or subtree. The degree (or valency) of a vertex in a (molecular) graph is equal to the number of edges meeting at this vertex. To obtain the N-tuple code, one has first to identify the vertices with the highest degree and select amongst them one that will result in a code that produces lexicographically the largest number. After the starting vertex is located, that vertex and adjacent edges are removed. The subtrees thus produced are examined. Typically, this means looking for the largest chain, and, if several chains of the same length appear, their codes are derived and combined in such a way that the result corresponds to the lexicographically highest number. The N-tuple code was the basis for the so-called *compact* codes[22] (called so because they use a limited number of digits for linearly encoding a given chemical structure) that have the potential to develop into a full new system of chemical coding. Compact codes have been used to encode[23-27] various classes of saturated and unsaturated acyclic molecules including acyclic compounds with heteroatoms, polycyclic molecules such as benzenoid hydrocarbons, annulenes, aza-annulenes, annulenoannulenes, aza-annulenoannulenes, cyclazines, aza-cyclazines, etc. In Figure 1 we give as an illustrative example the N-tuple code for a tree representing the carbon skeleton of 2,2,3-trimethylpentane and the labeling of its vertices induced by the N-tuple code.

The N-tuple code was also extended by Contreras et al.[28] to polycyclic structures and other molecular structures of increasing complexity. This code together with new approaches for symmetry analysis and cycle detection in molecular graphs was implemented in a modular way in the program CAMGEC2.[28]

Into this area of enumeration there also fall the efforts of Davidson[29-31] who used side-chain complexity minimization algorithms for naming and coding carbon skeletons of alkanes and ring-chain assemblies. His latest effort[32] is focused on fast canonical generation of the alkane series. Davidson's approach is based on the selection of the main chain of an alkane as the path that yields the least complex side chains without the maximum-length constraint that

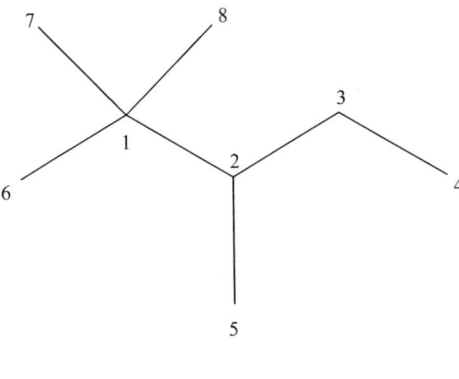

Figure 1 *A tree representing the carbon skeleton of 2,2,3-trimethylpentane, its N-tuple code and the labels of its vertices induced by the N-tuple code*

leads to an efficient algorithm representable as a nested binary tree. The largest side-chain required to specify an N-carbon alkane becomes $(N-1)/3$. This, for example, allows 3.8 million C_1–C_{22} alkanes to be coded for name translation in dictionary order, using an alphabet of 33 C_1–C_6 alkyl groups, also ranked by complexity. The generating process produces reversible isomer codes, making the computation rate in isomers per second inverse linear with N and much faster than reported rates for other structure-generators. Davidson called his alkane isomer codes (size + alkyl codes) the *ultimate* compact codes. We note that the number of isomers for C_{22} alkanes, given by Davidson (2267998),[32] is smaller than given by other authors (2278658).[8,33]

Classical enumerations of the constitutional isomers of various classes of hydrocarbons in the manner of Henze and Blair[34-40] are still being pursued. Henze and Blair derived recursive algorithms for enumerating the number of constitutional isomers of acyclic structures using the corresponding radicals. For example, they computed isomers of alkanes using alkyl radicals. However, Henze and Blair did not tackle the enumeration of constitutional isomers of cyclic structures. Lam from the Department of Chemical Pathology of the Chinese University of Hong Kong (China) undertook this task since he realized that alkyl polyradicals can be employed to enumerate isomers of cyclic compounds using ideas, and even numbers, of Henze and Blair. After showing that alkyl 1,1-biradicals can be used to enumerate the constitutional isomers of alkenes[41] (these numbers fully agree with the numbers obtained by von Knop *et al.*[7,8] using an algorithm based on the N-tuple code) and alkylcyclopropanes,[42] Lam[43] used alkyl 1,1,1-triradicals to enumerate isomers of alkylcyclobutadienes. For an alkyl triradical of carbon content N, the number of constitutional isomers is L_N, where also $L_N = S_{N-1}$ where S_{N-1} denotes the number of constitutional isomers of an alkyl group with $N-1$ carbons. The S_N numbers Lam took from Henze and Blair.[34]

Wang et al.[44] have dealt with enumeration of hydroxyl ethers. The considered class of compounds was limited to the so called *stable* hydroxyl ethers, with at most single oxygen atom bonded to any carbon atom. They presented generating functions for constitutional isomers, all stereo isomers and achiral stereo isomers of these compounds.

2.1.2 Benzenoid Hydrocarbons. – The interest in answering the ubiquitous question *How many benzenoid hydrocarbons are there?* continues. In the last forty years, this question was considered by many authors who proposed a galaxy of methods for generation and enumeration of benzenoid isomers[8,45,46] (in the mathematical literature this problem is known as counting hexagonal polyominoes[47,48] or hexagonal animals[49] and belongs to the cell-growth problems[4,50]). Graphs representing benzenoid hydrocarbons are often called in the literature polyhexes,[5,51] benzenoid graphs[5,52] and there is also a term polyhex hydrocarbons[53,54] in use. Polyhexes are planar graphs that may be obtained by any combination of regular hexagons such that any two of its hexagons have exactly one common edge or are disjoint. These schemes sometimes differ in whether they count polyhex structures which are not subgraphs of the honeycomb lattice (i.e., they sometimes do and sometimes do not include helicenic structures).

Among the first attempts in the chemical literature to enumerate benzenoid structures were efforts by Balaban and Harary[51] in 1968, Balaban et al.[55] in 1980 and von Knop et al.[56] in 1983. With increasing capacity of computers and an improvement of algorithms, enumerations of benzenoids have been extended to very large systems. After von Knop and his group published results for simply- and multiply-connected benzenoids with up to 16 hexagons,[53,57] using an algorithm based on the DAST (dualist angle-restricted spanning tree) code, several groups extended these enumerations, using their own algorithms, gradually building up to systems with up to 25 hexagons, e.g. refs. 58–63. All these efforts belong to the so-called *constructive* enumerations, that is, counting benzenoids by constructing each molecule using a suitable code. The *constructive* enumerations are useful for generating combinatorial libraries.

In order to extend the enumerative results to much higher numbers of hexagons, Vöge et al.[64] devised a new algorithm, based on methods used in statistical physics, which allows the computer enumeration of benzenoid systems (without constructing them; the so-called *nonconstructive* enumeration) with up to 35 hexagons. They proved that $b_h \approx \kappa^h$, where b_h is the number of benzenoid systems with h hexagons and κ is a growth constant. They also established the rigorous bounds for the growth constant: $4.789 \leq \kappa \leq 5.905$ and estimated that $\kappa = 5.16193016(8)$.

The work of Vöge et al.[64] made it possible to establish the asymptotic law by which b_h increases with h, for large values of h.[65] The numbers b_h for $h > 10$ turn out to be rather huge, e.g., $b_{11} = 141229$, $b_{16} = 359863778$, $b_{21} = 1012565172403$, and $b_{35} = 585100026562580180 6530$.[64] Using the asymptotic form one can calculate the approximate values of b_h for h greater than 35, if one is interested in these numbers. However, the exact computations were rather lenghty – Vöge et al.[64] needed about 10 weeks on a Compaq AlphaServer ES40, utilizing up to 5 GB of memory, to get the values of b_h with up to $h \leq 35$. Even so, this is considerably less time than one used by von Knop et al.[57] to get the values of b_h with up to $h \leq 16$ – they needed almost 92 days on a personal computer Siemens PCD3D (20 MHz, 386-AT).

It should be also mentioned that the number of known benzenoid hydrocarbons is about a thousand. However, the methods for synthesis of benzenoids have been improved[66,67] so that even very large benzenoids[68] are already available exhibiting unusually high stability[68,69] and many interesting properties (e.g., self-organization, conductivity).[70] There is also available a strategy, based on graph-theoretical and topological arguments, for designing all kinds of aromatic structures that includes benzenoid hydrocarbons and their derivatives.[71]

Klein and Misra[72] addressed a problem of enumerating poly-N-phenacenes, consisting of a cycle of benzene rings each fused in either *meta* or *para* fashions to the rings on either side of it. Here the fusion may be at opposite edges of a ring (termed *para*-fusion) or at next-neighbor edges of a ring (and termed *meta*-fusion). Thence a cycle of rings results, where the *meta*-fusions can be to non-neighbor edges either to the "left" or "right", as one proceeds around the cycle. Note here may be made of synthetic efforts toward preparing and characterizing such species.[73–77] The enumeration of the possible isomers is reminiscent of the Pólya approach, which entails enumeration of suitable symmetry classes of mathematical maps from a domain (usually of skeletal sites) to a range (of substituents). However for the present enumeration it turns out the relevant symmetry group acts both on the domain and range, so that some extension of the standard enumerative approach is needed. Thence a general symmetry-attentive method for enumeration is made, and applied for the cases where the number N of benzene rings is prime. The method is more general than prime N, this being the simplest case, where the relevant cyclic-related groups have simpler subgroup structures. The methodology is first presented on the simpler case of open-chain polyphenacenes, as earlier treated by Balaban and Harary.[51] Leading up to the general approach a systematic scheme to generate the various isomer labels is formulated, is noted to relate to a systematic 'nomenclature', and applied to cycles of up to 13 benzene rings (where there are over a 30000 isomers, of the 'regular' type). The method and applications incorporate the possibility of the cyclo-phenacene being 'regular' like an (untwisted) belt, or with a Möbius twist (either left- or right-handed), or even being multiply twisted. Some cursory attention is even directed to the possibility of knotted cyclo-polyphenacenes (possibly with various numbers of twists) and polyphenacenes that consist of a linking of multiple cycles.

2.1.3 Pólya Theory and Mark Tables. – The classification of substitutional isomers according to their symmetry realized as a result of the substitution has been a topic of some chemical interest for a little over 2 decades, with several chemical researchers utilizing methodology parts of which dates back to Burnside[78] around 1900. A number of researchers (Hässelbarth,[79] Mead,[80] Brocas, Gielen and Willem,[81] and Fujita[82]) utilized so-called *mark tables* to solve this type of problem, such tables conveniently characterizing the different possible subgroup structures within the parent permutationally represented symmetry of the skeleton into which the different substituents are being placed. Over the years books on this subject have appeared, by Fujita[83] and more recently by El-Basil,[84] merging the whole idea neatly with conventional Pólya theory of enumeration.[85] In 2002, a special issue of MATCH (vol. 46) was published dedicated to recent developments in this area. Thence the bulk of these papers may all be nicely reviewed together here. The introductory paper there by El-Basil[86] gives a short (dozen-page) history, with emphasis on the important role played by Pólya theory and mark tables in counting isomeric structures.

Dolhaine and Hönig[87,88] continue their work on counting isomers of complex organic compounds using the *Mathematica®* application program package *Isomers,*[89] which is the implementation of an efficient isomer enumeration algorithm based on the classical Pólya counting theorem.[85] They applied their program to obtain the number of possible tetramers of a selected subset of the nine known inositols (1,2,3,4,5,6-hexahydroxycyclohexanes). The question *How many tetramers of four specific inositols, namely neo-, muco-,* D-*chiro and* L-*chiro are there?* has been asked by Tomas Hudlicky from the University of Florida at Gainesville, whose group is making these compounds.[90,91] This question was directed to Dolhaine and his coworkers because Dolhaine et al.[92] applied in 1999 their program *Isomers* to a number of molecules of interest for organic chemists, including some oligomers of inositols. The answer to this question should also be of interest in the field of structural biology, where sugars and pseudosugars gain status as information-carrying macromolecules.[93] Doing combinatorial chemistry without knowledge of just how many combinations are to be expected, appears to be impratical, or at least incomplete. Dolhaine and Hönig[87] obtained these numbers – there are 24300 *linear* and 6480 *branched* tetramers. Therefore, there are 30780 tetramers possible which are made of the four specific and different (*neo-, muco-,* D-*chiro* and L-*chiro*) inositols.

In a subsequent publication, Dolhaine and Hönig[88] gave tables of all possible combinations and stereoisomers of the known nine inositols in their dimer, trimer and tetramer forms. The respective numbers of achiral compounds and cumulative estimations of the number of possible linear pentamers, hexamers, heptamers, octamers and nonamers are also listed. These calculations have been carried out by the *Mathematica – AddOn 'Isomers.m'* program which is available together with some help, free of charge (for academics only), at: http://www.cis.TUGraz.at/orgc/institut/softnew.htm.

Kerber[94] presents a mathematically elegant concise overview of the fundaments of enumeration theory under finite group action (following up on his mathematics book[95]). Kerber attends but briefly to the realized and potential chemical applications, so that his overview may be seen as encompassing an indication of further techniques of likely chemical use. In the following article Hässelbarth and Kerber[96] describe applications of enumeration under finite group action on a particular class of combinatorial libraries that arise from a symmetric parent compound by means of reactions with a given set of building blocks. Such libraries were described by Carell et al.,[97,98] and their enumeration by weight was made by van Almsick et al.,[99] using the Pólya theorem.[85] The enumeration by symmetry group is discussed by Hässelbarth[100] in 1986. In their paper Hässelbarth and Kerber[96] give a refinement of both these methods, enumerating such libraries by weight and by symmetry group using mathematical apparatus presented by Kerber in the preceding paper.[94] They present their methodology on previously considered libraries,[97,98] arising from xanthene and from a benzene skeleton with three acid-chloride groups.

Nemba and Balaban[101] obtained the number of chiral and achiral isomers of cycloalkanes C_NH_{2N} with M homomorphic alkyl groups. The M alkyl groups C_NH_{2N+1} are homomorphic if they are structurally and stereochemically identical when they are detached from the ring system. Nemba and Balaban[101] used the upgraded form of the enumeration technique described by Robinson *et* al.,[102] and produced numbers of enantiomeric pairs and achiral carbon skeletons of homopolysubstituted cycloalkanes $C_NH_{2N-M}(C_NH_{2N+1})_M$ for various values of $N \geq 3$ and $1 \leq M \leq N$.

In the mentioned special issue of MATCH, there are general methodological papers by Fujita and El-Basil[103] and by Lloyd.[104] A paper by Klein and Misra[72] is mentioned earlier in section 2.1.2 here. Another paper by Lloyd and Dolhaine[105] applies the mark-table methods outside of the area of isomer enumeration – enumerating *thermal cycles*.

Fujita[83] continues an immense near single-handed work on the development of symmetry-attentive combinatorial isomer-enumeration methodology and its application. In the present work, Fujita[106] reports exhaustive combinatorial enumeration of the square-planar complexes with achiral and chiral ligands under the point-group D_{4h}, where its mark table[86,103] and its inverse mark table are prepared and used to calculate the subduction of coset representations. The enumeration results are used to discuss equivalency under point-group symmetry, that is, enantiomeric relationships for chiral complexes and prochirality for achiral complexes. The alternative combinatorial enumeration of the square-planar complexes was carried out under the permutation group S_4, which is the symmetric group on four indices.[107] After defining the proper and improper permutations, the subgroups of S_4 are classified into stereogenic and astereogenic groups for the square-planar complexes. Then, equivalency under permutation-group symmetry is employed to clarify enantiomeric and diasteroisomeric relationships. Fujita's work based on the comparison between the action of a point-group and that of a permutation-group provides a novel and versatile approach to stereochemistry.

Fujita[108] also studied molecules of ligancy 4 that have been derived from allene, ethylene, tetrahedral, and square-planar skeletons to show that their symmetries are dually and distinctly controlled by point groups and permutation groups. He has shown the following: insomuch as the point-group symmetry controls the chirality/achirality of a molecule, sphericity in a molecule[109] and enantiomeric relationship between molecules,[109] the permutation-group symmetry controls the stereogenicity of a molecule, tropicity in a molecule and diastereomeric relationship between molecules. To characterize permutation groups, proper and improper permutations have been defined by comparing proper and improper rotations. These permutation groups were classified into stereogenic and astereogenic groups. After a coset representation (CR) of a permutation group has been ascribed to an orbit (equivalence class), the tropicity of the orbit has been defined in terms of the global stereogenicity and the local stereogenicity of the CR. As a result, the conventional stereogenicity has been replaced by the concept of local stereogenicity in this paper. Fujita coined the terms homotropic, enantiotropic, and hemitropic and has used them to characterize prostereogenicity. A molecule is defined as being prostereogenic if it has at least one enantiotropic orbit. Since this definition has been found to be parallel with the definition of prochirality, relevant concepts have been discussed with respect to the parallelism between stereogenicity and chirality in order to restructure the theoretical foundation of stereochemistry and stereoisomerism. The derivation of the skeletons has been characterized by desymmetrization due to the subduction of CRs. The Cahn-Ingold-Prelog (CIP) system[110] has been discussed from the permutational point of view to show that it specifies diastereomeric relationships only. The apparent specification of enantiomeric relationships by the CIP system has been shown to arise from the fact that diastereomeric relationships and enantiomeric ones overlap occasionally in case of tetrahedral molecules.

Fujita[111] has extended the sphericity concept, proposed for specifying stereochemistry in a molecule,[109] to a study of stereoisomerism among molecules. The novelty of this approach is to characterize the global symmetries of molecules as the *local symmetries of stereoisomerism*. Therefore, stereochemistry and stereoisomerism have been discussed on a common basis. Promolecules, which have been generated as stereochemical models of molecules by placing proligands (structureless ligands with chirality/achirality) on the vertices of a tetrahedral skeleton, have been analyzed by a permutation-group approach and by a point-group one. The skeleton has been considered to belong to the symmetric group of degree 4 and to the isomorphic point group T_d. The chirality fittingness derived from the sphericity concept has been applied to the characterization of local symmetries of promolecules, where two types of the Young tableaux have been compared. Thus, the Young tableaux of symmetry have been introduced to treat the ligand packing based on the chirality fittingness. These tableaux have been compared with the Young tableaux of permutation, which have taken no account of such chirality fittingness. The two types of the Young tableaux have been applied to the enumeration of tetrahedral isomers with chirality fittingness observance and without chirality fittingness. This enumeration has enabled the clarification of the quantitative aspect of the sphericity concept in characterizing isomer equivalence. Therefore, equivalent isomers under a point-group symmetry have been shown to construct an orbit of stereoisomers that is ascribed to a coset representation. Homomeric, enantiomeric, and diastereomeric relationships between stereoisomers have been discussed by means of homospheric, enantiospheric and hemispheric orbits of stereoisomers. Skeleton-based and ligand-based categories for enantiomers and diastereomers have also been discussed. The stereogenicity and the prostereogenicity of the Chan-Ingold-Prelog system[110] have been related to the Young tableaux of permutation.

Fujita and El-Basil[112] have developed a graphical method for generating one- and some two-dimensional group characters by using a reduced homomer set. The reduced homomer set was obtained by introducing the notion of negative graphs as corresponding to homomers with exchanged colors of vertices.

2.2 Kekulé Structure Count. – As remarked in our previous report[1] chemical work on counting Kekulé structures appears to be less wide-spread than a decade or two ago. It is however still studied by the mathematicians[113,114] and sometimes by chemists.[115] In graph-theoretical literature Kekulé structures are identified as 1-factors,[116] perfect matchings[52,113,117] or Kekulé patterns.[114,118,119] In 1985 von Knop *et al.*[45] published a book including a comprehensive listing of benzenoids of up to 9 hexagons along with their Kekulé-structure counts. Cyvin and Gutman[119] published a book in 1988 on methods for counting Kekulé structures in benzenoid hydrocarbons. Dias[120] a year earlier (1987) produced a handbook on benzenoid hydrocarbons that contained for each structure also its Kekulé number.

Misra and Klein[121] have further considered theoretical characteristics of cyclo-polyphenacenes, including the enumeration of their Kekulé structures. Such species always have Kekulé structures, and a comparable enumeration algorithm was made a while back by Cyvin et al.,[122] though they focused more on structures like kekulene[123] – that is, this earlier work focused on *unstrained* structures. The enumeration proceeds by way of a transfer-matrix technique,[5,124–128] such as discussed in our previous report.[1] This technique has most usually previously been applied to polymers with but a single monomer unit, whereas in the present case there are three monomer units (*para*-fused, left-*meta*-fused, and right-*meta*-fused) occurring in different sequences. As occasionally previously noted[127] the technique applies reasonably to such *disordered* polymers also. Misra and Klein[121] incorporate the Kekulé-structure counts into a reactivity index, which in application to open-chain polyphenacenes is found to agree well with an available set of experimental data.[129] Further they utilize a simple *combinatorial curvature* index to indicate the amount of strain in the structures, to obtain decent agreement with Dobrowolski's earlier SCF results[130] on all 52 of the regular cyclo-hexaphenacenes.

Torrens[115] considered the enumeration of Kekulé structures for alternant benzenoid hydrocarbons using the connection between the permanent of the vertex-adjacency matrix and the number of Kekulé structures:

$$\text{per } A = K^2 \qquad (3)$$

This relation holds only for alternant structures as early noted by Percus.[131] It was discussed in graph-theoretical terms and applied to alternant hydrocarbons by Cvetković et al.[132,133] In the above formula, K is the number of Kekulé structures and per A is the permanent of the vertex-adjacency matrix A of an alternant hydrocarbon (*i.e.* a bipartite graph,[4] or bicolorable structure[134]).

Permanents are generally more difficult to compute than determinants,[135] so that Torrens[115] proposed an algorithm for fast computation of the permanents of sparse matrices. Note, the vertex-adjacency matrix ([0,1] matrix) is for molecular graphs usually a sparse matrix. Torrens' approach based on an earlier algorithm reported by Cash,[136] applies the algorithm in an upgraded form to a suitable submatrix of the vertex-adjacency matrix. If we color an alternant structure with N sites by two colors in such a way that sites colored with the same color are never connected, then the sites can be divided in two groups each containing sites of the same color. The corresponding vertex-adjacency matrix is consequently given in terms of two submatrices B and B^T, each submatrix connecting between sites of different colors:

$$A = \begin{bmatrix} 0 & B \\ B^T & 0 \end{bmatrix} \quad (4)$$

where B^T is the transpose of B and 0 is the zero matrix. Then, the permanent of the adjacency matrix is given by:

$$\text{per } A = \text{per } B \text{ per } B^T = [\text{per } B]^2 \quad (5)$$

From (3) and (5) follows:

$$K = \text{per } B \quad (6)$$

In fact all these equations have been known for several years, e.g. refs. 132–134. Torrens' contribution then is a computer program based on (6) which is about seven times faster than the C-language program devised by Cash,[136] part of Torrens' program's advantage coming from the utilization of bitwise operations instead of integer comparisons. Still among the molecules considered by Torrens are two structures (pentalene, azulene) that are non-alternant hydrocarbons, though they have an alternant perimeter.

Lukovits et al.[137] have calculated the number of Kekulé structures of a special class of carbon nanotubes (CNT) with open ends which are also called tubulenes. CNTs are composed of cyclindrical graphene sheets consisting of sp^2 carbon atoms and may be thought of as hexagonal conjugated systems surrounded by a cloud of π-electrons. They were first prepared by Iijima[138] in 1991 as tiny tubes with about 1.5 nm of diameter and a length of several microns. A tube of this diameter contains about 20 hexagons along its circumference. CNTs possess conducting or semiconducting properties depending on the tube structure[139-142] and may be prepared by a laser technique[138] or by electrochemical procedures[143] or in some other way, e.g. refs. 144–148.

Lukovits et al.[137] have calculated the number of Kekulé structures in defect-free *armchair* tubulenes[149] using the transfer-matrix technique.[1,5,124-128] Other researchers have also been engaged in generating the number Kekulé structures of nanotubes and tubulenes, e.g. refs. 1, 150–153.

Lukovits et al.[137] considered polynaphthalenes and polyphenanthrenes and their macrocyclic analogs. The results they obtained indicate that the number of Kekulé structures is greater if the considered tubulenes are extended in the vertical (i.e., if they get longer) than in the horizontal direction. Moreover, experimental observation obtained by using scanning probe microscopy supports this prediction.[137] But here it should be emphasized that this conclusion is for the considered tubulenes, with the particular types of ends considered by Lukovits et al.[137] That is, the global Kekulé structure count can be vastly different depending on the ends of the tubes, because of a *long-range* order inherent in Kekulé structures. That the Kekulé-structure count is very sensitive to the ends is evident with the trivial example entailing the addition (or subtraction) of but a single vertex from the end of a Kekulé-structure-rich polymer, whence with an odd number of sites the count drops abruptly to 0. But a more general understanding is available.[154] The resultant general form of the count $K_{L,c}$ for a polymer of length L and circumference c appears asymptotically (for large L) as

$$K_{L,c} \approx A_c \Lambda_c^L \qquad (7)$$

where A_c and Λ_c are independent of L. Further Λ_c (the maximum eigenvalue of the transfer matrix) for a given type of tube ends up dependent on the ends of the polymer tube only in a rather simple discretized fashion. That is, of the infinite of possible ends there are but $\approx c$ different classes, each class giving a distinct value of Λ_c. Overall for general conjugated π-network polymers (non-zero) Λ_c falls in a range between 1 and σ^c, with σ independent of c and L as well as the particular monomer unit. For a particular polymer this full range from $1 \to \sigma^c$ is typically approximately uniformly filled (by $\approx c$ values). In any event the simple relation of Swinborne-Sheldrake et al.[155] correlating overall resonance energy with the Kekulé structure count that is employed by Lukovits et al.[137] should be used with caution for extended systems if one uses a global (i.e., end-sensitive) Kekulé structure count.

The sensitivity in selected properties of general graphitic fragments to the nature of different boundaries, has been studied sometime back,[156–158] as this relates to the tendency for unpaired electrons to appear in extended graphitic nano-structures. That is, with an unfavorable boundary which gives very few Kekulé structures, it becomes more favorable for major contributions from resonance structures with unpaired electrons at the boundary if the unpaired electrons are so placed that the remnant effective boundary is such as to support a greater number of Kekulé structures. Indeed it has been found (earlier on in Klein[159] and Klein and Bytautas,[160] and recently now by Ivanciuc et al.[161,162]) that the predictions made from such relatively simple resonance-theoretic considerations end up correlating quite accurately with the number of unpaired boundary-localized electrons predicted by suitable UHF solutions to Hubbard models. In these articles such comparisons have been made for about two-dozen different graphitic edges, much extending several earlier pieces of previously observed but unrationalized behaviors for the Hückel-model solutions at boundaries of graphite[163–165] or of large benzenoids.[157,158,166] Further in the work of Ivanciuc et al.[161,162] results have been enunciated for a general class of vacancy defects, and also in this case there has been found agreement in about two dozen cases with UHF solutions for the Hubbard model in regard to the occurrence of defect localized unpaired spins. Especially for the case of a single-site vacancy defect the results here again rationalize some earlier observations[167–170] made with simple tight-binding methods. Further there is a qualitative agreement with the results of scanning electron microscopy for two types of edges,[171] and for the single-site vacancy defect.[172–174]

2.3 Counting Walks. – Walks can be used for characterization[175–179] of (molecular) graphs, for quantification[180] of their complexity (or more precisely their labyrinthicity[181]) and for definition of various molecular descriptors.[177,182–184] A walk in a (molecular) graph is an alternating sequence of vertices and edges, such that each edge begins and ends with the vertices immediately preceding and following it.[4] The length of the walk is the number of edges in it. Repetition of vertices and/or edges is allowed in a walk. A random walk in a graph is naturally designated by a suitable probability measure, of which there are two natural ones. First, one probability measure entails starting walks from each vertex with equal likelihood and subsequent steps are such that each neighboring vertex is stepped to with equal probability, so that the probability of stepping from vertex i to j is $1/d_i$ where d_i is the degree of vertex i. Second, another conceivable probability measure takes each possible walk of a given length as equally likely. If a graph is regular (*i.e.*, having all vertices of the same degree), then the two resultant probability distributions for walks are the same – but not otherwise. The walks generated with the first probability measure are typically called *random*, and here we term the walks of the second case *equi-weighted*. A self-returning walk is a (random) walk starting and ending at the same vertex.

It is well-known[5,185,186] that the number of walks of length ℓ beginning at vertex i and ending at vertex j is given by the i,j-element of the ℓ-th power of the vertex-adjacency matrix A: $(A^\ell)_{ij}$. Then the number of self-returning walks of length ℓ is given by the i,i-element of the ℓ-th power of the vertex-adjacency matrix A: $(A^\ell)_{ii}$. It has been found, for example, thirty years ago, that the total number of self-returning walks of length ℓ which coincide with the trace of A^ℓ can be applied to the theory of total π-electron energy,[187] in as much as this gives moments of the eigenvalue distribution (as elaborated some later here). And more recently, self-returning-walk counts have been found to be applicable to systematic search of isocodal vertices in molecular graphs.[175,188–191] Isocodal vertices in a graph are those vertices that have the same numbers of self-returning walks for each length of walk.

The atomic walk count of order ℓ, denoted by $(awc)_\ell$, is the number of all possible walks of length ℓ which start at a specified vertex (atom) i and end at any vertex (atom) j:[180]

$$(awc)_\ell(i) = \sum_{j=1}^{N}(A^\ell)_{ij} \qquad (8)$$

where N is the number of vertices in a graph G.

The molecular walk count of order ℓ, denoted by $(mwc)_\ell$, is obtained by summing up all atomic walk counts of order ℓ:[182,192]

$$(mwc)_\ell = \sum_{i=1}^{N}(awc)_\ell(i) = \sum_{i=1}^{N}\sum_{j=1}^{N}(A^\ell)_{ij} \qquad (9)$$

The total walk count, denoted by twc, is the sum of all $(mwc)_\ell$ for $\ell = 1$ to $N-1$:[182,192]

$$twc = \sum_{\ell=1}^{N-1}(mwc)_\ell = \sum_{\ell=1}^{N-1}\sum_{i=1}^{N}\sum_{j=1}^{N}(A^\ell)_{ij} \qquad (10)$$

The total walk count can be obtained in a much simpler way utilizing a relationship between the Morgan extended connectivities[14] (*EC*) and matrix powers of the vertex-adjacency matrix. This relationship was first observed by Razinger,[193] who gave an emiprical example but offered no proof. Rücker and Rücker[180] proved that *EC*'s and $(awc)_\ell$'s are identical. Consequently, the *EC* of vertex i of the ℓ-th order can be obtained by iterative summation of the $(\ell-1)$-th order contributions of the neighbors of i and is equal to $(awc)_\ell(i)$. Thus, calculation of *twc* requires nothing but a sequence of addition steps. This elegant procedure, termed by Rücker and Rücker[180] the Morgan summation procedure, is illustrated in Figure 2. Rücker and Rücker[180] also prepared

(a) Graph representing the carbon skeleton of isopropylcyclopropane

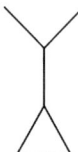

(b) Values of $(awc)_\ell(i)$ obtained by iterative summation over all neighbours, starting from vertex-degrees and a values of $(mwc)_\ell$

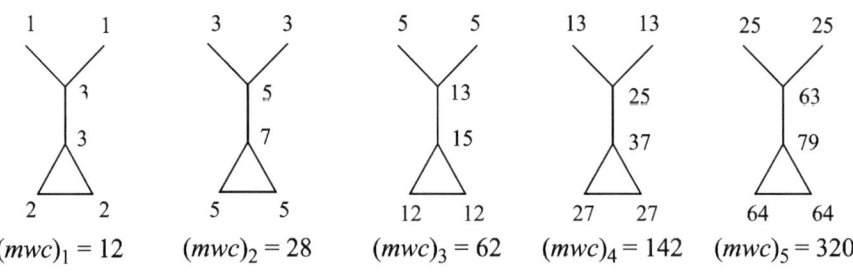

(c) The value of *twc*

$$twc = 12 + 28 + 62 + 142 + 320 = 564$$

Figure 2 *An illustrative example of the Morgan summation procedure*

program MORGAN (so named presumably to point to the pioneering work of Morgan[14]) based on the described procedure.

There is also a third method available for counting total walks. This method is based on the relationship between molecular walks of length ℓ, eigenvectors (c_{ri}; $i=1,2,\ldots,N$) and eigenvalues λ_r^ℓ of the vertex-adjacency matrices with various values of exponent ℓ.[192] A starting point is the eigenvalue equation:

$$Ac_r = \lambda_r c_r; \quad r = 1, 2 \ldots, N \tag{11}$$

Using the standard matrix-theoretical methods, from eq. (11) follows:

$$(A^\ell)_{ij} = \sum_{r=1}^{N} \lambda_r^\ell c_{ri} c_{rj} \tag{12}$$

Eq. (12) in combination with eq. (9) gives:

$$(mwc)_\ell = \sum_{r=1}^{N} \sigma_r \lambda_r^\ell \tag{13}$$

where

$$\sigma_r = (c_{r1} + c_{r2} + \cdots + c_{rN})^2 \tag{14}$$

If λ_r is a degenerate eigenvalue, then σ_r, as defined by (14), is not uniquely determined. If such a case occurs, then the sum of σ_r's over all degenerate eigenvectors is a true graph invariant σ_ρ' with ρ a label for distinct eigenvalues. Then in place of (13) one has $(mwc)_\ell = \sum_\rho \sigma_\rho' \lambda_\rho^\ell$. Formula (13) represents the spectral decomposition of the molecular-walk counts.[116,194,195]

Rücker and Rücker[183] reported total walk counts for saturated acyclics through polycyclic hydrocarbons with up to ten carbon atoms. They also reported walk counts for molecules containing multiple bonds and/or heteroatoms using the Morgan summation procedure.[181,196] Graphs which are used to represent this kind of molecule are *general* graphs,[197] that is, graphs in which multiple edges and/or loops are allowed. In *simple* graphs (as we earlier considered) multiple edges and loops are *not* allowed.[198] General graphs are also known as the vertex- and edge-weighted graphs.[5] Walks in general (weighted) graphs depend on weights given to vertices and edges.[181,196,199] For example, if a loop on vertex i is given a weight of 1, it means that it will give rise to additional walks such as the walk of length 1 from i to i. Once we decide the values for vertex- and edge-weights, the application of the Morgan summation procedure is straightforward. In Figure 3 we give the total walk count for a selection of molecules taken from Rücker and Rücker.[196]

In fact the bulk of the discussion concerning the counting of equi-weighted walks carries over with but slight modification to the generation of random

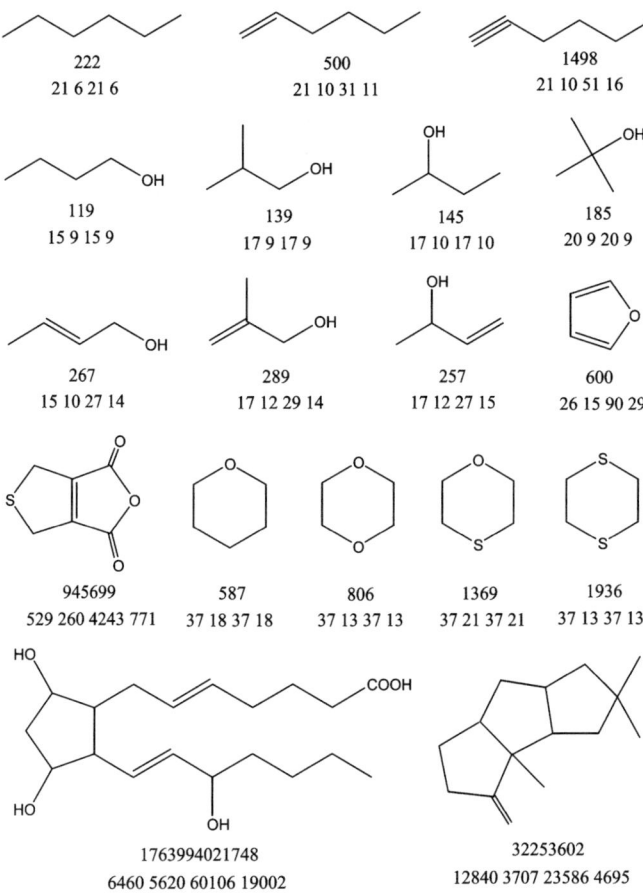

Figure 3 *Total walk counts (twc) for a selection of molecules. O and S are arbitrarily assigned weights 1 and 2, respectively. Four numbers in the row below twc are the total number of connected substructures (N_t), the number of distinct connected substructures (N_s), the total number of connected subgraphs (N_t') and the number of distinct connected subgraphs (N_s'). Data presented here are all taken from ref. 196*

walk sums. Random walks are generated by powers of a (Markov) matrix M with elements which are probabilities for the associated steps

$$(M)_{ji} = \begin{cases} 1/d_i, & j \neq i \\ 0, & \text{otherwise} \end{cases} \tag{15}$$

Then e.g. $(M^\ell)_{ji}$ is the probability for an ℓ-step random walk beginning at site i to end at site j. In analogy to the $(awc)_\ell (i)$ invariant one may consider a random-walk count:

$$rwc_\ell(i) = \sum_{j=1}^{N}(M^\ell)_{ij} \qquad (16)$$

which has the neat interpretation that $rwc_\ell(i)/N$ is the probability that one ends after ℓ steps at site i having started randomly at any site. The sum $\sum_{i=1}^{N} rwc_\ell(i)$, which would be the analog of $(mwc)_\ell$, is not so interesting though in that this is always $= N$, because this sum divided by N is evidently the probability of ending someplace after ℓ steps. (And thence also the analog of twc is trivial also.) Again the $rwc_\ell(i)$ may be neatly computed via eigensolutions to the matrix M, which however generally is non-symmetric, when the graph is not regular. But in this general case the eigensolutions are related to those of a symmetric matrix, as may be seen if one first starts out noting that M may be written as $M = AD^{-1}$ where D is the diagonal matrix of degrees d_i. Then we introduce the matrix:

$$H = D^{-1/2} M D^{1/2} \qquad (17)$$

where $D^{1/2}$ is the diagonal matrix of square-roots of degrees and $D^{-1/2}$ is its inverse. Evidently H is similar to M (*i.e.*, they are related by a similarity transformation), so that H has the same eigenvalues, and eigenvectors c'_r which are simply related to those c_r of M, thusly

$$c'_r = D^{-1/2} c_r \text{ and } c_r = D^{1/2} c'_r \qquad (18)$$

But evidently also $H = D^{-1/2} A D^{-1/2}$, so that H is symmetric and more standard matrix diagonalization routines may be employed, to yield the c_r and associated eigenvalues λ_r. Then *e.g.*,

$$(rwc)_\ell(i) = \sum_{r}^{N} (c_{ri} d_i^{-1/2}) \lambda_r^\ell (\sum_{j=1}^{N} c_{rj} d_j^{-1/2}) \qquad (19)$$

where the c_{ri} are components of c_r. In fact this matrix H is rather well-known, sometimes being termed the (*normalized*) *Laplacian* matrix of G, and there is much theory about it, *e.g.*, as reviewed in Chung.[200] This matrix is also interesting in that half the sum of its elements is just the Randić[201] connectivity index $^1\chi$. In general one may anticipate that many different graph invariants defined in terms of equi-weighted graphs may also have interesting and useful analogues defined in terms of random graphs. Some such invariants have been noted in by Klein[202] to be related to *sum rules* for resistance distances. But there are many graph invariants built for equi-weighted graphs where the random-walk analogues have not yet been explored. One example of this is the different invariants derived by Diudea[203,204] from his *walk matrices*.

Dress *et al.*[205] reported the number of walks in trees with up to 18 vertices. The total number of trees they considered was 205004 trees. They studied the harmonicity[206] of these trees and found that exactly three of the studied trees are harmonic, none almost harmonic, 11 superharmonic, and all other considered trees to be subharmonic. Dress and Gutman[207] also reported asymptotic results regarding the number of walks in a graph.

The concept of walks was extended by Lukovits and Trinajstić[208] to walks of zero and negative orders by using a backward algorithm based on the usual procedure to obtain the values of molecular walk counts. These authors called the Morgan summation procedure the Razinger algorithm, and have formalized it as follows. Let **1** be an N-dimensional (column) vector, all entries of which are equal to one. Let $\boldsymbol{a}_\ell$ be a (column) vector, the entries of which are the awc's of order ℓ for all vertices. Then the Razinger algorithm can be formally written as:[209]

$$\boldsymbol{a}_\ell = A^\ell \mathbf{1} = A^{\ell-1}\boldsymbol{a}_1 = \ldots = A\boldsymbol{a}_{\ell-1} \tag{20}$$

It is clear if matrix A^{-1} exists, the procedure can be inverted:

$$\boldsymbol{a}_{\ell-1} = A^{-1}\boldsymbol{a}_\ell \tag{21}$$

Formula (21) allows the concept of atomic walk counts to be extended to awc's of zero and negative orders. However, there are three possible subclasses of the inverse procedure: (I) The inverse of the vertex-adjacency matrix exists – this is the simplest case; (II) matrix A^{-1} does not exist, and $\boldsymbol{a}_0 = \mathbf{1}$ and (III) matrix A^{-1} does not exist, and $\boldsymbol{a}_0 \neq \mathbf{1}$. Lukovits and Trinajstić[208] have shown how to get walks of negative order for cases (II) and (III). Note that awc's and mwc's of negative order may assume non-integer and even negative values. In Figure 4 we give $(awc)_\ell$ and $(mwc)_\ell$ for several values of ℓ starting with $\ell = 1$ up to $\ell = -5$ for ethylcyclobutane.

Rücker and Rücker,[210] after learning about the work by Lukovits and Trinajstić,[208] derived a nonrecursive equation for walks of zero and negative orders in a graph (or molecule), utilizing their previous work on spectral decomposition.[192,195] They also developed a computer program for counting the atomic and molecular walks for both positive and negative orders. With these two papers[208,210] the concept of walk counts covers walking in both directions – forward and backward, and no walking at all.

2.4 Combinatorial Measures of Molecular Complexity. – Most authors agree that the concept of complexity is too complex to be defined precisely.[211–213] And it has been further proposed that complexity rather than being a non-numerical quantity is a partially ordered quantity – see, footnote 27 of the ref.[214] But for a chemist's purpose one may consider a system to be complex if it consists of a great number of components – the more components in a molecule, the more complex the molecule is.

We do not fully review here the variety of existing complexity indices and their dependencies on structural features (size, branching, cyclicity, multiple edges, heteroatoms, symmetry) of molecular graphs (they were recently

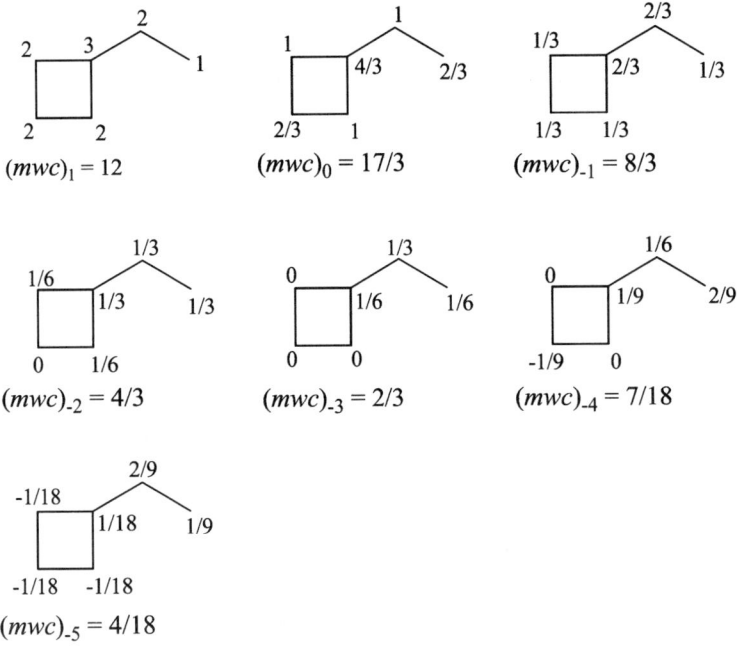

Figure 4 *The values of $(awc)_\ell$ and $(mwc)_\ell$ for $\ell = 1, 0, -1, -2, -3, -4, -5$ for a graph representing the carbon skeleton of ethylcyclobutane*

reviewed elsewhere[182,215]) and for more general (chemical, biological, ecological) networks[216] – we consider here only those combinatorial quantities that have been used to characterize molecular complexity *via* the corresponding graphs. We already mentioned the role of walks in characterizing graph (molecular) complexity.[181] Rücker and Rücker[181] also introduced the term *labyrinthicity* if the complexity of a structure is measured by walk counts alone, because the walk count neglects symmetry. On the other hand, some other authors strongly incorporate symmetry characteristics of molecules. For example, Randić proposed a symmetry-dependent complexity measure based on the concept of augmented vertex-degree.[217–219] The augmented degree of a vertex i in a (molecular) graph is the sum of its degree and degrees of all other vertices with weight $1/2^d$ depending on their distance d from the vertex i. The Randić complexity measure is then equal to the sum of the augmented degrees of vertices not equivalent by symmetry. It should be noted that the vertex-degrees have all kinds of nice combinatorial properties, *e.g.* refs. 220, 221, and they span the history of (chemical) graph theory from the beginnings (the Königsberg bridge problem, the handshaking lemma)[222] to our times, *e.g.* refs. 223–225. We should also remember that N-tuple code is based on vertex-degrees in a graph and its subgraphs.[7] Some other authors have used in the past, for example, the degree distribution for the construction and enumeration of constitutional isomers in the alkane series.[226,227]

Combinatorial properties that quite naturally correspond to a chemist's view of complexity expressed above are the number of kinds of connected subgraphs (substructures) N_s and the total number of connected subgraphs (substructures) N_t. A connected subgraph (substructure) is a subgraph (substructure) in which for every pair of vertices (atoms) there is a sequence of edges (bonds) connecting them. Connected subgraphs can be generated by removing one or more edges and/or vertices from a graph under the condition that what remains is in one piece. The graph itself is counted as a subgraph for formal reasons. In Figure 5 we give all subgraphs of a simple graph G representing the carbon skeleton of methylcyclobutane.

After the pioneering work of Gordon and Kennedy[228] on the use of subgraphs in chemistry, Bertz and Herndon[229] were the first to explore the idea of using subgraphs for measuring molecular complexity (and similarity). Bertz et al.[230,231] were first to use N_s and N_t numbers to study molecular complexity and synthetic complexity. Bertz[232-234] has also used these numbers to study the complexity of synthetic reactions and routes. Bonchev[235] independently used the N_t number for assessing molecular complexity.

A problem with the subgraph-counting numbers N_s and N_t is that their growth is explosive. For example, these numbers for a relatively simple molecule such as cubane are[182] 64 and 2441, respectively. Therefore, it is generally out of the question to construct all connected sugbraphs of graph by hand. Naturally a computer program would be of use to enumerate all nonisomorphic connected subgraphs, and consequently to calculate N_s and N_t numbers, up to a certain graph (molecular) size. Such a program has been prepared by Rücker and Rücker[236] and was named NIMSG (NonIsoMorphic SubGraphs). Their computer program first generates all connected subgraphs and then uses a combination of well-discriminating graph invariants to eliminate duplicates. Their program is applicable to simple graphs and to general graphs. Before this program, they prepared one that was applicable only to simple graphs (they called it SUBGRAPH).[237] Since a connected subgraph can be described as a linear array of labels of adjacent edges, just as a path can be described as a linear array of labels of adjacent vertices, the connected subgraphs are generated by a path-finding algorithm working on the graph's edge-adjacency matrix. The *edge*-adjacency matrix differs from the *vertex*-adjacency matrix in considering adjacencies of edges instead of adjacencies of vertices.[5] Rücker et al.[238] studied the limits of the computer program NIMSG within the comprehensive graph samples which serve as supersets of the graphs corresponding to saturated hydrocarbons, both acyclic with up to 20 carbon atoms and (poly)cyclic with up to 10 carbon atoms. Since a fast computer method for reliably discriminating all nonisomorphic graphs is not available, Rücker et al.[238] used in their work graph invariants of discriminating power as high as possible. They established that NIMSG, using as discriminatory graph invariants the combination of the Balaban index J[239] and the eigenvalues of the distance matrix,[240] is realiable to use within the domain of mono-, di-, tri-, and tetra-cyclic saturated hydrocarbon substructures with up to ten carbon atoms as well as of all alkane substructures with up to 19 carbon atoms.

4: Combinatorial Enumeration in Chemistry

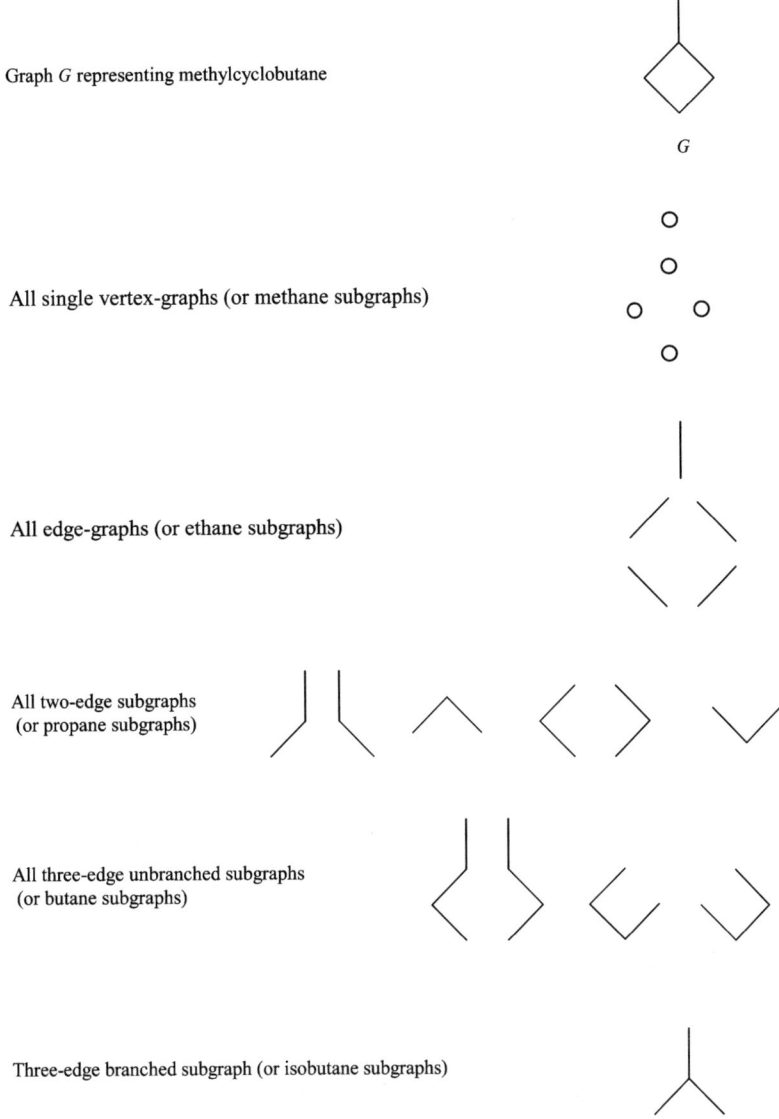

Graph G representing methylcyclobutane

All single vertex-graphs (or methane subgraphs)

All edge-graphs (or ethane subgraphs)

All two-edge subgraphs (or propane subgraphs)

All three-edge unbranched subgraphs (or butane subgraphs)

Three-edge branched subgraph (or isobutane subgraphs)

Figure 5 *All connected subgraphs of a graph G representing the carbon skeleton methylcyclobutane*

Therefore, the computer program NIMSG can be reasonably employed in chemistry whenever one is interested in computing substructures of a rather large number of saturated hydrocarbons which are themselves comprised of a modest number of vertices and edges.

When Rücker and Rücker[196] extended their program for the computation of N_s and N_t to unsaturated compounds represented by multigraphs, they had

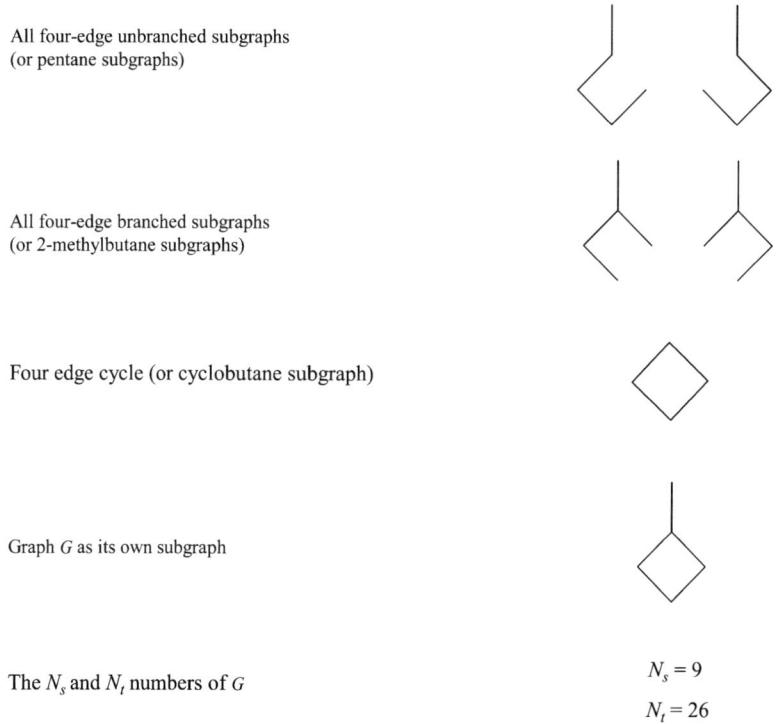

All four-edge unbranched subgraphs (or pentane subgraphs)	
All four-edge branched subgraphs (or 2-methylbutane subgraphs)	
Four edge cycle (or cyclobutane subgraph)	
Graph G as its own subgraph	
The N_s and N_t numbers of G	$N_s = 9$ $N_t = 26$

Figure 5 *Continued*

to distinguish between subgraph and substructure. The coincidence between subgraph and substructure, valid for simple graphs, does not hold for multigraphs. The difference between a substructure of an unsaturated molecule and a subgraph of the corresponding multigraph is that in the substructure all bonds still present have their multiplicities as they are in the parent molecule, whilst the subgraph may correspond to a less unsaturated analogue of the parent multigraph, that is, it may have a single or double edge where the multigraph has a double or triple edge. To illustrate this difference, we compare *n*-pentane and 1-pentene. In *n*-pentane (only the carbon skeleton is considered) one detects the following connected substructures: 5 methane substructures (single C atoms), 4 ethane substructures (C–C), 3 propane substructures (C–C–C), 2 butane substructures (C–C–C–C) and *n*-pentane (C–C–C–C–C) itself. Thus, the total number of connected substructures N_t is 15, whilst the number of distinct connected substructures N_s is 5. The total number of connected subgraphs N_t' and the number of distinct connected subgraphs N_s' are also 15 and 5, respectively, since *n*-pentane is a saturated molecule. In 1-pentene there are 5 methane substructures (single C atoms), 3 ethane substructures (C–C), 1 ethene substructure (C=C), 2 propane substructures (C–C–C), 1 propene substructure (C=C–C), 1 butane substructure (C–C–C–C), 1 butene (C=C–C–C) and 1-pentene itself. Hence, $N_t = 15$ and $N_s = 8$. The 5 methane substructures give 5 single-vertex subgraphs; the 3 ethane and ethene substructures give 5 single-edge subgraphs; the ethene substructure gives additionally 1 double-edge 2-vertex subgraphs; the 2 propane and propene substructures give 4 three-vertex linear subgraphs; the propene substructure gives additionally 1 3-vertex multigraph; the butane and butene substructures give 3 four-vertex linear subgraphs; the butene substructure gives additionally 1 4-vertex multigraph and the 1-pentene graph gives 2 five-vertex linear subgraphs and its own subgraphs. Thus, $N_t' = 23$ and $N_s' = 9$. The N_t, N_s, N_t' and N_s' indicate that 1-pentene is more complex than *n*-pentane. Rücker and Rücker[196] designed a computer program to construct and count all connected subgraphs and substructures and distinct connected subgraphs and substructures of general graphs and the corresponding unsaturated heterocompounds. They named this program BERTZ, presumably in honour of Steven H. Bertz who in 1981 published a pionieering paper entitled *The First General Index of Molecular Complexity*[241] and has continued to contribute to the present day to our understanding of molecular complexity and complexity of chemical reactions.[215,232-234] Program BERTZ is applicable to large sets of molecules, such as we find in combinatorial libraries, but in these cases because of the huge numbers of structures to process it is necessary to limit the maximal size of substructures (subgraphs) to a certain number of bonds (edges). In Figure 3 we also give N_t, N_s, N_t' and N_s' numbers for a set of molecules for which we listed their *twc* values – all the numbers are taken from Rücker and Rücker.[196]

There have been some further miscellaneous uses of complexity measures. Arteca[242,243] has discussed entanglement complexity for linear chain polymers. With the representation of a polymer conformation by the embedding of the linear-chain graph in 3-dimensions, a projection may be made into a 2-dimensional plane, whence the number of crossings there can be used as an entanglement complexity measure. Side-chain complexity enters into Davidson's lexicography[32] for alkanes. And further in biology and in physics there has been work on various complexity measures.

The number of spanning trees is another classic[244] combinatorial measure of graph complexity which has been used[245] for (poly)cyclic molecules. A spanning tree of a (molecular) graph G is a connected, acyclic subgraph that includes all the vertices of G.[4,5] Since acyclic graphs do not possess cycles, the spanning tree of an acyclic graph is identical to the graph itself. Therefore, the number of spanning trees can serve only as a measure of complexity in the case of (poly)cyclic graphs. Bonchev and coworkers[246] were the first authors who used spanning trees (they called them maximal trees) explicitly as a measure of complexity for chemical structures, whilst Brooks *et al.*[244] were first who defined the number of spanning trees of graphs as complexity for graphs. Independently, Gutman *et al.*[245] used the number of spanning trees to study the complexity of cyclic graphs. There are several methods for obtaining the number of spanning trees.[247] Here we will mention only the method based on the Laplacian matrix. The Laplacian matrix L of a graph G is defined as:[248]

$$L = D - A \qquad (22)$$

The Laplacian matrix is sometimes also called the Kirchhoff matrix[248] of a graph because of its role in the matrix-tree theorem,[116] implicit[249] in the work of Kirchhoff.[250] And sometimes L is called the *combinatorial* Laplacian, to distinguish it from the *normalized* Laplacian earlier noted in connection with random walks.

The Laplacian matrix is a real symmetric matrix, so that diagonalization of the Laplacian matrix of a graph (molecule) G with N vertices (atoms) gives N real eigenvalues λ_i, $i = 1, \ldots, N$. The smallest eigenvalue of the Laplacian spectrum $\lambda_i(L)$ is always 0, as a consequence of the special structure of the Laplacian matrix. The uses of the Laplacian matrix, its characteristic polynomial, its eigenspectrum, and related invariants have been explored in chemistry for at least the last decade.[184,247,251-262]

The number of spanning trees, denoted by t, of a (molecular) graph is:[248,263]

$$t = (1/N) \prod_{i=2}^{N} \lambda_i(L) \tag{23}$$

In regular graphs (that is, graphs in which every vertex has the same degree r),[4] the eigenvalues $\lambda_i(L)$ of the Laplacian matrix:

$$0 = \lambda_1(L) < \lambda_2(L) \leq \ldots \leq \lambda_N(L) \leq 2r \tag{24}$$

and of the vertex-adjacency matrix $\lambda_i(A)$:

$$r = \lambda_1(A) < \lambda_2(A) \leq \ldots \leq \lambda_N(A) \leq -r \tag{25}$$

are related by:[116]

$$\lambda_i(L) = r - \lambda_i(A) \tag{26}$$

In fullerenes whose graphs are regular graphs with $r = 3$ (since all their vertices are of degree 3), eq. (23) transforms by means of eq. (26) into:[264]

$$t = (1/N) \prod_{i=2}^{N} [3 - \lambda_i(A)] \tag{27}$$

This formula was used by Fowler[265] to study complexity of isomeric C_{40}, C_{60} and C_{70} fullerenes. He found that the most stable isomers of C_{40}, C_{60} and C_{70} fullerenes each have the greatest number of spanning trees and consequently the greatest complexity within their respective sets of 40, 1812 and 8149 candidates. Fowler[265] obtained similar results when he applied the Randić complexity (RC) formula in the normalized form to the same three sets of isomeric fullerenes:[217-219]

$$RC = (1/N)\sum_{i=1}^{N}\sum_{j=1}^{N} r_i/2^{d_{ij}} \qquad (28)$$

where r_i is the degree of the vertex i whilst d_{ij} is the distance in terms of the number of edges between vertices i and j. However, this formula produced less marked separation of isomers than eq. (27).

Mallion and Trinajstić[266] used two combinatorial quantities *spanning-tree density* and *reciprocal spanning-tree density* in an attempt to quantify the structural intricateness (complicatedness) of (poly)cyclic molecular graphs. They referred to it by the term *intricacy*. The term *intricate* graphs has already been used by some other authors.[267]

The spanning-tree density of a simple (molecular) graph G, denoted by *STD*, is defined as:

$$STD = t / {}^{E}C_{N-1} \qquad (29)$$

where ${}^{E}C_{N-1}$ is the number of ways of choosing any $N-1$ edges from the E available edges in G. *STD* represents the *probability* that if any set of $N-1$ edges in G is selected, and if the remnant set of $(E-N+1)$ edges in G is deleted, the resulting entity is a spanning tree.

In order to deal with numbers that are greater than 1, Mallion and Trinajstić[266] introduced the reciprocal spanning-tree density, denoted by *RSTD*, as:

$$RSTD = {}^{E}C_{N-1}/t \qquad (30)$$

The criterion of structural intricacy is – the bigger *RSTD* is, the more intricate the structure is.

For some classes of graphs *RSTD* can be given in closed form. For example, formulas for a complete graph K_N (that is, a regular graph G with N vertices and $r = N-1$) and a complete bipartite graph $K_{m,n}$ (that is, a complete graph which is made up of two sets of vertices — m and n — such that each vertex in the m set is joined to all and only to vertices in the n set; note that $N = m + n$) are given as:

$$RSTD(K_N) = {}^{(1/2)N(N-1)}C_{N-1} / N^{N-2} \qquad (31)$$

$$RSTD(K_{m,n}) = {}^{mn}C_{m+n-1} / m^{n-1} \times n^{m-1} \qquad (32)$$

Examples of the complete graph and the complete bipartite graphs are the *Kuratowski graph* K_5 and *Utilities graph* $K_{3,3}$, respectively. These graphs have been used to characterize planar graphs – a graph is planar if and only if it has *no* subgraph homeomorphic to K_5 or $K_{3,3}$.[268] The Kuratowski graph was used, for example, in the study of the rearrangement of tetragonal-pyramidal complexes,[269] whilst the Utility graph is grounded in a puzzle discussed by Dudeney in 1917 is his book *Amusements in Mathematics*[270] in which he collected a number of mathematically based puzzles. Dudeney presented a puzzle which can be visualized by the Utility graph under the title *Water, Gas and Electricity*, hence the name of the graph. The puzzle is to lay on water, gas and electricity pipes from three sources to each of three houses without any pipe crossing another. Also K_5 and $K_{3,3}$ were used in the discussion of the topological chirality of proteins.[271] A graph is topologically chiral if its embedding in Euclidean 3-space cannot be converted to its mirror image by a continuous deformation which avoids edge intersections. Nonplanarity is a necessary condition for topological chirality because a planar graph is achiral in 3-space.[272]

The intricacies of both K_5 ($RTSD = 1.68$) and $K_{3,3}$ ($RSTD = 1.56$) graphs are similar according to the *RSTD* criterion. However, the corresponding numbers of spanning trees are rather different: $t(K_5) = 125$ and $t(K_{3,3}) = 81$. The Kuratowski graph and the Utility graph are depicted in Figure 6.

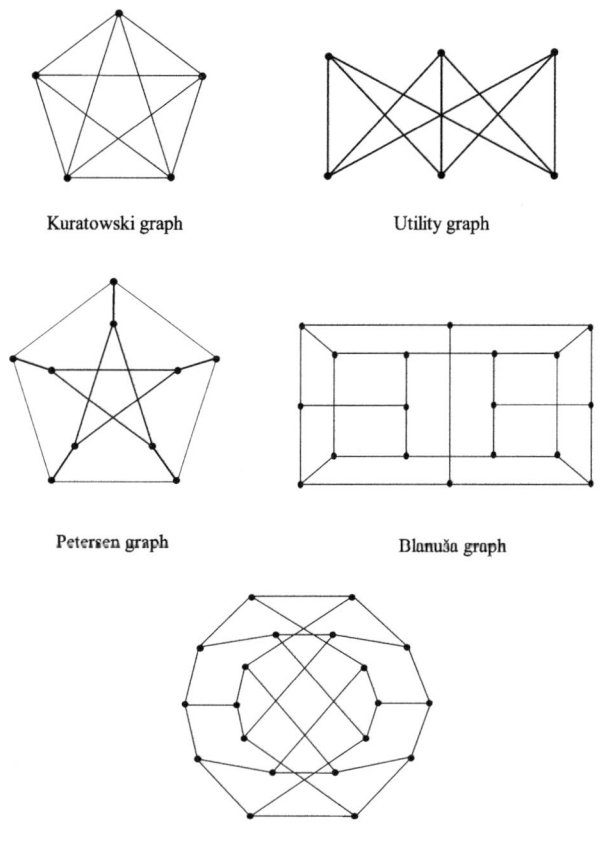

Figure 6 *A collection of chemically-important graphs*

Mallion and Trinajstić[266] calculated the values of *RSTD* for a number of interesting regular graphs of degree 3 which have found application in chemistry, such as: the Petersen graph[273] ($RSTD = 2.50$; $t = 2000$) (which depicts possible routes for the isomerisation of trigonal-bipyramidal complexes[274–276]); the Blanuša graph[277] ($RSTD = 3.01$; $t = 1037136$), which can be obtained by suitably combining two copies of the Petersen graph (the Blanuša graph has been introduced in the context of the four-color problem[4]); the Desargues-Levy graph[278] ($RSTD = 8.89$; $t = 6144000$) (which has found application in describing isomerisations, rearrangements[275,279–282] and nucleophilic displacement reactions[283]); and Schlegel graphs[284] corresponding to C_{60} ($RSTD = 3500$; $t = 3.75 \times 1020$) and C_{70} ($RSTD = 15000$; $t = 1.14 \times 1024$) fullerenes (fullerenes probably being the most interesting class of chemical compounds which have emerged in the last two decades of the 20th century, when wide-spread research began on fullerene chemistry starting with the discovery of the first fullerene C_{60}, named buckminsterfullerene[285]). The Petersen graph, the Blanuša graph and the Desargues-Levy graph are also given in Figure 6, while C_{60} and C_{70} fullerenes and their Schlegel graphs are depicted in Figure 7.

Mallion and Trinajstić[266] also considered Schlegel graphs of Platonic solids: the tetrahedron ($RSTD = 1.25$; $t(16)$; the cube ($RSTD = 2.06$; $t = 384$); the octahedron ($RSTD = 2.06$; $t = 384$); the icosahedron ($RSTD = 10.54$; $t = 5184000$); and the dodecahedron ($RSTD = 10.54$; $t = 5184000$). A number of chemical structures possessing skeletons resembling Platonic solids are known, *e.g.* refs. 276, 286. The reason why the *RSTD*- and *t*-values of the pair consisting of the

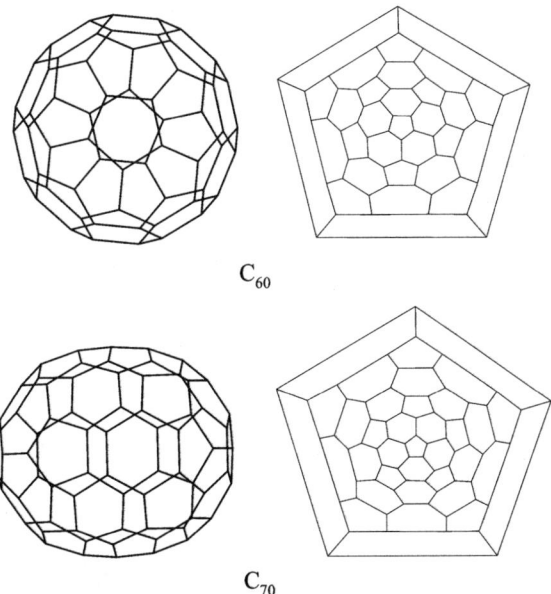

Figure 7 C_{60} and C_{70} fullerenes and their Schlegel graphs

cube and the octahedron as well as of the pair of the icosahedron and dodecahedron are the same is that each of these two pairs are duals[4] to each other.[287] The tetrahedron is its own dual.

2.5 Other Enumerations. – Cash and Dias,[288] in a continuation of Dias' work on developing methods for calculating eigenvectors, eigenvalues and resonance-structure counts for benzenoid radicals and diradicals, studied polyradicals. The resonance-structure counts of benzenoid polyradicals can be determined from the tail coefficients of their matching polynomials.[289] Coefficients of the matching polynomial are made up from only acyclic contributions and this polynomial is a key concept in the topological resonance energy (TRE).[290-292] The TRE theory is nowadays accepted by many as a reliable theory of aromaticity.[293]

Cash and Dias[288] used *Mathematica*®[294] to get the matching polynomials. In their calculation, they utilized a program by Salvador *et al.*[295] and the fragmentation procedure in the case of large structures by Babić *et al.*[296] The structure counts are large numbers for condensed benzenoid mono- and diradicals, but they become huge numbers in polyradicals. For example, the following are structure counts (given in brackets for only one diradical isomer of each formula) for the polycircumtriangulene diradical series: $C_{22}H_{12}$ (306), $C_{37}H_{18}$ (999), $C_{73}H_{21}$ (189792), $C_{121}H_{27}$ (153915460), whilst for the tetraradical series: $C_{46}H_{18}$ (273956), $C_{88}H_{24}$ (604073400), $C_{142}H_{30}$ (3859415491248). Dias[297,298] and Dias and Cash[299] have made further resonance-structure counts for different classes of benzenoid radicals.

Hosoya *et al.*[300] in continuation of their work on topological properties of polyhedral graphs, searched for twin graphs in polyhedral graphs with nine and ten vertices. Graphs corresponding to polyhedra are called polyhedral graphs.[5] Thence polyhedral graphs are planar, and their planar representations are often termed Schlegel diagrams.[284] Hosoya and co-workers[301] have previously defined topological twin graphs as pairs of highly similar nonisomorphic graphs that are isospectral with respect not only to the characteristic polynomial,[5,52,116] but also to the distance polynomial,[240] matching polynomial,[289] the Z-counting polynomial[302] and have identical values of the Wiener index,[303] Z-index[302] and many other topological indices.[5,52,184,256]

The polynomial $P(x) = |\lambda I - A|$ of the vertex-adjacency matrix A of a graph G is the *characteristic* polynomial of G. The set of eigenvalues of A (*i.e.*, the zeros of $P(x)$) is called the *spectrum* of G. Isospectral graphs[5,304-307] (sometimes also called cospectral graphs[4,308]) are graphs with *identical* characteristic polynomials and consequently *identical* spectra. Hosoya and his co-workers[300,301] prepared their own computer program to generate and enumerate polyhedral graphs. With this program they reproduced the known numbers of distinct polyhedral graphs with up to eleven vertices[309] and obtained the number of polyhedral graphs with 12 vertices (6384636 graphs). In Hosoya *et al*'s paper,[300] isospectral pairing was checked for the polyhedral graphs with nine vertices (2606 graphs) and ten vertices (32300 graphs). Among 2606 polyhedral graphs with nine vertices they found 53 isospectral pairs, one isospectral triplet and one isospectral quartet. They are the smallest isospectral triplet and quartet among the polyhedral graphs. Among 32300 polyhedral graphs with ten vertices they found many more isospectral pairs (622), nine isospectral triplets and two isospectral quartets. The isospectral pairs of polyhedral graphs with nine and ten vertices give rise to three and 15 pairs of topological twins. In Figure 8 we depict three pairs of the topological twin polyhedral graphs with nine vertices.

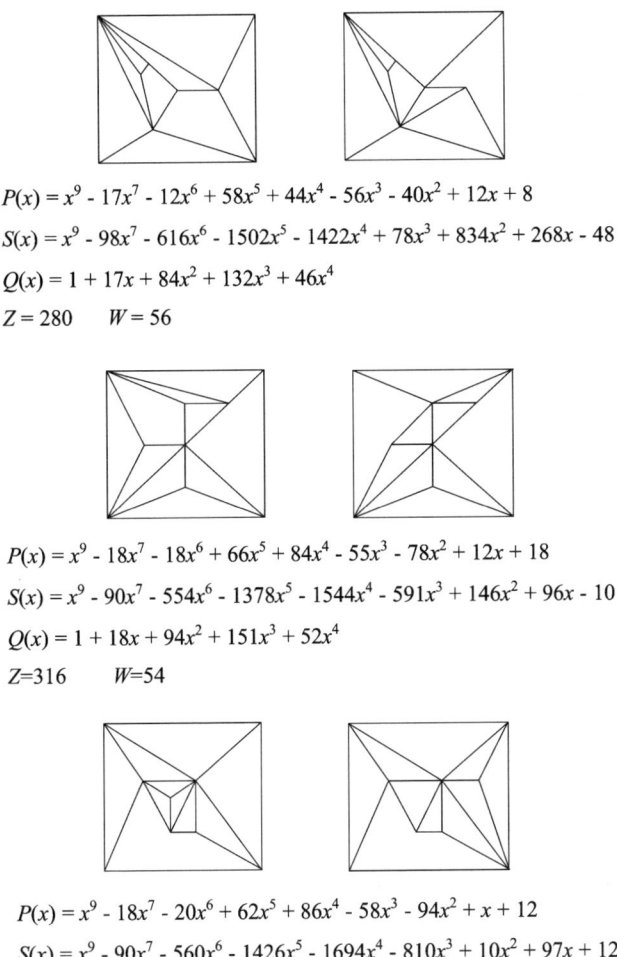

$P(x) = x^9 - 17x^7 - 12x^6 + 58x^5 + 44x^4 - 56x^3 - 40x^2 + 12x + 8$

$S(x) = x^9 - 98x^7 - 616x^6 - 1502x^5 - 1422x^4 + 78x^3 + 834x^2 + 268x - 48$

$Q(x) = 1 + 17x + 84x^2 + 132x^3 + 46x^4$

$Z = 280 \quad W = 56$

$P(x) = x^9 - 18x^7 - 18x^6 + 66x^5 + 84x^4 - 55x^3 - 78x^2 + 12x + 18$

$S(x) = x^9 - 90x^7 - 554x^6 - 1378x^5 - 1544x^4 - 591x^3 + 146x^2 + 96x - 10$

$Q(x) = 1 + 18x + 94x^2 + 151x^3 + 52x^4$

$Z = 316 \quad W = 54$

$P(x) = x^9 - 18x^7 - 20x^6 + 62x^5 + 86x^4 - 58x^3 - 94x^2 + x + 12$

$S(x) = x^9 - 90x^7 - 560x^6 - 1426x^5 - 1694x^4 - 810x^3 + 10x^2 + 97x + 12$

$Q(x) = 1 + 18x + 92x^2 + 148x^3 + 53x^4$

$Z = 312 \quad W = 54$

Figure 8 *Schlegel diagrams of the three pairs of the topological twin polyhedral graphs with nine vertices taken from ref. 300. $P(x)$ denotes the characteristic polynomial, $S(x)$ the distance polynomial, $Q(x)$ the Z-counting polynomial, Z is the Hosoya Z-index and W the Wiener index*

Rücker and Meringer[310] attempted to answer the following question: *How many organic compounds are graph-theoretically nonplanar?* The term *graph-theoretically nonplanar* means that the structural formula of such a compound cannot be drawn in a plane without at least two bonds crossing. It is interesting to note that most known compounds, even those that are geometrically quite clearly nonplanar such as helicenes, fullerenes and many other polycyclic compounds, correspond to planar graphs, *i.e.*, their structural formulas (considerably distorted, if necessary – see, for example, in Figure 7 Schlegel diagrams of C_{60} and C_{70} fullerenes and in Figure 8 Schlegel diagrams of six polyhedral graphs with nine vertices) can be drawn on a sheet of paper without two bonds crossing each other.

A justly famous graph-theoretical planarity criterion has been introduced and proved by Kuratowski.[268] This states that a graph is non-planar iff it has no subgraph homeomorphic to either K_5 or $K_{3,3}$. However, graph-theoretical (*gt*) planarity tests available as algorithms or computer programs do not use the Kuratowski criterion – these tests are based on the trials to embed the structure in a plane without crossing of edges, a procedure more efficient than finding subgraphs of unknown structure.[195] (That is, there are many possible subgraphs which are homeomorphic to either K_5 or $K_{3,3}$.) Rücker and Meringer[310] used the Lempel-Even-Cederbaum planarity test[311] written in C++ (a part of the Graph Template Library (GTL)[312]). They first generated, using computer program MOLGEN 4.0,[313] all connected simple undirected graphs of nine vertices ($N=9$) and tested them for *gt*-(non)planarity in classes of constant edge-numbers E, that have also constant numbers of cycles $C(=E-N+1)$. All graphs with $C \leq 3$ were found to be planar, all those of $C \geq 14$ were found to be nonplanar, and for $4 \leq C \leq 13$ the percentage of planar graphs among all such graphs monotonically decreases from 99.09 to 0.23%.

Next they performed a corresponding test on all simple 4-graphs (graphs with no vertex-degree higher than four – graphs that model carbon compounds) with $N=12$ vertices, in classes of constant E or C. And again, they found that all graphs with $C \leq 3$ were found to be planar, whilst for the classes with $4 \leq C \leq 13$ the percentage of planar graphs monotonically decreases from 98.80 to 0.84%.

Rücker and Meringer[310] performed two more tests – they considered all simple connected graphs with up to 12 vertices and all simple 4-graphs with up to 12 vertices. In the first case they found that beginning with $N=9$ the majority of simple connected graphs are nonplanar and in the second case that from $N=12$ the majority of simple 4-graphs are also nonplanar. Finally, Rücker and Meringer[310] considered all compounds in the MDL Beilstein file. Because of several technical, legal and financial obstacles, the *gt*-planarity test was carried out by D. Hounshell of MDL who ran the test on all Beilstein compounds in the MDL facilities (MDL, San Leandro, California). The *gt*-planarity test did not uncover compounds whose *gt*-nonplanarity was hitherto unknown. Rücker and Meringer[310] also tested peptides/proteins since it is known that *gt*-nonplanar peptides/proteins do exist.[271,314–316] *gt*-Nonplanar peptides/proteins were retrieved from the CAS Registry file. They found that a very few proteins are *gt*-nonplanar.

Barone et al.[317] presented a program to generate a virtual library of structures. They named their program GASP (Génération Automatique de Structures Polycycliques). The authors claim that their program can generate all polycyclic structures with a given number of rings of given sizes. It works like the chemist who draws a ring, then adds another ring and so on. GASP starts with a ring, *i.e.*, its connection (connectivity) table,[318] then it adds the second ring in all possible positions, then, on each generated structure it adds the third ring and so on. The authors presented the results obtained for several structures (steroid, taxane and triquinane skeletons). For example, in the case of steroid-like skeleton (4 rings of size 6,6,6,5) GASP generates 988 polycyclic structures – among them there are 48 structures in which all 4 rings are fused. Comparison with the Beilstein database (http://www.beilstein.com) revealed that among these 48 structures there are 21 new structures. Barone et al.[317] were induced to publish GASP by a paper of Brinkmann et al.[319] in which these authors produced an algorithm for fast constructive enumeration of polycyclic chains with arbitrary ring sizes. The algorithm has been implemented as a computer program named *chains.c*. Computations have been carried out on Pentium II 350 Mhz. For example, there are 1681 polycyclic chains with 10 rings of size six (time needed to get this number was 0.02 seconds). If one has 10 rings such that five are of size 5 and five of size 6, then there are 391251 possible polycyclic chains (time needed for this number was 1.58 seconds).

3 Concluding Remarks

Combinatorial enumeration has played an important role in the history of chemistry.[5,320–322] With the development and applications of combinatorial chemistry and (virtual and synthetic) combinatorial libraries catalysed by the development of powerful personal computers and portable software, combinatorial enumeration has become a chemist's, a biochemist's and a medicinal chemist's tool of everyday use.[2,3,323] Therefore, it is beyond the point to discuss the usefulness of combinatorial enumeration in chemical, biochemical and medicinal research. Nevertheless, we wish to point to some recent achievements that were possible due to combinatorial enumeration.

Lahana and coworkers[324] reported the rational design of immunosuppressive peptides without relying on information regarding their receptors or mechanisms of action. The design strategy was based on a virtual combinatorial library and the use of a variety of topological and shape descriptors with an analysis of molecular dynamics trajectories for the identification of potential drug candidates. Lahana et al.[324] generated the virtual combinatorial library for the peptides of the general form RXXXRXXXXY. This sequence has seven positions to mutate in order to create the library. The use of 35 amino acids – 20 natural and 15 unnatural – leads to 35^7 combinations (64 billion compounds), well above the present computing power. Amino acids were characterized by their physical and chemical properties and also by their topological indices. The number of 35^7 compounds in the library was reduced to 6^7 (279936) compounds by taking into account lipophilicity distribution, considered critical for the bioactivity studied. Screening the library of 279936 compounds, using two types of filters: static and dynamic, resulted in identification of 26 peptides satisfying all constraints. The bioactivity of these peptides was tested in a heterotopic mouse heart allograft model.[325] The compound predicted to be the most potent displayed an immunosuppressive activity about 100 times higher than the lead compound. This kind of approach has become indispensable in drug design research, the first step being to build the combinatorial library.

Until about 1992 combinatorial libraries were almost entirely restricted to peptides and oligonucleotides,[2] but in the last decade they have been extended to many kinds of compounds.[326,327] In this respect there is a very instructive recent article by Storm and Lüning[328] on using virtual combinatorial libraries for the efficient preparation of macrocycles. However, combinatorial libraries are still predominantly used in drug research.[323,329]

Important research related to combinatorial libraries concerns library screening.[330] Virtual screening of a combinatorial library may accelerate drug lead discovery by selecting subsets of compounds according to their structural similarity or dissimilarity toward compound collections of specific bioactivity. One way to virtually screen the combinatorial library is by means of structural descriptors. The process of virtual screening of a combinatorial library starts from a wide selection of reactants that are used to generate *in silico* a huge number of compounds, according to multicomponent reaction.[331] In the next step, for each chemical compound a comprehensive set of structural descriptors is calculated, followed by a dimensionality reduction by selecting from the descriptors set a chemical space that is relevant for the investigated biological target. Finally, the compounds for synthesis and the high-throughput screening are selected with a statistical algorithm that implements a similarity, diversity or drug-like paradigm. There are many kinds of molecular descriptors in use:[5,52,184,256,332–335] physicochemical or empirical (*e.g.*, logarithm of octanol-water partition coefficient, log P^{336}), constitutional (*e.g.*, number of hydrogen-bond donors, number of hydrogen-bond acceptors, number of rotatable bonds), graph invariants (e.g., cyclomatic number, walk counts), topological indices (e.g., Wiener index,[303] Hosoya index,[302] Zagreb indices,[337,338] Randić index,[201] Balaban index,[239] Harary index[339,340]), geometric (e.g., molecular volume, polar surface area), quantum (HOMO energy, atomic charges) and grid (various steric, electrostatic and lipophilic fields). Also there are rather general combinatorial questions concerning *data mining* and data manipulation in virtual combinatorial libraries. For a sampling of further recent references see.[341–346]

Ivanciuc and Klein[347] presented a simple and fast algorithm for building-block computation of a variety of Wiener-type indices for the virtual screening of very large combinatorial libraries. They short-cut the usual more detailed computational procedures for the indices for each individual compound. Also these two authors[348] have reported the computation of Wiener-type indices for virtual combinatorial libraries generated from the heteroatom-containing building-blocks. The approach is related to that used by Bytautas and Klein[349] in computing average Wiener numbers for immensely large (*i.e.*, $> 10^{24}$ structures). Ivanciuc[350] reported an efficient algorithm for the building-block computation of the Ivanciuc-Balaban indices[351] for the virtual screening of any large combinatorial library.

We should end our report by pointing to the motivation as to why Vollhardt et al.[352] synthesized bent [4]phenylene using their own words: *This compound is of interest because it constitutes the fifth and last [4]phenylene isomer* (here they refer to the paper by Gutman *et al.*[353] on the enumeration of the isomers of phenylenes), *and its preparation allows the completion of an experimental analysis of the comparative properties of this group of topomers, including data based on calculations.* [R]phenylenes (R = the number of benzene rigs) represent a subclass of [4N]annuleno-[4N+2]annulenes in which (formally) benzene (aromatic) rings alternate with cyclobutadiene (anti-aromatic) rings.[354] This kind of the structural arrangement of [R]phenylenes results in a number of interesting properties.[355] Gutman *et al.*[353] enumerated the [R]phenylene $C_{6R}H_{2R+4}$ isomers with up to $R = 12$. There are five $C_{24}H_{12}$ isomers, 122 $C_{42}H_{18}$ isomers, and the number of phenylene isomers goes up quickly giving 101161 $C_{72}H_{28}$ isomers. The preparation of higher linear phenylenes would be of interest, among other things, to prove or disprove theoretical ideas[356] about possibility that these compounds possess conducting properties.

Thence it seems that combinatorial enumeration in chemistry continues its historical tradition, with an increasing degree of activity, to an ever wider range of applications.

Acknowledgement

DB and NT acknowledge the support by the Ministry of Science and Technology of the Republic of Croatia (Grant Nos. 0098034 and 0098057) and DJK acknowledges the support of the Welch Foundation of Houston, Texas.

References

1. D.J. Klein, D. Babić and N. Trinajstić, in: *Chemical Modelling – Applications and Theory*, ed. A. Hinchliffe, Specialist Periodical Reports, The Royal Society of Chemistry, Cambridge, UK, 2002, Vol. 2, pp. 56–95.
2. W.A. Warr, in: *Encyclopedia of Computational Chemistry*, editor-in-chief P. von R. Schleyer, Wiley, Chichester, 1998, pp. 407–417.
3. *Handbook of Combinatorial Chemistry*, eds. K.C. Nicolaou, R. Hanko and W. Hartwig, Wiley-VCH, Weinheim, 2002.
4. F. Harary, *Graph Theory*, Addison-Wesley, Reading. MA, 2nd printing, 1971.
5. N. Trinajstić, *Chemical Graph Theory*, CRC Press, Boca Raton, FL, 2nd revised Edn., 1992.
6. J. von Knop, K. Szymanski, W.R. Müller, H.W. Kroto and N. Trinajstić, *J. Comput. Chem.*, 1987, **8**, 549–554.

7. J. von Knop, W.R. Müller, Ž. Jeričević and N. Trinajstić, *J. Chem. Inf. Comput. Sci.*, 1981, **21**, 91.
8. N. Trinajstić, S. Nikolić, J. von Knop, W.R. Müller, and K. Szymanski, *Computational Chemical Graph Theory: Characterization, Enumeration and Generation of Chemical Structures by Computer Methods*, Ellis Horwood Ltd., Simon & Schuster, Chichester, 1991.
9. I. Gutman, W. Linert, I. Lukovits and A.A. Dobrynin, *J. Chem. Inf. Comput. Chem.*, 1997, **37**, 349.
10. T. Došlić, *Croat. Chem. Acta*, 2002, **75**, 881.
11. A. Cayley, *Quart. J. Math.*, 1889, **23**, 376.
12. I. Lukovits and I. Gutman, *Croat. Chem. Acta*, 2002, **75**, 563.
13. I. Lukovits, *J. Chem. Inf. Comput. Sci.*, 1999, **39**, 563.
14. H.L. Morgan, *J. Chem. Doc.*, 1965, **5**, 107; A simplified interpretation of the Morgan indexing algorithm was devised in the following paper: V. Kvasnička and J. Pospichal, *J. Chem. Inf. Comput. Sci.*, 1990, **30**, 99.
15. I. Lukovits, *J. Chem. Inf. Comput. Sci.*, 2000, **40**, 361.
16. I. Lukovits, in: *Topology in Chemistry: Discrete Mathematics of Molecules*, eds. D.H. Rouvray and R.B. King, Horwood Publishing Ltd, Chichester, 2002, pp. 327–337.
17. D. Veljan, *Combinatorial and Discrete Mathematics*, Algorithm, Zagreb, 2001, p. 134.
18. D.I.A. Cohen, *Basic Techniques of Combinatorial Theory*, Wiley, New York, 1978, pp. 121–132.
19. R. Pauncz, *Alternant Molecular Orbital Method*, Saunders, Philadelphia, 1967.
20. T. Yamanouchi, *Proc. Physico-Math. Soc. Japan*, 1936, **18**, 436.
21. L. Pogliani, M. Randić and N. Trinajstić, *Int. J. Math. Educ. Sci. Technol.*, 2000, **31**, 811.
22. M. Randić, *J. Chem. Inf. Comput. Sci.*, 1986, **26**, 136.
23. M. Randić, *Croat. Chem. Acta*, 1986, **59**, 327.
24. M. Randić, S. Nikolić and N. Trinajstić, *J. Mol. Struct. (Theochem)*, 1988, **165**, 213.
25. S. Nikolić and N. Trinajstić, *Croat. Chem. Acta*, 1990, **63**, 155.
26. J. von Knop, W.R. Müller, K. Szymanski, S. Nikolić and N. Trinajstić, in: *Computational Chemical Graph Theory*, ed. D.H. Rouvray, Nova Sci. Publ., Commack, N.Y., 1990, pp. 9–32.
27. M. Randić, S. Nikolić and N. Trinajstić, *J. Chem. Inf. Comput. Sci.*, 1995, **35**, 357.
28. M.L. Contreras, J. Alvarez, M. Riveros, G. Arias and R. Rozas, *J. Chem. Inf. Comput. Sci.*, 2001, **41**, 964.
29. S. Davidson, *J. Chem. Inf. Comput. Sci.*, 1989, **29**, 151.
30. S. Davidson, *J. Chem. Inf. Comput. Sci.*, 1991, **31**, 417.
31. S. Davidson, *J. Chem. Inf. Comput. Sci.*, 1992, **32**, 215.
32. S. Davidson, *J. Chem. Inf. Comput. Sci.*, 2002, **42**, 147.
33. L. Bytautas and D.J. Klein, *J. Chem. Inf. Comput. Sci.*, 1998, **38**, 1063.
34. H.R. Henze and C.M. Blair, *J. Am. Chem. Soc.*, 1931, **53**, 3042.
35. H.R. Henze and C.M. Blair, *J. Am. Chem. Soc.*, 1931, **53**, 3077.
36. C.M. Blair and H.R. Henze, *J. Am. Chem. Soc.*, 1932, **54**, 1098.
37. C.M. Blair and H.R. Henze, *J. Am. Chem. Soc.*, 1932, **54**, 1538.
38. D.D. Coffman, C.M. Blair and H.R. Henze, *J. Am. Chem. Soc.*, 1933, **55**, 252.
39. H.R. Henze and C.M. Blair, *J. Am. Chem. Soc.*, 1933, **55**, 680.
40. H.R. Henze and C.M. Blair, *J. Am. Chem. Soc.*, 1934, **56**, 157.

41. C.-W. Lam, *J. Math. Chem.*, 1998, **23**, 421.
42. C.-W. Lam, *J. Math. Chem.*, 2000, **27**, 23.
43. C.-W. Lam, *J. Math. Chem.*, 2002, **31**, 333.
44. J. Wang, R. Li and S. Wang, *J. Math. Chem.*, 2003, **33**, 171.
45. J. von Knop, W.R. Müller, K. Szymanski and N. Trinajstić, *Computer Generation of Certain Classes of Molecules*, SKTH–Kemija u industriji, Zagreb, 1985, pp. 81–116.
46. I. Gutman and S.J. Cyvin, *Introduction to the Theory of Benzenoid Hydrocarbons*, Springer-Verlag, Berlin, 1989, pp. 33–49.
47. S.W. Golomb, *Polyominoes*, Scribner, New York, 1965.
48. W.F. Lunnon, in: *Graph Theory and Computing*, ed. R.C. Read, Academic Press, New York, 1972, pp. 87–100.
49. F. Harary, in: *Graph Theory and Theoretical Physics*, ed. F. Harary, Academic Press, London, 1967, pp. 1–41.
50. D.A. Klarner, *Can. J. Math.*, 1967, **19**, 851.
51. A.T. Balaban and F. Harary, *Tetrahedron*, 1968, **24**, 2505.
52. I. Gutman and O.E. Polansky, *Mathematical Concepts in Organic Chemistry*, Springer-Verlag. Berlin, 1986, pp. 59–62.
53. W.R. Müller, K. Szymanski, J. von Knop, S. Nikolić and N. Trinajstić, *J. Comput. Chem.*, 1990, **11**, 223.
54. N. Trinajstić, *J. Math. Chem.*, 1990, **5**, 171.
55. K. Balasubramanian, J.J. Kauffman, W.S. Koski and A.T. Balaban, *J. Comput. Chem.*, 1980, **1**, 149.
56. J. von Knop, K. Szymanski, Ž. Jeričević and N. Trinajstić, *J. Comput. Chem.*, 1983, **4**, 23.
57. J. von Knop, W.R. Müller, K. Szymanski, and N. Trinajstić, *J. Chem. Inf. Comput. Sci.*, 1990, **30**, 159.
58. D. Mašulović, R. Tošić, B.N. Cyvin and S.J. Cyvin, *MATCH Commun. Math. Comput. Chem.*, 1993, **29**, 165.
59. R. Tošić, D. Mašulović, I. Stojmenović, J. Brunvoll, B.N. Cyvin and S.J. Cyvin, *J. Chem. Inf. Comput. Sci.*, 1995, **35**, 181.
60. G. Caporossi and P. Hansen, *J. Chem. Inf. Comput. Sci.*, 1998, **38**, 610.
61. F. Chyzak, I. Gutman and P. Paule, *MATCH Commun. Math. Comput. Chem.*, 1999, **40**, 139.
62. G. Brinkmann, G. Caporossi and P. Hansen, *MATCH Commun. Math. Comput. Chem.*, 2001, **43**, 133.
63. G. Brinkmann, G. Caporossi and P. Hansen, *J. Chem. Inf. Comput. Sci.*, 2003, **43**, 842.
64. M. Vöge, A.J. Guttmann and I. Jensen, *J. Chem. Inf. Comput. Sci.*, 2002, **42**, 456.
65. I. Gutman, *Bull. Chem. Technol. Macedonia*, 2002, **21**, 53.
66. R.G. Harvey, *Polycyclic Aromatic Hydrocarbons*, Wiley-VCH, New York, 1997.
67. S. Hagen and H. Hopf, *Topics Curr. Chem.*, 1998, **196**, 47.
68. M.D. Watson, A. Fechtenkötter and K. Müllen, *Chem. Rev.*, 2001, **101**, 1267.
69. M. Randić, *Chem. Rev.*, 2003, **103**, 3449.
70. e.g., P. Phillips, *Nature*, 2000, **406**, 687.
71. Y. Sritana-Anant, T.J. Seiders and J.S. Siegel, *Topics Curr. Chem.*, 1998, **196**, 1.
72. D.J. Klein and A. Misra, *MATCH Commun. Math. Comput. Chem.*, 2002, **46**, 45.
73. F.H. Kohnke, A.M. Slavin, J.F. Stoddart and D.J. Williams, *Angew. Chem.*, 987, **99**, 941.

74. K. Yamamoto, T. Harada, Y. Okamoto, H. Chikamatsu, M. Nakazaki, Y. Kai, T. Nakao, M. Tanaka, S. Harda and N. Kasai, *J. Am. Chem. Soc.*, 1988, **119**, 3578.
75. A. Borchardt, A. Fuchicello, K.V. Kilway, K.K. Baldridge and J.S. Siegel, *J. Am. Chem. Soc.*, 1992, **114**, 1921.
76. R.M. Cory and C.L. McPhail, *Tetrahedron Lett.*, 1996, **37**, 1987.
77. O. Kintzel, P. Luger, M. Weber and A.-D. Schluter, *Eur. J. Org. Chem.*, 1998, 99.
78. W.S. Burnside, *Theory of Groups of Finite Order*, University Press, Cambridge, 2nd Edn., 1911.
79. W. Hässelbarth, *Theoret. Chim. Acta*, 1984, **66**, 91.
80. C.A. Mead, *J. Am. Chem. Soc.*, 1987, **109**, 2130.
81. J. Brocas, M. Gielen and R. Willem, *The Permutational Approach to Dynamic Stereochemistry*, McGraw-Hill, New York, 1983.
82. S. Fujita, *Theoret. Chim. Acta*, 1989, **76**, 247.
83. S. Fujita, *Symmetry and Combinatorial Enumeration in Chemistry*, Springer-Verlag, Berlin, 1991.
84. S. El-Basil, *Combinatorial Organic Chemistry – An Educational Approach*, Nova Sci. Publ., Huntington, N.Y., 1999.
85. G. Pólya, *Acta Math.*, 1937, **68**, 145; translation of this classical paper of combinatorial analysis appeared in G. Pólya and R.C. Read, *Combinatorial Enumeration of Groups, Graphs and Chemical Compounds*, Springer-Verlag, New York, 1987.
86. S. El-Basil, *MATCH Commun. Math. Comput. Chem.*, 2002, **46**, 7.
87. H. Dolhaine and H. Hönig, *MATCH Commun. Math. Comput. Chem.*, 2002, **46**, 71.
88. H. Dolhaine and H. Hönig, *MATCH Commun. Math. Comput. Chem.*, 2002, **46**, 91.
89. M. van Almsick, H. Dolhaine and H. Hönig, *MATCH Commun. Math. Comput. Chem.*, 2001, **43**, 153.
90. T. Hudlicky, K.A. Abboud, D.A. Entwistle, R. Fan, R. Maurya, A.J. Thorpe, J. Balonick and B. Myers, *Synthesis*, 1996, 897.
91. T. Hudlicky, K.A. Abboud, J. Balonick, R. Maurya, M.L. Stanton and A.J. Thorpe, *Chem. Comm.*, 1996, 1717.
92. H. Dolhaine, H. Hönig and M. van Almsick, *MATCH Commun. Math. Comput. Chem.*, 1999, **39**, 21.
93. e.g., K. Drickamer and J. Carver, *Curr. Opin. Struct. Biol.*, 1992, **2**, 653.
94. A. Kerber, *MATCH Commun. Math. Comput. Chem.*, 2002, **46**, 151.
95. A. Kerber, *Applied Finite Group Action*, Springer-Verlag, Berlin, 1998.
96. W. Hässelbarth and A. Kerber, *MATCH Commun. Math. Comput. Chem.*, 2002, **46**, 199.
97. T. Carrell, E.A. Wintner, A. Bashir-Hashemi and J. Rebek, Jr., *Angew. Chem. Int. Ed. Engl.*, 1994, **33**, 2059.
98. T. Carrell, E.A. Wintner, A.J. Sutherland, J. Rebek, Jr., Y.M. Dunayevskiy and P. Vouros, *Chem. Biol.*, 1995, **2**, 171.
99. M. van Almsick, H. Dolhaine and H. Hönig, *J. Chem. Inf. Comput. Sci.*, 2000, **40**, 956.
100. W. Hässelbarth, *Theor. Chim. Acta*, 1985, **67**, 339.
101. R.M. Nemba and A.T. Balaban, *MATCH Commun. Math. Comput. Chem.*, 2002, **46**, 235.
102. R.W. Robinson, F. Harary and A.T. Balaban, *Tetrahedron*, 1976, **32**, 355.

103. S. Fujita and S. El-Basil, *MATCH Commun. Math. Comput. Chem.*, 2002, **46**,121.
104. E.K. Lloyd, *MATCH Commun. Math. Comput. Chem.*, 2002, **46**, 215.
105. E.K. Lloyd and H. Dolhaine, *MATCH Commun. Math. Comput. Chem.*, 2002, **46**, 137.
106. S. Fujita, *Helv. Chim. Acta*, 2002, **85**, 2440.
107. R. Pauncz, *The Symmetric Group in Quantum Chemistry*, CRC Press, Boca Raton, FL, 1995.
108. S. Fujita, *J. Math. Chem.*, 2003, **33**, 113.
109. S. Fujita, *J. Am. Chem. Soc.*, 1990, **112**, 3390.
110. S. Cahn, C.K. Ingold and V. Prelog, *Angew. Chem. Int. Ed. Engl.*, 1966 **5**, 385.
111. S. Fujita, *Bull. Chem. Soc. Jpn.*, 2001, **74**, 1585.
112. S. Fujita and S. El-Basil, *J. Math. Chem.*, 2003, **33**, 255.
113. F. Zhang and W. Yan, *MATCH Commun. Math. Comput. Chem.*, 2002, **48**, 117.
114. W.C. Shiu, P.C.B. Lam, F. Zhang and H. Zhang, *J. Math. Chem.*, 2002, **31**, 405.
115. F. Torrens, *Int. J. Quantum Chem.*, 2002, **88**, 392.
116. D.M. Cvetković, M. Doob and H. Sachs, *Spectra of Graphs - Theory and Application*, 3rd revised & enlarged Edn., Barth-Verlag, Heidelberg, 1995.
117. L. Lovász and M.D. Plummer, *Matching Theory*, North-Holland, Amsterdam, 1986.
118. P.E. John, H. Sachs and M. Zheng, *J. Chem. Inf. Comput. Sci.*, 1995, **35**, 1019.
119. S.J. Cyvin and I. Gutman, *Kekulé Structures in Benzenoid Hydrocarbons*, Springer-Verlag, Berlin, 1988.
120. J.R. Dias, *Handbook of Polycyclic Hydrocarbons. Part A Benzenoid Hydrocarbons*, Elsevier, Amsterdam, 1987.
121. A. Misra and D.J. Klein, *J. Chem. Inf. Comput. Sci.*, 2002, **42**, 1171.
122. S.J. Cyvin, B.J. Cyvin, J. Brunvoll, H. Hosoya, F. Zhang, D.J. Klein, R. Chen and O.E. Polansky, *Monat. Chem.*, 1991, **122**, 435.
123. H.A. Staab and F. Diederich, *Chem. Ber.*, 1983, **116**, 3487.
124. D.J. Klein, T.G. Schmalz, G.E. Hite, A. Metropoulos and W.A. Seitz, *Chem. Phys. Lett.*, 1985, **120**, 367.
125. D.J. Klein, T.G. Schmalz and G.E. Hite, *J. Comput. Chem.*, 1986, **7**, 443.
126. D. Babić and A. Graovac, *Croat. Chem. Acta*, 1986, **59**, 731.
127. D.J. Klein, T.P. Živković and N. Trinajstić, *J. Math. Chem.*, 1987, **1**, 309.
128. A.N. Andriotis, M. Menon and D. Srivastava, *J. Chem. Phys.*, 2002, **117**, 2836.
129. G.M. Badger, *The Structure and Reactions of the Aromatic Compounds* University Press, Cambridge, 1954, pp. 89–90.
130. J.C. Dobrowolski, *J. Chem. Inf. Comput. Sci.*, 2002, **42**, 490.
131. J.K. Percus, *J. Math. Phys.*, 1969, **10**, 1881.
132. D. Cvetković, I. Gutman and N. Trinajstić, *Chem. Phys. Lett.*, 1972, **16**, 614.
133. D. Cvetković, I. Gutman and N. Trinajstić, *J. Chem. Phys.*, 1974, **61**, 2700.
134. N. Trinajstić, in: *Semiempirical Methods of Electronic Structure Calculation. Part A: Techniques*, ed. G.A. Segal, Plenum, New York, 1977, pp. 1–27.
135. H. Minc, *Permanents*, Addison-Wesley, Reading. MA, 1978.
136. G.G. Cash, *J. Math. Chem.*, 1995, **18**, 115.
137. I. Lukovits, A. Graovac, E. Kálmán, G. Kaptay, P. Nagy, S. Nikolić, J. Sytchev and N. Trinajstić, *J. Chem. Inf. Comput. Sci.*, 2003, **43**, 609.
138. S. Iijima, *Nature*, 1991, **354**, 5613.
139. R. Saito, G. Dresselhaus and M. Dresselhaus, *Physical Properties of Nanotubes*, Imperial College Press, London, 1998.

140. M. Terrones, W.-K. Hsu, H.W. Kroto and D.R.M. Walton, *Topics Curr. Chem.*, 1999, **199**, 189.
141. P. Avouris, *Acc. Chem. Res.*, 2002, **35**, 1026.
142. D.J. Klein, W.A. Seitz and T.G. Schmalz, *J. Phys. Chem.*, 1993, **97**, 1231.
143. G. Kaptay, I. Sytchev, J. Miklósi, P. Póczik, K. Papp, P. Nagy and E. Kálmán, in: *Advances in Molten Salts. From Structural Aspects to Waste Processing*, ed. M. Gaun-Escard, Begell, New York, 1999, pp. 257–262.
144. L. Gherghel, C. Kübel, G. Lieser, H.-J. Räder and K. Müllen, *J. Am. Chem. Soc.*, 2002, **124**, 13130.
145. C.N.R. Rao and A. Govindaraj, *Acc. Chem. Res.*, 2002, **35**, 998.
146. H. Dai, *Acc. Chem. Res.*, 2002, **35**, 1035.
147. L.P. Biro, Z.E. Horváth, L. Szalmas, K. Kertész, F. Wéber, G. Juhász, G. Radnóczi and J. Gyulai, *Chem. Phys. Lett.*, 2003, **372**, 399.
148. C. Velasco-Santos, A.L. Martínez-Hernández, A. Consultchi, R. Rodríguez and V.M. Castano, *Chem. Phys. Lett.*, 2003, **373**, 272.
149. A. Hirsch, *Angew. Chem. Int. Ed. Engl.*, 2002, **41**, 1853.
150. E.C. Kirby, *Croat. Chem. Acta*, 1993, **66**, 13.
151. H. Sachs, P. Hansen and M. Zheng, *MATCH Commun. Math. Comput. Chem.*, 1996, **33**, 169.
152. D.J. Klein and H. Zhu, *Discrete Appl. Math.*, 1996, **67**, 157.
153. P.E. John, *Croat. Chem. Acta*, 1998, **71**, 435.
154. D.J. Klein, T.P. Živković and R. Valenti, *Phys. Rev. B*, 1991, **43**, 723.
155. R. Swinborne-Sheldrake, W.C. Herndon and I. Gutman, *Tetrahedron Lett.*, 1975, 755.
156. W.A. Seitz, D.J. Klein, T.G. Schmalz and M.A. Garcia-Bach, *Chem. Phys. Lett.*, 1985, **115**, 139.
157. S.E. Stein and R.L. Brown, *Carbon*, 1985, **23**, 105.
158. S.E. Stein and R.L. Brown, *J. Am. Chem. Soc.*, 1987, **109**, 3721.
159. D.J. Klein, *Chem. Phys. Lett.*, 1994, **217**, 261.
160. D.J. Klein and L. Bytautas, *J. Phys. Chem.*, 1999, **93**, 5196.
161. O. Ivanciuc, L. Bytautas and D.J. Klein, *J. Chem. Phys.*, 2002, **116**, 4735.
162. O. Ivanciuc, D.J. Klein and L. Bytautas, *Carbon*, 2002, **40**, 2063.
163. M. Fujita, K. Wakabayashi, K. Nakada and K. Kusakabe, *J. Phys. Soc. Jpn*, 1996, **65**, 1920.
164. K. Wakabayashi, M. Fujita, K. Kusakabe and K. Nakada, *Czech. J. Phys.*, 1996, **46**, 1865.
165. K. Nakada, M. Fujita, B. Dresselhaus and M.S. Dresselhaus, *Phys. Rev. B*, 1996, **54**, 17954.
166. N. Tyutyulkov, G. Madjarova, F. Dietz and K. Müllen, *J. Phys. Chem. B*, 1998, **102**, 10183.
167. C.A. Coulson, M.A. Herraez, M. Leal, E. Santos and S. Senet, *Proc. Roy. Soc. (London) A*, 1963, **274**, 461.
168. C.A. Coulson and M.D. Poole, *Carbon*, 1964, **3**, 275.
169. K.H. Lee, M. Causá and S.S. Park, *J. Phys. Chem. B*, 1998, **102**, 6020.
170. M. Hjort and S. Stafström, *Phys. Rev. B*, 2000, **61**, 14089.
171. P.L. Giunta and S.P. Kelty, *J. Chem. Phys.*, 2001, **114**, 1807.
172. J.R. Hahn, H. Kang, S. Song and I.C. Jeon, *Phys. Rev. B*, 1996, **53**, 1725.
173. F. Atamny, O. Spillecke and R. Schlögl, *Phys. Chem. – Chem. Phys.*, 1999, **1**, 4113.

174. S.M. Lee, Y.H. Lee, Y.G. Hwang, J.R. Hahn and H. Kang, *Phys. Rev. Lett.*, 1999, **82**, 217.
175. M. Randić, *J. Comput. Chem.*, 1980, **1**, 386.
176. M. Barysz and N. Trinajstić, *Int. J. Quantum Chem.: Quantum Chem. Symp.* 1984, **18**, 661.
177. M. Razinger, *Theoret. Chim. Acta*, 1986, **70**, 365.
178. M. Randić, M. Barysz, J. Nowakowski, S. Nikolić and N. Trinajstić, *J. Mol. Struct. (Theochem)*, 1989, **185**, 96.
179. A.S. Shalabi, *J. Chem. Inf. Comput. Sci.*, 1991, **31**, 483.
180. G. Rücker and C. Rücker, *J. Chem. Inf. Comput. Sci.*, 1993, **33**, 683.
181. G. Rücker and C. Rücker, *J. Chem. Inf. Comput. Sci.*, 2000, **40**, 99.
182. S. Nikolić, N. Trinajstić, I.M. Tolić, G. Rücker and C. Rücker, in: *Complexity in Chemistry – Introduction and Fundamentals*, eds. D. Bonchev and D.H. Rouvray, Taylor & Francis, London, 2003, pp. 29–89.
183. G. Rücker and C. Rücker, *J. Chem. Inf. Comput. Sci.*, 1999, **39**, 788.
184. R. Todeschini and V. Consonni, *Handbook of Molecular Descriptors*, Wiley-VCH, Weinheim, 2000, pp. 480–486.
185. R.A. Marcus, *J. Chem. Phys.*, 1965, **43**, 2643.
186. Ref. 17, p. 251.
187. I. Gutman and N. Trinajstić, *Chem. Phys. Lett.*, 1973, **20**, 257.
188. J. von Knop, W.R. Müller, K. Szymanski, M. Randić and N. Trinajstić, *Croat. Chem. Acta*, 1983, **56**, 405.
189. G. Rücker and C. Rücker, *J. Chem. Inf. Comput. Sci.*, 1991, **31**, 422.
190. C. Rücker and G. Rücker, *J. Math. Chem.*, 1992, **9**, 207.
191. O. Ivanciuc and A.T. Balaban, *Croat. Chem. Acta*, 1996, **69**, 63.
192. I. Gutman, C. Rücker and G. Rücker, *J. Chem. Inf. Comput. Sci.*, 2001, **41**, 739.
193. M. Razinger, *Theoret. Chim. Acta* 1982, **61**, 581.
194. F. Harary and A.J. Schwenk, *Pacific J. Math.*, 1979, **80**, 443.
195. C. Rücker and G. Rücker, *J. Chem. Inf. Comput. Sci.*, 1994, **34**, 534.
196. G. Rücker and C. Rücker, *J. Chem. Inf. Comput. Sci.*, 2001, **41**, 1457.
197. R.J. Wilson, *Introduction to Graph Theory*, Oliver & Boyd, Edinburgh, 1972, p. 10.
198. D.E. Johnson and J.R. Johnson, *Graph Theory With Engineering Applications*, Ronald Press, New York, 1972, p. 45.
199. I. Lukovits, A. Miličević, S. Nikolić and N. Trinajstić, *Internet Electronic J. Mol. Design*, 2002, **1**, 388; http://www.biochempress.com.
200. F.R.K. Chung, *Spectral Graph Theory*, American Mathematical Society, Providence, Rhode Island, 1997.
201. M. Randić, *J. Am. Chem. Soc.*, 1975, **97**, 6609.
202. D.J. Klein, *Croat. Chem. Acta*, 2002, **75**, 633.
203. M. Diudea, *J. Chem. Inf. Comput. Sci.*, 1996, **36**, 535.
204. M. Diudea, *J. Chem. Inf. Comput. Sci.*, 1996, **36**, 833.
205. A. Dress, S. Grünewald, I. Gutman, M. Lepović and D. Vidović, *MATCH Commun. Math. Comput. Chem.*, 2003, **48**, 63.
206. S. Grünewald, *Appl. Math. Lett.*, 2002, **15**, 1001.
207. A. Dress and I. Gutman, *Appl. Math. Lett.*, 2003, **16**, 389.
208. I. Lukovits and N. Trinajstić, *J. Chem. Inf. Comput. Sci.*, 2003, **43**, 1110.
209. J. Figueras, *J. Chem. Inf. Comput. Sci.*, 1993, **33**, 717.
210. G. Rücker and C. Rücker, *J. Chem. Inf. Comput. Sci.*, 2003, **43**, 1115
211. M. Mitchell Waldrop, *Complexity*, Touchstone/Simon & Schuster, New York, 1992.

212. D. Bonchev and W.A. Seitz, in: *Concepts in Chemistry: A Contemporary Challenge*, ed. D.H. Rouvray, Wiley, New York, 1997, pp. 353–381.
213. *Complexity in Chemistry – Introduction and Fundamentals*, eds. D. Bonchev and D.H. Rouvray, Taylor & Francis, London, 2003.
214. S. Nikolić, I. M. Tolić, N. Trinajstić and I. Baučić, *Croat. Chem. Acta*, 2000, **73**, 909.
215. S.H. Bertz, *Complexity in Chemistry: Introduction and Fundamentals*, eds D. Bonchev and D. Rouvray, Taylor & Francis, London, 2003, pp. 91–156.
216. D. Bonchev, *SAR QSAR Environ. Res.*, 2003, **14**, 199.
217. M. Randić, *Croat. Chem. Acta*, 2001, **74**, 683.
218. M. Randić and D. Plavšić, *Croat. Chem. Acta*, 2002, **75**, 107.
219. M. Randić and D. Plavšić, *Int. J. Quantum Chem.*, 2003, **91**, 20.
220. R.J. Cook, *J. Comb. Theory*, 1979, **B 26**, 337.
221. M. Trusczynski, *J. Graph Theory*, 1984, **8**, 171.
222. L. Euler, *Commentarii Academiae Scientarum Imperialis Petropolitanae* 1736, **8**, 128; English translation appeared in *Sci. Am.*, 1953, **189**, 66.
223. M. Randić, *J. Chem. Inf. Comput. Sci.*, 2000, **40**, 627.
224. I. Lukovits, *MATCH Commun. Math. Comput. Chem.*, 2001, **44**, 279.
225. I. Lukovits, S. Nikolić and N. Trinajstić, *Chem. Phys. Lett.*, 2002, **354**, 417.
226. T.I. Bieber and M.D. Jackson, *J. Chem. Inf. Comput. Sci.*, 1993, **33**, 696.
227. M.D. Jackson and T.I. Bieber, *J. Chem. Inf. Comput. Sci.*, 1993, **33**, 701.
228. M. Gordon and J.W. Kennedy, *J. Chem. Soc. Faraday Trans. II*, 1973, **69**, 484.
229. S.H. Bertz and W.C. Herndon, in: *Artificial Intelligence Application in Chemistry*, eds. T.H. Pierce and B.A. Bohne, American Chemical Society, Washington, DC, 1986, pp. 169–175.
230. S.H. Bertz and W.F. Wright, *Graph Theory Notes of New York*, 1998, **35**, 32.
231. S.H. Bertz and T.J. Sommer, *Chem. Comm.*, 1997, 2409.
232. S.H. Bertz and C.M. Zamfirescu, *MATCH Commun. Math. Comput. Chem.*, 2000, **42**, 39.
233. S.H. Bertz, *New J. Chemistry*, 2003, **27**, 860.
234. S.H. Bertz, *New J. Chemistry*, 2003, **27**, 870.
235. D. Bonchev, *SAR QSAR Environ. Res.*, 1997, **7**, 23.
236. G. Rücker and C. Rücker, *J. Chem. Inf. Comput. Sci.*, 2001, **41**, 314; (Errata: *J. Chem. Inf. Comput. Sci.*, 2001, **41**, 865).
237. G. Rücker and C. Rücker, *MATCH Commun. Math. Comput. Chem.*, 2000, **41**, 145.
238. C. Rücker, G. Rücker and M. Meringer, *J. Chem. Inf. Comput. Sci.*, 2002, **42**, 640.
239. A.T. Balaban, *Chem. Phys. Lett.*, 1982, **89**, 399.
240. Z. Mihalić, D. Veljan, D. Amić S. Nikolić, D. Plavšić and N. Trinajstić, *J. Math. Chem.*, 1992, **11**, 223.
241. S.H. Bertz, *J. Am. Chem. Soc.*, 1981, **103**, 3599.
242. G.A. Arteca, *J. Chem. Inf. Comput. Sci.*, 2002, **42**, 326.
243. G.A. Arteca, *J. Chem. Inf. Comput. Sci.*, 2002, **43**, 63.
244. R.L. Brooks, C.A.B. Smith, A.H. Stone and W.T. Tutte, *Duke Math. J.*, 1940, **7**, 312.
245. I. Gutman, R.B. Mallion and J.W. Essam, *Mol. Phys.*, 1983, **50**, 859.
246. D. Bonchev, O.N. Temkin and D. Kamenski, *React. Kinet. Catal. Lett.*, 1980, **15**, 119.
247. S. Nikolić, N. Trinajstić, A. Jurić, Z. Mihalić and G. Krilov, *Croat. Chem. Acta*, 1986, **69**, 883.

248. B. Mohar, in: *MATH/CHEM/COMP 1988*, ed. A. Graovac, Elsevier, Amsterdam, 1989, pp. 1–8.
249. J.W. Moon, *Counting Labelled Trees*, Canadian Mathematical Monographs, Ottawa, 1970, chapter 5.
250. G. Kirchhoff, *Ann. Phys. Chem.*, 1847, **72**, 497; English translation is given in J. B. O'Toole, *I.R.E. Trans. Circuit Theory*, 1958, **CT5**, 4.
251. O. Ivanciuc, *Rev. Roum. Chim.*, 1993, **38**, 1499.
252. B. Mohar, D. Babić and N. Trinajstić, *J. Chem. Inf. Comput. Sci.*, 1993, **33**, 153.
253. N. Trinajstić, D. Babić, S. Nikolić, D. Plavšić, D. Amić and Z. Mihalić, *J. Chem. Inf. Comput. Sci.*, 1994, **34**, 368.
254. D.J. Klein, *MATCH Commun. Math. Comput. Chem.*, 1997, **35**, 7.
255. D.J. Klein and H.-Y. Zhu, *J. Math. Chem.*, 1998, **23**, 179.
256. *Topological Indices and Related Descriptors in QSAR and QSPR*, eds. J. Devillers and A.T. Balaban, Gordon & Breach, Amsterdam, 1999.
257. I. Gutman and D. Vidović, *Indian J. Chem.*, 2002, **41A**, 893.
258. I. Gutman, *MATCH Commun. Math. Comput. Chem.*, 2003, **47**, 133.
259. I. Gutman, D. Vidović and B. Furtula, *Indian J. Chem.*, 2003, **42A**, 1272.
260. D.J. Klein and M. Randić, *J. Math. Chem.*, 1993, **12**, 81.
261. M. Kunz, *MATCH Commun. Math. Comput. Chem.*, 1995, **32**, 221.
262. B.E. Eichinger, *Macromol.*, 1980, **13**, 1.
263. B. Mohar, in: *Graph Theory, Combinatorics and Applications*, eds. Y. Alavi, G. Chartrand, O.R. Ollermann and A.J. Scwenk, Wiley, New York, pp. 871–898.
264. Z. Mihalić and N. Trinajstić, *Fullerene Sci. Technol.*, 1994, **2**, 89.
265. P.W. Fowler, *MATCH Commun. Math. Comput. Chem.*, 2003, **48**, 87.
266. R.B. Mallion and N. Trinajstić, *MATCH Commun. Math. Comput. Chem.*, 2003, **48**, 97.
267. e.g., G. Rücker and C. Rücker, *J. Chem. Inf. Comput. Sci.*, 1991, **31**, 123.
268. K. Kuratowski, *Fund. Math.*, 1930, **15**, 271.
269. M. Randić, V. Katović and N. Trinajstić, in: *Symmetries and Properties of Non-Rigid Molecules: A Comprehensive Survey*, eds. J. Maruani and J. Serre, Elsevier, Amsterdam, 1983, pp. 399–408.
270. H.E. Dudeney, *Amusements in Mathematics*, Nelson, London, 1917.
271. C. Liang and K. Mislow, *J. Am. Chem. Soc.*, 1994, **116**, 3588.
272. K. Mislow, *Croat. Chem. Acta*, 1996, **69**, 485.
273. J. Petersen, *L'Intermédiaire des Mathématiciens*, 1898, **5**, 225.
274. J.D. Dunitz and V. Prelog, *Angew. Chem.*, 1968, **80**, 700.
275. M. Randić, *Croat. Chem. Acta*, 1977, **49**, 643.
276. R.B. King, in: *From Chemical Topology to Three-Dimensional Geometry*, ed. A.T. Balaban, Plenum Press, New York, 1990, pp. 343–414.
277. D. Blanuša, *Glasnik mat. fiz. astr.*, ser 2, 1946, **1**, 31.
278. H.S.M. Coxeter, *Bull. Am. Math. Soc.*, 1950, **56**, 413.
279. M. Gielen, in: *Chemical Applications of Graph Theory*, ed. A.T. Balaban, Academic Press, London, 1976, pp. 261–298.
280. M. Randić, D.J. Klein, V. Katović, D.O. Oakland, W.A. Seitz and A.T. Balaban, in: *Graph Theory and Topology in Chemistry*, eds. R.B. King and D.H. Rouvray, Elsevier, Amsterdam, 1987, pp. 266–284.
281. D.J. Klein, A. Graovac, Z. Mihalić and N. Trinajstić, *J. Mol. Struct. (Theochem)*, 1995, **341**, 157.
282. T. Živković, *Croat. Chem. Acta*, 1996, **69**, 215.
283. K. Mislow, *Acc. Chem. Res.*, 1970, **10**, 321.

284. H.S.M. Coxeter, *Regular Polytopes*, Dover, New York, 3rd Edn., 1973.
285. H.W. Kroto, J.R. Heath, S.C. O'Brien, R.F. Curl and R.E. Smalley, *Nature*, 1985, **318**, 162.
286. R.B. King, *Croat. Chem. Acta*, 2002, **75**, 447.
287. e.g., A. Rassat, *Angew. Chem.*, 2003, **115**, 632.
288. G.G. Cash and J.R. Dias, *J. Math. Chem.*, 2001, **30**, 429.
289. C.D. Godsil and I. Gutman, *J. Graph Theory*,1981, **5**, 137.
290. I. Gutman, M. Milun and N. Trinajstić, *MATCH Commun. Math. Comput. Chem.*, 1975, **1**, 171.
291. J.-i. Aihara, *J. Am. Chem. Soc.*, 1976, **98**, 2750.
292. I. Gutman, M. Milun and N. Trinajstić, *J. Am. Chem. Soc.*, 1977, **99**, 1692.
293. e.g., V.I. Minkin, M.N. Glukhovtsev and B.Y. Simkin, *Aromaticity and Antiaromaticity*, Wiley-Interscience, New York, 1994, pp. 14–19.
294. *Mathematica®* is a registered trademark of Wolfram Research, Inc., http://www.wolfram.com
295. J.M. Salvador, A. Hernandez, A. Beltran, R. Duran and Mactuis, *J. Chem. Inf. Comput. Sci.*, 1998, **38**, 1105.
296. D. Babić, G. Brinkmann and A. Dress, *J. Chem. Inf. Comput. Sci.*, 1997, **37**, 920.
297. J.R. Dias, *J. Phys. Org. Chem.*, 2003, **15**, 94.
298. J.R. Dias, *J. Chem. Inf. Comput. Sci.*, 2001, **42**, 686.
299. J.R. Dias and G.G. Cash, *J. Chem. Inf. Comput. Sci.*, 2001, **41**, 129.
300. H. Hosoya, K. Ohta and M. Satomi, *MATCH Commun. Math. Comput. Chem.*, 2001, **44**, 183.
301. H. Hosoya, U. Nagashima and S. Hyugaji, *J. Chem. Inf. Comput. Sci.*, 1994, **34**, 428.
302. H. Hosoya, *Bull. Chem. Soc. Jpn*, 1971, **44**, 2332.
303. H. Wiener, *J. Am. Chem. Soc.*, 1947, **69**, 2636.
304. A.T. Balaban and F. Harary, *J. Chem. Doc.*, 1971, **11**, 258.
305. I. Gutman and N. Trinajstić, *Topics Curr. Chem.*, 1973, **42**, 49.
306. W.C. Herndon, *Tetrahedron Lett.*, 1974, 671.
307. T. Živković, N. Trinajstić and M. Randić, *Mol. Phys.*, 1975, **30**, 517.
308. V.L. Collatz and U. Sinogowitz, *Abh. Math. Sem. Univ. Hamburg*, 1957, **21**, 63.
309. R.C. Read and R.J. Wilson, *An Atlas of Graphs*, Oxford University Press, New York, 1998.
310. C. Rücker and M. Meringer, *MATCH Commun. Math. Comput. Chem.*, 2002, **45**, 153.
311. K.S. Booth and G.S. Lueker, *J. Comput. System Sci.*, 1976, **13**, 335.
312. M. Forstner, A. Pick and M. Raitner, *Graph Template Library GTL*, see http://www.infosun.fmi.uni-passau.de/GTL/.
313. A. Kerber, R. Laue, T. Grüner and M. Meringer, *MATCH Commun. Math. Comput. Chem.*, 1998, **37**, 205; see also R. Grund, A. Kerber and R. Laue, *MATCH Commun. Math. Comput. Chem.*, 1992, **27**, 87.
314. B. Mao, *J. Am. Chem. Soc.*, 1989, **111**, 6132.
315. C. Liang and K. Mislow, *J. Am. Chem. Soc.*, 1995, **117**, 4201.
316. K. Mislow and C. Liang, *Croat. Chem. Acta*, 1996, **69**, 1385.
317. René Barone, Remi Barone, M. Arbelot and M. Chanon, *Tetrahedron*, 2001, **57**, 6035.
318. M.F. Lynch, J.M. Harrison, W.G. Town and J.E. Ash, *Computer Handling Chemical Structure Information*, Macdonald, London, 1971.
319. G. Brinkmann, A.A. Dobrynin and A. Krause, *MATCH Commun. Math. Comput. Chem.*, 2000, **41**, 137.

320. D.H. Rouvray, *Chem. Soc. Rev.*, 1974, **3**, 355.
321. D.H. Rouvray, in: *Chemical Graph Theory – Introduction and Fundamentals*, eds. D. Bonchev and D.H. Rouvray, Gordon & Breach, New York, 1991, pp. 1–39.
322. A.T. Balaban, in: *Chemical Graph Theory – Introduction and Fundamentals*, eds. D. Bonchev and D.H. Rouvray, Gordon & Breach, New York, 1991, pp. 177–234.
323. A.E.P. Adang and P.H.H. Hermkens, *Curr. Med. Chem.*, 2001, **8**, 985.
324. G. Grassy, B. Calas, A. Yasri, R. Lahana, J. Woo, S. Iyear, M Kaczorek, R. Floc'h and R. Buelow, *Nature Biotechnol.*, 1998, **16**, 748.
325. R. Buelow, P. Veyron, C. Clayberger, P. Pouletty and J.L. Touraine, *Transplantation*, 1995, **59**, 455.
326. R.A. Lewis, S.D. Pickett and D.E. Clark, *Rev. Comput. Chem.*, 2000, **16**, 1.
327. A. Ivanisevic, K.V. McCumber and C.A. Mirkin, *J. Am. Chem. Soc.*, 2002, **124**, 11997.
328. O. Storm and U. Lüning, *Chem. Eur. J.*, 2002, **8**, 793.
329. T. Wright, V.J. Gillet, D.V.S. Green and S.D. Pickett, *J. Chem. Inf. Comput. Sci.*, 2003, **43**, 381.
330. M.J. Blandamer, P.M. Cullis and P.T. Glesson, *Phys. Chem. Chem. Phys.*, 2002, **4**, 765.
331. R.W. Armstrong, A.P. Combs, P.A. Tempest, S.D. Brown and T.A. Keating, *Acc. Chem. Res.*, 1996, **29**, 123.
332. N. Trinajstić, M. Randić and D.J. Klein, *Acta Pharm.*, 1986, **36**, 267.
333. M. Randić, in: *Encyclopedia of Computational Chemistry*, Editor-in-chief P. von R. Schleyer, Wiley, Chichester, 1998, pp. 3018–3032.
334. M. Karelson, *Molecular Descriptors in QSAR/QSPR*, Wiley-Interscience, New York, 2000.
335. L. Poglianni, *Chem. Rev.*, 2000, **100**, 3827.
336. A. Leo, C. Hansch and D. Elkins, Chem. Rev., 1971, **71**, 525.
337. I. Gutman and N. Trinajstić, *Chem. Phys. Lett.*, 1972, **17**, 535.
338. S. Nikolić, G. Kovačević, A. Miličević and N. Trinajstić, *Croat. Chem. Acta*, 2003, **76**, 113.
339. D. Plavšić, S. Nikolić, N. Trinajstić and Z. Mihalić, *J. Math. Chem.*, 1993, **12**, 235.
340. O. Ivanciuc, T.-S. Balaban and A.T. Balaban, *J. Math. Chem.*, 1993, **12**, 309.
341. V.J. Gillet, W. Khatib, P. Willett, P.J. Fleming and D.V.S. Green, *J. Chem. Inf. Comput. Sci.*, 2002, **42**, 375.
342. L. Xue and J. Bajorath, *J. Chem. Inf. Comput. Sci.*, 2002, **42**, 757.
343. A. Schuffenhauer, J. Zimmermann, R. Stoop, J.-J. van der Vyver, S. Lecchini and E. Jacoby, *J. Chem. Inf. Comput. Sci.*, 2002, **42**, 947.
344. S. Putta, C. Lemmen, P. Beroza and J. Greene, *J. Chem. Inf. Comput. Sci.*, 2002, **42**, 1230.
345. J.W. Raymond and P. Willet, *MATCH Commun. Math Comput. Chem.*, 2003, **48**, 197.
346. Y. Pan, N. Huang, S. Cho and A.D. MacKerell, Jr., *J. Chem. Inf. Comput. Sci.*, 2003, **43**, 267.
347. O. Ivanciuc and D.J. Klein, *Croat. Chem. Acta*, 2002, **75**, 577.
348. O. Ivanciuc and D.J. Klein, *J. Chem. Inf. Comput. Sci.*, 2002, **42**, 8.
349. L. Bytautas and D.J. Klein, *J. Chem. Inf. Comput. Sci.*, 2000, **40**, 471.
350. O. Ivanciuc, *Internet Electronic J. Mol. Design*, 2002, **1**, 1; http://www.biochempress.com.

351. O. Ivanciuc, T. Ivanciuc and A.T. Balaban, *J. Chem. Inf. Comput. Sci.*, 1998, **38**, 395.
352. D.T.-Y. Bong, L. Gentric, D. Holmes, A.J. Matzger, F. Scherbag and K.P.C. Vollhardt, *Chem. Comm.*, 2002, 278.
353. I. Gutman, S.J. Cyvin and J. Brunvoll, *Monat. Chem.*, 1994, **125**, 887.
354. B.C. Berris, G.H. Hovakeemian, Y.-H Lai, H. Mestdagh and K.P.C. Vollhardt, *J. Am. Chem. Soc.*, 1985, **107**, 5670.
355. C. Dosche, H.-G. Löhmannsröben, A. Bieser, P.I. Dosa, S. Han, M. Iwamoto, A. Schleifenbaum, and K.P.C. Vollhardt, *Chem. Phys. Phys. Chem.*, 2002, **4**, 2156.
356. N. Trinajstić, T.G. Schmalz, T.P. Živković, S. Nikolić, G.E. Hite, D.J. Klein and W.A. Seitz, *New J. Chem.*, 1991, **15**, 27.

5
Electric Multipoles, Polarizabilities, Hyperpolarizabilities and their Magnetic Analogues

BY DAVID PUGH

1 Introduction

In the two previous reports[1] brief historical introductions to the development of theoretical studies of the electric multipoles and electric and magnetic molecular response functions, covering mainly the last 15 years, have preceded a review of the appropriate two year period. The present article attempts only to bring the review of the literature up to date by covering the period June 2001 to May 2003. As in the 2001 report the review of magnetic properties is confined to diamagnetic materials. Questions of spin-spin coupling and the splittings of levels in transition metal complexes etc. have not been treated. Even with these constraints it is no longer possible to identify or include all the papers in the literature where a multipole moment or hyperpolarizability has been calculated using a standard procedure as part of some more general study. The review is therefore necessarily selective and the choice of papers that have been discussed in greater detail is to some extent arbitrary. In particular 'state of the art' high level *ab initio* methods, when once established, tend to come rapidly into general use with comparatively minor variations in different studies. I have chosen to look at the results of a few such papers in detail in order to give a better idea of the procedures involved, but in many cases there are other publications, much more briefly referred to, where work of similar quality has been reported.

The physical quantities that are the subject of this review are almost all given general definitions in the Taylor series expansion for the change in molecular energy when uniform electric and magnetic fields (F and B respectively) act on a diamagnetic molecule:

$$\delta W = \mu_\alpha^0 F_\alpha - \frac{1}{2}\alpha_{\alpha\beta}F_\alpha F_\beta - \frac{1}{2}\chi_{\alpha\beta}B_\alpha B_\beta - \frac{1}{2}\xi_{\alpha\beta,\gamma}B_\alpha B_\beta F_\gamma - \frac{1}{6}\beta_{\alpha\beta\gamma}F_\alpha F_\beta F_\gamma$$

$$- \frac{1}{4}\eta_{\alpha\beta,\gamma\delta}B_\alpha B_\beta F_\gamma F_\delta - \frac{1}{24}\gamma_{\alpha\beta\gamma\delta}F_\alpha F_\beta F_\gamma F_\delta - \cdots\cdots\cdots \quad (1)$$

The equation includes the first three electric field response terms involving the dipole polarizability, α and the first and second hyperpolarizabilities, β and γ. The linear magnetizability is denoted by χ (which is also used for the macroscopic electrical susceptibilities in bulk materials) and terms dependant on the mixed electric/magnetic hypermagnetizabilities, ξ and η, which are relevant in the Cotton-Mouton effect, are also present. Not shown explicitly are terms derived from the higher permanent multipole moments, Θ, Φ, etc., which would be present in nonuniform fields, and terms dependent on the nuclear magnetic moments that give rise to the nuclear shielding factors. In table 1 a summary of the units commonly used for these quantities is provided. Theoretical papers, especially those using *ab initio* or density functional methods, often use atomic units. Conversions to the units commonly employed by experimentalists are given in some cases. The system of atomic units can be derived from the atomic units of energy, the Hartree, E_h, electric field (E_h/ea_0), and magnetic field ($\hbar/ea_0^2 = \dfrac{\sqrt{m_e E_h}}{ea_0}$), where a_0 is the Bohr radius.[2,3] If the energy contribution from one term in equation (1) is given the value, E_h and the above atomic units are substituted for the fields (omitting the numerical factor in the expansion) then the atomic unit for the corresponding response function is obtained.

2 Electrical Properties

2.1 High Level *Ab Initio* Calculations. – The studies reviewed in this section are characterized by the use of large polarizing basis sets often in conjunction with *post*- Hartree-Fock methods that make a serious attempt to include most of the electron correlation contributions. The coupled-cluster method

Table 1 *Units. The numerical quantities are the value of one atomic unit (au) in the units specified*

name	symbol	atomic unit	SI	other units
dipole moment	μ^0	ea_0	8.47835e–30/Cm	2.54175/debye
quadrupole moment	Θ	ea_0^2	4.48655e–40/Cm^2	
hexadecapole moment	Φ	ea_0^4	1.25636e–60/Cm^4	
dipole polarizability	α	$(ea_0)^2 E_h^{-1}$	1.64878e–41/C^{2}m^{2}J^{-1}	
1st hyperpolarizability	β	$(ea_0)^3 E_h^{-2}$	3.20636e–53/C^{3}m^{3}J^{-2}	8.63131e–33/cm^{5}esu^{-1}
2nd hyperpolarizability	γ	$(ea_0)^4 E_h^{-3}$	6.23538e–65/C^{4}m^{4}J^{-3}	
magnetizability	χ	$e^2 a_0^2/m_e$	7.89104e–29/JT^{-2}	7.89104e–30/cm^3
hypermagnetizabilty	ξ	$e^3 a_0^3/m_e E_h$	1.534562e–40/CmT^{-2}	
hypermagnetizability	η	$e^4 a_0^4/m_e E_h^2$	2.98425e–52/C^{2}m^{2}J^{-1}T^{-2}	

incorporating single and doubly excited configurations in the cluster calculation and then including triply excited configurations through perturbation theory, the CCSD(T) method, has become an established favorite for calculations at this level. Multiple reference self consistent field methods (MRSCF) including complete active subspace selection of CI's (CASSCF) and even full configuration interaction (FCI) for (very) small molecules are also used. The use of such methods is limited by computational restraints to fairly small molecules and the difficulty of obtaining accurate results increases rapidly for the higher order response functions, especially for some of the tensor components that are related to the asymmetry of the molecule.

2.1.1 Diatomic Molecules. – Diatomic molecules are susceptible to attack by theoretical methods that are impracticable for almost all polyatomics. Slater type atomic orbitals have been extensively used in the diatomic calculations since the absence of three and four center integrals makes this procedure comparatively efficient. An alternative approach is to use a finite difference method for solving the Schrödinger equation, which can lead to essentially exact Hartree_Fock solutions. In an attempt to demonstrate conclusively that a systematic procedure for the selection of Gaussian basis sets of increasing size can lead to a complete one electron basis, Kobus, Moncrieff and Wilson[4] have made a detailed comparison of dipole moments, polarizability and hyperpolarizability obtained from finite difference calculations and conventional GTO finite basis set procedures. Previous comparisons have related only to energies and orbitals of the unperturbed molecules. In the new work the finite difference method has been combined with a finite field perturbation to compute the dipole moment and the axial components of the polarizability and hyperpolarizability tensors at the Hartree-Fock level. The full tensors have also been calculated using a large series of GTO bases. The results for Hydrogen Fluoride, both at the equilibrium geometry and at other internuclear separations are reported in detail and equilibrium geometry results given for H_2, BH and LiH.

As is always the case in *ab initio* property calculations the choice of adequate basis sets lies at the heart of the problem. The authors have developed a procedure for constructing a sequence of basis sets of increasing size and including functions centered at the mid-point of the bond. The scheme may be summarized by the plan,

[F:sp;H:s]	→	[F:spd;H:sp]	→	[F:spdf;H:spd]
↓		↓		↓
[F:sp;H:s;bc:s]	→	[F:spd;H:sp;bc:sp]	→	[F:spdf;H:spd;bc:spd]
↓		↓		↓
[F:sp;H:s;bc:sp]	→	[F:spd;H:sp;bc:spd]	→	[F:spdf;H:spd;bc:spdf]

where the number of primitive gaussians is 30 for s-functions and 15 for p, d and f.

Proceeding along the rows of the diagram means adding higher angular momentum functions on the atoms and down the columns represents increasing the basis size at the mid-point of the bond (bc = bond center). The two matrices below give results for the two non-zero components of the first hyperpolarizability tensor β of hydrogen fluoride from the GTO calculation in a layout corresponding to the above plan (the molecule is aligned with the z-axis):–

$$\beta_{zzz}/\text{au} \qquad\qquad \beta_{xxx}/\text{au}$$

$$\begin{pmatrix} -18.7076 & -8.4558 & -8.4480 \\ -15.3388 & -8.4852 & -8.4494 \\ -10.4107 & -8.4900 & -8.4495 \end{pmatrix} \qquad \begin{pmatrix} -0.8282 & -0.0118 & -0.4591 \\ -1.0153 & -0.0059 & -0.4611 \\ -1.3800 & -0.0819 & -0.4610 \end{pmatrix}$$

Finite Difference: $\beta_{zzz}/\text{au} = -8.4501$

All quantities are in atomic units. The convergence of β_{zzz} towards the finite difference result on approaching the lower diagonal corner is convincing, although the effect of the bond-centered functions seems to be marginal. However, in the case of the dipole moment and the axial polarizability, agreement to about 6 decimal places is achieved but is attained in the last few places only with the inclusion of the bond-centered functions. Even at this level of calculation the convergence within the GTO HF calculation for the non-axial component of the first hyperpolarizability, β_{xxx}, is not entirely convincing. There is no finite difference result due to difficulties with numerical differentiation in the numerical finite field approach. It could be concluded that the problem of calculating reliable complete first hyperpolarizability tensors for the more difficult polyatomic case must still be intractable in some cases.

Shtoff et al.[5] have investigated the frequency-dependent γ-hyperpolarizability of LiH using an approach based on Floquet theory. The theory uses a combination of perturbation theory with the finite field method. Electron correlation is taken into account through the CISPI algorithm and the total wavefunction involves spectral and pseudospectral states and polynomial terms. The method should have general application to the calculation of higher order nonlinear molecular response functions.

Alkali metal hydrides (LiH, NaH and KH) are the subject of a comprehensive study by Avramopoulis and Papadopoulis[6] that treats the electronic and vibrational contributions to dipole moment, polarizability and β and γ hyperpolarizabilities. This work is intended to provide a benchmark study for these molecules and we discuss it here to exemplify the kind of results obtainable for the static response functions of small molecules using current state of the art methods. The calculations employ basis sets devised by Sadlej[7] and Sadlej and Urban[8], referred to as the Pol set. These are specified in the usual notation as,

H[10s6p4d/5s3p2d], Li[10s6p4d/5s3p2d], Na[13s10p4d/7s5p2d], K[15s13p4d/9s7p2d]

and described as medium size polarized basis sets, derived by the basis set polarization method which is related to the Hellman-Feynman theorem. In the case of LiH there are a number of high level calculations in the literature with which these calculations can be compared and it is of interest at this point to examine how well they have converged on settled values. In table 2 values of the electronic contributions to the averaged quantities, μ, α, β and γ, obtained from these basis sets and some other sets introduced by Pluta and Sadlej[9] and by Dunning and co-workers[10,11] are compared. All the values in table 2. have been calculated at CCSD(T) level. An earlier CCSD(T) calculation by Tunega and Noga[12] is also included. In table 3. the results for the principal components, α_{zz}, β_{zzz}, γ_{zzzz} are compared with those from a complete CI calculation by Jonsson et al.[13]

Agreement in the case of μ and α is entirely satisfactory. In the case of β the value with the very large Dunning basis is lowered by about 5% which is acceptable but much more drastically in the calculation of Tunega and Noga. In the case of γ both C and D differ from A and B by substantial percentages. Agreement between the CCSD(T) and full CI results with the same basis set is good in all cases. It therefore seems that, even in the case of LiH – the smallest stable heteronuclear diatomic – there are still uncertainties in the calculation of the static β and γ attributable to variations with basis set rather than to inadequate treatment of electron correlation.

Table 2 *Dipole moment and averaged (hyper)polarizabilities for LiH. All values are in atomic units*

	A	B	C	D
μ	−2.309	−2.320	−2.294	−2.2935
α	28.83	29.25	28.28	28.31
β	882.4	877.7	838.9	658.8
$\gamma \times 10^{-3}$	86.159	88.014	74.103	110.8

A: Avramopoulos and Papadopoulos;[6] B: Pluta and Sadlej;[9] C: Dunning and co-workers;[10,11] D: Tunega and Noga.[12]

Table 3 *Axial Components of LiH (hyper)polarizabilities. All values are in atomic units*

	A	E
α_{zz}	26.92	26.89
β_{zzz}	772.7	729
$\gamma_{zzz} \times 10^{-3}$	124.3	127

A: Avramopoulos and Papadopoulos;[6] E: Jonsson et al.[13]

Table 4 Polarizabilities and hyperpolarizabilities of alkali metal hydrides from reference 6. The variation in the averaged response functions as the level of electron correlation in the calculation is increased for the three alkali metal hydrides. All values are in atomic units

	SCF	MP2	CCSD(T)		SCF	MP2	CCSD(T)		SCF	MP2	CCSD(T)
Li				Na				K			
α	24.14	26.25	28.87	α	34.35	39.9	45.75	α	43.03	60.64	61.44
β	419.8	599.9	885.8	β	1146.8	1760.9	2944.8	β	1746.5	3225.6	6047.1
$\gamma \times 10^{-2}$	419.0	584.4	861.5	$\gamma \times 10^{-2}$	807	1249	2243.5	$\gamma \times 10^{-2}$	2012	3546	7022

It is clear from table 4, which shows data for all three hydrides, that the percentage change as more correlation is included is such as to invalidate the lower level values for β and γ as quantitative estimates – although they may be a useful indication of trends between different molecules. A quantitative estimate of the nonlinear response must therefore rely exclusively on the highly correlated approach. There is no indication in table 4. that convergence with degree of correlation has been achieved at the CCSD(T) level, but other results show that the same values are likely to be attained, probably in all cases, with a full CI using the same basis set. The usual assumption, that CCSD(T) will adequately cope with electron correlation, is at least partly justified and the remaining problems must arise from lack of completeness in the basis sets.

Tables 2–4. and the foregoing discussion apply to the electronic part of the calculation. The paper also provides a detailed analysis of vibrational contributions. An interesting feature is that the vibrational contributions have been decomposed in two ways: the more usual division into the zero-point vibrational averaging (ZPVA) contribution and the pure vibrational contribution (PVA) and alternatively into a nuclear relaxation (NR) part and a curvature (CURV) term associated with the changes in the shape of the potential surface. The authors show that the two decompositions lead to essentially the same results. The Bishop-Kirtman perturbation theory[14] has also been tested against the semi-numerical Numerov-Cooley[15] procedure. The basis set dependence has been investigated and the variations between the SCF, MP2 and CCSD(T) methods are discussed for the vibrational contributions. Again it appears that the highly correlated method is necessary to produce quantitative accuracy. Electronic and vibrational contributions to α, β and γ at the CCSD(T) level are shown in table 5.

In the case of KH it has been found that, generally, a treatment that includes correlation only for the 10 electron valence shell gives results that are close to those obtained when the correlation is extended over all 20 electrons.

Maroulis[2] has conducted an extensive finite field study of the nitrogen molecule. He first draws attention to the numerous applications in spectroscopy and experimental studies of intermolecular forces where a knowledge of the higher multipole moments and polarizability derivatives are required. In

Table 5 *Vibrational effects in the alkali metal hydrides (reference 6). All values are in atomic units*

	LiH	NaH	KH
α^e	28.87	45.75	61.44
α^{vib}	3.98	6.61	14.59
β^e	885.8	2944.8	6047.1
β^{vib}	−252.9	−456.7	−1015.7
$\gamma^e \times 10^{-2}$	861.5	2243.5	7022
$\gamma^{vib} \times 10^{-2}$	91.7	70.4	195.1

the case of nitrogen this is particularly important for atmospheric studies. The paper contains a useful bibliography of previous work on N_2 and quotes many experimental results for comparison with the calculations.

Maroulis systematically studies the basis set dependence (a discussion of his technique applied to a small polyatomic molecule will be given below) and also proceeds from the HF calculations to include correlation effects with increasing rigor through MP2, SDQ-MP4, MP4, CCSD, culminating in CCSD(T). The higher order quantities calculated are:-

- Θ the quadrupole moment
- Φ the hexadecapole moment
- $\overline{C}$ the mean quadrupole polarizability
- $\overline{\gamma}$ the second dipole hyperpolarizability
- $\overline{E}$ the dipole-octupole polarizability
- $\overline{B}$ the dipole-dipole-octupole hyperpolarizability

The first and second derivatives of these quantities with respect to the internuclear separation have also been calculated. All these quantities are related to various kinds of experimental measurement. The agreement with the experimental results is remarkably good. Table 6. gives a few examples.

A [15s12p9d7f] basis set was used for the SCF calculations and [10s7p6d4f] for the correlated methods. The gradients were calculated with a [7s5p4d2f] set.

The agreement between the two levels of correlated calculation and the experimental values is within the experimental error except for the two gradients, where the MP2 result is substantially different from the coupled cluster value, although the latter is still in agreement with experiment. In all cases, except for the average polarizability and the hexadecapole moment, the SCF values are significantly different from the correlated and experimental values.

Table 6 *Electric Field Response Functions for N_2 (adapted from ref 2). All values are in atomic units*

Property	SCF	MP2	CCSD(T)	Experiment
Θ	−0.9302	−1.1952	−1.1258	−1.08 ± 0.01
Φ	−7.40	−6.55	−6.75	−8.5 ± 2.0
$\overline{\alpha}$	11.5651	11.5351	11.7709	11.74
$\Delta\alpha$	5.1957	4.1753	4.6074	4.60 ± 0.10
$\overline{\gamma}$	715	879	927	917 ± 5
γ_{zzzz}	799	1176	1194	—
γ_{xxzz}	250	310	322	—
$\left(\dfrac{d\overline{\alpha}}{dR}\right)_e$	7.17	5.30	6.18	5.98 ± 0.2
$\left(\dfrac{d^2\overline{\alpha}}{dR^2}\right)_e$	4.05	−1.12	1.84	1.88 ± 0.78

The results substantiate the claim made by the author that reliable theoretical determinations of this type of property, useful for the interpretation of experimental results, can now be obtained for diatomic molecules. However, the failure of the rigorous SCF studies, for example, in the case of FH discussed above, to produce convincing convergence in the case of the non-axial component of β, is an indication that this conclusion has to be treated with caution, at least when applied to polar molecules.

More recently[16] the same methods have also been applied to a comparative study of the properties of N_2, P_2 and As_2, but direct experimental evidence for comparison is not available for phosphorus and arsenic.

Recent experimental work on nitrogen (Ritchie et al.[17]) has been concerned with the temperature variation of the electric-field gradient induced birefringence. New values of the molecular hyperpolarizability, and quadrupole moment are derived from the temperature independent contribution that can be extracted from the analysis of the data. These results are also in good agreement with the values in the above table.

The same systematic approach to finite field calculations has been applied to BCl[18], where roughly the same degree of internal consistency has been found in the variation of the calculated parameters through the range of approximations, but where there is no experimental evidence with which to compare the final predictions. Maroulis and Haskopoulos[19] have also a similar approach to calculate the interaction dipole and hyperpolarizability of the inert gas dimer, Ne...Ar.

In a continuing study of the inert gases, He and Ar and their electric properties, Rizzo et al.[20] have applied the results of recent *ab initio* calculations of the interatomic potentials and interaction-induced polarizabilities to make a fully quantum statistical calculation of the second virial coefficients for both gases.

The calculation of the properties of radicals presents special difficulties. Urban and co-workers[21,22] have discussed possible approaches to the treatment of the diatomic radicals, (O_2, CN and NO) and provided a review of their work using the Restricted Open Shell Hartree Fock (ROHF) method. The UHF method appears at first sight to be the more obvious choice for studies on radicals. It suffers from two deficiencies. The first, spin contamination of the wavefunctions, which do not represent pure spin states, becomes less important in high level calculations since the exact wavefunction is a pure spin state. The second is the great increase in computational costs consequent on use of spin-orbital basis sets. The approach favored by the authors combines the ROHF method with a CCSD(T) correlated calculation and at this level there is a substantial computational advantage in using the smaller spatial orbital basis of the restricted methods.

In reference (21) the dipole polarizabilities of the three radicals are calculated. Experimental values for α_{zz} and α_{xx} are available only for O_2 and NO and the authors are mainly concerned to evaluate the reliability of their results by a discussion of the convergence of the method and by comparison with other calculations. They are satisfied that results obtained with their largest (d-aug-cc-pVXZ and aug-cc-pV5Z) basis sets will be close to the CBS (Complete Basis Set) limit. In fact there is very little variation of the results with basis set at either the ROHF or the ROHF-CCSD(T) levels. This finding, is similar to that noted for dipole polarizability calculations in closed shell systems. A discussion of the accuracy of the CCSD(T) calculation for the inclusion of the correlation effects applies the T1 diagnostic method, which evaluates the norm of the singly excited amplitudes, and concludes that the procedure is accurate. Vibrational contributions are calculated and are found to amount to about 0.3% of the electronic value for both oxygen components but to reach 1% and 2% for the axial components in the other radicals. The principal results are summarized in table 7.

In reference (22) the ROHF-CCSD(T) method is more extensively reviewed and the work extended to the calculation of dipole moments and hyperpolarizabilities for O_2 and CN. Much of this paper is concerned with electron affinities which are not directly relevant to this review. The static second hyperpolarizability tensor, γ_{ijkl}, has been evaluated using the same methods and basis functions as those described above. The dipole moment of the CN radical is found to vary very little with basis set, but changes from 0.9061 in the ROHF calculation to 0.559 in the final ROHF/CCSD(T) result. With all basis sets from aug-cc-pVTZ to d-aug-cc-pV5Z the CCSD(T) value is within about 2% of the quoted experimental value of 0.57 ± 0.03.

2.1.2 Small Polyatomic Molecules. – Xenides and Maroulis[23] have extended their studies of the open and cyclic forms of ozone to make a comparison with similar effects in sulphur dioxide. Theoretical work finds that in addition to the open structure (SO ≈ 1.43, OO ≈ 2.4, OSO ≈ 118°) there should exist a cyclic form, (SO ≈ 1.69, OO ≈ 1.50, OSO ≈ 52.7°), similar to the two forms of ozone. A comparison of the polarizabilities has been made, the most striking result of which is the increase of $\bar{\beta}$ from 30 au in the open form to 110 au in the cyclic form. Symmetry decrees that in the case of ozone $\bar{\beta}$ is zero but it is found that the nonzero components of α and β decrease for ozone in comparison with the open structure values, whereas in SO_2 they increase. In further work on protonated ozone (O_3H^+) they find that there are substantial differences between the dipole polarizabilities of the open and cyclic forms.

Karamanis and Maroulis[24] have investigated fluorodiacetylene (F–C≡C–C≡C–H), the type of polar substituted diacetylene that might be the basis of high second order optical response when substituted into a polymeric structure. The paper may be regarded as typical of current high level *ab initio* calculations of the static properties of small polyatomic molecules. The reconciliation of the requirement for very large basis sets, to accommodate the distortions of the electron distribution, and the application of highly correlated methods, is again the central feature of the study.

Maroulis's system of constructing basis sets for polarization studies has been described in detail in previous publications.[25] The outline structure is that, starting from a Dunning (9s5p/9s5p/4s) substrate for F/C/H and adding polarization and diffuse functions, three sets of increasing size, B0 = [5s3p2d/5s/3p2d/3s2p], B1 = [5s3p3d1f/5s3p3d1f/3s3p1d] and B2 = [5s3p3d2f/5s3p3d2f/

Table 7 Small Radical Polarizabilities (adapted from ref 21) Experimental values are shown in brackets immediately below the corresponding CCSD(T) result. It is the view of the authors that the DODS (distributed oscillator strength distribution) values are the most reliable. Where they are the only values available these are given without error estimates. In some cases values from refractivity (extrapolated to zero frequency) and dielectric constant measurements are also available. The average of all values has been taken and the estimated errors are merely an indication of the spread of these results. See ref. 21 for further details. All values are in atomic units

	α_{zz}		α_{xx}		α_{yy}		$\bar{\alpha}$	$\Delta\alpha$
	ROHF	CCSD(T)	ROHF	CCSD(T)	ROHF	CCSD(T)	CCSD(T)	CCSD(T)
O$_2$	20.79	15.03 (15.37 ± 0.1)	7.49	8.19 (8.21 ± 0.04)			10.47 (10.61 ± 0.02)	6.84 (7.16 ± 0.10)
CN	20.06	25.45	17.53	16.20			19.28	9.25
NO	14.84	15.36	8.76	9.24	9.17	9.98	11.53	5.74
		15.24		9.67		9.67	11.58 ± 0.05	5.57

3s3p2d] are developed. For the smallest set calculations are performed at SCF, MP2, MP3, SQD-MP4, MP4, CCSD and CCSD(T) levels but only by the SCF and SCF/MP2 procedures with the two larger basis sets.

Some typical results are displayed in a condensed form in table 8. In the full tables in the paper it is generally found that the MP2 and MP4 results are very similar while MP3 and SQD-MP4 show differences from MP2 amounting to about 10%. The overall convergence with increasingly rigorous treatment of correlation is well represented by the sequence SCF-MP2-CCSD(T).

In all cases the convergence of the property with increasing size of basis set is convincing. In all cases other than the dipole polarizability there is a substantial change from the SCF results when correlation is introduced at the MP2 level, but the change from MP2 to CCSD(T) is much smaller.

It is perhaps pertinent to ask what conclusions a non-specialist in the construction of basis sets and correlated methods would come to simply from inspecting the tables. As far as the dipole moment is concerned, the two correlated methods, with all basis sets, give results in the range 0.18 au to 0.23 au and discounting the SCF results one might have a realistic estimate of the dipole as (0.20 ± 0.02) au. It is probably not possible to improve on this by attempting to guess the change in the CCSD(T) value for B1 and B2. There is an experimental value available for the dipole moment, which is given as (0.2049 ± 0.0007). If this experimental precision is justified there is therefore still room for meaningful improvement in the calculations and it would be interesting to see what the B1 and B2 CCSD(T) values were. Nevertheless the theory does already seem to be able to provide a useful quantitative estimate of the dipole.

As has been found in many cases the diagonal or average components of the dipole polarizability are accurately given at the SCF level and convergence with basis set and correlation is very satisfactory. An inspection of the B2/MP2 and B0/CCSD(T) results for the axial component of the first hyperpolarizability might lead to the conclusion that the value was very close to –320, but the trend across the B0 line implies a final B2/CCSD(T) value of about –310. Again, there is enough consistency for the prediction to be quantitatively useful in assessing any future experimental results. In the case of β_{zxx} the divergence between the B2/MP2 and B0/CCSD(T) values might lead to an estimate of -40 ± 4, but extrapolating the trend across the B0 line to B1 and B2 one might predict a value of about –42 with more closely defined error limits.

There is also in the paper comprehensive data on the second hyperpolarizability. This is of less immediate interest since the particular applications of the polar substituted diacetylene lies in the β value. Generally the γ components behave similarly to the β components in that the on axis and diagonal values show better consistency than other elements.

Maroulis[26] has also recently published a similar study of the haloethynes, $H-C \equiv C-X$, $X = F$, Cl, Br and I. He finds good agreement with experimental α values with very little contribution from electron correlation effects, but for β only fair agreement for the first three halogens and unsatisfactory values for the iodine substituent.

Table 8 *Dipole and (hyper)polarizabilities for Fluorodiacetylene (from reference 24). Atomic units are used*

μ	SCF	MP2	CCSD(T)
B0	0.3451	0.1989	0.2256
B1	0.3367	0.1868	—
B2	0.3312	0.1913	—
α_{zz}	SCF	MP2	CCSD(T)
B0	85.99	85.90	86.04
B1	86.03	86.15	—
B2	86.22	86.16	—
β_{zzz}	SCF	MP2	CCSD(T)
B0	−242.28	−330.95	−319.70
B1	−235.66	−323.10	—
B2	−236.77	−320.56	—
β_{zxx}	SCF	MP2	CCSD(T)
B0	−27.79	−38.81	−36.01
B1	−33.18	−44.87	—
B2	−33.33	−44.24	—

Maroulis and Haskopoulos[27] have calculated interaction induced dipole moments for CO_2 inert gas interactions using MP2 and Coupled Cluster methods with large basis sets. The most stable configurations of the intermolecular complexes are found to be **T** shaped. Similar methods have been utilized by Wang et al.[28] to investigate the interaction hyperpolarizabilities of Ar-HF complexes.

The finite field approach has also been employed by Ohta et al.[29] in an attempt to provide a reliable estimate of the $\bar{\gamma}$ value for carbon tetrachloride. The authors emphasize the potential importance of liquid CCl_4 as a standard for measurement of second order susceptibilities and hyperpolarizabilities. They argue that there should be fewer uncertainties associated with the use of carbon tetrachloride than for dipolar liquids such as CS_2, probably the most frequently used standard. In the case of CS_2, when the experimental measurement involves the use of a low frequency field as in the case of EFISH or OKE techniques, second order effects dependent on the reorientation of the molecular dipoles play a major part and mask the third order ($\chi^{(3)}$, γ) response. The standard is then susceptible to variations (for example, related to the laser pulse width) induced by effects operating on a time-scale comparable with the orientational relaxation times. Earlier theoretical work at the HF level has now been extended to include the usual range of correlated procedures culminating in a CCSD(T) calculation. The α and β tensors have also been investigated at the same levels. One difficulty is that the molecular geometry changes appreciably when the structures are optimized at different levels. The HF α values with 6-31G+pdd and 6-31G(3d)+pd basis sets (68.2 and 69.4 au) are in rather better agreement with the experimental (70.4–70.9) values than the CCSD(T) values (73.4, 73.1). In all these cases and in the β and γ calculations, the geometry has been optimized at the HF/6-31G level, which gives the CCl bond length as 1.823A rather than the experimental 1.767A. Only in the case of α have different geometries been investigated giving CCl = 1.765 and α = 65.8au at HF/6-31G(3d)+pd and CCl = 1.767, α = 69.5 at MP2/6-31G(3d)+pd, worsening the HF result but greatly improving the correlated one. It therefore seems likely that the overestimate with the CCSD(T) method is due to the incorrect geometry. The value of β for this molecule, which is effectively close to being spherically symmetric, is small and notoriously hard to calculate. There is, however, some evidence of convergence in the range 10–20 au, but the correct sign is only achieved for 6-32G+pd and larger basis sets. Again, in the case of γ, there is a large jump in the values between the 6-31G* and larger basis sets, but a reasonably convincing convergence to a value of γ_S = 13.8 for CCSD(T)/6-31G(3d)+pd. Converted into the usual experimental units this gives 6.9×10^{-36}esu in comparison with the experimental (OKE) value of 10.6×10^{-36}esu. An allowance of 1660 au for dispersion to 800 nm, based on earlier TDHF work and an estimate of 2420 au for the vibrational effect from preliminary HF/MP2 studies, brings the theoretical value to 17 900au or 9.0×10^{-36}esu, which is within 10% of the experimental value. Although definitive results are still to be obtained for dispersion and vibrational contributions, it seems probable that a reasonably complete understanding of the nonlinear response of this simple liquid is now attainable and should considerably enhance the reliability of determinations of nonlinear material properties. One caveat is that there still has to be an absolute standard used to calibrate the CCl_4 liquid measurements and the usual standard, quartz, has been sufficiently ill-determined to require revision (Mito et al.[30], see also Mito and Moriwaki[31]) as recently as 1995. Almost all comparisons of absolute *ab initio* calculations with experiment depend ultimately on the absolute quartz frequency doubling value.

Papadopoulos and Avramopoulos[32] are engaged in a program to elucidate the gas and condensed phase properties of acetonitrile. The static and dynamic α, β and γ have been calculated for the isolated molecule using correlated methods up to CCSD(T) for single HF reference state, and multireference calculations have also been carried out at the MP2 level. Geometry and basis set effects and vibrational contributions have been calculated. It is found that all methods give similar values for the electronic parts of α and γ, but that for β the highly correlated CCSD(T) results differ substantially from those obtained at the MP2 level. Using the highest level correlated methods, liquid phase susceptibilities are computed using the dipolar Onsager reaction-field approximation. The linear response functions – relative permittivity and refractive index – are in excellent agreement with experiment while the $\chi^{(3)}$ values are found to be acceptable.

Gubskaya and Kusalik[33] have investigated the nonlinear response of liquid water. A finite field method using point charge perturbations rather than uniform fields has been used to calculate dipole and quadrupole moments, the dipole polarizability and the principal components of hyperpolarizabilities for H_2O in the gas and liquid phases. The liquid environment is represented by a local field generated by a set of fixed charges, and three different models corresponding to different arrangements of the charges are examined. The values are compared with those obtained for the gas phase. The calculations are at the MP2 and MP4 levels with balanced, moderately sized basis sets. A marked increase in the first hyperpolarizability when the molecule is in the liquid phase is noted. Other field gradient polarizabilities are calculated for the first time.

Bartkowiak et al.[34] have investigated the effect of the solute/solvent interaction in polar solvents on the first hyperpolarizability of urea. The environmental effects were represented by the quantum-mechanical Langevin dipoles/Monte-Carlo method and also by super molecule calculations. The calculated β values are compared with EFISH measurements.

Yamada et al.[35] have treated the pyridinium cation-chloride anion pair and found that the component of the γ-hyperpolarizability in the direction joining the anion and cation drastically increases as the ion pair distance increases.

Romaniello and Lelj,[36] have used TD-DFT, including relativistic corrections, to study metal/dithiolene complexes where the metals are Ni, Pd and Pt. The largest first hyperpolarizability, found with one of the Pt complexes gave value of $40.3 \times 10^{-30} cm^5 esu^{-1}$.

2.1.3 Larger Molecules. – For medium sized molecules that are often the subjects of experimental measurements of nonlinear response functions, there has been an increasing emphasis on Density Functional Methods. The single determinant DFT function in principle includes correlation effects so that excitations to the associated virtual orbitals should lead to states that will be close to the actual excited states of the molecule. The possibility exists of utilizing the TDHF procedure or even a direct sum over states calculation to find the frequency dependent response functions. Adaptation of the density functionals to cater for response function calculations is found to be necessary if reasonable results are to be obtained, which can be construed as an indication of the ultimately semi-empirical nature of the density functional methods – although the semi-empirical element is superimposed on basis sets that are certainly more realistic than those employed in older forms of semi-empirical procedure.

Gruning et al.[37] have discussed the application of DFT with various modified functionals to the calculation of the time-dependent response properties of 21 small molecules. Corrections to the standard local density (LDA) and generalized gradient (GGA) approximations that take special account of the shape of the molecule are known to be necessary in response function work, often linked to the substantial contributions that would be expected (in, for example, sum- over-states theory) from highly dispersed bound states and continuum states. The authors argue that the shape corrections are not exclusively concerned with the long-range asymptotic part of the potential. They consider two shape dependent potentials, the gradient-regulated asymptotic connection procedure applied to the Becke-Perdew potential (BP-GRAC) and a procedure based on the statistical averaging of model orbital potentials (SAOP). The performance of these potentials in the calculation of the dipole polarizability, the Cauchy S_4 coefficient and the static average hyperpolarizability is analysed and compared with that of the LDA, GGA BP, BYLP and B3LYP functionals without shape corrections. The superiority of the modified potentials is demonstrated.

Hieringer et al.[38] have carried out a study of electronic excitations and the γ-hyperpolarizability of the five-ring heterocyclics, furan, thiophene, selenophene and tellurophene using time-dependent density functional theory. Basis set saturation, exchange-correlation potentials and relativistic effects are discussed. It is suggested that dimer interactions may provide an explanation for the large deviations between recent *ab initio* and experimental results.

Salek et al.[39] have systematically derived equations for the application of time-dependent density functional theory from the Ehrenfest variational principle and from the quasienergy principle. They apply the methods using hybrid density functionals, including exchange-correlation functionals at the GGA level and exact HF exchange, to the first hyperpolarizability of 4-nitroaniline and a porphyrin of the push-pull type, obtaining good agreement with experiment.

An alternative derivation of density functional response theory is given by Heinze et al.[40] The action integral formalism, which can lead to violations of the causality requirement, is avoided by basing the derivation directly on the Runge-Gross theorem. Equations that can be used to treat time-dependent DFT in the frequency domain are derived and efficient noniterative methods for calculating dynamic response functions are presented. The theory is applied to investigate the effect of substituents on the frequency-dependent hyperpolarizability of benzene and stilbene.

Soscún et al.[41] have investigated the dipole polarizability and the γ-hyperpolarizability of benzene using coupled perturbed HF and TDHF methods to access both static and dynamic properties. Correlation effects were included at MP2 level and through DFT using the B3LYP functional. The largest basis set was STO/6-31 + G(d*,p) with the d- orbital exponent varied to maximize the static benzene γ value at the TDHF level. With these refinements excellent agreement with the experimental results was achieved.

The 3-methyl-4-nitropyridine-1-oxide molecule (POM) was identified as being of special interest for its second order properties. The presence of the NO and nitro groups at opposite ends of the ring leads to an approximate cancellation of dipole contributions in the ground state, but a fairly large dipole moment arises in the predominant excited electronic state. The usual semiquantitative interpretations of the first hyperpolarizability of donor/acceptor conjugated structures then accounts for the relatively high β value while the absence of an appreciable ground state dipole is thought to favor noncentrosymmetric crystallization[42,43]. An examination of the pyridine *N*-oxide molecule and its substituents using modern *ab initio* and DFT methods is of interest in assessing the theoretical validity of the semi-empirical ideas that lead to the extensive work carried out on POM. Soscún et al.[44] have investigated the structure, dipole moment, dipole polarizability, first hyperpolarizability and second hyperpolarizability of pyridine N-oxide using HF + MP2 and CCSD ab initio and BLYP and B3LYP DFT methods with basis sets increasing in size to 6-311++G(3d,3p) and Sadlej sets. Only static properties have been calculated. The results were compared with those of similar calculations on the pyridine molecule. The values obtained for the dipole moments and dipole polarizabilities are consistent amongst themselves and with experimental results where available and with other theoretical work. The B3LYP/6-311++G(3d,3p) method gives 2.28 and 4.25 Debye for the dipole moments of pyridine and pyridine N-oxide to be compared with the experimental 2.22 and 4.24–4.32 Debye respectively. In the case of the dipole polarizability, $\bar{\alpha}$, the calculated values from B3LYP/6-311++G(3d,3p) and B3LYP/Sadlej are respectively 63.79 au and 64.50 au to be compared with values ranging from 61 to 64 au in pure liquid pyridine or in solution. Results for the β-hyperpolarizability show much less consistent behavior and a CCSD calculation gives values that have substantially greater absolute magnitude for the principal components in both molecules as compared to the other methods of calculation. It is established, however, that the oxide has a hyperpolarizability about an order of magnitude greater than that of pyridine. In terms of the two level sum over states model the substantial dipole in the ground state of the oxide is now largely removed by the charge transfer leading to the excited state, so that there is a large change in dipole moment associated with the virtual excitation. The effect would be enhanced in POM due to the acceptor property of the nitro group. In the case of the oxide it is observed that the agreement between the CCSD and MP2 calculations (β_{zzz} = –409.1 au and –486.7 au respectively) is much better than the agreement of either with the B3LYP result ((β_{zzz} = –256.4 au), providing a further example of the failure of the standard DFT methods in push-pull compounds.[45,46]

Yamada et al.[47] have calculated the second hyperpolarizabilities (γ) of cationic 3- or 4- substituted pyridine derivatives using ab initio methods including CCSD(T). The magnitude of γ for the pyridinium cation is found to be smaller than that of pyridine; but it is found that the value for the cation changes dramatically on substitution whereas neutral pyridine does not show such large substitution effects.

There have been many reports of exceptionally large β-hyperpolarizabilities in ionic or radical structures and there are general reasons for believing that a structure finely balanced between a covalent and zwitterionic form may have a very large response to a perturbing field. The separation of conjugated π-electron ionic donor and acceptor systems by a σ-bonded spacer provides a method of stabilizing such structures and also introduces interesting effects in the response functions that may be exploitable in devices. Sitha et al.[48] have performed coupled perturbed HF calculations with 6-31G** basis sets on a heterocyclic zwitterionic molecule with such a σ-spacer separating the donor and acceptor and calculate that the static first hyperpolarizability has a value of 240×10^{-30} cm^{5}esu^{-1} Substitution of electron withdrawing groups on the acceptor ring can increase this value to 3969×10^{-30} cm^{5}esu^{-1}. The size of the response is linked to the large charge separation leading to large changes in dipole moment, while the excitation energy and oscillator strength of the associated virtual transition are low. There is evidence from MP2 calculations on small molecules that electron correlation will further enhance the effect. Rao and Bhanuprakash[49] have extended this work using semi-empirical methods.

Borbulevych et al.[50] have investigated the structure of N,N-dimethyl-4-nitroaniline and its derivatives experimentally through X-ray diffraction and theoretically by *ab initio* calculations. They relate the spatial arrangement of the methyl groups relative to the plane of the ring to differences in the relative importance of the quinoid form and the first hyperpolarizabilities of the ortho and meta substituted compounds.

Thanthiriwatte and de Silva[51] report *ab initio* results at the HF level on novel donor-acceptor substituted fluorenyl derivatives (see eg structures I and II). They establish that much larger β-hyperpolarizabilities are found as compared with biphenyl and point out that this is due to the increased conjugation in the fluorenyls which are constrained to be planar whereas biphenyls can rotate about the central bonds. The *ab initio* results are consistent with qualitative interpretations of the effects of the various donor and acceptor groups and changes in the molecular orbital structure.

Yang et al.[52] have used MP2/6-31+G(d) calculations to compare the static β-hyperpolarizabilities of planar and twisted sesquifulvalenes. It is found that the dominant longitudinal component is five times larger in the twisted form where it is negative with respect to the dipole moment axis. The enhancement in the twisted structure is related to the phenomenon of sudden polarization in polar hydrocarbons and leads to β-values that are of the same order as those found in push-pull conjugated polyenes of similar size.

2.1.4 Vibrational Effects. – Work that includes vibrational effects in the calculation of the response functions continues to become more prominent. While very large rapid electronic effects predominate in the functions that describe phenomena involving only optical frequency fields (SHG, THG etc.), in other cases, where lower frequencies are present, it has been established that vibrational effects are sometimes of equal or greater importance.

In earlier work Luis, Kirtman[53,54] and collaborators introduced field induced vibrational co-ordinates (FICs) to simplify the calculation of the nuclear

I
7-nitro-9H-fluoren-2-ylamine

II
2-(7-amino-9H-fluoren-2-ylmethylene)-malonitrile

relaxation component of the vibrational response. The FICs as originally defined formed a small complete set in terms of which an exact account of the nuclear relaxation induced by a particular applied field configuration could be given. The derivation assumed, however, that any optical fields involved in the nonlinear process could be treated in the infinite frequency approximation. (In terms of a sum over states approach this would be equivalent to replacing energy denominators of the form $\omega \pm \omega_{nm}$ by ω) Clearly this approximation cannot give an account of the effect of frequency dispersion. The theory has been generalized by Luis et al.[55] who have introduced frequency-dependent field induced co-ordinates (FD-FICs). Provided that the optical frequency is outside the infra-red region these can be used with a harmonic approximation to obtain accurate results for the electro-optic Pockels and Kerr effects.

Another approach to the selection of significant vibrational co-ordinates for use in the calculation of vibrational response functions has been made by Torii[56]. He introduces the concept of intensity carrying modes (ICMs), derived from derivatives of the dipole, polarizability and first hyperpolarizability. Again, he discusses the construction of a small ICM vector space on which the vibrational response functions can be calculated. The theory is applied to a push-pull polyene, a neutral polyene and a streptocyanine dye cation and produces a satisfactory picture of the polarization properties of these molecules.

Quinet et al.[57] have devised a procedure for computing analytical second derivatives of the frequency-dependent first hyperpolarizability with respect to vibrational co-ordinates, for use with the TDHF approximation. The scheme has been used to calculate the first order zero-point vibrational adjustment (ZPVA) correction for H_2O, NH_3 and CH_4. It is found that the frequency dispersion coefficients are similar to those of the electronic effect except in the case of CH_4.

Torrent-Sucarrat et al.,[58] using HF and MP2 methods, have made calculations of the longitudinal response functions for a set of seven typical medium sized conjugated nonlinear optical molecules. Static electronic, static zero-point vibrational average (ZPVA) and static and dynamic pure vibrational contributions have been included. Higher order pure vibrational terms are treated through 4th or 5th order and field-induced co-ordinates are utilized with a special finite field technique to carry out the correlated treatment. It is found that, while the ZPVA terms are comparatively small, the pure vibrational terms are commonly of the same order or larger than the static electronic term and major contributions come from the higher order vibrational terms. The authors suggest that the initial convergence of the perturbation series in mechanical and electrical anharmonicity requires further investigation.

Torrent-Sucarrat et al.[59] have investigated the choice of basis set on the electronic and vibrational contributions to the nonlinear optical properties of three typical π-conjugated organic molecules. Their conclusions reinforce those reported by other workers – to the effect that the 6-31G basis set often fails to provide even quantitative accuracy while the 6-31+G(d) set is sufficient for semi-quantitative work. MP2 produces a substantial proportion of the electronic correlation effects, as defined by QCISD calculations, but the comparative values of the electronic and vibrational contributions at MP2 level are not consistently of semi-quantitative accuracy.

Maroulis[60] has applied his FF Coupled Cluster technique to derive formulae for the variation of the static polarizability and second hyperpolarizability with the bond length for the symmetric stretching of CO_2 and claims reasonable agreement with recent experimental work.

2.2 Semi-empirical Calculations. – Goller et al.[61] have conducted a systematic investigation of the sum-over-states (SOS) method as used to calculate second order hyperpolarizabilities. They have made detailed studies of the γ-hyperpolarizability of two prototype molecules, 4-nitroaniline (PNA) and 1,3,5,7-all-*trans*-octatetraene (OTE), which are respectively typical of the donor-acceptor aromatic and polyenic structures of interest for optoelectronic applications. There has already been a very large amount of semi-empirical and *ab initio* work published on this kind of structure but, as the authors point out, in the case of the semi-empirical studies the use of the SOS method with rather arbitrary numbers of configurations is unsatisfactory, especially for third order effects (i.e for 2nd order hyperpolarizabilities). The method used by these authors can be defined as AM1/SOS/CI. The AM1 semi-empirical Hamiltonian belongs to the NDDO (Neglect of Diatomic Differential Overlap) class. Much earlier SOS work has employed PPP or versions of CNDO where the approximations used to evaluate the integrals are more drastic. They also attempt to deal with the CI method, used in conjunction with the SOS expansion, in a more systematic way than previous authors. In the SOS method the expansion of the response functions is over the calculated excited states so that all the eigenvalues and eigenvectors of the CI matrix are required. This limits the systematic application, even of CISD, to very small molecules. The authors re-emphasize the well-established result that for γ calculations it is essential to include doubly excited configurations – otherwise the effect of negative terms in the response is grossly over-estimated. Their favored procedure is to select a complete active subspace (CAS) that is at least adequate to include as many microstates as there are $\pi\pi^*$ excitations. For many medium sized molecules, of interest in optoelectronics, this criterion is satisfied by working with a set of 400 microstates (CISD/400). The convergence of the SOS expansion with increasing number of configurations is explored for PNA and OTE and the results of the CISD/400 approximation compared with that from CISD. In these cases the results are in good agreement. A review of other theoretical calculations of γ(THG) and of experimental results is also included. Some overall trends are noted. For PNA the THG and DFWM γ values are found experimentally in nearly non-polar solvents to be in the range 15–20 $\times 10^{-36}$ cm^{7}esu^{-2} (in methanol rising to 48 $\times 10^{-36}$ cm^{7}esu^{-2}) while all calculated values, semi-empirical and *ab initio* are much lower (1–7 $\times 10^{-36}$cm^{7}esu^{-2}). The CISD/400 and CISD results in this paper are in the same lower range. The discrepancy may be attributable to solvent effects. In the case of OTE the range of previous theoretical determinations has had a much better overlap with the gas phase experimental results (7.5–9.1 × 10^{-36} cm^{7}esu^{-2}). However many of the calculations give higher values and the value obtained here for γ(THG) is 17 × 10^{-36} cm^{7}esu^{-2} from CISD/400.

While this treatment has not brought the experimental and theoretical values into better accord it has, nevertheless, some attractive features in comparison to previous approaches. The paper contains an analysis of the dipole moments, the electronic excited state energies and transition moments. Although the method has been parameterized only on the dipole moment data, the electronic spectrum obtained by the NDDO method is in very good agreement with those excitation bands that can be identified in the experimental spectra. This increases confidence that the frequency variation of the response function, which has been calculated over a range of frequencies, will be meaningful. The study of the convergence of various approximations to CISD is also convincing and the authors state explicitly in their conclusions that methods that cannot cater for a large enough CAS will not lead to reliable results, in the sense that the calculated values will not have reached a stable limit. This rules out the application of the method to molecules substantially larger than the two under consideration. (It should perhaps be recalled that the widespread use of the SOS method to treat the β-hyperpolarizabilities of much larger push-pull molecules etc. is viable because the dominant effects are clearly identifiable with the contributions of singly excited configurations.). A further feature of interest is the analysis of the effects of molecular geometry on the hyperpolarizabilities. These are found to be substantial and the authors recommend that DFT/ B3LYP structural optimizations should be carried out before applying the SOS method.

Machado and da Gama[62] have used a AM1/TDHF procedure to investigate the γ-hyperpolarizability of donor-acceptor substituted polyenic derivatives. They permute a number of donor/acceptor combinations with a conjugated bridging group of three or four double bonds to optimize the γ-values for THG at 1.91 micron. They predict that the most effective response is to be obtained from the combination of the phenylamine donor with the dicyanomethylene acceptor (III). They also consider steric effects and check that their method of calculation produces reasonable agreement with a series of substituted polyenes investigated experimentally by Puccetti et al.[63]

Yang and Jiang[64] have studied the γ-hyperpolarizabilities of squaraines. They have used the AM1/FF method to calculate static values and AM1/ TDHF to obtain various forms of the dynamic hyperpolarizability that are related to experimental nonlinear phenomena. It has been known for some time that some centrosymmetric squaraines have large negative γ values. In this paper derivatives of 2,4 diphenyl squaraine (IV) are studied.

By choosing 2,4-diphenyl squaraine as the prototype, the para and ortho substitution effects on the magnitude and sign of γ were investigated. The molecule with the powerful electron donating groups at the para positions exhibits the large negative γ value that originally aroused interest in the properties of the squaraines, while the less powerful donors lead to a smaller value. When the donors are very weak or replaced by electron withdrawing

X=H, OH; R=NH$_2$, CH$_3$, CHO, NO$_2$.

IV

groups, the value may become positive and, with powerful electron withdrawing groups in the para positions, γ may become large and positive. The introduction of hydroxy groups in the ortho positions polarizes the carbonyl groups and makes the central acceptor more powerful so that the negative γ values are enhanced. The change in the sign of γ is of particular interest in connection with laser beam propagation through the medium.

The FF-AM1 method is the basis of a study of some push-pull benzene, thiophene and *trans*-butadiene molecules by Mandal et al.[65]

Sudharsanam et al.[66] have synthesized unsymmetrical donor-acceptor aryl disulphides for which a moderately high β has been measured and which have good transparency to below 420 nm. Calculations show that the molecules have an asymmetric charge distribution about the disulphide bond, which is responsible for their high β values. The design of nonlinear optical chromophores with multiple disulphide bonds should lead to compounds with good second order nonlinearity combined with transparency in the visible.

Bartkowiak et al.[67] have extended work on second order processes in push-pull conjugated molecules with bond alternation to consider the case of two-photon absorption (TPA). The cross-section for TPA between an initial state, g, and final state, f, is given (before various averaging processes over polarization etc.) by,

$$S_{gf} = \sum_k \left(\frac{\langle g|\mu_1 \cdot r|k\rangle \langle k|\mu_2 \cdot r|f\rangle}{\omega_{kg} - \omega} + \frac{\langle g|\mu_2 \cdot r|k\rangle \langle k|\mu_1 \cdot r|f\rangle}{\omega_{kg} - \omega} \right) \quad (2)$$

In earlier work Zalesny et al.[68] have shown that after angular averaging, where there is a strongly predominant $g \rightarrow CT$ charge transfer excitation, the TPA absorption depends on the quantity,

$$\frac{\langle g|x|CT\rangle \Delta\mu}{\omega_{gCT}} \quad (3)$$

which is essentially the same as the two state model expression for SHG.

The paper discusses relations between the averaged TPA cross-section, δ, and the first and second hyperpolarizabilities, β and γ in conjugated molecules, examines the validity of the above relationship, and addresses the question as to how each of them varies with the bond length alternation parameter (BLA). The static β and γ values are calculated by the finite field method and the frequency-dependent TPA factors by a SOS/CIS method. In both cases the GRINDOL[69] variant of the INDO method provides the underlying semi-empirical parametrization. This method is claimed to be less heavily parametrized than either the simpler CNDO theories, which tend to be reparametrized for particular classes of molecule, or the MOPAC/MNDO/AM1 approach where the integrals are obtained from more elaborate empirical analysis. The main conclusions are exhibited in plots of the variation of β,γ and the TPA parameter, δ as a function of the BLA. An example, for structure V is shown in fig 1. The other molecules show similar behaviour. There has been a good deal of work on explaining the trends for hyperpolarizability v. BLA in terms of simple two and three state models(see for example[70]) and an attempt is made to extend these theories to the TPA response, finding formula connecting β,γ and δ with the excitation energy of the two state model. The inadequacy of a two state model for treating the third order effect producing γ (since double excitations are not part of a two state model) raises questions about this procedure, but in compounds showing strong donor-acceptor behaviour it may be that the singly excited configuration produced by excitation to the charge-transfer state is dominant.

On a more formal basis it can be shown that TPA is related to the imaginary part of the γ(-ω;ω,ω,-ω)-tensor. Nakano et al.[71] have developed a method for finding the spatial contributions of electrons to γ(-ω;ω,ω,-ω) involving an expression for the γ density. The imaginary part can be calculated for each virtual excitation and the method can also be applied to other dynamic response properties. A π-electron model of *trans*-stilbene and its donor substituted derivative, 4,4'-diamino-*trans*-stilbene has been used as an example of the application of the method of calculation of the imaginary part of γ.

V

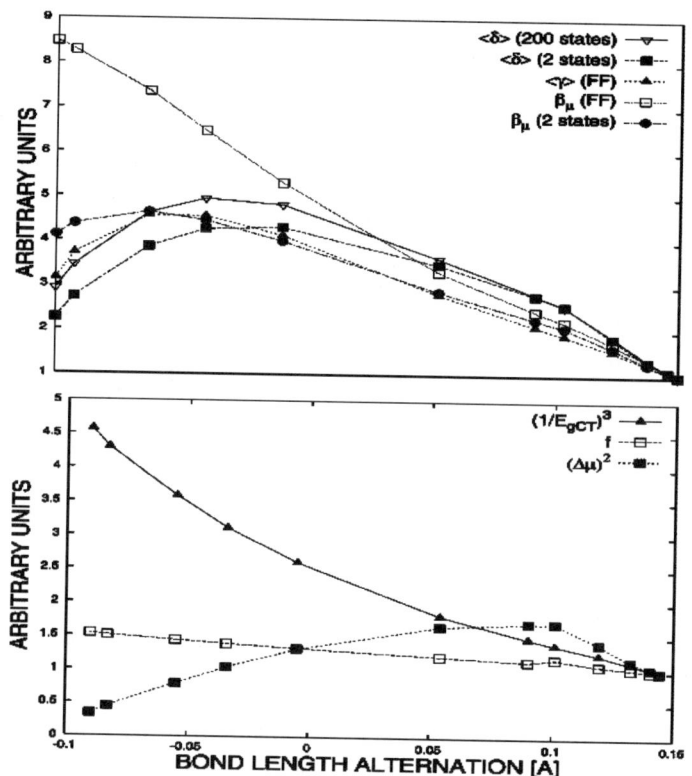

Figure 1 From ref. 67. ▽: *TPA parameter calculated with 200 single excited configurations.* ■ *TPA parameter from 2-state model.* ▲ *γ from FF.* □ *β(EFISH) FF.* ● *β(EFISH) 2-state. The lower diagram shows the variation of the excitation energy, the oscillator strength and the change in dipole moment for the two state model*

Meng et al.[72] have investigated the ground-state geometry and γ-hyperpolarizability of anthraquinone derivatives using a finite field technique. The calculated γ_{zzzz} values are considerably affected by substituents. Measurement of the third order response of 1,4-diamino-2,3-diphenoxy-anthraquinone by degenerate four-wave mixing gives a high value for the $\chi^{(3)}$ of the dye in accordance with the high calculated γ value of 5.71×10^{-31} cm^{7}esu^{-2}.

Diazabutadienes and hexatrienes have been subjected to a combined experimental and theoretical study by Dworczak et al.[73] and Tomonari and Ookubo[74] have used a simplified SOS model to analyse the effect of tetra-substitution on fused ring compounds.

Faustino and Petrov[75] have attempted to calculate the β-hyperpolarizability for 4-nitroaniline (pNA) molecule with a fraction of the molecules in an excited state. The results are compared with hyper-Rayleigh measurements.

Lee et al.[76] have attempted to relate the charge transfer in push-pull molecules to the electronegativity and polarizability. Hillebrand et al.,[77] using semi-empirical optimization of structures and *ab initio* dipole calculations have designed a new class of 2-(2'-hydroxyphenyl)benzoxazoles (HBO) chosen to combine high first hyperpolarizability with photothermal stability and fast excited state intramolecular proton transfer (ESIPT) and conclude that the HBO family of dyes is worthy of further study in this connection.

Sen *et al.*[78] have proposed a simplified Huckel-like model for donor-acceptor molecules, which takes into account electron correlation.

Diaz *et al.*[79] use MNDO, AM1 and PM3 semi-empirical methods to calculate the β-hyperpolarizability and dipole moment of 2-substituted 4-methylene-4H-oxazol-5-ones. Yartsev and Singh[80] have developed a two-dimensional site model to investigate push pull molecules in different conformations; Lukes *et al.*[81] use AMI and ZINDO/S to treat bisthienyls with five membered bridges; Zhou and Feng[82] employ INDO/S and AM1 in an investigation of chains formed from *sym*-tribenzotridehydro annulene and Zhou *et al.*[83] have used density functional theory and the ZINDO method to calculate the TPA cross-sections and γ-hyperpolarizabilities of multi-branched stilbene derivatives.

Daul *et al.*[84] have used semi-empirical and DFT methods to calculate the first and second hyperpolarizabilities of a set of ten (M)-tetrathia-[7]-helicenes.

Abbotto *et al.*[85] have used *ab initio* methods to study a family of azinium-(CH=CH-thienyl)-dicyanomethanido chromophores that show marked zwitterion/quinoid changes in the ground and excited states. The consequent large charge transfer leads to very large values for the first hyperpolarizability. Chromophores with 1,3-dithiol-2-ylidene as the electron donor have been investigated using AM1 structure optimization and INDO/CI for the electronic properties.[86]

Yang and Champagne[87] have used *ab initio* and semi-empirical techniques to analyse the off-diagonal tensor components of the β-hyperpolarizability of λ-shaped molecules. They rationalize the results in terms of a valence bond charge transfer model.

Meng *et al.*[88] have used the FF formulation to calculate the γ-hyperpolarizabilities of anthraquinone derivatives in conjunction with experimental studies.

2.3 Solvent and other Environmental Effects. – Moran *et al.*[89,90] have supplemented experimental measurements of linear absorption spectra and the Resonance Raman effect in 'push-pull' chromophores in various solutions with numerical simulations of the spectra using time-dependent wave-packet propagation methods. This method allows the effect of excited-state geometry changes, vibrational and solvent effects to be incorporated in the simulation. The frequency dispersion of β is also calculated in each solvent by a time-domain form of the two state model that incorporates solvent reorganization, inhomogeneous broadening and the vibronic structure of the charge-transfer state. It is shown that accurate extrapolation to zero frequency of β-values, measured by doubling near infra-red frequencies, requires a realistic description of the excited state as the wavelength approaches a two-photon resonance. It is pointed out that these effects may be particularly important at the high chromophore concentrations that are necessary for device application and are always present in crystals. It may also be essential to consider the extrapolation procedure more carefully than has been usual when comparing measurements with static hyperpolarizability calculations.

Zhu et al.[91] have used DFT with a number of functionals to make a study of the effects of the field, the basis set and the cavity size on the molecular polarizabilities and hyperpolarizabilities of substituted benzenes in liquid or solution. The conductor-like screening model (COSMO) has been used. An optimized model is then used to produce a set of calculated values for the polarizabilities and hyperpolarizabilities which show good agreement with the experimental values. A comparison with other theoretical results indicated that, provided the same solvation model is used, variations between different theoretical methods are comparatively small.

Nandi et al.[92] have employed a scaled self-consistent reaction field (SSCRF) method[93], previously introduced by them, to study the solution values of the static and dynamic β-hyperpolarizabilities of three donor/acceptor conjugated molecules, pNA, trans, 5-dimethylamino-5'-nitro-2,2'-bithiophene (DNBT) and 1,1-dicyano-6-amino-hexatriene (DCH). The calculations are based on structures optimized by the AM1 method in the absence of solvent, and excitation energies for use in a two state model have been calculated by CNDO/S-CI modified by the SSCRF procedure, using 200 singly excited configurations. The finite field method has also been used for the static quantities. The paper contains a very useful set of references to recent work, especially on pNA, including references relating to EFISH and Hyper-Rayleigh scattering measurements in different solvents. Comparisons between theory and experiment are not given in any detail, but the authors draw attention to the fact that serious discrepancies still exist. As a starting point for a study of the current state of theory and experiment on one of the most fundamental problems in organic NLO (the prototype push-pull systems in solution) the paper should prove very useful.

Jorgensen et al.[94] describe a method for obtaining the quadratic response functions for heterogeneously solvated molecules. The method applies to cases where the solvated molecules are on or near the surface of a metal, represented as a perfect conductor. The solvent is simulated by a continuous dielectric medium. The scheme operates at both uncorrelated and correlated *ab initio* levels and has been used to investigate the β-hyperpolarizability and TPA matrix elements for heterogeneously solvated CO. The results are found to depend strongly on the heterogeneous solvent configuration.

Mestechkin[95] has used TDHF theory with a semi-empirical method to calculate the β-hyperpolarizabilities of push-pull molecules that are of interest in connection with Langmuir-Blodgett films. Intermolecular interactions in the films are included. Internal fields are calculated exactly for the ideal film structures through Madelung sums.

2.4 Polymers.

A general review of nonlinear optic polymeric materials, covering the theoretical principles used in the design of chromophores and the methods of incorporating them in polymers to be used in particular kinds of optoelectronic device, has been provided by Dalton[96]. Much of the book is devoted to the methods of developing a molecular polymeric material with given NLO properties into a form suitable for incorporation into a device and to a description of the kinds of device which might usefully employ such a material. The importance of obtaining the correct mix of molecular properties is, however, clearly explained for each case and the volume may be of considerable interest in directing the attention of molecular theoreticians to potentially useful projects.

Jacquemin et al.[97] have used the INDO formalism with the SOS method and with a real space finite field approach to calculate the β tensor of polymethineimine oligomers. The aim has been to find an explanation for the reversal of the sign of the longitudinal β component as the chain length is increased. The INDO calculations agree well with previous ab initio results and a four state SOS model explains the sign reversal in terms of competition from two contributions of opposite sign, which relate to bond charge and bond polarization phenomena. Jacquemin et al.[98] have also investigated polyacetylene/polymethineimine copolymers and find that, as a result of the interplay between delocalization and asymmetry, there exists an optimal composition, estimated as between 50% and 66% of polyacetylene, for obtaining the maximum value of β.

Work on the hyperpolarizabilities of polymers – especially conjugated chains – continues to attract a great deal of interest. A number of studies include both electronic and vibrational effects, using the formalisms that have been developed in the last decade or so to treat the vibrational contributions. Calculations on oligomers are supplemented by methods that make use of infinite one dimensional translational symmetry and employ 'crystal' orbitals (Bloch sums). As far as NLO applications are concerned it is the 'hyperpolarizability density' that is relevant and one of the main objectives of calculations on series of molecules with increasing chain length is to find an optimum length to maximize this quantity. In some cases a simple saturation effect is found, in which case all long polymers would be equally effective beyond a saturation length, but in many others there appears to be a finite length that maximizes the response.

Weniger and Kirtman[99] have described new methods of extrapolating results obtained on oligomers of increasing size to the infinite chain limit. Calculations for the ground state energy of polyacetylene are presented which demonstrate the usefulness of the methods.

Moore and Yaron[100] have applied the PPP (Pariser, Parr, Pople) π-electron semi-empirical method with singly excited CI to the infinite polyacetylene chain. At this level of calculation the lowest excited states are excitons, but the authors find that the results of Huckel theory, where the nonresonant γ-hyperpolarizability is proportional to the third power of the bandwidth and inversely proportional to the sixth power of the optical gap, still hold to a very good approximation. These results give strong support to the Huckel description of long chains. The implication, familiar from solid state physics, is that the states of the long chain (both band states and exciton states) can be successfully represented by re-normalized delocalized orbitals, which, because of their nearly infinite range, interact very weakly. In the shorter chains of the series it is well established that doubly excited and higher configurations must be included to produce the correct results. Further work is necessary to establish how the transition from the short chain to long chain regimes occurs.

Jacquemin et al.[101] have worked at *ab initio* HF MP2 level with 6-31G basis sets to calculate static and dynamic longitudinal β values, including electronic and vibrational effects, for donor/acceptor substituted polyacetylene oligomers up to 16 unit cells. An optimal length is found and the optimum value of β is increased by a factor of 2 when the correlated method is used. Frequency dispersion also has a major effect. Champagne et al.[102] have investigated the effect of doping with alkali metals on γ for polyacetylene chains with up to 70 carbon atoms.

The chain length dependence of the hyperpolarizability and its dependence on bond alternation in linear H-(BN)$_n$-H oligomers has been investigated by Jacquemin et al.[103] using *ab initio* HF and MP2 methods. Criteria for optimum hyperpolarizability density have been given.

The convergence of the exchange integral lattice sum is one of the critical computational features of one dimensional lattice work. Jacquemin et al.[104] have shown that for various stereoregular polymers, including conjugated chains, exponentially decreasing contributions are found, although in some cases convergence is very slow. A new crystal orbital coupled-perturbed HF method has been used by Kirtman et al.[105] to investigate the static longitudinal α and β values of polydiacetylene, polybutatriene and interacting pairs of polyacetylene chains. Gu et al.[106,107] also report dynamic calculations using this method.

Push-pull polyenes of various lengths, regarded as objects of investigation in their own right or as oligomers of polyacetylene, still arouse considerable interest as the precursors of structures that might have NLO applications. The effects of substitutions[108,109] and chain flexibility[110,111] on the large β hyperpolarizability in push-pull (end groups substituted with donor and acceptor) polyenes have been treated by several simplified models and third order effects are also relevant in the context of other NLO effects.[112,113]

Small doubly charged polyacetylene oligomers have been investigated by Oliveira et al.[114] using a number of basis sets with correlation included up to MP2 level.

Balakina et al.[115] report calculations on the β tensor for epoxy oligomers with a chromophore in the main chain and Philip and Sreekumar[116] have used a two-level solvatochromic method and other computations to interpret the second harmonic response of chiral polyesters.

Diaz et al.[117] have used the PM3 hamiltonian from the MOPAC package to calculate the hyperpolarizabilities of polyconjugated carbazolyl-oxazolones and find that there is a low dependence of β on the number of double bonds, implying that the comparatively large effect was due to the rigidity of the chain and the influence of the carbazole moiety.

2.5 Dendritic Structures. – In order to exploit the very high nonlinear response of some organic structures at the molecular level it is necessary to produce properly ordered macroscopic structures containing the active molecular components. For second order properties the macroscopic behaviour must be non centrosymmetric, a requirement that imposes a severe constraint on the development of suitable materials. Much effort has been expended on non-centrosymmetric crystals, poled polymers and Langmuir-Blodgett films. Another possibility is to grow dendrimers that have a suitably noncentrosymmetric arrangement.

One of the most important uses of the theory is to try to simulate the effect of the solvent on the shape of the dendritic growths.

Yamaguchi et al.[118,119], building on earlier synthetic and structural work by Yokoyama et al.[120] have made experimental and theoretical studies of four generations of dendrites of nitro-azobenzene derivatives grown in polar and nonpolar solvents (see fig 2.). A schematic representation of the dendrites, grown in polar solvents, where the structures show greatest deviation from an overall spherical arrangement, is given in fig (3). CNDO/SOS calculations of the β tensors of the dendrites predict that the value for the fourth generation is 21.7 times that of the first generation (98×10^{-30} esu), a trend that is in good qualitative agreement with measured values. Fujita et al.[121] and Nakano et al.[122] have investigated the γ hyperpolarizability of dendritic structures.

Nomura et al.[123] have also used the CNDO/S SOS procedure with singly excited configurations to calculate α, β and γ for phenylacetylene dendrimers. The static γ value is shown to depend quadratically on the number of acetylene chains, which suggests that two chromophores are involves in a coupled response in the third order effect. Nomura and Shibuya[124] have also analyzed the response of a model in which a set of linear oscillators are coupled dendrimerically. It is shown, by numerical calculation, that the total oscillator stength increases linearly with the number of oscillators while the third order

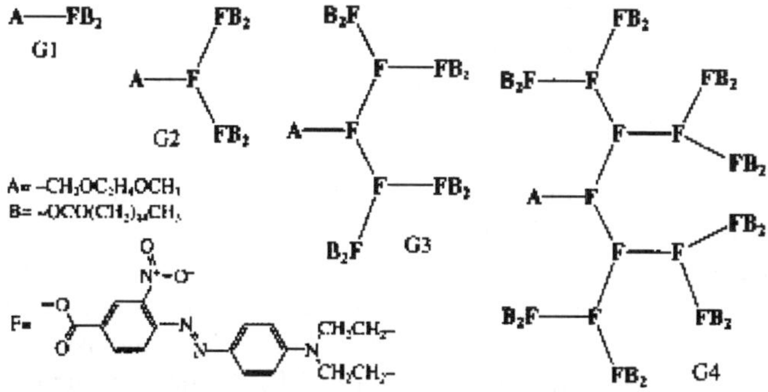

Figure 2 Schematic development of four generations of dendrites (ref. 119)

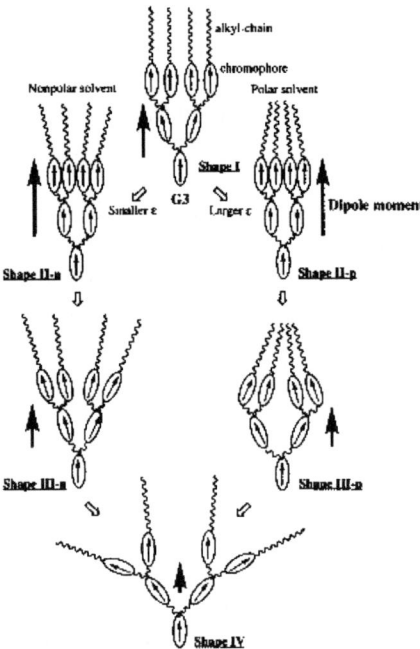

Figure 3 *Structures of dendrimers grown in polar solutions (ref. 119)*

polarizability (γ) increases much more steeply. Lin et al.[125] have used DFT/B3LYP theory to study the potential energy surface and the hyperpolarizabilities of the core (G0) and first generation (G1) dendrimers of polyamidoamide (PAMAM) with the medium sized polarized GTO/CGTO basis sets. A calculation of the inversion barrier indicates that folding of the branches is very easy and it is found that in most cases β for G1 is increased when the molecular structure is distorted from its equilibrium arrangement. It is suggested that these results will provide structural guidelines for the optimization of β by changing dendrimer shape in different solvents.

2.6 Octupolar Molecules. – Zyss and collaborators continue their work on the β tensor in molecular and crystal structures that do not conform to the standard 'one-dimensional' donor-acceptor model that has provided the underlying motivation for almost all the earlier work in this area. In particular they are concerned with octupolar systems where symmetry dictates that the dipole moment is zero, but where there may be substantial tensor components of the quadratic hyperpolarizability. Ledoux and Zyss[126] have reviewed the field. Bartholomew et al.[127] have considered dipolar and octupolar contributions, distinguishing between 'through-bond' and 'through-space' intramolecular charge transfer contributions. Their theoretical conclusions are verified by polarized harmonic light scattering and EFISH measurements. A five-state model has been developed by Cho et al.[128] to interpret the nonlinear optical properties of tetrahedral donor-acceptor octupolar molecules and *ab initio* calculations have been carried out to verify the predictions of the model and to evaluate the π-electron contributions. Boomerang-shaped octupolar molecules derived from triphenylbenzene, which is also octupolar, have been investigated by Brunel *et al.*[129] Le Bozec *et al.*[130] have examined octupolar tris(bipyridyl) metal complexes, Horiuchi *et al.*[131] have investigated the organic/inorganic crystal 2-amino-5-nitropyridinium chloride and Di Bella *et al.*[132] have used the INDO/SCI-SOS method to study Schiff base complexes with large off-diagonal β components. Cho *et al.* have discussed the relationship between the TPA cross-section and β in the light of experimental measurements for octupolar oligomers containing form 2 to 12 molecules of 1,3,5-tricyano-2,4,6-tris (styryl) benzene derivatives and Zojer *et al.*[133] have used *ab initio* methods to calculate the TPA cross sections for conjugated organic molecules with a quadrupolar charge transfer arrangement.

The two state model, originally introduced for the archetypal donor-acceptor complex, has been extended to a three state model to cope with the interpretation of γ in simpler molecules and now a five state model is sometimes used to interpret the more complex active structure in two dimensional and octupolar systems. Lee *et al.*[134] employ a four-state valence-bond three-charge transfer model to investigate two photon absorption and the quadratic hyperpolarizability in four representative series of octupolar molecules. Experimental investigation of polychlorotriphenylmethyl radicals using the harmonic light scattering technique has been supported by a discussion of the evolution of β in systems with D_{3h} and C_{2v} symmetry.[135]

2.7 Crystals. – Munn *et al.*[136] have reviewed the theory underlying the calculation of local fields in crystals and their effect on the linear and nonlinear susceptibilities. In relating the macroscopic susceptibilities to the molecular polarizabilities and hyperpolarizabilities it has been found necessary in previous work to consider each molecule as an assembly of sub-molecules. This method is discussed and it is concluded that, provided sufficiently small sub-molecules are used, good results can be obtained. Fully distributed response, including charge transfer within molecules, can approach experimental accuracy.

Ostroverkhov *et al.*[137,138] have reviewed the criteria that must be satisfied in nonpolar crystals to obtain large nonlinear response functions. The arrangement of chromophores to maximize the contributions from the 'off-diagonal' elements, which are the only non-zero components of $\chi^{(2)}$ in such crystals, is discussed. Such elements are described as 'Kleinman disallowed' since the Kleinman symmetry rules, which apply exactly at zero frequency, would predict that they vanish. The authors conclude that λ-shaped molecules with C_{2v} or similar symmetry are suitable candidates.

Mang *et al.*[139] have carried out *ab initio* calculations on chains of lithium formate monohyrate dimers ($HCOOLi.H_2O$) and have extrapolated the values to obtain an estimate of the quadratic NLO properties of the crystal.

2.8 Clusters. – An assessment of *ab initio* methods (using FF with correlation up to CCSD(T)) in the calculation of the polarizability and hyperpolarizability of Si_4 clusters has been provided by Maroulis and Pouchan.[140]

Ab initio studies of urea clusters have been reported by Wu *et al.*[141,142]. They assess the validity of oriented gas models for molecular crystals by comparison with the results obtained for the clusters and propose a shell-structure model to improve the internal field factor calculations for the crystals. They show that the effect of hydrogen bonding in the clusters, which partially invalidates the additivity implied by the oriented gas model, is underestimated by about 15% at the HF level and must be treated by a method that includes electron correlation. Yakovlev *et al.*[143] have investigated the change in the response of $(TeO_2)_3$ as compared with TeO_2 and find that there is a steep enhancement of the γ-hyperpolarizability. They find that the linear polarizabilities of SO_2, SeO_2 and TeO_2 are of the same order of magnitude while the second hyperpolarizability of TeO_2 is rather greater thatn that of SeO_2 and exceeds that of SO_2 by an order of magnitude.

The relationship between the response of a molecular crystal and the polarizability and hyperpolarizabilities of the individual molecules treated in isolation is central to the understanding of the crystal properties and to the design of more useful materials by the molecular engineering approach. The refinement of methods for local field calculations(see reference 136) is one way to tackle this problem, but still makes the assumption that the response of the sub-units are independent, in the sense that each sub-unit has an intrinsic polarizability or hyperpolarizability through which it responds to the local fields. If the virtual excitations associated with the perturbation by the external field are not completely localized on one molecule (or sub-molecule) this concept becomes somewhat blurred, although not necessarily invalid. There may be an 'effective' molecular response function which includes such effects, but it will certainly not be the same as that pertaining to the isolated molecule. Alternative approaches may use crystal orbital methods- but in solid state physics it is not easy to include correlation effects when beginning from infinitely delocalized Bloch functions- or attempt to extrapolate to the infinite crystal from finite cluster calculations. Guillaume *et al.*[144] have adopted the latter method in a study on clusters of 2-methyl-4-nitroaniline (MNA in NLO jargon). The TDHF/AM1 method has been used to calculate the hyperpolarizability for the linear stacks of molecules produced by displacements along the crystal axes. It is found that the crystallographic *a*-axis stacks produce a large increase in effective hyperpolarizability due to electrostaic interactions between the molecules, whereas there is a decrease in the effective hyperpolarizability for the *b* and *c*-axis stacks. The effects are magnified in the frequency-dependent functions and can be related to shifts in the excitation energies and transition dipoles to the low lying excited states that dominate the uv-spectrum. A large number of intermolecular charge transfer states were found and although the transition moments to these states are small for the *b* and *c* axis stacks, they have a non-negligible effect on the hyperpolarizability along the *a*-axis. These charge transfer effects will presumably be very difficult to accommodate within a local field model.

Botek and Champagne[145] have employed a scheme for treating the static and dynamic response within the time-dependent HF regime in which the effects are decomposed into contributions from atomic orbitals. For each molecule the total property value consists of an intrinsic part from the orbitals belonging to that molecule and a mixed contribution that accounts for the intermolecular interactions. The method has been applied to clusters of all *trans*-hexatriene.

Wu et al.[146] have made theoretical and simulation studies of nonlinear optical crystals in the form of metal clusters. Promising amounts of SHG are predicted.

The averaged dipole polarizabilities of heterodimers of water with small organic molecules have been calculated by Fileti et al.[147] at the MP2 level using the aug-cc-pVDZ basis set. The calculations have been extended to interacting clusters in an attempt to explain the intensities and depolarization ratios in Rayleigh scattering experiments.

Nakano et al.[148] have used a FF method to investigate the effect of π–π intermolecular interactions on the γ-hyperpolarizability of $C_5H_7^+$ cations when these are stacked along the normal to the monomer plane. The results are interpreted in terms of the phase relationships between the interacting frontier π-orbitals along the stack.

2.9 Theoretical Developments. – Lipinski[149] has investigated the differences in properties calculated from the energy expansion and the dipole expansion in cases where the Hellmann-Feynman theorem does not hold. He finds that the differences may be very large and that elements calculated by the dipole expansion in these cases are not correctly related by Kleinman symmetry, leading to the conclusion that the energy expansion is more reliable.

Quinet and Champagne[150] describe two analytical procedures based on the TDHF method for evaluating the first derivatives of the first hyperpolarizability with respect to atomic Cartesian co-ordinates. The methods are applied to a study of the frequency dispersion of the first derivatives of the first hyperpolarizability with respect to the vibrational normal co-ordinates of H_2O, NH_3 and CH_4 and to the determination of hyper-Raman intensities within the double harmonic oscillator approximation.

The same authors have implemented a general time-dependent Hartree-Fock scheme using the $2n+1$ rule to calculate the sum frequency generation first hyperpolarizability. The method has been applied to estimate the second order nonlinear optical response of 4-nitroaniline with one frequency fixed in the ultraviolet and one tunable infra-red frequency. The scheme has also been employed to determine the anisotropic component of the first hyperpolarizability of chiral molecules including helicenes.

Kuzyk[151] has applied his sum rule arguments relating to upper limits for molecular susceptibilities to the diagonal part of the second order susceptibility tensor as measured by EFISH and to a variety of third order measurements. The results confirm that all off-resonant susceptibilities fall within the predicted limits. He extends the arguments to the susceptibilities measured in hyper-Rayleigh scattering (HRS) and discusses the reasons for some exceptions where the response is greater than the expected limiting value. Clays[152] has examined the role of multi-photon fluorescence, which can contribute to the signal generated in experiments intended to lead to the determination of higher order susceptibilities. He shows that, if allowance is made for the fluorescence, those hyperpolarizabilities that appeared to exceed the limits imposed by Kuzyk's theory, when corrected, fall almost exactly on the upper limit.

Ducere et al.[153] have assessed the procedure known as carbomerization as a method of devising organic materials with enhanced first hyperpolarizabilities. Carbomer structures are obtained from any compound by introducing dicarbon, C_2, units into each bond of the structure. General arguments lead to the conclusion that the resulting compound will have symmetry and conjugation properties that are similar to those of the parent compound. The authors refer to the work of Marder et al.[154,155] where it was shown that the first hyperpolarizability in a molecule containing a given conjugated bridge between donor and acceptor groups was maximized for a particular donor/acceptor strength and could not be further improved by using more powerful donor/acceptor couples. They argue that the carbomerization procedure scales up the whole molecular structure, donor, bridge and acceptor, while preserving the relationships established in the parent optimized structure. They have demonstrated the procedure in a study of the completely and partially carbomerized structures derived from pNA. The numerous conformers are investigated through molecular mechanics and the hyperpolarizability through the ZINDO/SOS method. An increase in β by a factor of about 30 is found in favorable cases, but the synthesis of the carbomerized compounds may be problematic.

Simpson et al.[156] argue that it would be advantageous in the study of chromophores of low symmetry to choose an internal co-ordinate system aligned with the charge transfer axis of the dominant virtual transition. The number of significant β components is then greatly reduced.

Zhao et al.[157,158] have developed a dynamic Lie algebra formulation for the calculation of hyperpolarizabilities and applied it, using a model Hamiltonian, to calculate, from the density matrix, statistically averaged values for the hyperpolarizabilities of the para-substituted benzenes.

Weber and Daul[159] have devised a scheme for improved coupled perturbed Hartree-Fock and Kohn-Sham convergence acceleration which can be applied to the calculation of molecular polarizabilities and hyperpolarizabilities.

Larsen et al.[160] have reformulated Hartree-Fock theory for time-independent molecular properties with perturbation-dependent basis sets. The formulation refers directly to the atomic orbital basis and is based on an exponential parametrization of the one-electron atomic orbital density matrix. Formulae for molecular energy derivatives up to fourth order are given.

3 Magnetic Properties

3.1 Magnetizability and Nuclear Shielding. – To appreciate the general structure of the magnetic theory it should be recalled that the magnetic part of the molecular electronic hamiltonian (written schematically) is,

$$H' = \sum_n \left[\frac{i\hbar e}{2m} \mathbf{B}_n \cdot \left(\hat{l}_n + \hat{l}_{sn} \right) + \frac{e^2}{2m} \mathbf{A}_n^2 \right] \tag{4}$$

where $\mathbf{B}_n$ and $\mathbf{A}_n$ are the magnetic field and potential at the nth electron, $\hat{l} = \frac{i\hbar e}{2m}(\mathbf{r} \times \nabla)$ is the orbital angular momentum operator and l_s the spin operator. The magnetic fields produced by the nuclear moments, $\mathbf{M}_N$, will contribute to $\mathbf{B}_n$ and $\mathbf{A}_n$ so that both the magnetizability and the nuclear shielding constants are obtained from the linear response to the perturbation, H'. The rotational g-tensor is also implicitly defined by this level of response, provided the variation of properties with the rotational quantum number, J, can be adequately approximated. If the method of calculation satisfies the magnetic analogue of the Hellman-Feynman theorem then the magnetizability, shielding factors and g-tensor can all be obtained as second derivatives of the energy, W, expressed as a function of the field, nuclear moments and rotational quantum number. The second order energy is,

$$W^{(2)} = \frac{e^2}{2m} \langle 0|A^2|0 \rangle + \sum_{k \neq 0} \frac{\langle 0|H'|k \rangle \langle k|H'|0 \rangle}{(W_0 - W_k)} \tag{5}$$

where the two terms lead respectively to the diamagnetic and paramagnetic contributions to the magnetizability and the shielding constants. The sum is over all excited states and H' in the matrix elements is essentially the angular momentum operator. The explicit expressions for the magnetizability, the shielding constants and the rotational g-tensor, although they vary somewhat with the type of quantum chemical approximation used, always consist of a comparatively simple term derived from the diamagnetic effect and a paramagnetic term that involves a sum over states. These ubiquitous features of the calculations have a determining influence on the type of approximation used.

The relationship between the magnetizability and magnetic shielding has been explored by Janowski and Wolinski[161]. While it is relatively easy to calculate reasonable magnetizabilities the theoretical determination of screening factors and chemical shifts to the accuracy required for useful application in NMR work has proved to be much more difficult and requires the use of highly correlated wavefunctions. A relation between the shielding parameter and the magnetizability has first been established from the analysis of numerical results and subsequently it has been shown that the integrated shielding factor is simply proportional to the magnetizability, the constant of proportionality being -2/3 times the vacuum permeability. This result could be very useful in normalizing theoretical estimates of shielding factors to make them more compatible with the experimental values.

Benchmark calculations on small molecules, using highly correlated methods, have been carried out over the last decade and the methods have been reviewed by Helgaker et al.[162] a few years ago. The MRSCF *ab initio* approach favored in this work has produced results that are generally in very good agreement with experiment, but the level of calculation that appears to be necessary to reproduce the required level of spectroscopic accuracy in the shielding constants is clearly impracticable for applications to larger molecules. An adequate representation of the energy levels and matrix elements that appear in the paramagnetic terms in the above formulae requires the use of highly correlated methods and the computationally easier single reference HF or TDHF methods are not adequate.

The form of the equations does give an indication, however, that DFT methods might be convenient. The Kohn-Sham equations of DFT produce molecular eigenvalues and 'orbitals' in which correlation effects are implicitly included, so that, if the density functional were exact, the paramagnetic term summations would also be exact. An attempt to implement the DFT approach (briefly described in the previous SPR) was made by Wilson et al.[163–166] They devised two empirically adjusted hybrid density functionals based on the well-known B3LYP and B97 approximations.[167,168] The energy levels and matrix elements calculated with these functionals were used in the generalized gradient approximation (GGA) formulae for the magnetic response functions. This procedure is inconsistent in that since the hybrid density functional includes a contribution from the HF exchange operator, which is not multiplicative in the electron density, the GGA formulae for the response functions are not strictly applicable. Nevertheless good agreement with coupled cluster calculations and experimental data was achieved. More recently a new procedure has been devised by Wilson and Tozer.[169] They utilize the method of Zhao et al.[170] to obtain orbitals and eigenvalues from given electron densities. The electron densities may have been calculated by any *ab initio* method but the relation between the orbitals and eigenvalues and the Kohn-Sham Hamiltonian, which depends on the given electron density, should be consistent with density functional theory. The Kohn-Sham orbitals and eigenvalues obtained in this way will be consistent with the multiplicative operators that lead to the simple non-coupled formulae for the magnetic response functions. Wilson and Tozer have refined the procedure in that they allow the electron density to change during the iterative determination of the orbitals and use a larger basis set (compared to that used in the original *ab initio* electron density determination) during the process. The larger basis set is probably essential to achieve a reasonable degree of gauge invariance through the LORG method, which is the one employed. A disadvantage of this procedure is that, for larger molecules, the initial *ab initio* calculation of the electron density at an adequate correlated level will become impracticable. Wilson and Tozer have explored the use of standard multiplicative DFT functionals to produce the initial electron density. This latter procedure is rather difficult to distinguish from a straightforward application of DFT using a multiplicative approximation. However the Zhao Morrison Parr (ZMP) method for determining the orbitals and eigenfunctions from an (arbitrary) electron density introduces a constraint on the asymptotic behaviour of the orbitals which ensures that they satisfy physical requirements.

The DFT approximate treatments (in particular) depend on the choice of functionals which are partly of empirical origin and one cannot make definitive statements on theoretical grounds as to which approximation should be more acceptable. Extensive comparisons with experimental data are given in the references. As can be seen from these comparisons it is still proving difficult with any theoretical method to predict chemical shifts to within 1ppm, which would be desirable if tables of theoretical values are to become a generally useful source of information in connection with NMR spectroscopy. In this context the question of vibrational effects also becomes significant. Ruud et al.[171] have continued their work on magnetic vibrational effects and have investigated the effects of zero-point vibrational motion on the nuclear magnetic shielding constants for a number of organic molecules. Anharmonic contributions from the potential energy surface and harmonic contributions from the curvature of the property surface are included. The vibrational corrections in the case of hydrogen shielding are found to be in the range +0.50 to −0.70 ppm, so that in extreme cases errors of more than 1ppm may occur as a result of ignoring these corrections. The authors find, however, that, in the case of hydrogen, the corrections are approximately transferable from one molecule to another. Results for hydrogen shielding in 35 molecules are given. No such transferability has been found for nuclei other than hydrogen. In further studies of vibrational effects the same authors present the results of *ab initio* magnetizability calculations[172] on more than 60 molecules. They find that vibrational corrections to isotropic magnetizabilities are generally negligible except where there are aromatic or anti-aromatic ring systems when the effect my amount to 1% due to the contribution of the component normal to the molecular plane. The magnetizability anisotropy, however, may often contain vibrational contributions of 5–10%. It is also shown that the additivity of magnetizabilities (Pascal's rule) breaks down in the case of fluorine containing compounds.

A number of publications are concerned with aromaticity indices and ring currents and their relation to the chemical shifts. Indices, such as the aromatic stabilization energy (ASE) and Schleyer's nucleus independent chemical shift (NICS) are obtained, usually from *ab initio* and DFT calculations. The anisotropy and exaltation (due to aromatic ring currents) of the magnetizability is also calculated. Rodriguez-Otero et al.[173] have used *ab initio* MP2 and B3LYP DFT methods with 6-31G(d,p) basis sets to investigate lumiflavin; Mills[174] has studied the antiromaticity of fluorenylidene compounds; Poater et al.[175] interpreted the local aromaticites of polycyclic hydrocarbons and fullerenes; Stepien et al.[176] have used B3LYP.6-322+G** DFT calculations to invetigate substituted fulvenes. De Proft et al.[177] analyse the relative aromaticity of the three singlet benzyne isomers, 1,2-, 1,3- and 1,4- didehydrobenzenes.

Jartin et al.[178] have made a detailed study of the magnetic properties of s-indacene, regarded as an archetypal paraoptic, non-aromatic molecule. An intense paramagnetic flow in the electron distribution was identified which causes strong upfield chemical shifts for protons attached to hexagonal and pentagonal rings. The continuous transformation of the origin of co-ordinates (CTOCD) method which has been systematically exploited by Lazzeretti and co-workers is used in this study.

Sebastiani and Parrinello[179] have used DFT and molecular dynamics to evaluate the chemical shifts in water under normal and supercritical conditions. They have also devised a new method for computing NMR chemical shifts in periodic media.[180]

3.2 Hypermagnetizability and the Cotton-Mouton Effect.

– The Cotton-Mouton effect – birefringence induced by a static magnetic field -continues to attract a good deal of attention. The effect is third order and is the magnetic analogue of the optical Kerr effect. In analogy with the latter electric field effect the observed Cotton- Mouton effect in a gas or liquid depends on an indirect contribution resulting from the re-orientation of the molecules in the strong magnetic field and also on the distortion of the electron cloud *via* the mixed electric field–magnetic field hyper-magnetizablity/polarizability, η, defined (with the appropriate frequencies to be added) through the expansion of the energy for a diamagnetic material,

$$W(E,B) = W_0 - \mu_\alpha^{(0)} F_\alpha - \frac{1}{2}\alpha_{\alpha\beta} F_\alpha F_\beta - \frac{1}{2}\chi_{\alpha\beta} B_\alpha B_\beta$$
$$- \frac{1}{6}\xi_{\alpha\beta,\gamma} B_\alpha B_\beta F_\gamma - \eta_{\alpha\beta,\gamma\delta} B_\alpha B_\beta F_\gamma F_\delta - \cdots \quad (6)$$

where F and B are respectively the electric and magnetic fields and the summation convention over repeated indices is implied.

Double differentiation with respect to the magnetic or electric fields leads to expressions for the field dependent magnetizabilities. The technique currently favored for computation is to evaluate the electric field-dependent susceptibility by analytical methods when a finite electric field has been applied, and then to extract the coefficients from equation (6) by numerical differentiation. In the former procedure gauge invariant methods must be used.

$$\chi_{\alpha\beta}^E = -\left[\frac{\partial^2 W(B,E)}{\partial B_\alpha B_\beta}\right]_{B=0} = \chi_{\alpha\beta} + \xi_{\alpha\beta,\gamma} F_\gamma + \frac{1}{2}\eta_{\alpha\beta,\gamma\delta} F_\gamma F_\delta + \cdots \quad (7)$$

Experimentally the Cotton-Mouton coefficient for a gas is defined through the equation,

$$\Delta n = C_{CM} \lambda B^2 \quad (8)$$

where Δn is the induced difference in the refractive indices producing the birefringence. A molar constant, $_mC$, that includes some numerical factors arising from the expansion coefficients in equation (7) is given by,

$$_mC = \frac{2\lambda V_m C_{CM}}{27} \qquad (9)$$

A statistical mechanical calculation for the reorientational effect in rigid diatomic molecules then connects the macroscopic constant to the molecular response functions through the equation,

$$_mC = \frac{2\pi N}{27}\left\{\Delta\eta + \frac{1}{5kT}\left(\alpha_{\alpha\beta}\chi_{\alpha\beta} - \frac{1}{3}\alpha_{\alpha\alpha}\chi_{\beta\beta}\right)\right\} \qquad (10)$$

where,

$$\Delta\eta = \frac{1}{5}\left(\eta_{\alpha\beta,\alpha\beta} - \frac{1}{3}\eta_{\alpha\alpha,\beta\beta}\right) \qquad (11)$$

is the anisotropic part of the mixed hypermagnetizability and the other, temperature dependent, term is a product of the anisotropies in the linear polarizability and magnetizability.

The inert gases provide a testing ground for accurate *ab initio* procedures for the calculation of the more unusual molecular response functions and there has been considerable previous work on polarizabilities, hyperpolarizabilities, magnetizabilities and hypermagnetizabilities, especially in argon and helium. Rizzo et al.[181] have carried out a detailed study of the hypermagnetizability and Cotton-Mouton effect in helium. While the Kerr effect is present in all substances, no matter what the symmetry, the Cotton-Mouton effect is subject to some restrictions. In a gas of non-interacting spherical particles, as a consequence of its dependence on the anisotropic part of the hyperpolarizability, it would not be observed. In helium, therefore, the effect is entirely due to the interaction between the atoms and is related to the interatomic potential and also, since it depends on the mean interatomic distance, to the pressure. In this study the interaction induced anisotropic polarizability, magnetizability and hyperpolarizability are calculated and the pressure dependence of the Cotton-Mouton effect expressed in terms of virial coefficients. The calculations use high level extended basis sets with gauge invariant (GIAO) orbitals and a complete CI procedure. With the GIAO's it is not possible to determine the hyperpolarizability analytically so that the numerical procedure, which produces only the static hyperpolarizability must be used. The major part of the Cotton-Mouton effect arises however from the product term involving the anisotropic polarizability and magnetizability, both of which quantities have been calculated over a range of frequencies. The paper will perhaps serve as a benchmark for the calculation of small higher order magnetic effects. Jonsson et al.[182] have used the multiconfigurational SCF wave functions and coupled cluster methods to calculate the hypermagnetizability and its anisotropy for CO_2, N_2O, OCS and CS_2 from which the Cotton-Mouton effect can be evaluated. Vibrational effects are also assessed and it is suggested that they might be more important than the electronic contribution. Cappelli et al.[183] have attempted to calculate the same quantities for the same set of molecules using the computationally less demanding density functional theory. The agreement between the two sets of calculations and with the earlier experimental Cotton-Mouton measurements of Klug and Huttner[184] is satisfactory.

In a study closely integrated with the analysis of experimental results obtained at the Grenoble High Magnetic Field Laboratory Cappelli et al.[185] have calculated the dipole polarizability and magnetizability tensors at 632.8 nm (He/Ne laser wavelength) and the static hypermagnetizability anisotropy for furan, thiophene and selenophene, using DFT. The polarizable continuum model (PCM) is employed to describe the systems in a condensed phase. The temperature dependent Cotton-Mouton constant for the three molecules has thus been obtained for gaseous, pure liquid and solution phases and compared with results obtained with experimental results. The performance of the DFT and PCM models can thus be assessed against recent experimental data.

Caputo and Lazzeretti[186] have attempted to calculate the second hypermagnetizabilities of three small molecules at the Hartree-Fock level of accuracy. They treat the problem of gauge invariance in a number of ways, both in the common origin approach where the size of the basis set, if sufficiently close to completeness, should overcome the gauge invariance problem; and by a number of continuous transformation methods. In the latter the diatomic contribution is formally removed at each point in space by using coordinate dependent gauge transformations. Permutation symmetry criteria give an indication that the results are of near Hartree-Fock quality.

References

1. D. Pugh in *Chemical Modelling*, ed A. Hinchliffe, Specialist Periodical Reports, Royal Society of Chemistry, Vol 1, 2000, 1; Vol 2, 2001, 293.
2. G. Maroulis, *J. Chem. Phys.*, 2002, **118**, 2673.
3. S. Coriani, A. Rizzo, K. Ruud and T. Helgaker, *Mol. Phys.*, 1996, **88**, 931.
4. J. Kobus, D. Moncrieff and S. Wilson, *J. Phys.*, 2001, **B34**, 5127.
5. A.V. Shtoff, M. Rerat and S.I. Gusarov, *Eur. Phys.J.*, 2001, **15**, 199.
6. A. Avramopoulis and M.G. Papadopoulis, *Mol. Phys.*, 2002, **100**, 821.
7. A.J. Sadlej, *Collect. Czech. Chem. Commun.*, 1988, **53**, 1995.
8. A.J. Sadlej and M. Urban, *J. Mol. Struct (Theochem).*, 1991, **234**, 147.
9. T. Pluta and A.J. Sadlej, *Chem. Phys. Lett.*, 1998, **297**, 391.
10. T.H. Dunning Jr., *J. Chem. Phys.*, 1989, **90**, 1007.
11. R.A. Kendall, T.H. Dunning Jr. and R.J. Harrison, *J. Chem. Phys.*, 1992, **96**, 6796.
12. D. Tunega and J. Noga, *Theoret. Chem. Accounts*, 1998, **100**, 78.
13. D. Jonsson, P. Norman and H. Agren, *J. Chem. Phys.*, 1996, **105**, 6401.

14. D.M. Bishop, *Adv. Chem. Phys.*, 1998, **104**, 1.
15. V.E. Ingamells, M.G. Papadopoulos and A.J. Sadlej, *Chem. Phys. Lett.*, 2000, **316**, 1645.
16. G. Maroulis and D. Xenides, *J. Phys. Chem. A*, 2003, **107**, 712.
17. G.L.D. Ritchie, J.N. Watson and R.I. Keir, *Chem. Phys. Lett.*, 2003, **370**, 376.
18. G. Maroulis, *J. Phys. B*, 2001, **34**, 3727.
19. G. Maroulis and A. Haskopoulos, *Chem. Phys. Lett.*, 2002, **358**, 64.
20. A. Rizzo, C. Hattig, B. Fernandez and H. Koch, *J. Chem. Phys.*, 2002, **117**, 2609.
21. M. Medved, M. Urban, V. Kello and G.H.F. Diercksen, *J. Mol. Structure (Theochem)*, 2001, **547**, 219.
22. P. Neogrady, M. Medved, I. Cernusak and M. Urban, *Mol. Phys.*, 2002, **100**, 541.
23. D. Xenides and G. Maroulis, *J. Chem. Phys.*, 2001, **115**, 7953.
24. P. Karamanis and G. Maroulis, *J. Mol. Struct. (Theochem)*, 2003, **621**, 157.
25. G. Maroulis, *J. Chem. Phys.*, 1998, **108**, 5432; 1999, **111**, 583.
26. G. Maroulis, *J. Comput. Chem.*, 2003, **24**, 443.
27. G. Maroulis and A. Haskopoulos, *Chem. Phys. Lett.*, 2001, **349**, 335.
28. B.Q. Wang, Z.R. Li, D. Wu and C.C. Sun, *J. Mol. Struct.(Theochem)*, 2003, **620**, 77.
29. K. Ohta, S. Yamada, T. Tanaka and K. Kamada, *Mol. Phys.*, 2003, **101**, 315.
30. A. Mito, K. Hagimoto and C. Takahashi, *Nonlinear Optics*, 1995, **13**, 3.
31. A. Mito and H. Moriwaki, *Proc. SPIE*, 1999, **3610**, 188.
32. M.G. Papadopoulos and A. Avramopoulos, *J. Phys. Chem. A*, 2003, **107**, 3907.
33. A.V. Gubskaya and P.G. Kusalik, *Mol. Phys.*, 2001, **99**, 1107.
34. W. Bartkowiak, R. Zalesny, M. Cowal and J. Leszczynski, *Chem. Phys. Lett.*, 2002, **362**, 224.
35. S. Yamada, K. Yamaguchi and K. Ohta, *Synth. Met.*, 2003, **137**, 1419.
36. P. Romaniello and F. Lelj, *Chem. Phys. Lett.*, 2003, **372**, 51.
37. M. Gruning, O.V. Gritsenko, S.J.A. van Gisbergen and E.J. Baerends, *J. Chem. Phys.*, 2002, **116**, 9591.
38. W. Hieringer, S.J.A. van Gisbergen and E.J. Baerends, *J. Phys. Chem.*, 2002, **A106**, 10380.
39. P. Salek, O. Vahtras, T. Helgaker and H. Agren, *J. Chem. Phys.*, 2002, **117**, 9630.
40. H.H. Heinze, F. Della Sala and A. Gorling, *J. Chem. Phys.*, 2002, **116**, 9624.
41. H. Soscún, J. Hernándezm, R. Escobar, C. Toro-Mendoza, Y. Alvarado and A. Hinchliffe, *Int. J. Quant. Chem.*, 2002, **90**, 497.
42. J. Zyss, D.S. Chemla and J.F. Nicoud, *J. Chem. Phys.*, 1981, **74**, 4800.
43. J. Zyss, D.S. Chemla and J.F. Nicoud, in: 'Organic Materials for Nonlinear Optics', eds. D.S. Chemla and J. Zyss, Academic Press, Orlando, USA, 1987, **1**, 23.
44. H. Soscún, O. Castellano, Y. Bermúdez, C. Tro-Mendoza, A. Marcano and Y. Alvarado, *J. Mol. Struct.(Theochem)*, 2002, **592**, 19.
45. S.J.A. van Gisbergen, J.G. Sniders and E.J. Baerends, *J. Chem. Phys.*, 1998, **109**, 10657.
46. B. Champagne, E. Perpete, D. Jacquemin, S.J.A. van Gisbergen, E.-J. Baerends, C. Soubra-Ghaoui, K.A. Robins and B. Kirtman, *J. Phys. Chem.* 2000, **A104**, 4755.
47. S. Yamada, K. Yamaguchi and K. Ohta, *Mol. Phys.*, 2002, **100**, 1839.
48. S. Sitha, J.L. Rao, K. Bhanuprakash and B.M. Choudry, *J. Phys. Chem.*, 2001, **A105**, 8727.

49. J.L. Rao and K. Bhanuprakash, *Synthetic Metals*, 2003, **132**, 315.
50. O.Y. Borbulevych, R.D. Clark, A. Romero, L. Tan, M.Y. Antipin, V.N. Nesterov, B.H. Cardelino, C.E. Moore, M. Sanghadasa and T.V. Timofeeva, *J. Mol. Struct.*, 2002, **604**, 73.
51. K.S. Thanthiriwatte and K.M.N. de Silva, *J. Mol. Struct.(Theochem)*, 2002, **617**, 169.
52. M.L. Yang, D. Jacquemin and B. Champagne, *PCCP*, 2002, **4**, 5566.
53. J.M. Luis, B. Champagne and B. Kirtman *Int. J. Quantum Chem.*, 2000, **13**, 293.
54. J.M. Luis, M. Duran, B. Champagne and B. Kirtman, *J. Chem. Phys.*, 2000, **113**, 5203.
55. J.M. Luis, M. Duran and B. Kirtman, *J. Chem. Phys.*, 2001, **115**, 4473.
56. H. Torii, *J. Comp. Chem.*, 2002, **23**, 997.
57. O. Quinet, B. Kirtman and B. Champagne, *J. Chem. Phys.*, 2003, **118**, 505.
58. M. Torrent-Sucarrat, M. Sola, M. Duran, J.M. Luis and B. Kirtman, *J. Chem. Phys.*, 2002, **116**, 5363.
59. M. Torrent-Sucarrat, M. Sola, M. Duran, J.M. Luis and B. Kirtman, *J. Chem. Phys.*, 2003, **118**, 711.
60. G. Maroulis, *Chem. Phys.*, 2003, **291**, 81.
61. A.H. Goller, S. Erhardt, and U.W. Grummt, *J. Mol. Struct.(Theochem)*, 2002, **585**, 143.
62. A.E.D. Machado and A.A.S. da Gama, *J. Mol. Struct.(Theochem)*, 2003, **620**, 21.
63. G. Puccetti, M. Blsnchard-Desce, I. Ledoux, J.-M. Lehn and J. Zyss, *J. Phys. Chem.*, 1993, **97**, 9385.
64. M.L. Yang and Y.S. Jiang, *PCCP*, 2001, **3**, 4213.
65. K. Mandal, T. Chattopadhyay, P.K. Nandi and S.P. Bhattacharyya, *Indian J. Chem. A*, 2003, **42**, 449.
66. R. Sudharsanam, S, Chandrasekaran and P.K. Das, *J. Mater. Chem.*, 2002, **12**, 2904.
67. W. Bartkowiak, R, Zalesny and J. Leszczynski, *Chem. Phys.*, 2003, **287**, 103.
68. R. Zalesny, W. Bartkowiak, S. Styrcz and J. Leszczynski, *J. Phys. Chem. A.*, 2002, **106**, 4032.
69. J. Lipinski, *Int. J. Quantum Chem.*, 1988, **34**, 423.
70. M. Barzoukas, C. Runser, A. Fort and M. Blanchard-Desce, *Chem. Phys. Lett.*, 1996, **257**, 531.
71. M. Nakano, H. Fujita, M.Takahata and K. Yamaguchi, *Chem. Phys. Lett.*, 2002, **356**, 462.
72. Q.H. Meng, W.F. Yan, M.J. Yu and D.Y. Huang, *Dyes Pigment.*, 2003, **56**, 145.
73. R. Dworczak, W.M.F. Fabian, C. Reidlinger, A. Rumpler, J. Schachner and K. Zangger, *Spectroc. Acta A-Molec. Biomolec. Spectr.*, 2002, **58**, 2135.
74. M. Tomonari and N. Ookubo, *Chem. Phys. Lett.*, 2002, **351**,431.
75. W.M. Faustino and D.V. Petrov, *Chem. Phys. Lett.*, 2002, **365**, 170.
76. J.Y. Lee, K.S. Kim amd B.J. Mhin, *J. Chem. Phys.*, 2001, **115**, 9484.
77. S. Hillebrand, M. Segala, T. Buckup, R.R.B. Correia, F. Horowitz and V. Stefani, *Chem. Phys.*, 2001, **273**, 1.
78. R. Sen, D. Majumdar, K.K. Das and S.P. Bhattacharyya, *Ind. J. Chem. A.*, 2001, **40**, 804.
79. J.L. Diaz, B. Villacampa, F. Lopez-Calahorra and D. Velasco, *Chem. Mat.*, 2002, **14**, 2240.
80. V.M. Yartsev and M. R. Singh, *Chem. Phys.*, 2002, **276**, 293.

81. V. Lukes, M. Breza and V. Laurinc, *J. Mol. Struct.(Theochem)*, 2002, **582**, 213.
82. Y.F. Zhou and S.Y. Feng, *Solid State Commun.*, 2002, **122**, 307.
83. X. Zhou, A.M. Ren, J.K. Feng and X.J. Liu, *Chem.Phys. Lett.*, 2002, **362**, 541.
84. C.A. Daul, I. Ciofini and V. Weber, *Int. J. Quantum Chem.*, 2003, **91**, 297.
85. A. Abbotto, L. Beverina, S. Bradamante, A. Facchetti, C. Klein, G. A. Pagani, M. Redi-Abshiro and P. Wortmann, *Chem.-Euro. J.*, 2003, **9**, 1991.
86. X.J. Liu, J.K. Feng, A.M. Ren and J.Z. Sun, *Chem. J. Chin. Univ.-Chin.*, 2003, **24**, 488.
87. M.L. Yang and B. Champagne, *J. Phys. Chem. A*, 2003, **107**, 3942.
88. Q.H. Meng, W.F. Yan, M.J. Yu and D.Y. Huang, *Dyes Pigment.* 2003, **56**, 145.
89. A.M. Moran, D.S. Egolf, M. Blanchard-Desce and A.M. Kelley, *J. Chem. Phys.*, 2002, **116**, 2542.
90. A.M. Moran, A.M. Kelley and S. Tretiak, *Chem. Phys. Lett.*, 2003, **367**, 293.
91. W.H. Zhu, G.S. Wu and Y.S. Jiang, *Int. J. Quantum Chem.*, 2002, **86**, 347.
92. P.K. Nandi, T. Chattopahyay and S.P. Bhattacharyya, *J. Mol. Struct. (Theochem)*, 2001, **545**, 119.
93. P.K. Nandi, K.K. Das and S.P. Bhattacharyya, *J. Mol. Struct.(Theochem)*, 1999, **466**, 155.
94. S. Jorgensen, M.A. Ratner and K.V. Mikkelsen, *J. Chem. Phys.*, 2001, **115**, 8185.
95. M.M. Mestechkin, *Opt. Commun.*, 2001, **198**, 199.
96. L. Dalton, 2002, 'Nonlinear optical polymeric materials: from chromophore design to commercial application', vol **158** of *Advances in Polymer Science*.
97. D. Jacquemin, D. Beljonne, B. Champagne, V. Geskin, J.L. Bredas and J.M. Andre, *J. Chem. Phys.*, 2001, **115**, 6766.
98. D. Jacquemin, B. Champagne and J. M. Andre, *Macromolecules*, 2003, **36**, 3980.
99. E.J. Weniger and B. Kirtman, *Comput. Math. Appl*, 2003, **45**, 189.
100. E.E. Moore and D. Yaron, *J. Phys. Chem. A*, 2002, **106**, 5339.
101. D. Jacquemin, B. Champagne, E.A. Perpete, J.M. Luis and B. Kirtman, *J. Phys. Chem. A*, 2001, **105**, 9748.
102. B. Champagne, M. Spassova, J.B. Jadin and B. Kirtman, *J. Chem. Phys.*, 2002, **116**, 3935.
103. D. Jacquemin, E.A. Perpete and B. Champagne, *Phys. Chem. Chem. Phys.*, 2002, **4**, 432.
104. D. Jacquemin, J.G. Fripiat and B. Champagne, *Int. J. Quantum Chem.*, 2002, **89**, 452.
105. B. Kirtman, B. Champagne, F.L. Gu and D.M. Bishop, *Int. J. Quantum Chem.*, 2002, **90**, 709.
106. F.L. Gu, Y. Aoki and D. M. Bishop, *J. Chem. Phys.*, 2002.
107. F.L. Gu, D.M. Bishop and B. Kirtman, *J. Chem. Phys.*, 2001, **115**, 10548.
108. W.H. Zhu, G.S. Wu and Y.S. Jiang, *Int. J. Quantum Chem.*, 2002, **86**, 390.
109. M.B. Zuev, S.E. Nefediev and J.L. Bredas, *Pol. J. Chem.*, 2002, **76**, 1211.
110. S. Sugliani, M. del Zoppo, G. Zerbi and C.F. Shu, *Chemical Physics*, 2001, **271**, 127.
111. M. del Zoppo, S. Sugliani and G. Zerbi, *Synthetic Metals*, 2001, **124**, 167.
112. T.L. Fonseca, M.A. Castro, C. Cunha and O.A.V. Amaral, *Synthetic metals*, 2001, **123**, 11.
113. U. Gubler, S. Concilio, C. Bosshard, I Biaggio, P. Gunter, R.E. Martin, M.J. Edelmann, J.A. Wytko and F. Diederich, *Appl. Phys. Lett.*, 2002, **81**, 2322.
114. L.N. Oliveira, O.A.V. Amaral, M.A. Castro and T.L. Fonseca, *Chem. Phys.*, 2003, **289**, 221.

115. M.Y. Balakina, M.F. Ilyazov and M.B. Zuev, *Pol. J. Chem.*, 2002, **76**, 1199.
116. B. Philip and K. Sreekumar, *J. Polym. Sci. A-Polym. Chem.*, 2002, **40**, 2868.
117. J.L. Diaz, B. Villacampa, F. Lopez-Calahorra and D. Valasco, *Tetrahedron Lett.*, 2002, **43**, 4333.
118. y. Yamaguchi, Y. Yokomichi, S. Yokoyama and S. Mashiko, *J. Mol. Struct. (Theochem)*, 2001, **545**, 187.
119. Y. Yamaguchi, Y. Yokomichi, S. Yokoyama and S. Mashiko, *J. Mol. Struct (Theochem).*, 2002, **578**, 35.
120. S. Yokoyama, T. Nakahara, A. Otomo and S. Mashiko, *Thin Solid Films*, 1998, **331**, 248.
121. H. Fujita, M. Nakano, M. Takahata, S. Kiribayahi and K. Yamaguchi, *Mol. Cryst. Liquid Cryst.*, 2001, **371**, 261.
122. M. Nakano, H. Fujita, M. Takahata and K. Yamaguchi, *J. Chem. Phys.*, 2001, **115**, 6780.
123. Y. Nomura, T. Sugishita, S. Narita and T. Shibuya, *Bull. Chem. Soc. Japan*, 2002, **75**, 481.
124. Y. Nomura and T. Shibuya, *J. Chem. Soc. Japan*, 2002, **71**, 767.
125. C.S. Lin, K.C. Wu, R.J. Sa, C.Y. Mang, P. Liu and B.T. Zhuang, *Chem. Phys. Lett.*, 2002, **363**, 343.
126. I. Ledoux and J. Zyss, *Comptes Rendus Physique*, 2002, **3**, 407.
127. G.P. Bartholomew, I. Ledoux, S. Mukamel, G.C. Bazan and J. Zyss, *J. Am. Chem. Soc.*, 2002, **124**, 13480.
128. M.H. Cho, S.Y. An, H. Lee, I. Ledoux and J. Zyss, *J. Chem. Phys.*, 2002, **116**, 9165.
129. J. Brunel, A. Jutand, I. Ledoux and J. Zyss, *Synth. Met.*, 2001, **124**, 195.
130. H. Le Bozec, T Renouard, M. Bourgault, C, Dhenaut, S. Brasselet, I. Ledoux and J. Zyss, *Synth. Met*, 2001, **124**, 185.
131. N, Horiuchi, F. Lefaucheux, A. Ibanez, D. Josse and J. Zyss, *J. Opt. Soc. Am. B- Opt. Phys.*, 2002, **19**, 1830.
132. S. Di Bella, I. Fragala, I. Ledoux and J. Zyss, *Chem. – Eur. J.*, 2001, **7**, 3738.
133. E. Zojer, D. Beljonne, T. Kogej, H. Vogel, S.R. Marder, J.W. Perry and J.L. Bredas, *J. Chem. Phys.*, 2002, **116**, 3646.
134. W.H. lee, H. Lee, J.A. Kim, J.H. Choi, M.H. Cho, S.J. Jeon and B.R. Cho, *J. Am. Chem. Soc.*, 2001, **123**, 10658.
135. I. Ratera, S. Marcen, S. Montant, D. Ruiz-Molina, C. Rovira, J. Veciana, J.F. Letard and E. Frysz, *Chem. Phys. Lett.*, 2002, **363**, 245.
136. R.W. Munn, M.G. Papadopoulos and H. Reis, *Pol. J. Chem.*, 2002, **76**, 155.
137. V. Ostroverkhov, O. Ostroverkhova, R.G. Petschek, K.D. Singer, L. Sukhomlinova and R.J. Tweig, *IEEE J. Sel. Top. Quantum Electron.*, 2001, **7**, 781.
138. V. Ostroverkhov, K.D. Singer and R.G. Petschek, *J. Opt. Soc. Am. B-Opt. Phys.*, 2001, **18**, 1858.
139. C.Y. Mang, K.C. Wu, C.S. Lin, R.J. Sa, P. Liu and B. Zhuang, *Opt. Mater*, 2003, **22**, 353.
140. G. Maroulis and C. Pouchan, *PCCP*, 2003, **5**, 1992.
141. K.C. Wu, R.J. Sa, J. Li, B.T. Zhuang, C.S. Lin, X.H. Chen, S.Q. Peng and Z.F. Zhou, *Chin. J. Struct. Chem.*, 2001, **20**, 319.
142. K.C. Wu, J.G. Snijders and C.S. Lin, *J. Phys. Chem. B*, 2002, **106**, 8954.
143. D.S. Yakovlev, A.P. Mirgorodskii, A.V. Tulub and B.F. Shchegolev, *Opt. Spectrosc.*, 2002, **92**, 449.

144. M. Guillaume, E Botek, B. Champagne, F. Castet and L. Ducasse, *Int. J. Quantum Chem.*, 2002, **90**, 1378.
145. E. Botek and B. Champagne, *Chem. Phys. Lett.*, 2003, **370**, 197.
146. K.C. Wu, S.H. Chen, J.G. Snijders, R.J. Sa, C.S. Lin and B.T. Zhang, *J. Cryst. Growth*, 2002, **237**, 663.
147. E.E. Fileti, R. Rivelino and S. Canuto, *J. Phys. B*, 2003, **36**, 399.
148. M. Nakano, S. Yamada, M. Takahata and K. Yamaguchi, *J. Phys. Chem. A.*, 2003, **107**, 4157
149. J. Lipinski, *Chem. Phys. Lett.*, 2002, **363**, 313.
150. O. Quinet and B. Champagne, *J. Chem. Phys.*, 2002, **117**, 2481.
151. M.G. Kuzyk, *IEEE J. Sel. Top. Quantum Electron.*, 2001, **7**, 774.
152. K. Clays, *Opt. Lett.*, 2001, **26**, 1699.
153. J.M. Ducere, C. Lepetit, P.G. Lacroix, J. L. Heully and R. Chauvin, *Chemistry of Materials*, 2002, **14**, 3332.
154. S.R. Marder, D.N. Beratan and L-T. Cheng, *Science*, 1991, **252**, 103.
155. S.R. Marder, C.B. Gorman, B.G. Tieman and L-T. Cheng, *J. Am. Chem. Soc.* 1993, **115**, 3006.
156. G.J. Simpson, J.M. Perry and C.L. Ashmore-Good, *Phys. Rev. B.*, 2002, **66**.
157. X. Zhao, D. Guan, X.Z. Yi, G.B. Xu and M.H. Jiang, *Int. J. Quantum Chem.*, 2003, **93**, 335.
158. X. Zhao, D. Guan, X.H. Yi, Y.F. Xu, B.H. Ge and M.H. Jiang, *Chem. Phys.*, 2003, **287**, 21.
159. V. Weber and C. Daul, *Chem. Phys. Lett.*, 2003, **370**, 99.
160. H. Larsen, T. Helgaker, J. Olsen and P. Jorgensen, *J. Chem. Phys.*, 2001, **115**, 10344.
161. T. Janowski and K. Wolinski, *J. Chem. Phys.*, 2002, **117**, 1994.
162. T. Helgaker, M. Jaszunski and K. Ruud, *Chem. Rev.*, 1999, **99**, 293.
163. P.J. Wilson, R.D. Amos and N.C. Handy, *Chem. Phys. Lett.*, 1999, **312**, 475.
164. P.J. Wilson, R.D. Amos and N.C. Handy, *Mol. Phys.*, 1999, **97**, 757.
165. P.J. Wilson, R.D. Amos and N.C. Handy, *PCCP*, 2000, **2**, 187.
166. P.J. Wilson, R.D. Amos and N.C. Handy, *J. Mol. Struct (Theochem)*, 2000, 335.
167. A.D. Becke, *J. Chem. Phys.*, 1993, **98**, 5648.
168. H.L. Schmider and A.D. Becke, *J. Chem. Phys.*, 1998, **108**, 9624.
169. P.J. Wilson and D.J. Tozer, *J. Mol. Struct*, 2002, **602**, 191.
170. Q. Zhao, R.C. Morrison and R.G. Parr, *Phys. Rev A.*, 1994, **50**, 2138.
171. K. Ruud, P.O. Astrand and P.R. Taylor, *J. Am. Chem. Soc.*, 2001.
172. K. Ruud, P.O. Astrand and P.R. Taylor, *J. Phys. Chem. A.*, 2001, **105**, 9926.
173. J. Rodriguez-Otero, E. Martinez-Nunez, A. Pena-Gallego and S. A. Vasquez, *J. Org. Chem.*, 2002, **67**, 6347.
174. N.S. Mills, *J. Org. Chem.*, 2002, **67**, 7029.
175. J. Poater, X. Fradera, M. Duran and M. Sola, *Chem-Eur. J.*, 2003, **9**, 1113.
176. B.T. Stepien. T.M. Krygowski and M.K. Cyranski, *J. Org. Chem.*, 2002, **67**, 5987.
177. F. De Proft, P.V. Schleyer, J.H. van Lenthe, F. Stahl and P. Geerlings, *Chem-Eur. J.*, 2002, **8**, 3402.
178. R.S. Jartin, A. Ligabue, A. Soncini and P. Lazzaretti, *J. Phys. Chem. A.*, 2002, **106**, 11806.
179. D. Sebastiani and M. Parrinello, *PCCP*, 2002, **3**, 675.
180. D. Sebastiani and M. Parrinello, *J. Phys. Chem. A.*, 2001, **105**, 1951.
181. A. Rizzo, K. Ruud and D.M. Bishop, *Molecular Physics*, 2002, **100**, 799.

182. D. Jonsson, P. Norman, H. Agren, A. Rizzo, S, Coriani and K. Ruud, *J. Chem. Phys.*, 2001, **114**, 8372.
183. C. Cappelli, B. Mennucci, J. Tomasi, R. Cammi and A, Rizzo, *Chem. Phys. Lett.*, 2001, **346**, 251.
184. Klug and Huttner, *Chem. Phys. Lett.*, 1984, **90**, 207.
185. C. Cappelli, A. Rizzo, B. Mennucci, J. Tomasi, R. Cammi, G.L.J.A. Rikken, R. Mathevet and C. Rizzo, *J. Chem. Phys.*, 2003, **118**, 10712.
186. M.C. Caputo and P. Lazzeretti, *J. Chem. Phys.*, 2002, **116**, 9611.

6
Simulation of the Liquid State

BY KARL P. TRAVIS

1 Introduction

The last few years have heralded a rapid growth in the use of simulation as a tool for investigating the properties of liquids. What was once the preserve of the chemist or physicist has now become a standard tool widely used in multidisciplinary fields such as materials science and chemical engineering. Indeed, it is the expansion of the use of simulation in these areas in particular that has lead to a shift of emphasis away from theoretical studies, towards particular applications. An explanation for this change may be found from an analysis of the sources of research funding, which increasingly originates from industry. However, my personal opinion is that the drive towards application-based research has brought about unquestionable benefits. Simulation and experiment were once seen as separate entities, seldom coming together. Indeed, experimentalists and practitioners of simulation are infamous for not collaborating. In applications-lead research, both of these groups naturally come together. There is no doubt in my mind that the increasing use of simulation by chemical engineers and materials scientists has furnished the simulation community with a whole new set of interesting problems and phenomena with which to apply the existing simulation techniques, but also has acted as a spark to catalyse new ways of using simulation, and new techniques being introduced. Rather than signalling the death of fundamental research, this shift in emphasis will ultimately rejuvenate interest in basic fundamental science.

One area that has seen tremendous activity as a result of these changes is the area of multiscale simulation. Molecular simulation has broadened its realm of applicability to cover greater length and time scales, particularly the mesoscale. Techniques such as Dissipative Particle Dynamics and Lattice Boltzmann have been increasingly developed and used to model complex, multicomponent fluids. Interest in molecular scale simulation has also increased, catalysed by the exciting developments in nanotechnology such as the production of nano-machines, and small scale storage devices. Theoretical developments have been no less interesting; research carried out over the last decade or so on dynamical systems has yielded some remarkable discoveries such as a relationship between transport properties and dynamical chaos. Combining non-equilibrium statistical mechanical theory with experiment has lead to a suggestion that there is a fundamental limitation to the miniaturization of small engines.[1]

In writing this review it has been assumed that the reader is familiar with the basic principles behind the main simulation methodologies. The simulation literature is well served by a number of excellent text books which cover the fundamentals. See for example refs. 2–9.

2 Transport Properties

Transport phenomena in liquids include viscosity, thermal conduction and diffusion. These phenomena are important as they describe how a liquid relaxes back to equilibrium following application of a mechanical or thermal perturbation. Knowledge of the behaviour of the corresponding transport properties with thermodynamic state, and with external fields, is necessary for achieving the aim of a unified theory of liquids. The ability to compute these transport coefficients from molecular simulation is also advantageous in engineering applications where standard tools such as computational fluid dynamics rely on this information, which is not always accessible from experiments. Consequently, simulation of transport in liquids is a very active area of research.

2.1 Theories of Diffusion. – Theories of liquids based on the hard sphere fluid as reference state have been highly successful when applied to static properties of equilibrium fluids. Unfortunately, the same degree of success has not been achieved for transport coefficients.[10]

Kinetic theory gives an expression for the diffusion coefficient at infinite dilution as,[11]

$$D_0 = \frac{3}{8\rho\sigma^2}\left(\frac{kT}{\pi m}\right)^{1/2} \tag{2.1}$$

where ρ is the hard sphere number density, m the hard sphere mass, T is temperature, σ is the hard sphere diameter, and the subscript 0 refers to infinitely low density. Attempts to extend kinetic theory to liquid densities are plagued by the correlations that occur between the positions and velocities of particles in a dense fluid. Enskog developed kinetic theory within the approximate framework of uncorrelated collision sequences, extending it to finite density. He obtained what essentially amounts to a correction factor for the hard sphere diffusivity:

$$D_E = D_0\left(\frac{1.01896}{g(\sigma)}\right) \tag{2.2}$$

where $g(\sigma)$ is the radial distribution function evaluated at contact. The Enskog theory gives a reasonable estimate of the self-diffusion coefficient over the whole hard sphere phase diagram.[12]

Fluids composed of non-hard spheres can be treated by empirical 'perturbation' methods. The density and temperature dependence of the self-diffusion coefficient is taken to be that of a hard sphere fluid. The trick is to find an equivalent hard sphere diameter for the particles in question. Structured molecules can be dealt with by introducing a roughness factor.[13]

As an alternative to kinetic theory treatments of diffusion, various free volume theories have been put forward.[14-18] In the Cohen-Turnbull approach,[18] molecules are imagined to diffuse through space by a series of jigglings into voids (free volumes) created by neighbouring molecules under density fluctuations. This theory leads to an expression for the self-diffusion coefficient of a liquid

$$D = ga(v)\bar{c} \exp(-\alpha v^*/v_f) \tag{2.3}$$

in which v_f is the free volume per molecule, v^* is a critical volume just large enough to allow another molecule to move in, g is a geometric factor, $a(v)$ is roughly the diameter of the cage created in the liquid, and $\bar{c}$ is the gas kinetic speed. The pre-exponential factor can be approximated by the expression for the self diffusivity of dilute hard spheres (eqs. 2.1 and 2.2).[19]

The difficulty lies in obtaining accurate values of the free volume over a sufficiently wide range of density and temperature. Rah and Eu have developed a method of obtaining free volumes with reasonable accuracy from a generic van der Waals equation of state.[19] The generic van der Waals equation of state is obtained from the virial equation of state for fluids composed of molecules which interect with a potential energy comprising of both an attractive part and a repulsive part. Exact statistical mechanical expressions were derived for the generalized van der Waals parameters by Eu and Rah.[20] The same authors computed the free volume from the cavity function via Monte Carlo simulations and hence were able to calculate self-diffusion coefficients for liquid argon and liquid methane within the Cohen-Turnbull theory.[19] The authors point out their method of calculating diffusion coefficients allows a simple means of calculating other transport coefficients, since these may be expressed in terms of diffusion or self-diffusion coefficients together with the equilibrium pair correlation functions.[21-26]

Rah and Eu have extended their treatment of the free volume theory of self-diffusion to include mutual diffusion in binary liquid mixtures.[27] They note that the theory may readily be extended to more complex mixtures.

Harmandaris et al. have studied diffusion of the binary alkane systems C_5–C_{78}, C_{10}–C_{78} and C_{12}–C_{60} in which the lighter hydrocarbon was the solvent.[28] They obtained equilibrated starting configurations via a new MC routine which includes scission and fusion moves.[29] Using molecular dynamics they obtained values for the self-diffusion coefficients of each component in the mixture for a series of compositions. They compared the MD results with the predictions of a free volume theory for polymer-penetrant systems, a combined Rouse diffusant and chain end free volume theory, and with experimental data. They found that both component self-diffusivities increase when the weight fraction of solvent increases, a fact which they attribute to the increase in total free volume. Free volume theory was found to significantly underestimate the solvent diffusivity coefficient in the dilute regime whereas the combined theory was able to describe the MD results over the entire range of concentrations semi-quantitatively. The MD results were found to agree well with experiment in the C_{12}–C_{60} systems.

Drozdov and Tucker have developed a microscopic model for diffusion in near critical supercritical solvents. This model, which they called multiple time scale generalized Langevin equation (MTSGLE), which is based on the GLE, but extended to include slow changes in the solvent response caused by underlying density inhomogeneities.[30] Data for the local solvent environment surrounding the diffusing particles over time were obtained via MD. Application of their method to a near critical supercritical fluid showed that diffusion coefficients could be observed at intermediate times which were substantially different from in the bulk (times longer than the velocity relaxation time but shorter than the local density relaxation time.

Tong and co-workers calculated self-diffusion coefficients for non-associating fluids at moderate and high density using the hard sphere chain (HSC) equation.[31] This approach is found to give results which give a discrepancy with experimental data of 5% for pressures less than 300 MPa and temperatures greater than 100 K. They also developed a method for calculating the same quantity for associating fluids. This latter method is a combination of SAFT theory with HSC. The non-spherical associating molecules are treated as chains of tangent HS segments with an associating site. They obtain an equation for D_s in polyatomic associating fluids which is a product of a hydrogen bonding and non-hydrogen bonding term. Using their equation, they were able to reproduce the experimental D_s value with 7.5% discrepancy on average for water, alcohols and HF over a wide temperature and pressure range including supercritical water.

Free volume and kinetic theories of transport are somewhat old and well established. More recent developments are based on the remarkable idea that transport phenomena are connected with the chaotic properties of reversible dynamical systems. Much research effort is being directed into studying dynamical systems and their properties. One important property of dynamical systems is the Kolmogorov-Sinai entropy. The Kolmogorov-Sinai entropy can be obtained from the sum of all positive Lyapunov exponents:

$$h_{KS} = \sum_{\lambda_l > 0} \lambda_l \qquad (2.4)$$

and is an invariant quantity for a dynamical system which measures the average rate of information loss.

Shin and co-workers have studied the Lyapunov instability of a fluid composed of rigid diatomic molecules by molecular dynamics.[32] They analysed the spectra of Lyapunov exponents, the Kolmogorov-Sinai entropy, and the associated tangent space vectors as a function of the anisotropy dependent density. They found that the largest contribution to the instability of the phase space trajectory arises from the translational momentum variables, while in two-dimensions, the major contribution comes from the angular momentum variables. The instability contribution stemming from translational momentum was found to depend on both density and bond length whereas that from angular momentum depends only on bond length. In a more recent publication, the same authors looked at the relationship between the Kolmogorov-Sinai entropy, h_{KS} and the self-diffusion coefficient, D for hard sphere fluids and WCA fluids.[33] Based on theoretical arguments they proposed the relationship between these two quantities

$$\frac{h_{KS}}{Nv} \propto \left(\frac{D}{v\sigma^2}\right)^\eta \qquad (2.5)$$

where N is the number of particles, v the collision frequency in Enskog theory, σ the particle diameter, and η is a parameter which varies depending on the type of interaction potential function, but is independent of temperature and density. The simulation data they obtained could be fitted with high precision for both HS and WCA fluids yielding values of the exponent $\eta = 0.17$ and 0.44 respectively.

At equilibrium, the entropy of a classical N-particle system as stated by Gibbs is

$$S(t) = -k_B \int d\Gamma f(\Gamma) \log f(\Gamma) \qquad (2.6)$$

where $f(\Gamma)$ is a time-independent equilibrium distribution function and $d\Gamma$ is the classical phase space volume element. Boltzmann showed that for dilute gases, this equation predicts a monotonic increase in the entropy of an isolated gas as it relaxes towards equilibrium. Evans has shown that the Gibbs entropy diverges to negative infinity for a non-equilibrium steady state of a dense many body system subject to a time-independent external field.[34] In the absence of a thermostat, the Gibbs entropy is predicted to be a constant, a result known to Gibbs![35] Gibbs resolved this difficulty by introducing a coarse-grained entropy which can be shown to obey a generalized H-theorem and cannot decrease.

Evans and Rondoni have discussed at length the entropy of non-equilibrium steady states.[36] This publication extends some of the ideas put forward by one of the authors over a decade earlier in a textbook on non-equilibrium statistical mechanics.[4] The argument put forward by Evans and Rondoni in the present work is that if the Gibbs entropy is expanded in a series involving 1, 2, ... body terms (Greens expansion), the divergence of the Gibbs entropy arises only from terms involving integrals whose dimension is higher than, approximately, the Kaplan-Yorke dimension of the steady state attractor. The lower order terms are finite and sum in the weak field limit to the local equilibrium entropy of linear irreversible thermodynamics. The entropy must therefore be calculated from within the lower dimensional, accessible phase space rather than the full phase space.

Evans and Searles have reviewed the most up to date theoretical developments and simulation data relating to the so-called fluctuation theorem (FT).[37] The FT gives an analytical expression for the probability of observing Second Law violating dynamical fluctuations in thermostatted dissipative non-equilibrium systems.

Norman and Stegailov have discussed the concept of a dynamical memory time, t_m, and its relationship to the fluctuation of energy and the Lyapunov exponent.[38] They argue that molecular dynamics simulations contain hidden stochastic features and it is these features that are responsible for some of the successes of the molecular dynamics method.

2.2 Computational Methodology. – Major advances in computing transport properties by simulation have been made in the last 20 years. Perhaps the most significant development has been synthetic NEMD methodology. The best known algorithm of this type is the Sllod algorithm which was developed for studying planar Couette flow.[4] Since its introduction, new NEMD algorithms continue to be developed for calculating all manner of transport coefficients. Equilibrium routes to the transport coefficients, while in many cases vastly inefficient compared to NEMD, still have an important contribution to make, and hence are still the subject of much research effort.

As an example of a recent development in equilibrium methods of calculating transport coefficients, Hess and Evans have introduced a new method of obtaining the viscosity of a liquid. In place of the usual Green-Kubo (GK) method,[39] which involves an integral of an appropriate time correlation function, the new technique calculates this integral without the need to calculate the time correlation function. Instead, the transport coefficient is calculated from time averages of fluctuations in the relevant dynamical quantity. Hess and Evans used this method to evaluate the shear and bulk viscosity of a simple fluid. As the authors did not make any claims regarding their method being superior to the GK or NEMD routes to these transport coefficients, it appears the only advantage of this method at present is the ability to resolve the kinetic and potential contributions to a transport coefficient.

Hess has compared 4 different methods of obtaining the shear viscosity of a liquid via molecular dynamics simulations.[40] These methods: the Green-Kubo, Einstein, Sllod and what was termed a periodic perturbation method, but is essentially the sinusoidal transverse force method, were judged on accuracy, applicability and efficiency. Hess rediscovered the well known fact that the non-equilibrium methods are much more efficient than the equilibrium routes. However, Hess also found that when these methods are applied to complex fluids such as SPC water, the equilibrium methods are not only inefficient, but they are problematic when electrostatic interactions are included. For these fluids, it was concluded the periodic perturbation method was preferred.

Ratanapisit and co-workers have made detailed Green-Kubo calculations of a variety of transport coefficients including shear viscosity, self-diffusion and thermal conductivity using different symplectic integration algorithms.[41] The idea behind this work was to compare these algorithms for accuracy, stability and efficiency. Green-Kubo calculations of transport coefficients are a good test case for such a comparison since the accuracy of the method relies on long simulations. They found that algorithms which have poor energy conservation lead to significant uncertainties in transport property calculations. The symplectic algorithms were found to be superior to the non-symplectic algorithms such as the 4th order Runge Kutta and 4th order Gear algorithms.

Lee and co-workers discussed new approaches to the calculation of the shear viscosity and thermal conductivity of liquids based on modified Green-Kubo and Einstein relations.[42] Rather than computing the shear viscosity, say, from an integral of the stress correlation function, their approach is to use a mean square displacement of the stress in the spirit of the Einstein approach. This method is claimed to improve the statistics on this transport coefficient by effectively considering stress as a single particle property.

Todd has reviewed the field of non-equilibrium molecular dynamics (NEMD) with the intention of demonstrating the methods and their usefulness to the community of computational fluid dynamicists.[43] The use of NEMD as an aid to designing new polymeric materials is one of several topics covered in this article.

One essential component of any NEMD algorithm is the thermostatting mechanism. Without a thermostat, a system driven away from equilibrium would heat up irreversibly, with no steady state being achieved. Deterministic thermostats such as the Gaussian thermostat or Nosé-Hoover thermostat are usually employed to fix the kinetic energy of the system, and hence maintain a constant kinetic temperature. The difficulty with this approach is that one must have an *a priori* knowledge of the local streaming velocity in order that the thermal components of momenta alone are acted upon by the thermostat. For molecules, the situation is more complex since the local streaming velocity at an atomic site must be known. The development of the configurational thermostat and its use in NEMD simulations has removed many of these difficulties. Improvements in this area continue to be made. For instance, Lue et al.[44] have extended the configurational thermostat to enable it to be used for the NEMD of molecular fluids. They compared this type of thermostat with a standard kinetic thermostat for a range of molecules of differing length and flexibility and found quantitative agreement at low strain rates and qualitative agreement at high strain rates. They point out that the configurational thermostat is advantageous since no *a priori* knowledge of the atomic streaming velocity is required.

Delhommelle and Evans have used both a configurational thermostat and a centre-of-mass (com) kinetic thermostat to conduct isothermal homogeneous shear flow simulations of straight chain alkanes (C_{10} and C_{20}).[45] They found that both thermostatting schemes give similar results at low shear rates but that at reduced shear rates in excess of 0.5, the response of the internal degrees of freedom strongly depends on the thermostatting mechanism. They also show that equivalent results for the two thermostatted simulations can be obtained provided that the configurational temperature is fixed at a significantly higher value than the com kinetic temperature.

2.3 Bulk Liquid Transport Properties. – interest in bulk liquid transport properties remains as strong as ever judging by the number of publications in this area. The majority of simulations have been concerned with the calculation of diffusion coefficients. Part of the reason for this is the ease with which self-diffusion coefficients can be obtained from molecular dynamics simulations. Transport diffusion on the other hand, is perhaps one of the hardest transport properties to work with due to issues with defining frames of reference and flux definitions, which mean one must exercise extreme care when comparing diffusion coefficients with experiment. Much confusion has arisen in the literature and many errors have made over the years as a consequence of this.

While it is generally accepted that diffusion processes slow down dramatically as the vapour-liquid critical point is approached, there is still much disagreement over the exact behaviour of the self-diffusion coefficient in this region. Interpretation of different sets of experimental data for diffusion in pure fluids has lead to conflicting results. From the results of one study it can be concluded that self-diffusivity approaches zero at the critical point,[46] in another that it approaches infinity at the critical point,[47] while in yet another, it remains finite, showing no anomalous behaviour.[48] De and co-workers have discussed these findings from the point of view of dynamic scaling theory.[49] They conclude that self-diffusivity should be non-vanishing at the critical point, while transport diffusivities in both pure fluids and binary mixtures are predicted to vanish at their respective critical points.

Drozdov and Tucker looked at the behaviour of the coefficient of self-diffusivity close to the critical point of a pure fluid by conducting molecular dynamics simulations of a pure, Lennard-Jones fluid in the critical region ($T_r \approx 1.05$, $0.4 < \rho_r < 1.6$).[50] Employing a new method of calculating the diffusion coefficient, they observed a weak anomalous behaviour (decrease in self-diffusivity) at near critical densities which they attributed to the enhancement of the mean local density and broadening of the local density distribution. They noted that experimental measurements of diffusion for napthalene[51] and acetone[52] also found anomalous behaviour, but also conceded that a fairly extensive set of measurements for pure methane[53] failed to see anomalous behaviour. Cherayil,[54] using renormalized kinetic theory, determined that the self-diffusivity exhibits an anomalous decrease in the vicinity of the critical point, in agreement with the simulations of Drozdov and Tucker.[50] However in a comment paper, Harris[55] criticises the experimental data which purports to show anomalous diffusion and cited by Drozdov and Tucker in support of their simulation results.[50] Harris pointed out the subtle differences between tracer diffusion, self-diffusion and mutual diffusion that are often confused in the literature, but need to be appreciated for the correct interpretation of experimental diffusion studies. Furthermore, he pointed to a different set of experimental measurements of self-diffusion in pure fluids near critical points which found no evidence for anomalous diffusion.[48,56,57] In their response to the comments of Harris, Drozdov and Tucker point out that while diffusion experiments suggesting anomalous self-diffusion near the critical point were unreliable, they did not contradict its existence either.[58]

It is known that nematic liquid crystals exhibit dynamical anisotropy: the self-diffusion of molecules in a direction parallel to the director is greater than perpendicular. Darinskii and co-workers[59] studied this anisotropy in a model liquid crystalline system composed of semi-flexible polymer chains. Their model consisted of soft spheres connected by springs. The chain flexibility was adjusted through the spring stiffness or by addition of extra springs, allowing them to study the effect of flexibility on the anisotropy. In the isotropic phase they observed that the ratio of translational to rotational diffusion coefficients decreased with decreasing flexibility. In the nematic phase they observed anisotropy in translational diffusion and that this anisotropy shows a universal dependence on the order parameter.

Ali and co-workers have used mode coupling theory to calculate tracer diffusivities in a binary Lennard-Jones fluid mixture corresponding to an infinite dilution of one of the components.[60] They found that diffusion in the mixture is not solely dictated by binary collision as it would be if Enskog theory was correct, nor is solely determined by the transverse current of the solvent à la Stokes-Einstein. Rather, it is a combination of both these effects coupled with a significant contribution from density fluctuations. The theoretical formalism developed by these authors is able to predict the correct mass and size dependence of tracer diffusivity as well as going over to the correct hydrodynamic limit for larger mass and size, as predicted by simulation.

Schober investigated the isotope effect in a binary Lennard-Jones fluid using molecular dynamics simulation.[61] Using mass differences of 20–40% he found that at fixed density, there was a large reduction in the isotope effect for both species. At fixed temperature, the same effect was observed for the larger majority atoms, indicating that density change was the main driving force behind the isotope effect.

Dhole et al.[62] have developed a microscopic approach to cross diffusivity in a binary fluid mixture based on a continued fraction approach originally developed by Mori[63] (the cross diffusivity is that component of the mutual diffusivity that arises from correlations between the velocities of different molecules). Using their theory, Dhole and co-workers were able to explain interesting phenomena such as an increase in cross diffusivity with an increase in the mass ratio or concentration from first principles.

Thermal diffusion or the Ludwig-Soret effect describes the process by which temperature gradients in multicomponent liquid mixtures can lead to separation of the constituents. The transport coefficient describing this cross coupling is known as the Soret coefficient. Very few studies have been carried out to determine the magnitude of this transport coefficient. Being a cross coupling phenomenon, thermal diffusion is a weak effect and thus requires very sensitive methods in order to determine the transport property. Perronace et al. have recently conducted experimental measurements of the Soret coefficient for n-pentane and n-decane mixtures via thermal diffusion forced Rayleigh scattering (TDFRS).[64] In addition, they also used several different molecular simulation methods to obtain the same quantity, including equilibrium molecular dynamics, the subtraction method, boundary driven NEMD and synthetic NEMD. They found good agreement between simulation and experiment. They also found that both the subtraction method and the EMD Green-Kubo routes are hopeless for evaluating this weak transport phenomena.

Moon et al.[65] looked at reorientational motion in a dilute solution of linear 'tracer' molecules in a solvent that exhibits liquid, plastic and crystalline phases. Surprisingly, they found that the transition from liquid to plastic phase results in less hindered tracer rotation.

Adams and Siavosh-Haghighi studied the structure and rotational dynamics of supercritical CO_2 at a temperature 1% above the critical temperature and over a wide range of densities about the critical density.[66] They found density inhomogeneities and hence regions of modest local density enhancement in this fluid over a wide range of bulk densities. They also found that the total rotational friction and rotational diffusion constant are poor probes of such enhancement.

Apart from diffusion, computation of the various viscosity coefficients such as bulk and shear viscosity, are the subject of many publications. Bertolini and Tani have calculated the wavevector dependent bulk viscosity as well as the bulk and longitudinal viscosities of liquid argon by molecular dynamics.[67,68] They found that k-dependent bulk viscosity is positive for all wavevectors. Negative values for this coefficient were shown to arise from an incorrect extension to finite wavevectors of a relation linking bulk, shear and longitudinal viscosity coefficients. Cheung et al. used molecular dynamics to calculate the rotational viscosity of a nematic liquid crystal.[69] Such information is important with respect to the switching times of liquid crystal displays.

Theory based models for predicting liquid mixture viscosity are important in fields such as process engineering. Although not strictly speaking simulation, such is the importance of these models that we shall briefly mention a couple of papers that were published in this area. These references should be of interest to anyone concerned with using molecular simulation to calculate mixture viscosities.

Novak introduced an improved theory based model for liquid mixture based viscosity which is based on the Eyring mixture viscosity model and the NRTL model for describing deviations from ideality.[70] His model, when generalized to multicomponent mixtures is given by

$$\ln(\eta_m V_m) = \sum_i x_i (\eta_i V_i) - \Delta G^E / RT \qquad (2.7)$$

where η_m and V_m are respectively, the mixture viscosity and molar volume, x_i the mole fraction of component i, η_i and V_i the respective viscosity and molar volume of species i, R the gas constant, T the temperature and ΔG^E is the excess Gibbs free energy. The term involving the excess free energy represents non-ideal contributions to the model. Using this model, Novak compared predicted mixture viscosities with available data for polymer solvent systems. The predictions were found to be in good agreement with the available data. Song *et al.* used a similar mixing rule to that of Novak, but derived using different theory based models.[71] The essential difference between the two mixing rules lies in the treatment of the nonideality term. In the former case this is handled by an excess free energy term, while in the work of Song *et al.*, it is handled through a set of binary symmetric and antisymmetric mixing parameters.

Sigurgeirsson and Heyes have recently performed a wide ranging set of molecular dynamics simulations of hard sphere fluids using large system sizes.[72] They calculated the coefficients of self-diffusion, bulk and shear viscosity, and thermal conductivity for a large number of packing fractions up to the glass transition (0.57 packing fraction). Using their data together with data from the literature they tested various analytical formulae that have been proposed for calculating these transport coefficients. They found that the expression for the packing fraction dependence of the self-diffusivity proposed by Speedy,[73] fitted the simulation data quite well for systems of about 500 particles. Furthermore, they found that the packing fraction dependence of the other transport coefficients could be well represented by the simple formula $X/X_0 = 1/[1-(\xi/\xi_1)]^m$ within the equilibrium fluid range $0 < \xi < 0.493$ (where X is a transport coefficient, X_0 its value in the limit of zero density, ξ_1 and m are constants). The self-diffusion coefficient was shown to possess a significant 'hydrodynamic' contribution to its value at intermediate densities, implying that its value exceeds its Enskog prediction over a wide density range.

Faussurier and Murillo have employed the Gibbs-Bogolyubov inequality in order to establish a mapping between the Yukawa system and both the hard sphere and 1-component plasma reference systems.[74] They computed the transport coefficients (λ, D_s, η) of the Yukawa fluid, using known properties of the reference systems. The method can be applied to mixtures and also could be extended to other systems and properties.

Hoyt and Asta have studied diffusion in liquid silver and gold.[75] The simulated values for these properties were found to be in good agreement with experimental data. Margulis *et al.* studied ion diffusion in the ionic solvent 1-buthyl-3 methylimidazonlium hexafluorophosphate ([bmim][PF_6]).[76] They observed that the ions display complex dynamics with at least 2 different time scales for diffusion.

2.4 NEMD. – Non-equilibrium molecular dynamics is useful for studying the behaviour of liquids away from equilibrium. In this regime all the transport coefficients are strictly speaking transport properties since they become functions of the non-equilibrium driving force, such as the strain rate. The strain rate dependence of the shear viscosity is the subject of an ongoing debate. The majority of simulation studies find in favour of mode coupling predictions but recent work in the last few years has questioned its validity. Marcelli *et al.* have revisited the issue of determining the strain rate dependence of the shear viscosity, pressure and energy of a simple fluid.[77] Mode coupling theory predicts that all three properties are non-analytic functions of the strain rate. In contrast to this theory, and to most earlier simulations, they found that the pressure and energy varied as the 2nd power of the strain rate but were unable to unequivocally determine the functional dependence of the shear viscosity.[78] Using two- and three-body potentials, they determined that these functional dependencies were attributed solely to the two-body potential.

The realm of applicability of the Navier-Stokes equations is another area which has received much attention in recent years. Morriss et al. compared results from simulations of 2D convective flow with truncated solutions of the Navier-Stokes equations.[79] The Navier-Stokes equations were found to be quite accurate at predicting the transient development of the flow if the correct value of the viscosity was used. However, they observed breakdown of the hydrodynamics at high values of the field.

Understanding the behaviour of alkanes under shear flow is important in the lubrication industry. Developments in synthetic NEMD and the availability of massively parallel supercomputers have made the simulation of alkane flow fairly routine. A good example of this type of applied simulation is that of McCabe et al. who used both EMD and NEMD to calculate the shear viscosity of 9-octylheptadecane at a couple of temperatures.[80] Using the values of shear viscosity obtained from these simulations, they obtained values for the viscosity number (VN). The VN is used in the lubrication industry as a measure of a lubricants performance with respect to its viscosity variation with temperature. They found good agreement between simulated and experimental VNs.

Behaviour of liquid crystals under shear is another area of interest in NEMD. McWhirter and Patey have carried out NEMD simulations of a ferroelectric nematic liquid under planar Couette flow.[81] They found that the orientational order decreases with increasing shear rate as a result of the director dynamics being unstable under the shear flow. They observed critical orientations of the director for which it rotates with the greatest velocity, speculating that at these points, the individual dipoles cannot rotate fast enough compared to the director, giving rise to a time lag and hence destroying orientational order.

The structure of liquids under flow has also received some attention in the simulation literature. Lutsko looked at the effect of velocity correlations on the pair distribution function (PDF) of a hard sphere fluid undergoing shear flow.[82] He constructed a non-equilibrium generalized mean-spherical approximation for the PDF at finite separations that showed good agreement with simulation data. Hess studied the structure and non-equilibrium flow behaviour of simple and complex fluids by analysing their pair distribution functions and orientational distribution functions.[83]

Dufty looked at shear stress correlations in hard and soft sphere fluids (the latter modeled with a r^{-n} potential in which n is an integer).[84] The singular contribution to the hard sphere response function can be reproduced with large n values. He proposed a simple model for the stress correlation function at finite n based on the required HS limiting form.

3 Phase Equilibria

Location of the phase boundaries of a substance is of prime importance in fields such as materials science. The use of simulation to obtain phase diagrams has a long history but in 1987, the method was revolutionised by Panagiotopoulos's invention of Gibbs Ensemble Monte Carlo (GEMC).[85] Due to short comings in GEMC, simulation studies were restricted to equilibria involving vapour and liquids which were not too dense. The GEMC has been used to great effect in locating gas-liquid critical points, although histogram reweighting methods are now a much better tool for this purpose. Simulation of coexisting phases involving solids requires that simulation be used in conjunction with classical thermodynamics while simulation of the structure and dynamics of interfaces requires more direct methods be employed. The developments in this area, restricted to bulk systems, are discussed in this section.

Gelb and Müller have developed a new method for locating phase coexistence points using MD simulations.[86] Their technique involves using a relatively large simulation cell under NVT conditions and then quenching the system into a two-phase region, whereupon it spontaneously separates into two coexisting phases. The different phases are identified post-run by collecting histograms of the local density and/or composition of the pure phase regions. While this method explicitly includes the interface between coexisting phases, it does not suffer from the drawbacks of earlier direct methods due to the large system sizes employed, facilitated by efficient parallel MD algorithms and fast modern supercomputers.

Lekkerkerker and Oversteegen have tested two different levels of the free volume approximation to the potential of mean force for the phase diagram of an asymmetric mixture of hard spheres with a diameter ratio of 0.1.[87] Reasonable agreement was obtained with direct computer simulations.

Goujon and co-workers have used a direct Monte Carlo scheme to study the equilibrium properties of the n-pentane liquid-vapour interface.[88] Using appropriate local long-range corrections to the configurational energy within the Metropolis scheme and an algorithm allowing two maximum displacements chosen randomly, they obtained critical densities, temperatures and vapour pressures in good agreement with experiment.

Stubbs and co-workers have used configurational-bias MC to investigate the properties of alkane, alcohol and water mixtures.[89] Despite using state-of-the-art transferable potentials (OPLS/TIP4P) they obtained only qualitative agreement with experiment. Zhuravlev et al. used coupled-decoupled configurational bias MC (CD-CBMC) simulations in the Gibbs ensemble to determine the vapour-liquid coexistence curves for n-triacontrane and squalane.[90] They found branching and molecular weight lead to deviations from the law of corresponding states. They found evidence that partial collapse of the triacontane isomers (self-solvation) is the likely origin of these deviations.

A variant on the usual Gibbs ensemble method are the virtual Gibbs ensembles (VGEs) which belong to a broader class of techniques called pseudo-ensembles.[91-93] In the VGE method, virtual or 'faked' moves are carried out in order to partially decouple the Gibbs ensemble correlated fluctuations. A volume move on box I and box II might be accepted but only carried out on box I, similarly for a particle transfer. Shetty and Escobedo[94] have employed a specialized VGE to simulate vapour-liquid, vapour-solid, and liquid-solid equilibrium in single and multicomponent systems.

Calculating the solid region of a phase diagram requires additional tools to Gibbs ensemble MC. One such additional tool is Gibbs-Duhem integration. Lamm and Hall used a combination of Gibbs-Duhem integration and semi-grand canonical MC simulation to calculate temperature-composition and pressure-temperature phase diagrams for binary Lennard-Jones mixtures[95] They examined both solid-liquid and solid-vapor phase equilibria. Costa and co-workers have switched the free energy of the solid, liquid and vapour phases of the Girfalco model of C60 using a combination of MC simulation and thermodynamic integration (in the Girfalco model, spherical cage structures are replaced by LJ spheres in which the potential due to individual atoms is smeared out).[96] They confirmed the existence of a stable liquid phase for the system which exists only over a narrow range of temperatures and pressures, and obtained the full phase diagram of this substance. An alternative approach to modelling phase equilibria involving fullerenes has been followed by Kahl et al., who reformulated the self consistent Ornstein-Zernike approximation (SCOZA) based on the Wertheim-Baxtor formalism for solving the mean spherical approximation for a hard-core-multi Yukawa-tail fluid.[97] This reformulation allows systems to be treated which can be represented by a linear combination of Yukawa tails. They tested their theory against GEMC simulation for the critical point of model fullerenes (C_{60} and C_{70}) using the Girfalco potential, obtaining good agreement. They also tested the theory for a system with an explicit density dependent interaction. Albo and Müller have briefly reviewed the methods and intermolecular potentials appropriate for the molecular simulation of solids in supercritical fluids.[98]

Kettler et al. investigated the effect of long range Coulombic interactions of the VLE properties of polar and associating fluids.[99] This work follows on from their previous work on simple fluids in which they found that the fluid structure is primarily determined by the short range repulsions. For polar and associating fluids they find a similar result. Their approach involved decomposing a realistic potential (LJ and Coulombic) into a short range part and a residual part, and then conducting Gibbs ensemble simulations to study the VLE of polar and associating fluids. This result should prove useful in perturbation theory, in which the Coulombic interactions are treated as a perturbation to a short range reference.

Panayiotou developed a model based on density gradient theory, combined with an equation of state model of fluids and their mixtures to estimate the interfacial tension of pure fluids and their mixtures.[100] The model, which is applicable to non-polar fluids and hydrogen bonded fluids gave good agreement with experiment for pure fluids and mixtures of non-polar and weakly polar fluids, but only qualitative agreement for associating mixtures. Interfacial tension was also studied by Bresme, who used the Ornstein-Zernike integral equation to calculate the surface tension of colloid-fluid spherical interfaces.[101] Katsov and Weeks applied their general theory of non-uniform fluids[102-106] to the liquid-vapour interface of the LJ fluid.[107] The theory agreed well with simulation data for the interface profile and surface tension.

The dynamical behaviour of liquid-liquid and liquid-vapour interfaces has also been investigated by a number of workers. da Rocha and co-workers used MD to investigate the structures of the binary dense CO_2–H_2O interface.[108] They observed an interface which is molecularly sharp with distortions from a flat interface due to the presence of capillary waves induced by thermal fluctuations. Shen and Debenedetti developed a kinetic theory of homogeneous bubble nucleation.[109] Good agreement was observed for the predicted size of the critical bubble with classical nucleation theory.

Joule-Thomson inversion curve (JTIC) is the locus of states where fluid temperature is invariant upon isenthalpic expansion. Knowledge of the JTIC is used in the design and operation of throttling processes. The JTIC is difficult to obtain directly by experiment and so computer simulation is a useful tool in obtaining the JTIC. Colina et al. have obtained accurate JTIC for CO_2 using two different approaches based on NpT MC simulations.[110] The two approximations involved the compressibility and thermal expansivity routes.

Histogram reweighting is beginning to replace GEMC as the standard tool for MC simulation of phase coexistence with interfaces. Virnau et al. investigated the phase behaviour of a hexadecane-CO_2 mixture using this approach.[111] In this work the authors used a course-grained model of the hexadecane and CO_2, the former being modeled as a bead-spring chain.

Robles and Lopéz developed an equilibrium thermodynamic approach for obtaining the liquid-glass transition line in the reduced temperature versus reduced density plane for a monatomic Lennard-Jones fluid.[112] They used a modified classical perturbation theory of liquids developed by themselves[113] which makes use of a rational function approximation for the Laplace transform of the radial distribution function of the hard sphere fluid. The theory requires only an equation of state for the hard sphere system. They obtained good agreement with MD simulation for the liquid-glass transition line.

Wilding and Magee performed MC simulations of the phase behaviour in single component systems of particles interacting via a core-softened potential.[114]

4 Supercooled Liquids and Glasses

Many liquids, if cooled so rapidly that they are prevented from undergoing a freezing transition, may enter what is known as a supercooled regime. The supercooled state is metastable. Upon further cooling, the fluid viscosity increases by several orders of magnitude and the substance becomes an amorphous solid, lacking in long range molecular order, which is called a glass. The temperature at which the substance changes from being a fluid to a glass is known as the glass transition temperature, T_g. The glass transition temperature is not a well-defined property, for it depends on the rate of cooling. By convention, T_g is defined as the temperature at which the viscosity is of the order of 10^{13} Poise.

A basic understanding of the dynamics of supercooled liquids is still lacking. Consequently, a great deal of research has been undertaken to gain a better understanding of supercooled liquids and the approach to the glass transition. The application of computational statistical physics to the field of supercooled liquids and glasses has been discussed in an excellent review by Glotzer and co-workers.[115] The review lists all the unanswered questions in this field and shows how computational statistical physics has made significant contributions to aiding our understanding of liquid behaviour near the glass transition, particularly with respect to the observation of heterogeneous dynamics. An extensive, critical review of the glass transition of polymer melts has been undertaken recently by Binder et al.[116] Binder's review covers the various theories, experimental results and computer simulations that have been proposed and conducted. A selection of publications in the area of supercooled liquids that have appeared during the period of *this* review are described in the next two sections.

4.1 Phenomenology. – The glass transition is not a true phase transition in the thermodynamic sense since the formation of a glass depends on the experimental cooling rate. In glass forming liquids the shear viscosity increases smoothly over many orders of magnitude on cooling towards and beyond the glass transition temperature. At this point, the configurational relaxation times are in the range of a few minutes to a few hours. The thermal behaviour of the shear viscosity emphasises the kinetic aspects of the glass transition. An empirical law known as the Vogel-Fulcher-Tammann (VFT) law has been found to give a good fit to the viscosity of many systems over a wide temperature range. The VFT equation is given by

$$\eta = \eta_0 \exp\left[-E_\eta / k_B (T - T_0)\right] \qquad (4.1)$$

where η_0 is a reference viscosity, T_0 is the relaxation-time divergence temperature (a characteristic temperature lying below T_g) and E_η can be thought of as an activation energy for viscous flow at $T \gg T_0$. This equation can be re-cast in Arrhenius form by defining a temperature-dependent activation energy. By plotting the variation of the logarithm of viscosity versus reciprocal temperature, linearity implies Arrhenius behaviour while non-linearity indicates so-called super-Arrhenius behaviour.

Aside from the kinetic aspects of the glass transition, 2nd order thermodynamic quantities show anomalous behaviour near T_g. This behaviour coincides with the fact that the entropy of the supercooled liquid is approaching that of the corresponding crystalline phase. The temperature at which they become equal is called the Kauzmann temperature and is described as a "thermodynamic crisis". The Kauzmann paradox refers to the fact that a *thermodynamic* crisis at T_g is being avoided by a *kinetic* phenomenon – namely the crossing between the time scales of the system and of the experimental apparatus. Stillinger and co-workers have discussed the so-called Kauzmann paradox in some detail.[117] Citing data from real substances, theoretical modelling and numerical simulation in the temperature-pressure plane, they demonstrate that a Kauzmann locus $T_K(p)$ can occur for many materials and that third-law violations do not arise. They find no evidence for the concept of an 'ideal glass transition' at positive temperature. Tokuyama focussed his attention on the colloidal glass transition.[118] From a theoretical analysis of experimental data, Tokuyama has shown that the non-singular behaviour of the thermodynamic-like quantities near the glass transition is closely related to the non-singular behaviour of the kinetic coefficients such as the self-diffusion coefficient. Similar behaviour is expected for the viscosity which should lead to a continuous liquid-glass transition. Tokuyama hence predicts that neither non-divergence of the characteristic relaxation times nor an entropy crisis should occur. Kamath *et al.* define a configurational entropy as the difference between the entropy of the liquid and that of a hypothetical glass at the same temperature.[119] Their simulations show that this quantity displays a distinct change in its temperature dependence in the vicinity of T_A, the crossover temperature, below which caged dynamics becomes significant. These authors claim that their results add weight to the argument that vitrification may be driven by an underlying thermodynamic transition, rather than being purely kinetic in origin. The kinetic versus thermodynamic argument shows no signs of abating.

The scaling of viscosity data implied by the VFT law has lead to a useful classification of glass forming liquids into "strong" and "fragile". Strong glass formers are typically open network liquids such as SiO_2 and GeO_2, which display Arrhenius behaviour corresponding to $T_0 \ll T_g$. Fragile glass formers on the other hand, are those characterised by Coulombic interactions such as $ZnCl_2$ or van der Waals interactions such as in aromatics with many π electrons. These glass formers are also characterised by super-Arrhenius behaviour and a large jump in their specific heat at T_g. Strong liquids show Arrhenius behaviour and a small jump in specific heat at T_g.

Mansfield modified the 1-D Ising model in an external field to allow for a *P*-fold degenerate high energy state.[120] The model reproduces several empirical properties of glass-forming liquids such as: rapid fall off of energy and entropy upon cooling (Kauzmann paradox), the temperature dependence of the relaxation time follows the Vogel law, the spectrum of relaxation times broadens upon cooling in agreement with the stretched exponential law, and spatially heterogeneous dynamics are observed. The model suggests several assumptions about real liquids such as the fact that supercooled liquids possess two separate equilibrium "phases"; an "ideal" liquid and an "ideal" glass phase.

At temperatures well above the glass transition temperature, diffusion occurs by a flow mechanism. At temperatures well below T_g, diffusion occurs by a hopping mechanism. The cross over from one mechanism to the other as the glass transition is approached is not well understood, nor is the exact nature of the hopping mechanism.

Pang and co-workers used MD with the embedded atom method to study dynamical properties of supercooled copper.[121] By characterising several time correlation functions for the undercooled liquid metal they observed a slowing down in the coefficient of self-diffusion and a corresponding increase in structural relaxation time compared with the normal liquid state. Within the typical α-relaxation regime the structural rearrangement acts as a leading factor to induce the incipient nucleation, underlying the existence of a transition in dynamical heterogeneities. Their results confirm that the temperature dependence of the α-relaxation time serves as a direct and reliable measure of the glass-forming ability for real and model systems.

Schober used molecular dynamics to calculate the pressure dependence of the diffusion coefficient in a binary Lennard-Jones glass.[122] The pressure dependence of this quantity can be used to obtain what is known as an activation volume for diffusion. In a 1-component liquid, the temperature dependence of the diffusion coefficient follows an Arrhenius law:

$$D(T) = D_0 \exp(-H/kT) \qquad (4.2)$$

where D_0 is a pre-exponential factor and H is the activation enthalpy. Using the thermodynamic identities: $V = \partial G/\partial p$ and $G = H - TS$, together with Eq. 4.2, one obtains the activation volume for diffusion involving a single jump process

$$V_{act} = -kT\left[\frac{\partial \ln D}{\partial p}\right]_T + kT\left[\frac{\partial \ln D_0}{\partial p}\right]_T \qquad (4.3)$$

High values of V_{act} in crystals are taken to be indicative of activated diffusion via thermal vacancies. In amorphous materials, one can similarly calculate an apparent activation volume,

$$\hat{V}_{act} = -kT\left[\frac{\partial \ln D}{\partial p}\right]_T \qquad (4.4)$$

Low values of V_{act} in glasses can be taken as indicating a collective diffusion mechanism. High values are somewhat ambiguous. For liquids, the diffusion coefficient (just like the viscosity) can be fitted by the Vogel-Fulcher-Tammann (VFT) law

$$D^{VFT}(T) = D_0^{VFT} \exp\left[-E^{VFT}/k(T-T^{VFT})\right] \qquad (4.5)$$

By using Eq. 4.5 togther with 4.4, one obtains an expression for the apparent activation volume

$$\hat{V}_{act}^{VFT} = -kT \left[\frac{\partial \ln D_0^{VFT}}{\partial p} - \frac{1}{k(T-T^{VFT})} \frac{\partial E^{VFT}}{\partial p} - \frac{E^{VFT}}{k^2(T-T^{VFT})^2} \frac{\partial T^{VFT}}{\partial p} \right]_T \quad (4.6)$$

Mode Coupling theory on the other hand, predicts a scaling law for the diffusion coefficient

$$D^{MCT}(T) = D_0^{MCT}(T-T_c^{MCT})^\gamma \quad (4.7)$$

from which it follows that the apparent activation volume is

$$\hat{V}_{act}^{MCT} = -kT \left[\frac{\partial \ln D_0^{MCT}}{\partial p} - \gamma \frac{1}{k(T-T_c^{MCT})} \frac{\partial T_c^{MCT}}{\partial p} + \ln(T-T_c^{MCT}) \frac{\partial \gamma}{\partial p} \right]_T \quad (4.8)$$

Both the MCT and VFT expressions for the diffusion coefficient extrapolate to zero. However, the MCT diffusion coefficient extrapolates to zero at T_c^{MCT}, while for VFT, it is T_0^{VFT}. The former temperature is well above the glass transition temperature, while the latter is well below it. Within the numerical uncertainties, both expressions are too similar to be differentiated due to the crossover to the glass. However, the expressions for the activation volumes, which give singularities above and below the glass transition respectively, can be clearly distinguished. Schober calculated apparent activation volumes for both components of the binary mixture and found that they fell into 4 temperature regimes: A high value of around 0.6 atomic volumes for the hot liquid, a plateau value of about 0.3 atomic volumes and then a steep rise close to the MCT critical temperature, followed by a fall in value back to 0.3 average atomic volumes in the glassy state. The peak in activation volume at the critical temperature was found to agree with the prediction of MCT.

Rinaldi *et al.* performed molecular dynamics simulations of a supercooled liquid comprising of 3-site molecules parameterised so as to represent *ortho*-terphenyl, a well studied glass forming liquid.[130] Both self- and collective correlation functions were calculated for a large range of wavevectors and then compared with the predictions of mode coupling theory. The MCT equations they compared to their MD data, despite applying only to atomic liquids, were found to give good agreement with the site dynamics except in the wavevector regime centred around 9 nm^{-1}, where the centre-of-mass dynamics dominates.

Keyes *et al.* have extended the Random Energy Model (REM) and used it to analyse the phenomenology of fragile glass forming liquids.[131] The model parameters were obtained by fitting molecular dynamics results for the LJ fluid with the theory.

4.2 Structural Models for Supercooled Liquids. – These models are based on the molecular order in the supercooled liquid and employ Stillinger and Weber's inherent structure (IS) formalism.[132,133] The formalism separates the dynamics of liquids into vibrational motion within local potential energy minima ("inherent structures") and structured transitions between different local minima. Liquid properties can then be considered as the sum of "structural" contributions that are evaluated at the local energy minima visited by the system, and "vibrational" contributions associated with atomic motion within these local minima.

Lacks undertook an inherent structure analysis of the viscosity of a binary Lennard-Jones mixture.[134] The liquid viscosity at various strain rates was obtained from non-equilibrium molecular dynamics simulations, while the local minima in the energy landscape were determined by steepest-descent quenches from instantaneous atomic configurations. Lacks found that the viscosity can be separated into two contributions: a structural contribution associated with the energy minima that the system visits, and a vibrational contribution associated with displacements within the energy minima. Disappearance of high-stress energy minima gives rise to strain-induced relaxations and thence shear thinning. For strongly supercooled fluids which have a shear stress that tends to be dominated by the structural contribution, the viscosity is predicted to vary as the inverse of strain rate. This result is in agreement with experimental observations on supercooled liquids.[135,136]

Dzugutov *et al.* also carried out MD simulations of a simple Lennard-Jones liquid in the supercooled regime in an attempt to gain further insight into the supercooled dynamics, particularly the spatial variation of the atomic mobility and the breakdown of the Stokes-Einstein relation.[137] They found that during the supercooling process, a transition to deeper minima in the energy landscape gives rise to the formation of icosahedrally structured domains, characterised by slow diffusion, and which grow upon cooling. At the same time, there is an accompanying sharp slowing down in the structural relaxation relative to diffusion. Dzugutov *et al.* concluded that this effect could not be explained by the spatial variation in atomic mobility.

Tsige and Taylor also used MD to study the glass transition temperature in poly(methyl methacrylate).[123] By separating out the vibrational motion of atoms, monomers and molecules from the overall displacements from their initial positions, they were able to establish that the diffusive motion displays a power law variation with time, with an exponent that varies continuously from 0.5 below T_g to 1 at high temperatures. By analysing the self part of the van Hove correlation functions for both hydrogen atoms and monomers they conjectured that the so called 2nd peak which is usually interpreted in terms of hopping processes, is in fact due to rotation of methyl groups.

Colmenero *et al.* used a combination of MD and quasi-elastic neutron scattering (QENS) to study self motion of main chain hydrogens in the α-relaxation regime of the glass former, polyisopropene.[124] They found evidence for a crossover from a Gaussian to a non Gaussian regime which they show to be due to an anomalous jump diffusion model with a distribution of jump lengths. The model gives rise to a time dependent non-Gaussian parameter.

The measurement and characterisation of the spatially heterogeneous dynamics on approach to the glass transition has traditionally been discussed in terms of the two-point time dependent van Hove correlation function, $G(r,t)$. This function provides information on particle caging upon cooling, but is unable to provide local information about correlated motion and dynamical heterogeneity. Lacevic et al.[125] have instead used the fourth-order time-dependent density correlation function introduced by Dasgupta et al.[126,127] to measure higher order spatiotemporal correlations of the density of a supercooled binary Lennard-Jones fluid. They found that the characteristic length scale of this function has a maximum as a function of time which increases steadily beyond the characteristic length of the static pair correlation function in the temperature range approaching T_c^{MCT} from above. This length scale can therefore provide a measure of the dynamical spatial heterogeneity of the liquid in the α-relaxation regime.

Vollmayr-Lee and co-workers used MD to study a binary Lennard-Jones mixture at temperatures below the kinetic glass transition.[128] Using a definition of particle mobility based on the amplitude of fluctuation around its average position, they studied in detail the spatial distribution of the 5% of particles deemed to be the most and least mobile. It was found that this sub-population of atoms were distributed very heterogeneously in that both mobile and immobile particles formed clusters. They suggested that the heterogeneity has its origins in the size of solvent cage surrounding mobile/immobile particles, being smaller in the former, and larger for the latter.

Mosa et al. studied the interaction of the relaxation processes with density fluctuations for o-terphenyl in both the liquid and supercooled liquid phases using molecular dynamics.[129] They found evidence of a secondary relaxation, which is vibrational in origin, and relates to phenyl-phenyl stretching. Their results support experimental evidence of a fast vibrational relaxation regime in o-terphenyl, although the experimental and simulated timescales differ by a factor of 6. The authors attributed this large discrepancy to neglect of quantum effects in their model, the intramolecular potential used and to the method of performing the simulations in which the fluid is perturbed in an unphysical manner compared to experiment.

Doliwa and Heuer looked at jump motion among potential energy minima of a Lennard-Jones glass former by Langevin molecular dynamics.[138] They looked at the spatial and temporal aspects of hopping and determined an effective jump width and mean waiting time as a function of temperature. They show that diffusion can be pictured as a random walk among metabasins in the energy landscape and that the whole temperature dependence comes from the distribution of waiting times, which decays algebraically as $\tau^{-1/2}$ at very short times and $\tau^{-\alpha}$ for long times with $\alpha \approx 2$ near T_c. They suggest the existence of long-lived metabasins can explain the slowing down of molecular motion in their supercooled model liquid.

Chelli et al. performed molecular dynamics simulations of the fragile glass former, metatoluidine in order to study the structure and dynamics of the liquid in the supercooled state.[139] They followed the liquid glass transition by calculating the molar specific heat as a function of temperature. By analysing the various contributions to the molar specific heat, they were able to ascertain that it is essentially the configurational contribution that determines the behaviour of this quantity at the liquid-glass transition. On the basis of their results and analysis, Chelli *et al.* proposed a mechanism for the liquid-glass phase transition which is based on the structural rearrangement of clusters, resulting from large scale rotations of the aromatic rings. Their simulations suggest the existence of a critical temperature, in support of MCT.

Barbieri *et al.* used molecular dynamics simulations to help interpret experimental data on the density and thermal behaviour for CaO–ZrO_2–SiO_2 glasses.[140] Their aim was to determine the role played by the different atoms present in the glass formulation and to correlate the structural modifications to the macroscopic properties of this class of glasses. They determined that ZrO_2 acts as a network former up to molar contents of 5–10%. On the other hand, CaO was found to work as a network modifier; addition of large amounts of CaO leads to clustering of ions.

Mode coupling theory of the glass transition is able to offer an explanation for the existence of a crossover from Arrhenius to super-Arrhenius behaviour in fragile glass formers. However since strong glass formers by definition do not show a crossover between the temperature dependencies in viscosity (or at best have only a weak one) it is not clear whether such liquids have a temperature regime in which MCT can be applied. Horbach and Kob used MD to study relaxation dynamics of a viscous melt of silica, an archetypal strong glass former in a bid to check whether MCT of the glass transition is able to describe the dynamics at low temperature.[141] In this work they analysed the time and temperature dependence of the various intermediate scattering functions for different wave vectors (coherent and incoherent). These functions are good probes of the slowing down of the system since they change their shape from single exponential at high temperature, to a two-step relaxation at low temperature. They found that in the temperature regime where the relaxation times of the dynamics are mesoscopic, many aspects of the dynamics of this strong glass former can be rationalized very well by MCT.

Murarka and Bagchi have used MD to understand the relaxation mechanism in a strongly non-ideal glass-forming binary mixture (Kob-Anderson model).[142] They analysed the dynamics of tagged pairs of atoms (AA, BB and AB types) by calculating time dependent partial pair distribution functions (angular and radial). They found clear differences in the relative motion of the three types of pairs, AA, BB and AB. The relative diffusion constants of the BB pair is almost twice that of the AA pair. The authors suggest this is evidence of a largely hopping mechanism for B, while an occasional one for A. Murarka and Bagchi also generalized the concept of a non-Gaussian parameter used to describe deviations from Gaussian behaviour in the motion of an individual atom to one measuring deviations from Gaussian behaviour of relative motion of atoms. At intermediate times, a significant deviation from Gaussian behaviour of pair distribution functions was observed with differing degrees for the three pair types. They also applied the 1-component mean field theory of Haan and found it could successfully describe the dynamics of the

AA and AB pairs but not BB. They explain this in terms of large anharmonic motions of B particles in a weak effective potential.

Harbola and Das have investigated the nature of transverse fluctuations in a supercooled binary liquid in the context of the self-consistent mode coupling model.[143] They found a crossover in the relaxation behaviour of the transverse current correlation functions for the isotropic liquid, from the propagating shear waves to diffusive shear mode at a critical wave number. This critical wave number decreased with increasing density indicating growing correlations. They also observed anomalous stretching of the frequency dependent shear modulus with density.

Kim and co-workers studied dynamical rotational heterogeneity in a supercooled molecular liquid, CS_2.[144] They characterised this heterogeneity via the distribution of individual molecular contributions to the anisotropic polarizability correlation function. They found a crossover from Arrhenius to super-Arrhenius behaviour of the rotational correlation times.

Munakata derived the dynamic density functional theory and Smoluchowski equation for an interacting many body system by applying a Markovian approximation to the Mori-Fujisaka nonlinear generalized equation, *i.e.* a kinetic theory approach.[145] Both approaches were found to be useful in studying glass transitions and dynamics in supercooled liquids.

Starr *et al.* used a Voronoi construction to analyse the local geometry of a simulated glass-forming polymer melt.[146] They found that the distributions of Voronoi volume and asphericity were universal properties of dense liquids. They also found that the average free volume along a path of constant density extrapolates to zero at the same temperature that the extrapolated relaxation diverges and can be related to the Debye-Waller factor.

Generation of equilibration configurations for use in simulations of the slow dynamics of supercooled liquids is a difficult task. De Michele and Sciortino made a comparison of two methods of generating such configurations to see which method lead to the fastest equilibration time.[147] They compared molecular dynamics with the parallel tempering (PT) technique.[148,149] The latter method was developed for dealing with the slow dynamics of disordered spin systems. In a nutshell, the PT algorithms simultaneously simulate a set of identical non-interacting replicas of the system, each at a different temperature. Pairs of replicas swap their temperatures according to an MC procedure. They found that PT does not increase the speed of equilibration of the slow configurational degrees of freedom.

Brinkmann and Teichler conducted constant strain rate and constant shear stress NEMD simulations of a model metallic glass $Ni_{0.5}Zr_{0.5}$.[150]

Huerta and Naumis used MC to simulate the competition between molecular relaxation and crystallization times in the formation of a glass.[151] They found evidence in support of Phillips' conjecture that the glass transition is related to rigidity due to the lack of a pathway to crystallization.

Habasaki and Hiwatari conducted MD simulations of the ion conducting glass, lithium metasilicate ($LiSiO_3$).[152] They observed dynamic heterogeneity for both this system and the $LiKSiO_3$ system.

5 Confined Liquids

The properties of liquids under confinement can differ markedly from those of the corresponding bulk fluid, particularly when the defining length scale is of the order of a few molecular diameters. Modification of the fluid properties is a direct result of competition between the intermolecular forces operating between the fluid molecules, and the intermolecular forces acting between the fluid molecules and the molecules that constitute the confining material. This competition can lead to the formation of a non-uniform fluid density profile, a shift in the thermodynamic equilibrium properties and, a modification of the transport properties.

Changes such as these are exploited in many key industrial processes such as in catalysis, separation science, and gas storage. The latter process is currently the subject of heated debate surrounding the use of carbon nanotubes as potential hydrogen storage devices for use in hydrogen fuelled automobiles.[153]

5.1 Structure. – Liquids confined by solid walls are known to display layering in the direction perpendicular to them. Simulation investigations of these layering effects are discussed.

Wang and Fichthorn looked at the influence of molecular structure on confined LJ, *n*-decane and 2,2-dimethyloctane fluids.[154] The LJ and *n*-decane fluids were observed to form layers parallel to the surfaces while the 2,2-dimethyloctane formed a 'pillared-layered' structure consisting of both parallel and perpendicularly aligned molecules. They observed a step change in the number of molecules as pore width is varied for LJ and *n*-decane, but not for the asymmetric 2,2-dimethyloctane. They also observed oscillatory behaviour of the force, viscosity and translational diffusion. Pocheron *et al.*[155] have looked at the behaviour of ultrathin films of alkanes under confinement by GCMC. They found no evidence of any discontinuities in the average numbers of particles with wall separation.

Porcheron and co-workers used configurational bias MC to study the properties of a thin film of *n*-butane, sandwiched between corrugated walls.[156] By changing the degree of corrugation they observed a change in the solvation pressure profile from a damped oscillatory profile to nonoscillatory behaviour as the characteristic length of the corrugations (furrows) became comparable to the typical dimensions of a butane molecule.

Cui and co-workers have investigated the properties of thin films of alkanes confined between mica-like walls.[157,158] Their work involved using molecular dynamics simulations of confined dodecane in which both the slit width and strength of the wall-liquid interaction were varied. They observed a significant increase in the density of the alkane film as the degree of confinement increased such that the film comprised of 6 molecular layers. The effective alkane density at this point was greater than the bulk freezing density and behaved in a solid like manner. The authors suggest this might provide a microscopic mechanism for the observed increase by up to 6 orders of magnitude of the viscosity of confined thin films observed in experiments using the surface force apparatus.[159–161]

Jin and Wang[162] used GCMC to simulate the adsorption of HCFC-22 (modelled with a Stockmayer potential) in slit-like activated carbon adsorbents in an attempt to find the optimum parameters for the adsorption recovery of this substance.

5.2 Phase Equilibria. – Capillary condensation in porous adsorbents is an example of a phase transition which has been well studied. Capillary condensation is the formation of a liquid phase at a pressure lower than the bulk saturated vapour pressure due to the effects of surface tension. In porous adsorbents there is a continuous progression from multilayer adsorption to capillary condensation. An excellent review published in 1999 by Gelb *et al.* covers many aspects of phase separation in porous materials, including capillary condensation.[163]

Shi and co-workers[164] used multiple histogram re-weighting to examine capillary condensation, prewetting and layering transitions for confined fluids. Capillary condensation for methane in graphite slit-pores was found to exhibit 2D behaviour. Crossover of the effective exponent for the width of the coexistence curve from 2D Ising-like behaviour further away from the critical point to mean field near the critical point was observed.

Schoen used GCMC simulations to study a LJ fluid confined between furrowed walls.[165] For a fixed wall separation, the phase behaviour (in particular capillary condensation) was found to depend sensitively on the substrate corrugation. Gentner *et al.* carried out molecular dynamics simulations of a liquid composed of chains of atoms confined between two atomistic walls moving in opposite directions in order to study the dynamics of wetting.[166] Macroscopically, dynamic wetting is measured by the variation in contact angle with time. They changed the strength of the liquid-solid interaction in the simulations to see what effect it had on the wetting dynamics. They found that the steady state contact angle changed along with the flow field near the wetting line. Two regimes were identified: a hydrodynamic regime which described the low speed results, and a molecular-kinetic regime applicable to a large range of speeds.

Vishnyakov and co-workers studied the critical properties of the Lennard-Jones fluid in slit-like pores of differing width by Gibbs ensemble MC and the lattice gas model.[167] Both methods gave similar results for vapour-liquid phase diagrams. They found a linear dependence of the critical temperature on inverse pore width and that T_c may depend strongly on the strength of the solid-fluid interactions. The lattice gas model showed non-monotonic dependence of the critical density on pore width. Continuing the theme of liquid-vapour equilibria, Potoff and Siepmann carried out GCMC simulations with histogram reweighting and a mixed field analysis to determine the vapour-liquid phase diagrams and critical parameters for a series of straight and branched chain alkanes physisorped on a flat gold surface.[168] The systems were found to exhibit 2-dimensional Ising critical behaviour. The ratio of the 2-D to the 3-D critical temperature was found to depend weakly on chain length for the *n*-alkanes. Branched isomers were found to have higher critical temperatures than their linear counterparts, which stands in contrast to the behaviour of the related bulk liquid.

An understanding of the effects of confinement on freezing of liquids is essential in applications such as lubrication, fabrication of nanomaterials and nanotribology. For a full review of this area, the reader is once again referred to Ref. 163. Sliwinska-Bartkowiak and co-workers reported experimental measurements of the melting and freezing behavior of fluids in nanoporous media: controlled pore glass, Vycor and MCM-41.[169] These experimental findings back up the results of earlier simulation studies of freezing transitions in slit and cylindrical shaped pores in which it was found that the confined fluid freezes into a single crystalline structure for average pore diameters greater than 20 fluid diameters, a frustrated crystalline structure for pore sizes between 15 and 20 fluid diameters, and an amorphous solid below 15 molecular diameters. Radhakrishnan *et al.* have rationalised the diverse and often contradictory data on freezing behavior of small molecules in pores.[170] They have shown that the behavior can be classified and understood in terms of two main parameters: (1) The relative strength of fluid-wall to fluid-fluid attractive interaction, α, and (2) the reduced pore width $H^* = H/\sigma_{ff}$, where σ_{ff} is the Lennard-Jones diameter of the fluid-fluid interaction. By conducting molecular simulations of confined fluids together with Landau free energy calculations to locate true thermodynamic phase transitions, they were able to construct global phase diagrams for slit pore fluids in terms of the parameters H^* and α. In a further publication, Radhakrishnan *et al.* performed molecular simulations of simple fluids, confined in slit-shaped carbon pores[171] in order to study the nature of the solid phase found from earlier studies of freezing in pores. They found that the fluids exhibit crystal-hexatic and hexatic-liquid transitions consistent with Kosterlitz-Thouless-Halpern-Nelson-Young theory. The temperature range for stabilization of the hexatic phase is significantly widened under confinement. The transitions cross over from being continuous for a single adsorbed layer, to weakly first order for 2 layers.

Koga used MD to study the phase behaviour of water confined between hydrophobic surfaces at nanoscale distances.[172] Melting and freezing were observed when one surface was brought close to the other at fixed temperature and bulk pressure. Solvation forces were found to be discontinuous upon freezing and melting and exhibited strong hysteresis.

As well as liquid-vapour and liquid-solid phase transitions in pores, there has been some interest in studying order-disorder transitions such as the isotropic-liquid crystal transition. Quintana *et al.*[173] looked at the effects of confinement on the phase diagram of a Gay-Berne fluid via MC simulation. Confinement leads to a widening of the smectic-isotropic coexistence region together with an increase in the triple point temperature. No evidence was found for confinement induced stabilization of the smectic phase. Erickson and co-workers[174] used lattice DFT to study order-disorder phase transitions in adsorbed confined fluids with directional interactions. The model predicts that ordering can lead to formation of long chains – a phenomenon that has been observed experimentally in magnetorheological fluids in a magnetic field. Klapp and Schoen have conducted MC simulations of a dipolar soft sphere fluid confined in a nanoscopic slit-pore.[175] They observed that a transition from a globally isotropic fluid to a globally polarized fluid occurs at significantly lower pressure and density than in the bulk fluid. They suggest that their observations provide evidence that confinement can support the onset of ferroelectric order.

Shen and Monson have touched on the issue of the effect of framework flexibility on confined liquid properties.[176] They found that framework flexibility can have a significant effect on adsorption/desorption isotherms and fluid phase coexistence for a highly porous (95%) network similar to that of silica aerogels.

Brennan and Dong[177] used GEMC and Gibbs-Duhem integration to study the phase equilibria of a 1-component fluid confined in random porous media. They found an extreme sensitivity of the phase diagram on the microscopic structure of the porous samples.

Lee et al.[178] have used GCMC to investigate the components of the force between parallel charged surfaces in an electrolyte.

5.3 Glass Transition. – One of the ideas originating from research into the glass transition is that the slowing down of the structural relaxation is caused by an underlying correlation length ξ, which increases as the temperature is lowered towards the glass transition temperature. If such a correlation length does in fact exist, then it will be reduced by spatial confinement as soon as ξ approaches the typical dimension of the restrictive geometry. A shift in the value of T_g is therefore expected for glass forming liquids in confined geometries.

Gallo et al. performed molecular dynamics simulations of a supercooled Lennard-Jones binary mixture in a disordered array of soft spheres.[179] They found that the MCT predictions for bulk liquids also apply to confined liquids. However, they also found that the range over which MCT theory is valid is smaller in the confined regime and that hopping processes are greatly enhanced above T_c in the confined liquid compared to the bulk.

Kranbuehl and co-workers studied the effect of confinement geometry on the glass transition temperature.[180] They carried out MC simulations of freely jointed polymer chains spatially confined by 1-dimensional (between parallel plates), 2-dimensional (in a tube) and by 3-dimensions (in a sphere) using repulsive wall potentials. They observed that the glass transition temperature decreases with increasing degree of confinement, i.e. lowering of the confined dimension. The onset of this decrease occurs first for the spherical confined geometry which also gives rise to the largest change in T_g. Two-dimensional tubes showed constraint effects at larger values of the constrained length than in 1-dimensional pores. Their results are in accord with experimental studies of constraint geometry effects on T_g.

McCoy and Curro hypothesised that the shift in T_g for confined polymers is a result of the inhomogeneous density profile of the liquid.[181] They assumed that T_g of the confined polymer can be approximated by T_g of the corresponding bulk material at a density equal to the average confined density. Models based on this hypothesis gave results in agreement with experiment.

5.4 Diffusion. – Transport in porous catalysts and adsorbents occurs mainly by diffusion, which is often rate determining. A detailed understanding of diffusion in porous solids is therefore of great importance for optimization of these processes, and simulation is a powerful tool in this respect.

The widely accepted model for diffusion in porous solids is based on the dusty gas model of Mason and co-workers.[182–184] This model, introduced some years ago, yields the correct limiting zero density behaviour consistent with the results of Knudsen flow experiments. The key point about the model is that the solid itself is treated as if it were a component in the diffusing mixture. Using linear, irreversible thermodynamics (or equivalently, Stefan-Maxwell formalism), one can then obtain a set of flux equations for each *fluid* component in the mixture (the flux of the membrane particles drops out through choosing the laboratory frame as a frame of reference). Furthermore, the flux expressions contain a viscous flow term which arises when a pressure gradient is present. For flow in wide pores, the viscous mode of transport is expected to dominate, whereas at small pore widths, viscous flow will be negligible, giving rise to predominantly diffusive transport. At some intermediate regime, both modes will occur together with neither being particularly dominant.

Despite its successes, the dusty gas model is far from satisfactory. Several objections may be raised, particularly when it is applied to flow in a single pore. In this case one often sees the flux expression written as a sum of viscous and diffusive terms in many publications, with the tacit assumption that there are two modes of transport which cannot be separated. It is possible to design a simulation method in which an effective pressure gradient is applied without an accompanying density gradient – this is the usual Poiseuille flow algorithm (see ref 185 for example). In this case it is clear that only viscous flow may occur in the direction of the pressure gradient (for the isothermal case). In principle, this Poiseuille flow algorithm provides a means of calculating the viscous part of the total flux when a chemical potential gradient is used to drive the flow à la the dual control volume grand canonical molecular dynamics (DCV GCMD) method. Indeed, Travis and Gubbins proposed just such a method.[186] However, it is now thought that this method is incorrect, at least for the 1-component fluid case.[187] The current thinking is that there is only one mode of transport i.e. viscous for this case. It is impossible to separate two manifestations of the same phenomenon. By treating the flux as comprising an effective diffusive and a viscous component, a diffusion coefficient emerges which has some peculiar properties. For example, if the pore width is increased, one can show that this transport "diffusion coefficient" increases without limit rather than approaching zero as it should do for the bulk fluid. Indeed, a diffusive flux is found to obey Darcey's law, which strongly suggests that what is being measured is not diffusion, but rather viscous flow.

Confusion continues in the diffusion literature. Bhatia and Nicholson have used three different simulation techniques to study the transport of a simple fluid in a cylindrical nanopore: DCV GCMD, equilibrium molecular dynamics and so-called forced flow NEMD (essentially the Poiseuille flow algorithm). To their surprise, these authors discovered that all three methods yielded the same value for the transport coefficient provided diffusely reflecting walls were used in the pore model. The surprise comes from the fact that the DCV GCMD simulation supposedly simulates diffusion and viscous flow simultaneously whereas the forced flow simulation simulates pure viscous flow. Their results are perfectly explainable if one accepts the idea that for a single component fluid in single pore, the isothermal transport is viscous only. Progress may only be made if we abandon the idea of including the solid as part of the diffusing mixture, and develop a new theory from the framework of irreversible thermodynamics.

The industrial separation of air into it major components using molecular sieving carbon (MSC)[188] has provided a good application to motivate many simulation studies. It is not difficult to see why since the quest for an understanding of the exact separation mechanism (which is believed to be due to differences in intra-pore diffusivity of oxygen and nitrogen) has been referred to as the "Holy Grail" of separation science.

The most recent attempt at unravelling the mechanism has been made by Travis, who has shown that the mechanism could be explained in terms of loss of rotational degrees of freedom.[189] In this work, a systematic simulation study was undertaken in which the self- and transport diffusivities of nitrogen and oxygen were compared in a range of graphite slit-pores of differing widths. It was found that once the pore width became narrow enough to prevent free rotation of nitrogen molecules about one of their axes, yet still wide enough to allow oxygen molecules to rotate largely unhindered, a high diffusive selectivity resulted in favour of oxygen (see figure 1). Evidence for hindered rotation was obtained from angular velocity autocorrelation functions (see figure 2).

The previous study concerned only intracrystalline resistances to transport, pore entry/exit effects being ignored. It has long been speculated that it is the existence of a significant potential energy barrier at the entrance to a pore which may explain the typical order of magnitude difference between diffusion coefficients obtained via PFG NMR measurements and those from uptake measurements. The former measures only intracrystalline diffusivity, while the latter implicitly includes the effects of potential barriers at pore entrances/exits.

Ahunbay *et al.* studied the diffusion of methane in a single crystal membrane of silicalite using DCV GCMD in an attempt to determine the contribution made by surface resistance (entry and exit effects).[190] They found that this contribution is larger and of longer range than might be expected. Using a model in which the resistances are assumed to be additive they found that the exit (desorption) resistance saturated at about 50 nm of crystal thickness while the adsorption resistance is predicted to saturate at about 150 nm. Ahunbay *et al.* concluded that intra-crystalline resistance was still the dominant contribution for the applications they were interested in.

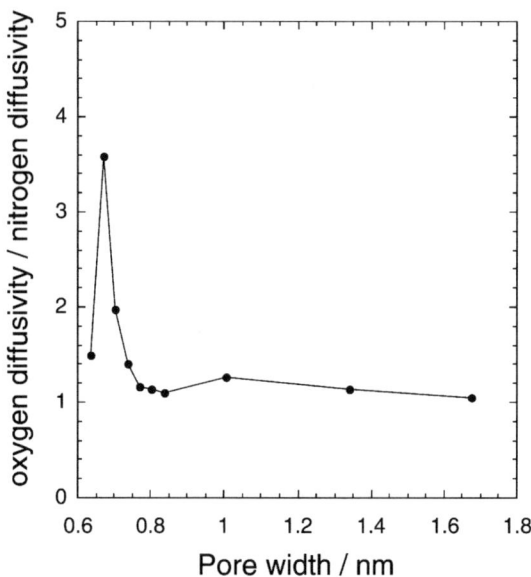

Figure 1 *Plot of the diffusive selectivity versus pore width obtained from EMD simulations of oxygen and nitrogen confined in graphitic slit-pores. The diffusive selectivity is defined as the ratio of the coefficients of self-diffusion for the corresponding species*

There have been many studies of the spatial dependence of viscosity in highly confined fluids, but relatively few studies of the spatial dependence of diffusion coefficients. Recently, MacElroy and co-workers have looked at this issue. They combined the results of NEMD and EMD simulations of confined fluids with Pozhar-Gubbins theory in order to study self-diffusion of dense fluids within a crystalline solid.[191] Their results show that the spatial dependence of transport parameters should be taken into consideration to reliably predict the diffusion fluxes within zeolitic systems. In particular they found that the self-diffusivity increased significantly in magnitude over nanometer length scales in the model pores studied.

Diffusion in non-isotropic media typified by the zeolite class of materials, must be described by a diffusion tensor,[188] the principal components of which are not necessarily equal. However, the principal components of the diffusion tensor are often found to be interdependent. For MFI-type zeolites for example, the diffusivities in the main crystallographic directions are found to obey the following relationship[192,193]

$$\frac{c^2}{D_z} = \frac{a^2}{D_x} + \frac{b^2}{D_y} \tag{5.1}$$

in which a, b and c are the unit cell extensions in the x, y and z directions. MFI type zeolites are characterised by the presence of straight and zigzag channels

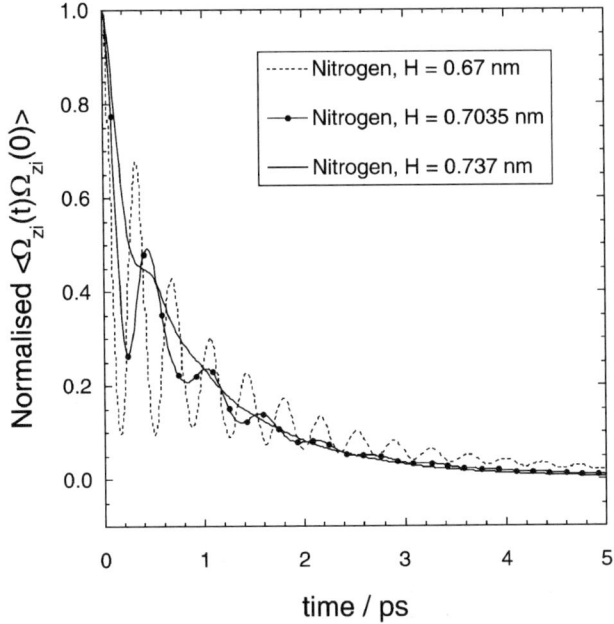

Figure 2 *Angular velocity autocorrelation function for nitrogen molecules confined in slit-shaped carbon pores of differing width. The function involves a correlation of the component of angular velocity about the z-principle axis (perpendicular to the bond). The oscillatory behaviour evident at pore widths of about 0.7 nm and below indicates librational motion of the molecules*

running along the x and y directions. Diffusive movement of an adsorbate molecule in the z-direction is restricted and not continuous. Instead, the molecule diffuses in this direction by subsequent moves in the x-direction and, occasionally, the y-direction. The above equation was derived by assuming that subsequent moves between intersections are statistically independent of each other. Diffusion in MFI zeolites in the main conforms to Eq. 5.1. However, deviations from this relationship imply a memory dependent diffusion process.

Diffusion anisotropy is not restricted to zeolites. Su and co-workers have studied the diffusion of monolayer octamethylcyclotetrasiloxane and cyclohexane films confined between atomically structured uncharged mica surfaces using molecular dynamics and GCMC.[194] They found that diffusion parallel to the walls is anisotropic. They also found that the direction and magnitude of the diffusion depends on the relative alignment of the surfaces.

In ordinary diffusion, the mean square displacements are proportional to the time. In some circumstances, molecules are unable to pass each other due to geometrical constraints, such as encountered in the narrow channels of many zeolites. Mean square displacements in these cases are found to vary with the square root of time and the process is termed single file diffusion. Simulation studies of single file diffusion continue to generate interest. Adhangale and Keffer for example, used MD to look at the behaviour of polyatomic ethane molecules adsorbed in AlPO4–5 and observing both ordinary and single file diffusion in these 1D nanoporous materials.[195] The transition from ordinary diffusion to single file diffusion was found to occur for a methyl group diameter of 4.75Å.

Mon and Percus have looked at single file diffusion of fluids confined in narrow cylindrical pores.[196] They found that the diffusion coefficient D_{xx}, depends on a hopping time as $D_{xx} \propto (\tau_{hop})^{-1/2}$, which they regard as a universal relationship.

Vasenkov and Kärger have combined the results of Dynamic Monte Carlo (DMC) with analytical calculations to give a complete description of the time dependence of tracer exchange in single file systems.[197] They found that depending on the elapsed time, the tracer exchange can display different time scaling related to transport modes of either normal or single-file diffusion.

Diffusion in zeolites remains a popular research theme due to the industrial importance of these materials and to the fact that their structures are well characterised, and therefore straightforward to simulate. Skoulidas and Sholl studied the self- and transport diffusivity of several light gases: CH_4, CF_4, He, Ne, Ar, Xe and SF_6 in silicalite at room temperature.[198] They found that self-diffusivity decreases with increased pore loading, whereas transport diffusivity increases with increased loading. They found that in general, both the Darken approximation of a constant, corrected diffusivity and the Lattice Gas approximation of a constant transport diffusivity, are incorrect.

Tunca and Ford have extended multidimensional transition state theory to obtain escape rates of adsorbate molecules from zeolite cages.[199] The improved method takes into account the occupancy of neighbouring cages amongst other things.

Fritzche and Kärger carried out molecular dynamics simulations of the diffusion of methane in silicalite-1 and analysed their results in terms of memory effects.[200] They found good agreement between their data and analytical treatments of such memory effects.

Diffusion in carbon nanotubes is another popular research theme motivated by their many (arguably over-hyped) potential applications. Düren et al.[201] used DCV GCMD to investigate the effect of changing molecular properties on adsorption and diffusion fluxes of binary fluid mixtures in carbon nanotubes. They found agreement with Grahams law for the whole range of molecular weights and LJ parameters investigated except in cases where there is strong interaction between the components of the mixture.

While many studies of confined diffusion in slit-pores have concentrated on carbon as the substrate, other surfaces such as metallic ones are now attracting attention. For example, Zhang and co-workers have used DCV GCMD to study the transport diffusion of water-methanol mixtures through slit pores composed of Au(111) surfaces decorated with methyl and hydroxyl self-assembled monolayers.[202] They observed for the first time in a simulation the convex and concave interfaces of fluids transporting across slit pores under a chemical potential gradient. They found that the nature of the interfaces depends on the fluid-fluid and fluid-solid interactions. In addition, they found that the methanol flux was higher than that of water in hydrophobic pores leading them to speculate that hydrophobic pores could be employed to separate methanol-water mixtures.

Finally, from an engineering perspective, it is material properties such as porosity which are the parameters of importance when considering their effect on diffusion. Dominguez and Rivera therefore investigated the effects of porosity on diffusion.[203] The solid matrices were prepared by different methods yielding different porosities *e.g.* MD of binary mixtures followed by etching out one of the components, or by using the final configurations of a constant density MD run of a 1-component liquid. Their results showed that the diffusion coefficient depended on the kind of matrix used: matrices having the highest porosity and diffusion coefficient were the ones prepared using attractive particle interactions and without the use of a template.

5.5 Rheology. – Theory, backed up by simulation has shown that the viscosity of a liquid confined in a pore of just a few molecular diameters thickness is non-uniform.[204,185] This suggests that the Navier-Stokes equations break down at molecular length scales. This breakdown can be traced to two main sources: (1) the assumption of constant shear viscosity implicit in Newton's law of viscosity, and (2) the boundary conditions applied.

The standard boundary condition for fluid flow along a surface is the no-slip condition, which, for a fluid at rest, implies zero tangential fluid velocity at the wall. There has been much debate in the literature over the years about the accuracy of the no-slip boundary condition. Some simulation studies of confined fluids suggested that a degree of slip is encountered.[205] However, the work of Travis and Gubbins found that while slip does occur for highly confined simple fluids in slit pores, the Navier length is of the order of the particle diameter.[185] These results can be taken to read that the no-slip condition is correct given that the position of the solid-liquid interface can only be defined to within a particle diameter in any case. For non-simple fluids, experimental results demonstrate that slip does occur and that the Navier length can be of the order of a micrometer. de Gennes has recently discussed the origin of slip in fluids giving rise to large Navier lengths.[206] He postulates the existence of a thin gaseous film between the fluid and the wall to explain this. Spikes and Granick have recently developed a new equation for Newtonian fluid flow in the presence of slip at a solid surface.[207] In their model, slip only occurs when a critical shear stress is reached, and once slip begins, it takes place at a constant slip length. Their model is able to reconcile the results from different experimental studies.

The variation of viscosity with degree of confinement is not well understood and hence has attracted much attention in the simulation literature. Apart from those simulations which have looked at the breakdown of the Navier-Stokes hydrodynamics and attempted to extend it to smaller lengthscales, there has recently been an attempt to model the confined viscosity from the kinetic theory angle. Zhang and co-workers have proposed two correlation models for describing the viscosity of simple fluids in pores based on Chapman-Enskog theory and its extension to liquid densities by Heyes.[202] These theories for bulk fluids are extended to cover pore fluids by making use of the approximate empirical observations concerning how viscosity varies with density, temperature and pore width. They obtained the following correlation model for the reduced pore fluid viscosity, η_p

$$\ln\left(\frac{\eta_p^*}{\eta_b^*}\right) = \frac{L(\rho^*)^c}{T^*}\left(\frac{1}{H^*-1}\right)^n \quad (5.2)$$

where L, c and n are empirical constants, η_b is the reduced bulk fluid viscosity, H^*, T^*, ρ^* are the reduced pore width, temperature and density respectively. Claiming a lack of experimental data to use for testing their model, they generated their own simulation data for this purpose. Perhaps surprisingly given the crudeness of the approach and various liberties taken with the calculation of confined viscosities, the model results agreed reasonably well with the simulated data for a range of temperatures, densities and pore widths, but became increasingly poor as the reduced pore width decreased below a value of 3.

Tovbin and co-workers have used a different approach to the usual NEMD/EMD techniques for studying fluid flows in narrow pores.[208] Their approach involves using a lattice-gas model which includes the effects of molecular interactions in a quasi-chemical approximation. The model consists of a set of equations which are described by the authors as being quasi-hydrodynamic. Using this method they analysed the dynamic modes of the flow of argon gas in carbon slit-pores. The information they obtained from their simulations concerning the redistribution of molecules and their velocities is claimed by the authors to be comparable to that obtainable from MD simulations with the added advantage of faster computation.

Horbach and Binder have performed EMD and NEMD simulations of amorphous silica (a network forming liquid) confined between atomic walls.[209] They observed a pronounced layering of the melt near the walls. The layering was found to be irregular in character, a fact which they attributed to the preferred orientational ordering of SiO_4 tetrahedra near the wall. Strong structural rearrangements in the ring size distribution were found to be present at the walls when the melt was subjected to Poiseuille flow.

Motivated by the need to develop a better understanding of viscous behaviour of confined simple fluids, Porcheron and Schoen have carried out microcanonical equilibrium MD simulations of simple fluids confined in slit pores.[210] To analyse the viscous behaviour they focussed attention on the propagation of density modes in the confined fluid in the hydrodynamic regime. They achieved this by calculating the intermediate scattering function, $F(k_\parallel,t)$ and its associated memory function, $M(k_\parallel,t)$. They derived an expression for $F(k_\parallel,t)$ in the hydrodynamic limit in terms of the thermal diffusivity, D_T, sound attenuation coefficient Γ, the in-plane adiabatic velocity of sound, $v_\parallel$ and ratio of heat capacities at constant transverse stress and volume, γ. Values of these material properties were then obtained by fitting the equation to their simulation data. The memory function approach turns out to be the most useful since the other approach, involving $F(k_\parallel, t)$, yields material properties which depend on wave vector.

Using a solvent primitive model in which both ions and solvent molecules are soft core spheres and the polar nature of the solvent is taken implicitly as a background with a given dielectric constant, Tang *et al.* carried out EMD and NEMD (constant electric field) simulations to calculate the diffusion coefficient and electric conductivity of ions in neutral cylindrical pores, with diameters 3–15 times the ion diameter.[211] From EMD the diffusion coefficients of the ions and solvent were found to decrease with decreasing pore radius or increasing packing fraction of solvent particles. NEMD obtained conductivity showed a similar trend. They calculated ionic conductivities from EMD results using the Nernst-Einstein relation and compared these with NEMD data. Poor agreement for very small pores, the EMD results being lower than NEMD ones was found.

Lutsko and Dufty investigated the long ranged correlations in a fluid undergoing planar poiseuille flow.[82] They demonstrate that an exact relationship between the density autocorrelation function and the density-momentum correlation function implies that the density autocorrelation decays faster than $1/r$, in contradistinction to mode coupling theory. They derived an analytic non-perturbative mode-coupling model which predicts a crossover from $1/r$ at small r to a stronger asymptotic power law decay with a characteristic length scale which is macroscopic.

5.6 Chemical Reactions. – An improved understanding of the effects of confinement on the rates of chemical reactions is important in catalysis. However, surprisingly few simulation studies have been carried out to determine this effect.

Turner and co-workers have investigated the influence of confinement on chemical reactions taking place in a porous solid.[212] They employed a new methodology in their simulation studies in which transition state theory is combined with the Reactive Monte Carlo (RxMC) method of Johnson *et al.*[213] Whilst their method is straightforward in principle, the difficulty lies in knowing the structure of the transition state and the activation energy *a priori*. Nevertheless, using this method, Turner and co-workers simulated the decomposition of hydrogen iodide, finding good agreement with experiment for the rate constant over a pressure range from 1–300 bar. They observed a significant enhancement in the rate constant (compared with the bulk fluid) when the reaction took place within a carbon slit pore and carbon nanotubes. Significantly, the rate constant was found to increase by a factor of 47 for the (8,8) nanotubes at 200 bar.

Travis and Searles have studied the effects of confinement on the rates of isomerization in *n*-butane adsorbed in slit pores of varying width.[214] They have found that the rate of isomerization decreased with increasing degree of confinement, relative to a bulk fluid of the same mean density and temperature.

5.7 Water Adsorption. – The presence of water in porous carbons is quite often detrimental to many applications involving these adsorbents. The presence of water can significantly inhibit the capacity and selectivity of industrial porous carbon adsorbents during the removal of organic and inorganic contaminants. Small amounts of water can also reduce the diffusion rates of alkanes in microporous solids by an order of magnitude. The exact mechanism giving rise to these observations remains elusive.

Brennan and co-workers have used Grand Canonical Monte Carlo simulations to study the adsorption of water in realistic models of both activated and nonactivated porous carbons, with the aim of elucidating this mechanism.[215] They found that water adsorption increases as the activated site density increases and that the presence of water vapour in porous carbons has a significant affect on the connectivity of the available pore space. Brennan and co-workers have also reviewed the experimental, semi-empirical and simulation studies of water confined in porous carbons.[216]

5.8 Applications of Density Functional Theory. – Classical density functional theory (DFT) methods provide an alternative approach to the study of confined fluids over the usual molecular simulation and macroscopic thermodynamic approaches. Essentially DFT yields an equilibrium density profile from a minimisation of a free energy functional. Adsorption isotherms for example, can be obtained at a fraction of the computational cost of the GCMC method. Because DFT methods are so useful in obtaining information about phase behaviour and structure of confined fluids we include a brief survey of publications in this area. A recent review of various DFT methods in use for the statistical mechanics of materials has been completed by Oxtoby.[217]

Schmidt developed a DFT for fluids in porous media which treats both solid matrix and adsorbate components on the level of the density profiles.[218] This is a major simplification over the standard approach using an external potential to describe the matrix particles. The method was tested on confined fluids by comparison with computer simulation results. This method has great potential for studying equilibrium properties of confined fluids such as freezing in pores.

Nguyen and co-workers have used DFT (at the close packed limit) together with a lattice gas model to develop a model for predicting the temperature dependent variation in capacity with pore size.[219] They applied their model to methane in graphitic slit pores. The model predictions were found to be in excellent agreement with the results of GCMC simulations (provided the adsorbed phase did not exceed three layers). Using characteristic dimensions extracted from the model, information on the changing microstructure of the pore fluid can be obtained. The weakness in the model is that it is difficult to extend it to beyond 3 layers.

Sweatman used weighted DFT to study prewetting of a Lennard-Jones fluid.[220] The intrinsic free energy functional is separated into repulsive and attractive contributions. A hard sphere functional is used for the repulsive functional and weighted DFT method for the attractive part. This method was compared with the mean field DFT, GCMC and a theory of Velasco and Tarazona. Sweatman's results show that weighted DFT for attractive forces may offer a significant increase in accuracy over the other theories.

Ravikovitch and co-workers undertook a systematic comparison of non-local density functional theory (NLDFT) of confined fluids with MC simulations using as an example a Lennard-Jones fluid in slit and cylindrical nanopores of varying width.[221] They employed 2 versions of NLDFT, the smoothed density approximation (SDA) and the fundamental measure theory (FMT). Having found poor agreement between the DFT and MC results for the bulk phase diagrams of the LJ fluid, they employed different fluid-fluid potential parameters for DFT as opposed to MC by separately fitting the DFT and simulated phase diagrams for the bulk fluid to experimental data on the vapour-liquid equilibria of systems of interest. They concluded that within reasonable limits, NLDFT with properly chosen parameters of intermolecular interactions is capable of quantitatively predicting the confined fluid structure at solid surfaces and in pores, adsorption isotherms, and conditions of phase equilibria e.g. spinodal transitions.

Kim et al.[222] tested 3 different DFT approaches against MC simulation of hard discs in circular cavities. DFT using the Rosenfield approximation gave the best agreement.

Patra and Ghost developed a self-consistent density functional approach for the structure of non-uniform fluid mixtures in which the 2nd-order direct correlation function and the bridge function of the corresponding uniform fluid mixture are used as input.[223] The new approach is used to predict the structure of binary HS fluid mixtures near a hard wall. Yu and Wu developed a new DFT for inhomogeneous mixtures of polymeric fluids by combining Rosenfeld's fundamental measure theory for excluded volume effects with Wertheim's first order thermodynamic perturbation theory for chain connectivity.[224] The theory was tested against MC simulations of adsorption isotherms with good agreement.

We mention in passing an interesting development by Kamat and Keffer.[225] They have developed a statistical mechanical lattice model of adsorption that requires only 4 parameters to describe the adsorbent. This model is useful for obtaining adsorption isotherms in arbitrary zeolites and other nanoporous materials at a fraction of the computational cost of the MC method.

6 Water and its Solutions

Water and solutions in which water is the solvent have always attracted a great deal of interest in the simulation community. However, in spite of this, accurate modelling of water has presented a major challenge. No single model of water so far developed is able to account for all of water's anomalous properties. The more complicated the model, the more computational effort is typically required in a simulation. This has lead to the introduction of simplified models of water which contain most of the essential physics but can only qualitatively reproduce some of water's many interesting properties. A selection of some of the papers published is discussed.

Stanley and co-workers have reviewed recent literature in which statistical mechanical ideas have been used to try understand the puzzles surrounding the static and dynamic properties of liquid water.[226] The conclusion of their review is that local tetrahedral order in liquid water is of prime importance in understanding the statics while the number of diffusive directions in the potential energy landscape is of main concern in the dynamics.

Garberoglio and Vallauri have studied the short time dynamics of hydrogen bonded liquids to understand the role played by strong directional bonds.[227] Using an instantaneous normal mode analysis approach to the liquid dynamics they were able to identify the particular motions relevant in the dynamics at various frequencies. They found that the high frequency regime is most influenced by the strong directional hydrogen bonds (H-bonds) in all the systems they considered (water, hydrogen fluoride and methanol). At low frequencies the relative motions of nearest neighbours was found to be orthogonal to the direction of the H-bonds.

Water is known to display anomalous behaviour in the pressure-dependence of self-diffusion and orientational relaxation of its molecules at low temperature. Such behaviour is usually explained in terms of the changes in water-water hydrogen bonds caused by the application of pressure. Anomalous self-diffusion also occurs in solutions in which water is the solvent and the solute is polar. Chowdhuri and Chandra undertook MD simulations of water containing a single molecule of methanol and acetonitrile in order to determine the pressure dependence of dynamical properties under ambient and supercooled conditions.[228] At room temperature, tracer diffusion coefficients of both methanol and acetonitrile molecules decreased monotonically with increasing pressure. However, at the supercooled temperature of 258 K, diffusion coefficients first increased, then decreased with increasing pressure. The increase in diffusivity is large for methanol, but small for acetonitrile. They found a high degree of correlation between pressure-induced changes in H-bonded properties and anomalous changes in dynamic properties.

Evans has developed a theory based approach to studying the rotational dynamics of water.[229] The water model employed in the theory is that of a spherical top with a hard, convex tetrahedral surface. Collisions between these 'water' molecules exchange linear and angular momentum as per usual 2-body kinematics but rotational energy transfer is diminished from the hard body value (*i.e.* an energy bottleneck) to reflect the areas of the water molecule surface that are blocked due to H-bonds.

Verdaguer and Padro used MD to probe the microscopic dynamics in water and ionic aqueous solutions. They calculated velocity cross correlations between ions in aqueous NaCl and observed that the difference between results for Na^+ and Cl^- ions are associated mainly with the different sizes of the ions.[230]

Matsumoto and co-workers report MD simulations of the freezing of water. Using very long trajectories, they observed water crystallization.[231] They found that ice nucleation occurs after a sufficient number of relatively long-lived hydrogen bonds have developed spontaneously in a particular location to form an initial nucleus. This nucleus then changes in size and shape until it reaches a stage where rapid expansion permits crystallization of the entire system. This work is very significant but it comes at the expense of using several months of supercomputer time, and clearly illustrates the scale of the task when trying to understand the anomalous properties of water.

Many models of water have been proposed over the years but none are successful at reproducing all the anomalous properties of this important molecule. The development of new models has continued with differing degrees of success. Honda has developed an interaction potential model for H-bonded liquids using non-empirical reference energies.[232] It is a pair potential function of the overlap integrals between molecules and of the Coulomb interactions between partial charges on atomic sites. Due to a lack of many body interactions, the potential is incapable of reproducing liquid structures without tuning of parameters. The model is thought not to be applicable to mixtures. A somewhat different approach has been taken by Nezbeda, who has attempted to develop a simplified model of water for reasons of computational efficiency. Nezbeda performed MD simulations of aqueous solutions of electrolytes using this simple model, which accounts explicitly for the molecular structure of water, but which does not incorporate any long range Coulombic interactions.[233] Perhaps surprisingly, he found that the residual entropy is qualitatively similar to the same quantity evaluated for a system employing long-range Coulombic interactions. Kolafa et al. extended this investigation to cover dipolar and quadrupolar fluids.[234] They found that spatial arrangement of the molecules is only marginally affected by long ranged forces. The effect of electrostatic interactions is significant at short separations but the overall structure of short range and full systems is similar, as is their dielectric constants.

Predota and co-workers examined different algorithms for obtaining structural and thermodynamic dipolar properties of a polarizable water model: The adiabatic nuclear and electronic sampling method (ANES) and the pair approximation for polarization interaction (PAPI).[235]

Gubskaya and Kusilik have developed a means of obtaining the average total dipole moment for the water molecule from experimental refractive index data augmented with MD simulation data.[236]

7 Mesoscale Simulations

Mesoscale refers to length scales ranging from 10^{-8} to 10^{-6} m which cover industrially important materials such as colloids. These length scales have traditionally been outside the realm of molecular simulations, with the few exceptions employing massively parallel supercomputers to follow the motion of a very large number of atoms. Mesoscale techniques such as Brownian Dynamics (BD), Smooth Particle Hydrodynamics (SPH), Stokesian Dynamics (SD), and more recently, dissipative particle dynamics (DPD) and Lattice Boltzmann (LB), were developed as a means of bridging this gap by coarse graining the molecular detail. The two main mesoscale techniques that are currently receiving a great deal of attention in the simulation community are DPD and LB. Each of these methods has its own attractions, as well as weakness. At present the LB technique suffers from numerical instability problems while for DPD, the mapping of molecular parameters to the mesoscale parameters is not well established. Nevertheless, these mesoscale methods are becoming standard tools which are now offered as part of commercial modelling packages. The CCP5 organisation,[237] which has traditionally been associated with molecular simulation, is now engaged in a project to develop a suite of mesoscale algorithms in the spirit of the highly popular DL_POLY code.[238] The forthcoming DL_MESO will be custom built to exploit grid technology, and should lead to an ever greater number of practitioners working at the mesoscale. The number of applications of LB and DPD is growing rapidly and includes modelling concrete flow,[239] blood flow[240] and amphiphilic fluids,[241] *etc.* The number of publications is too numerous to review here so I will focus mainly on a handful of examples for each method in which improvements have been made to the technique.

7.1 Dissipative Particle Dynamics. – DPD is a mesoscopic simulation technique introduced by Hoogerbrugge and Koelman.[242] It is intended to simulate hydrodynamic behaviour as well as the rheological properties of complex fluids, including multiphase systems. Like MD, the simulation consists of a set of discrete particles which are acted upon by forces. However, in DPD, there are 3 separate forces: (1) a dissipative force, (2) a random force, and (3) a conservative force (similar to MD). The particles don't necessarily represent individual atoms or molecules but rather the position and momentum of fluid elements. The DPD method is useful for modelling suspensions *e.g.* colloids, ceramic slurries and concrete.

Standard implementations of DPD are based on simulations in which the conservative force is a soft-repulsive interaction. The main drawback with DPD is that there is no simple way in which to derive the soft-particle model from a realistic molecular model. Use of such a soft-repulsive model is made at the expense of correct phase behaviour. Its use becomes untenable for soft matter at the liquid-vapour or solid-vapour interface. In answer to this shortcoming, Liew and Mikami have devised a 'soft-attractive-and-repulsive' pair potential model.[243] This model is based on a combination of a Morse-like function with a damping smoothing function, $w(r_{ij})$,

$$U(r_{ij}) = \varepsilon \left(\left\{ 1 - \exp\left[-\beta \left(r_{ij}/r_{\min} - 1\right)\right] \right\}^2 - 1 \right) w(r_{ij}) \tag{7.1}$$

6: Simulation of the Liquid State

$$w(r_{ij}) = \begin{cases} 1.0 & \text{for } r_{ij} < r_{min} \\ \left[1 - \left(\dfrac{r_{ij} - r_{min}}{r_{cut} - r_{min}}\right)^n\right]^n & \text{for } r_{min} \leq r_{ij} \leq r_{cut} \\ 0.0 & \text{for } r_{ij} > r_{cut} \end{cases} \quad (7.2)$$

where ε determines the depth of the potential minimum which is located at r_{min}, n is a small integer while β is a dimensionless parameter. When $\beta = 6$, the potential becomes very similar to the Lennard-Jones 12-6 potential.

Pagonabarraga and Frenkel introduced another new way of implementing DPD.[244] In their method, the conservative interaction depends on the local excess free energy rather than on the interparticle separation. The advantage of this approach is that it is possible to fix beforehand the desired thermodynamic properties of the system (at the mean field level). This method is often referred to as the multibody DPD, or MDPD for short. Trofimov and co-workers have improved the multibody DPD model, providing a correction for particle correlations in strongly non-ideal systems. They tested it on a number of single-component examples and also generalized MDPD to multicomponent systems.[245]

Español and Revenga have developed a new hybrid particle model which is a thermodynamically consistent version of smoothed particle hydrodynamics (SPH) and a version of DPD, and therefore captures the best features of both these methods.[246] This new model is a discrete version of Navier Stokes equations that includes thermal fluctuations and therefore allows SPH to be applied to mesoscopic scales where these fluctuations are important. In this model, particle volume is introduced as a thermodynamic variable. This modification of the original DPD formulation means that the conservative forces are now given in terms of pressure forces, allowing one to introduce arbitrary equations of state such as van der Waals into the model. According to the authors, using their model, it is possible to study thermal effects in the generation of bubbles and droplets in non-equilibrium liquid-vapour co-existence. Another significant advantage of this new model over the original DPD model is that in the new model, a direct connection between the model parameters and the transport coefficients of the fluid is given.

7.2 Lattice Boltzmann. – LB is a computational method based upon the Boltzmann equation. It considers a typical volume element of fluid to be composed of a collection of particles that are represented by a particle velocity distribution function for each fluid component at each grid point. The collision rules are designed such that the time average motion of the particles is consistent with the Navier-Stokes equations. It can handle complex boundary conditions and complex geometries.

Succi and co-workers have reviewed the development of the Lattice Boltzmann method with a particular focus on the role played by the *H* theorem in forcing compliance of the method with the second law of thermodynamics.[247]

Halliday *et al.* introduced position and time dependent "source" terms into the microscopic evolution equation of a lattice Boltzmann fluid and thereby usefully transformed the emergent dynamics of the lattice fluid.[248] They demonstrated their modifications by employing a source term which recovered the cylindrical polar coordinate form of the Navier-Stokes equations as the macroscopic equations of the lattice fluid.

Hirabayashi *et al.* have applied a new LB model in which they extended the LB scheme for magnetohydrodynamics to the fractal geometry in magnetic fluids in an alternating current magnetic field.[249] The scheme is said to be suitable for simulating various behaviours of magnetic fields under the influence of the internal angular momentum.

Ispolatov and Grant have developed a viscoelastic flow version of LB. They tested the model on a number of physical systems and found that it is described by the continuous Navier-Stokes equations with Maxwell viscoelastic term.[250]

Chin and Coveney have used a LB model to examine the spinodal decomposition of a two-dimensional binary fluid.[251] The model showed agreement with the Cahn-Hilliard theory during interface formation at very early times. They observed a scaling law with exponent 2/3 for late time domain growth, which is also seen in other models, as well as breakdown in scaling for certain sets of simulation parameters.

Sofonea and co-workers have developed a LB model in order to describe competition between surface tension and dipolar interactions in magnetic fluids.[252]

8 Simulation Methodology

An extensive array of tools is now available for the simulation of liquids. The main techniques of MC and MD have been extended to cover a multitude of ensembles, while NEMD algorithms are now available to simulate liquids under a variety of non-equilibrium conditions. A great deal of research effort has been directed towards the development of these tools with the aims of improving computational efficiency (neighbour lists, Ewald sums, *etc*), exploiting new developments in computer science (parallel computing, grid technology, *etc*), extending the realm of applicability of a given algorithm (*e.g.* configurational bias MC) and fusing two dissimilar methods together (*e.g.* the development of dual control volume methods). Efforts have also been made to exploit the advantages of particle based simulation methodologies in order to tackle problems involving times and lengths at the mesoscale (*e.g.* DPD and LB). There have a number of interesting developments in the last three years, some of these having been documented elsewhere in this review.

8.1 Reverse Monte Carlo. – Reverse Monte Carlo (RMC) is a general method of structural modelling based on Metropolis MC in which experimental data is used in place of an intermolecular potential. In a nutshell, RMC consists of a sequence of random trial moves (translations and rotations) of molecules, beginning from an initial configuration. A radial distribution function is calculated after each trial move and then Fourier transformed to give a structure factor. The structure factor is compared with one obtained from experimental data, the source of which could be X-ray diffraction, neutron scattering, NMR, or electron diffraction experiments. The trial configuration is accepted if the difference between its structure factor and the experimental one is less than the previous comparison involving the old configuration. Once the measure of the structure factor deviations reaches a minimum, a 3-dimensional representation of the real structure is obtained.

Most applications of RMC have been based on constructing model solids and hence lie outside of this review. However, RMC has been used to model liquids. An excellent review of the RMC method has been written by McGreevy,[253] and includes a discussion of the variations on standard RMC and its many applications. A more recent article by McGreevy and Zetterström raises some important questions concerning when to use, and when not to use RMC.[254]

Pusztai et al. have used the RMC method to gain structural information for chemically associating fluids.[255] They used molecular dynamics simulations to generate their "experimental input data". More recently, Pusztai and co-workers used RMC to calculate the static dielectric constant in ambient water, concluding that it fluctuates too widely to be useful as an extra constraint in RMC simulations.[256] A dynamic analogue of the RMC method was introduced by Tóth and Baranyai.[257] In their scheme, particles are moved around in the simulation box according to a molecular dynamics algorithm. The potential, from which the forces are derived, is purely fictitious, and obtained from differences in the square of the model and experimental structure factors. Their method is applicable to liquids as well as solids. Tóth used this method to obtain pair potential parameters for 1 and 2-component Lennard-Jones fluids.[258]

8.2 New Monte Carlo Algorithms. – New MC algorithms appear on an almost daily basis and several have already been described in earlier sections. Some other noteworthy examples are now described.

Bourasseau et al. introduced a new type of transfer move for use in Gibbs ensemble MC simulations involving cyclic alkanes.[259] Their scheme is based on a similar one used in GCMC proposed by Errington and Panagiotopoulos[260] (called reservoir bias transfer). The basic idea is to move a cyclic molecule as a whole unit by first undertaking a search for 'holes' by several test insertions of a single force centre of suitable size, with appropriate correction for the bias introduced.

Mbamala and Pastore looked at the issue of sampling efficiency of configuration space in MC simulations.[261] They compared single particle moves with multi-particle moves and also explored the effect of changing the acceptance ratio from its widely accepted value of 0.5. After performing a number of MC simulations using hard spheres they found no advantage in using the multi-particle moves over single particle moves. However, they did find that optimum sampling may be achieved from an acceptance ratio of 0.1 rather than 0.5.

Smith and Lísal introduced a new technique for conducting MC simulations at fixed total internal energy, U, or enthalpy, H.[262] Such a technique is potentially useful for adiabatic flash calculations for non-reacting pure fluids and mixtures at fixed enthalpy and pressure (P) and adiabatic flame temperature calculations at fixed (U,V) or (H,P) for example. The essence of the technique is the introduction of a new MC 'move' which fixes the value of the derivative $\partial \beta A / \partial x$ where βA is the dimensionless Helmoltz energy (and β has its usual meaning in statistical thermodynamics), and x is either the number of particles of species i, volume or β (letting x be N_i for example, recovers the GCMC method). It is worth mentioning in passing that an MD algorithm has also been developed by Kioupis and Maginn which enables the direct simulation of isenthalpic pressure changes and isobaric enthalpy changes.[263]

The aggregation-volume-bias Monte Carlo (AVBMC) algorithm was introduced recently by Chen and Siepmann as a method for carrying out MC simulations of associating fluids which overcomes difficulties in sampling due to bond formation. Essentially AVBMC adds a new MC move: an intrabox swap.[264] Attempts are made to swap two randomly selected molecules (one acting as swap target) upon which atoms are moved into or out of the bonding region of the target. In this more recent publication by the same authors, an extension of the technique is presented, which leads to a much improved sampling for super-strongly associating fluids.[265]

Boulougouris et al. developed a method of calculating the chemical potential of chain molecules from molecular simulation based on the staged deletion of a test particle.[266] This method results in large savings in CPU time compared to the conventional Widom test particle insertion route to the chemical potential.

Yan et al. developed an MC method based on a density-of-states sampling for study of arbitrary statistical mechanical ensembles in a continuum.[267] The density-of-states is estimated from random walks in the 2D space of particle number and energy.

8.3 Intermolecular Potentials. – The development of intermolecular potentials is a very active area of research and justice cannot possibly be done to this field in a review of this nature. However, several papers deserve mention due to their importance in the area of simulation of liquids.

The first of this select batch of publications concerns the development of potentials for the interaction between an atom and a solid surface and is therefore applicable to simulations such as those involving fluids adsorbed in porous materials. In this paper, Patil et al.[268] extend the atom-atom potential model of Tang and Toennies[269] to describe the atom-surface potential. They obtained an analytic expression for such a potential from a pairwise sum model including appropriate damping terms for the atom-surface attractive dispersion series.

Masaki et al. developed an effective pair potential for simulating molten tin.[270] The method they used is the one introduced by Reatto and co-workers in

1986.[271] This method is an iterative predictor-corrector method in which the predictor is the modified hypernetted-chain equation and the corrector is the simulation. While this technique is not too recent, it is worth giving a brief outline of it here.

One starts by noting that in integral equation theory, the effective pair potential, $\phi(r)$ is given by

$$\beta\phi(r) = g_{\exp}(r) - 1 - c_{\exp}(r) - \ln g_{\exp}(r) + B(r) \tag{8.1}$$

where

$$c_{\exp}(r) = \frac{1}{(2\pi)^3 n}\int\left(1 - \frac{1}{S_{\exp}(Q)}\right)\exp(-i\mathbf{Q}\cdot\mathbf{r})d\mathbf{Q} \tag{8.2}$$

and $\beta = 1/kT$, n is the number density, $B(r)$ is the bridge function, $g_{\exp}(r)$ is the experimental radial distribution function and $c_{\exp}(r)$ is the experimental direct correlation function. Both $g_{\exp}(r)$ and $c_{\exp}(r)$ can be calculated from the experimental static structure factor. However the bridge function cannot. The solution is to use the hard sphere bridge function to arrive at a zeroth order approximation to the pair potential. The bridge function is then obtained from a molecular dynamics simulation in which the zeroth order pair potential is used. The simulated radial distribution function can then be used to correct the bridge function and hence obtain a better estimate for the pair potential. The procedure is continued until successive pair potentials differ by an acceptably small amount. Masaki et al. used their molten tin potential to calculate self-diffusion coefficients over a wide range of temperatures.[270] The simulation results compared well with experimental diffusion coefficients obtained under zero gravity.

In earlier work, Hess and Kroger demonstrated that the density dependence of the energy and pressure of the WCA fluid and fcc solid could be reproduced by a short-range repulsive polynomial potential.[272] In more recent work published during the period covered by this review, the same authors extended this approach by introducing a simple polynomial function that has both a repulsive and relatively short ranged part.[273] Such polynomial potentials are computationally advantageous especially in situations where one must calculate 2nd derivatives of the potential e.g. in the computation of configurational temperature. The potential function used by these workers is given by

$$\Phi(r) = \frac{512}{27}\Phi_0\left(1 - \frac{r}{r_0}\right)\left(3 - 2\frac{r}{r_0}\right)^3, \quad r \leq 1.5\,r_0$$
$$= 0 \quad \text{for} > 1.5\,r_0 \tag{8.3}$$

where Φ_0 is the well depth analogous to ε in the Lennard-Jones potential and r_0 has the same physical meaning as the σ parameter. The cut-off is smooth such that the first and second derivatives vanish at the truncation point. Hess and Kroger evaluated the thermophysical properties of the model using MD and NEMD, comparing the results with analytical results. Good agreement was obtained.

Vrabec and co-workers have used a LJ plus point quadrupole model (2CLJQ potential) to represent a large range of molecules including noble gases, CO_2, I_2, and some organic molecules.[274] They parameterised this potential for each substance in turn by comparing vapour-liquid equilibria data from NpT MC calculations with experimental data of pure fluids. The model fluids yield vapour-liquid equilibria with an accuracy of 3% for the vapour pressure and 0.5% for the saturated liquid density, and 2% for the enthalpy of vapourisation. The authors note that these models will be useful for predicting vapour-liquid equilibria of fluid mixtures.

Chelli et al. extended the chemical potential equalization (CPE) method to include atomic dipolar charge distributions.[275] Their approach allows for a CPE parameterisation which is then capable of realistically reproducing the polarization response of linear and planar molecules.

Marcelli et al. found that the relationship between 2- and 3-body interactions discovered in earlier MC simulation work on equilibrium fluids is reproduced in NEMD simulations of argon.[276] They suggest this might offer a computationally efficient means of obtaining the effects of 3-body interactions in NEMD simulations form a conventional 2-body simulation.

8.4 Multiscale Methods. – For reasons previously mentioned, multiscale modelling/simulation is currently a hot topic and a large degree of research effort is being devoted to developing multiscale techniques. The main developments have been described in Section 7. Other key developments are described below.

Medvedev has used the Voronoi-Delaunay technique which uses the geometrical constructions: Voronoi polyhedra and Delaunay simplices, to yield information on the structure of complex inhomogeneous systems.[277] This technique allows one to study local order of atoms and the extended structure correlations of atoms as well as voids between atoms, and, as such, is well suited to study of structures consisting of a multitude of length scales. Medvedev applied this method to a study of the intermediate range order to show that the behaviour of the so-called prepeak in the structure factor is defined by a motif of spatial distribution of voids in the model.

Louis et al.[278] developed a method for coarse-graining polymers in a good solvent as single particle "soft colloids" interacting with density independent or density dependent interactions. They also showed how to extend their method to cover polymers in poor solvent and mixtures of different length polymers. Such coarse graining has allowed simulations which would have been virtually impossible with an explicit polymer model.

Stuart and co-workers have developed a method of performing variable-timestep molecular dynamics integration.[279] The method involves an iterative algorithm which selects the largest timestep consistent with the desired simulation accuracy (defined in terms of energy conservation). The method is useful for simulations involving disparate timescales.

8.5 Miscellaneous Developments. – Here we mention several developments that apply mainly to dynamic simulations.

Galea and Attard employed constraints to yield non-Hamiltonian equations of motion which could be used to drive non-equilibrium flows.[280] The constraints can be used to fix the kinetic energy of a system without the need for periodic velocity rescaling, which is necessary when Gauss' principle of least constraint is used to obtain isokinetic equations of motion. This new method looks promising and has been successfully tested on a system undergoing bulk Couette flow. However, a test of constraint driven Couette flow in the absence of the usual Lees-Edwards boundary conditions (fluid is confined between smooth parallel walls and the constraints are used to fix the momenta of fluid particles in such a way that a linear velocity profile is established) gave a measured velocity profile that deviated from linearity. The authors explained these deviations as being due to the smooth (non wetting) walls employed. Clearly, more careful studies are required.

Barenbrug et al. developed a means of reducing the error in a Brownian Dynamics (BD) simulation involving spherical particles which can encounter hard-body interactions such as in adsorption or chemical reactions.[281] In the usual implementation of BD in these situations, overlaps are checked for after each time step. However, many overlap events are missed due to the simplification of diffusional motion, leading to overall simulation errors proportional to the square root of time. By taking proper account of the hard-body interactions during each time step, Barenbrug et al. have reduced the error to one which is proportional to time rather than the square root of time.

Ikeshoji et al.[282] have developed expressions for the microscopic pressure tensor suitable for use in MC and MD simulations of heterogeneous systems involving planar and spherical interfaces.

Lopez-Lemus and Alejandre carried out MD of LJ fluids using lattice sums to calculate the dispersion interactions.[283] Their method is useful for simulations of inhomogeneous fluids where the interface is physically present since in these cases, the long-range correction is not easily included.

Siperstein et al.[284] discussed the implications of ignoring long range corrections due to potential truncation for simulations of confined liquids. They point out that the error incurred in assuming isotropy in the radial distribution function beyond the cut off (as done in bulk phase simulations) is small compared to the magnitude of the long-range correction.

References

1. G.M. Wang, E.M. Sevick, E. Mittag, D.J. Searles and D.J. Evans, *Phys. Rev. Lett.*, 2002, **89**, 050601.
2. J.M. Haile, *Molecular Dynamics Simulation. Elementary Methods*, J. Wiley & Sons, New York, 1992.
3. W.G. Hoover, *Computational Statistical Mechanics*, Elsevier, 1991.
4. D.J. Evans and G.P. Morriss, *Statistical Mechanics of Nonequilibrium Liquids*, Academic Press, London, 1990.

5. M.P. Allen and D.J. Tildesley, *Computer Simulation of Liquids*, Oxford University Press, 1987.
6. D.C. Rapaport, *The Art of Molecular Dynamics Simulation*, Cambridge University Press, 1995.
7. R.J. Sadus, *Molecular Simulation of Fluids: Theory, Algorithms and Object-Orientation*, Elsevier, Amsterdam, 1999.
8. D. Frenkel and B. Smit, *Understanding Molecular Simulation*, Academic Press, San Diego, 2002.
9. J.R. Dorfman, *An Introduction to Chaos in Nonequilibrium Statistical Mechanics*, Cambridge University Press, 1999.
10. D.M. Heyes, *The Liquid State: Applications of Molecular Simulations*, Wiley, 1998.
11. K.D. Hammonds and D.M. Heyes, *J. Chem. Soc. Farad. Trans. 2*, 1988, **84**, 705.
12. B.J. Alder, D.M. Gass and T.E. Wainwright, *J. Chem. Phys.*, 1970, **53**, 3812.
13. K.R. Harris, *Mol. Phys.*, 1994, **77**, 1153.
14. A.J. Batschinsky, *Z. Phys. Chem.*, 1913, **84**, 644.
15. H. Eyring, *J. Chem. Phys.*, 1936, **4**, 283.
16. T.G. Fox and P.J. Flory, *J. Appl. Phys.*, 1950, **21**, 581.
17. A.K. Doolittle, *J. Appl. Phys.*, 1951, **22**, 1471.
18. M.H. Cohen and D. Turnbull, *J. Chem. Phys.*, 1959, **31**, 1164.
19. K. Rah and B.C. Eu, *J. Chem. Phys.*, 2001, **115**, 2634.
20. B.C. Eu and K. Rah, *Phys. Rev. E*, 2001, **63**, 031203.
21. K. Rah and B.C. Eu, *Phys. Rev. Lett.*, 1999, **83**, 4566.
22. K. Rah and E.C. Eu, *Phys. Rev. E*, 1999, **60**, 4105.
23. K. Rah and B.C. Eu, *J. Chem. Phys.*, 2000, **112**, 7118.
24. K. Rah and B.C. Eu, *J. Chem. Phys.*, 2001, **115**, 9370.
25. K. Rah and B.C. Eu, *J. Chem. Phys.*, 2001, **114**, 10436.
26. K. Rah and B.C. Eu, *J. Chem. Phys.*, 2002, **117**, 4386.
27. K. Rah and B.C. Eu, *J. Chem. Phys.*, 2002, **116**, 7967.
28. V.A. Harmandaris, D. Angelopoulou, V.G. Mavrantzas and D.N. Theodorou, *J. Chem. Phys.*, 2002, **116**, 7656.
29. E. Zervopoulou, V.G. Mavrantzas and D.N. Theodorou, *J. Chem. Phys.*, 2001, **115**, 2860.
30. A.N. Drozdov and S.C. Tucker, *J. Phys. Chem. B*, 2001, **105**, 6675.
31. Q.-Y. Tong, G.-H. Gao, M.-H. Han and Y.-X. Hu, *Int. J. Thermophys.*, 2002, **23**, 635.
32. Y.-H. Shim, D.-C. Ihm and F.-K. Lee, *Phys. Rev. E*, 2001, **64**, 041106.
33. D. Ihm, Y.-H. Shin, J.-W. Lee and E.K. Lee, *Phys. Rev. E*, 2003, **67**, 027205.
34. D.J. Evans, *Phys. Rev. A*, 1985, **32**, 2923.
35. J.W. Gibbs, *Elementary Principles in Statistical Mechanics*, Yale University Press, Boston, 1902.
36. D.J. Evans and L. Rondoni, *J. Stat. Phys.*, 2002, **109**, 895.
37. D.J. Evans and D.J. Searles, *Adv. in Phys.*, 2002, **51**, 1529.
38. G.E. Norman and V.V. Stegailov, *Com. Phys. Comm*, 2002, **147**, 678.
39. S. Hess and D.J. Evans, *Phys. Rev. E*, 2001, **64**, 011207.
40. B. Hess, *J. Chem. Phys.*, 2002, **116**, 209.
41. J. Ratanapisit, D.J. Isbister and J.F. Ely, *Fluid Phase Equilibria*, 2001, **183–184**, 351.
42. S.H. Lee, D.K. Park and D.B. Kang, *Bull. Korean Chem. Soc.*, 2003, **24**, 178.
43. B.D. Todd, *Comp. Phys. Comm.*, 2001, **142**, 14.

44. L. Lue, O.G. Jepps, J. Delhommelle and D.J. Evans, *Mol. Phys.*, 2002, **100**, 2387.
45. J. Delhommelle and D.J. Evans, *J. Chem. Phys.*, 2002, **117**, 6016.
46. G. Cini-Castagnoli and A. Longhetto, *Physica*, 1970, **49**, 153.
47. J.S. Duffield and M.J. Harris, *Berichte Busenges-Gesellschaft*, 1976, **80**, 157.
48. P. Etesse, A.M. Ward, W.V. House and R. Kobayashi, *Physica B*, 1993, **183**, 45.
49. S. De, Y. Shapir and E.H. Chimowitz, *Chem. Eng. Sci*, 2001, **56**, 5003.
50. A. Drozdov and S.C. Tucker, *J. Chem. Phys.*, 2001, **114**, 4912.
51. Y.V. Tsekhanskaya, *Russ. J. Phys. Chem.*, 1971, **45**, 744.
52. H. Nishiumi, M. Fujita and K. Agou, *Fluid Phase Equilibria*, 1996, **117**, 356.
53. N.J. Trappeniers and P.H. Oosting, *Phys. Lett.*, 1966, **23**, 445.
54. B.J. Cherayil, *J. Chem. Phys.*, 2002, **116**, 8455.
55. K.R. Harris, *J. Chem. Phys.*, 2002, **116**, 6379.
56. K.R. Harris, *Physica A:*, 1978, **93**, 593.
57. P.W.E. Peereboom, H. Luigjes and K.O. Prins, *Physica A*, 1989, **156**, 260.
58. A.N. Drozdov and S.C. Tucker, *J. Chem. Phys.*, 2002, **116**, 6381.
59. A. Darinskii, A. Zarembo, N.K. Balabaev, I.M. Neelov and F. Sundholm, *PCCP*, 2003, **5**, 2410.
60. S.M. Ali, A. Samanta and S.K. Ghosh, *Chem. Phys. Lett*, 2002, **357**, 217.
61. H.R. Schober, *Sol. State Comm.*, 2001, **119**, 73.
62. K. Dhole, A. Samanta and S.K. Ghosh, *J. Chem. Phys.*, 2002, **116**, 7081.
63. H. Mori, *Prog. Theor. Phys.*, 1965, **34**, 399.
64. A. Perronace, C. Leppla, F. Leroy, B. Rousseau and S. Wiegand, *J. Chem. Phys.*, 2002, **116**, 3718.
65. C. Moon, G.S. Pawley and J. Crain, *Mol. Phys.*, 2001, **99**, 2037.
66. J.E. Adams and A. Siavosh-Haghighi, *J. Phys. Chem. B*, 2002, **106**, 7973.
67. G. Marcelli, B.D. Todd and R.J. Sadus, *Phys. Rev. E*, 2001, **63**, 021204.
68. D. Bertolini and A. Tani, *J. Chem. Phys.*, 2001, **115**, 6285.
69. D.L. Cheung, S.J. Clark and M.R. Wilson, *Chem. Phys. Lett*, 2002, **356**, 140.
70. L.T. Novak, *Ind. Eng. Chem. Res.*, 2003, **42**, 1824.
71. Y. Song, P.M. Mathias, D. Tremblay and C.C. Chen, *Ind. Eng. Chem. Res.*, 2003, **42**, 2415.
72. H. Sigurgeirsson and D.M. Heyes, *Mol. Phys.*, 2003, **101**, 469.
73. R.J. Speedy, *Mol. Phys.*, 1987, **62**, 509.
74. G. Faussurier and M.S. Murillo, *Phys. Rev. E*, 2003, **67**, 046404.
75. J.J. Hoyt and M. Asta, *Phys. Rev. B*, 2002, **65**, 214106.
76. C.J. Margulis, H.A. Stern and B.J. Berne, *J. Phys. Chem. B*, 2002, **106**, 12017.
77. G. Marcelli, B.D. Todd and R.J. Sadus, *Fluid Phase Equilibria*, 2001, **183-184**, 371.
78. K. Kawasaki and J.D. Gunton, *Phys. Rev. A*, 1973, **8**, 2048.
79. G.P. Morriss, T. Byrnes and D.R.J. Monaghan, *Mol. Phys.*, 2002, **100**, 2377.
80. C. McCabe, S. Cui and P.T. Cummings, *Fluid Phase Equilibria*, 2001, **183-184**, 363.
81. J.L. McWhirter and G.N. Patey, *J. Chem. Phys.*, 2002, **117**, 8551.
82. J.F. Lutsko and J.W. Dufty, *Phys. Rev. E*, 2002, **66**, 041206.
83. S. Hess, *Int. J. Thermophys.*, 2002, **23**, 905.
84. J.W. Dufty, *Mol. Phys.*, 2002, **100**, 2331.
85. A.Z. Panagiotopoulos, *Mol. Phys.*, 1987, **61**, 813.
86. L.D. Gelb and E.A. Müller, *Fluid Phase Equilibria*, 2002, **203**, 1.
87. H.N.W. Lekkerkerker and S.M. Oversteegen, *J. Phys. Condens. Matt.*, 2002, **14**, 9317.

88. F. Goujon, P. Malfreyt, A. Boutin and A.H. Fuchs, *J. Chem. Phys.*, 2002, **116**, 8106.
89. J.M. Stubbs, B.Chen, J.J. Potoff and J.I. Siepmann, *Fluid Phase Equilibria*, 2001, **183–184**, 301.
90. N.D. Zhuravlev, M.G. Martin and J.I. Siepmann, *Fluid Phase Equilibria*, 2002, **202**, 307.
91. M.Mehta and D.A. Kofke, *Mol. Phys.*, 1995, **86**, 139.
92. S.K. Nath, F.A. Escobedo and J.J. de Pablo, *J. Chem. Phys.*, 1998, **108**, 9905.
93. F.A. Escobedo, *J. Chem. Phys.*, 1999, **110**, 11999.
94. R. Shetty and F.A. Escobedo, *J. Chem. Phys.*, 2002, **116**, 7957.
95. M.H. Lamm and C.K. Hall, *Fluid Phase Equilibria*, 2002, **194–197**, 197.
96. D. Costa, G. Pellicane, M.C. Abramo and C. Caccamo, *J. Chem. Phys.*, 2003, **118**, 304.
97. G. Kahl, E. S.-Paschinger and G. Stell, *J. Phys. Condens. Matt.*, 2002, **14**, 9153.
98. S. Albo and E.A. Müller, *J. Phys. Chem. B*, 2003, **107**, 1672.
99. M. Kettler, I. Nezbeda, A.A. Chialvo and P.T. Cummings, *J. Phys. Chem. B*, 2002, **106**, 7537.
100. C. Panayiotou, *Langmuir*, 2002, **18**, 8841.
101. F. Bresme, *J. Phys. Chem. B*, 2002, **106**, 7852.
102. J.D. Weeks, K. Katsov and K. Vollmayr, *Phys. Rev. Lett.*, 1998, **81**, 4400.
103. K. Katsov and J.D. Weeks, *Phys. Rev. Lett.*, 2001, **86**,
104. K. Katsov and J.D. Weeks, *J. Phys. Chem. B*, 2001, **105**, 6738.
105. K. Vollmayr-Lee, K. Katsov and J.D. Weeks, *J. Chem. Phys.*, 2001, **114**, 416.
106. J.D. Weeks, *Annu. Rev. Phys. Chem.*, 2002, **53**, 533.
107. K. Katsov and J.D. Weeks, *J. Phys. Chem. B*, 2002, **106**, 8429.
108. S.R.P. da Rocha, K.P. Johnston, R.E. Westacott and P.J. Rossky, *J. Phys. Chem. B*, 2001, **105**, 12092.
109. V.K. Shen and P.G. Debenedetti, *J. Chem. Phys.*, 2003, **118**, 768.
110. C.M. Colina, M. Lísal, F.R. Siperstein and K.E. Gubbins, *Fluid Phase Equilibria*, 2002, **202**, 253.
111. P. Virnau, M. Müller, L.G. MacDowell and K. Binder, *Comp. Phys. Comm.*, 2002, **147**, 378.
112. M. Robles and M. Lopéz de Haro, *Euro. Phys. Lett.*, 2003, **62**, 56.
113. M. Robles and M. Lopéz de Haro, *PCCP*, 2001, **3**, 5528.
114. N.B. Wilding and J.E. Magee, *Phys. Rev. E*, 2002, **66**, 031509.
115. S.C. Glotzer, Y. Gebremichael, N. Lacevic, T.B. Schroder and F.W. Starr, *Comp. Phys. Comm.*, 2002, **146**, 24.
116. K. Binder, J. Baschnagel and W. Paul, *Prog. Polym. Sci*, 2003, **28**, 115.
117. F.H. Stillinger, P.G. Debenedetti and T.M. Truskett, *J. Phys. Chem. B*, 2001, **105**, 11809.
118. M. Tokuyama, *Physica A*, 2002, **315**, 321.
119. S. Kamath, R.H. Colby and S.K. Kumar, *J. Chem. Phys.*, 2002, **116**, 865.
120. M.L. Mansfield, *Phys. Rev. E*, 2002, **66**, 016101.
121. H. Pang, Z.H. Jin and K. Lu, *Phys. Rev. B*, 2003, **67**, 094113.
122. H.R. Schober, *Phys. Rev. Lett.*, 2002, **88**, 145901.
123. M. Tsige and P.L. Taylor, *Phys. Rev. E*, 2002, **65**, 021805.
124. J. Colmenero, A. Arbe, F. Alvarez, M. Monkenbusch, D. Richter, B. Farago and B. Frick, *J. Phys. Condens. Matt.*, 2003, **15**, S1127.
125. N. Lacevic, F.W. Starr, T.B. Schroder, V.N. Novikov and S.C. Glotzer, *Phys. Rev. E*, 2002, **66**, 030101.

126. C. Dasgupta, A.V. Indrani, S. Ramaswamy and M.K. Phani, *Euro. Phys. Lett.*, 1991, **15**, 307.
127. C. Dasgupta, A.V. Indrani, S. Ramaswamy and M.K. Phani, *Euro. Phys. Lett.*, 1991, **15**, 467.
128. K. Vollmayr-Lee, W. Kob, K. Binder and A. Zippelius, *J. Chem. Phys.*, 2002, **116**, 5158.
129. S. Mossa, G. Monaco and G. Ruocco, *Euro. Phys. Lett.*, 2002, **60**, 92.
130. A. Rinaldi, F. Sciortino and P. Taraglia, *Phys. Rev. E*, 2001, **63**, 061210.
131. T. Keyes, J. Chowdhary and J. Kim, *Phys. Rev. E*, 2002, **66**, 051110.
132. F.H. Stillinger and T.A. Weber, *Science*, 1984, **225**, 983.
133. F.H. Stillinger, *Science*, 1995, **267**, 1935.
134. D. Lacks, *Phys. Rev. Lett.*, 2001, **87**, 225502.
135. J.H. Simmons, R.K. Mohr and C.J. Montrose, *J. Appl. Phys.*, 1982, **53**, 4075.
136. R.L. Sammler, *J. Rheol*, 1996, **40**, 285.
137. M. Dzugutov, S.I. Simdyankin and F.H.M. Zetterling, *Phys. Rev. Lett.*, 2002, **89**, 195701.
138. B. Doliwa and A. Heuer, *Phys. Rev. E*, 2003, **67**, 030501.
139. R. Chelli, G. Cardini, P. Procacci, R. Righini and S. Califano, *J. Chem. Phys.*, 2002, **116**, 6205.
140. L. Barbieri, V. Cannillo, C. Leonelli, M. Montorsi, P. Mustarelli and C. Siligardi, *J. Phys. Chem. B*, 2003, **107**, 6519.
141. J. Horbach and W. Kob, *Phys. Rev. E*, 2003, **64**, 041503.
142. R. Murarka and B. Bagchi, *Phys. Rev. E*, 2003, **67**, 041501.
143. U. Harbola and S.P. Das, *J. Chem. Phys.*, 2002, **117**, 9844.
144. J. Kim, W.-X. Li and T. Keyes, *Phys. Rev. E*, 2003, **67**, 021506.
145. T. Munakata, *Phys. Rev. E*, 2003, **67**, 022101.
146. F.W. Starr, S. Sastry, J.F. Douglas and S.C. Glotzer, *Phys. Rev. Lett.*, 2002, **89**, 125501.
147. C. De Michele and F. Sciortino, *Phys. Rev. E*, 2002, **65**, 051202.
148. K. Hukushima and K. Nemoto, *J. Phys. Soc. Jpn*, 1996, **65**, 1604.
149. K. Hukushima, H. Takayama and H. Yoshino, *J. Phys. Soc. Jpn*, 1998, **67**, 12.
150. K. Brinkmann and H. Teichler, *Phys. Rev. B*, 2002, **66**, 184205.
151. A. Huerta and G.G. Naumis, *Phys. Rev. B*, 2002, **66**, 184204.
152. J. Habasaki and Y. Hiwatari, *J. Therm. Anal. Cal*, 2002, **69**, 1005.
153. R. Dagani, *Chem. Eng. News*, 2002, **80**, 25.
154. J.-C. Wang and K.A. Fichthorn, *Colloids and Surfaces A*, 2002, **206**, 267.
155. F. Porcheron, B. Rousseau and A.H. Fuchs, *Mol. Phys.*, 2002, **100**, 2109.
156. F. Poncheron, M. Schoen and A.H. Fuchs, *J. Chem. Phys.*, 2002, **116**, 5816.
157. S.T. Cui, P.T. Cummings and H.D. Cochran, *Fluid Phase Equilibria*, 2001, **183–184**, 381.
158. S.T. Cui, P.T. Cummings and H.D. Cochran, *J. Chem. Phys.*, 2001, **114**, 7189.
159. H.W. Hu, G.A. Carson and S. Granick, *Phys. Rev. Lett.*, 1991, **66**, 2758.
160. S. Granick, *Science*, 1991, **253**, 1374.
161. J. Klein and E. Kumacheva, *Science*, 1995, **269**, 816.
162. W. Jin and W. Wang, *J. Chem. Phys.*, 2001, **114**, 10163.
163. L.D. Gelb, K.E. Gubbins, R. Radhakrishnan and M. Sliwinska-Bartkowiak, *Rep. Prog. Phys*, 1999, **62**, 1573.
164. W. Shi, X. Zhao and J.K. Johnson, *Mol. Phys.*, 2002, **100**, 2139.
165. M. Schoen, *Colloids and Surfaces A*, 2002, **206**, 253.
166. F. Gentner, G. Ogonowski and J. De Coninck, *Langmuir*, 2003, **19**, 3996.

167. A. Vishnyakov, E.M. Piotrovskaya, E.N. Brodskaya, E.V. Votyakov and Y.K. Tovbin, *Langmuir*, 2001, **17**, 4451.
168. J.J. Potoff and J.I. Siepmann, *Langmuir*, 2002, **18**, 6088.
169. M. Sliwinska-Bartkowiak, G. Dudziak, R. Gras, R. Sikorski, R. Radhakrishnan and K.E. Gubbins, *Colloids and Surfaces A*, 2001, **187–188**, 523.
170. R. Radhakrishnan, K.E. Gubbins and M. Sliwinska-Bartkowiak, *J. Chem. Phys.*, 2002, **116**, 1147.
171. R. Radhakrishnan, K.E. Gubbins and M. Sliwinska-Bartkowiak, *Phys. Rev. Lett.*, 2002, **89**, 076101.
172. K. Koga, *J. Chem. Phys.*, 2002, **116**, 10882.
173. J. Quintana, E.C. Poire, H. Dominguez and J. Alejandre, *Mol. Phys.*, 2002, **100**, 2597.
174. J.S. Erickson, G.L. Aranovich and M.D. Donohue, *Mol. Phys.*, 2002, **100**, 2121.
175. S.H.L. Klapp and M. Schoen, *J. Chem. Phys.*, 2002, **117**, 8050.
176. J. Shen and P.A. Monson, *Mol. Phys.*, 2002, **100**, 2031.
177. J.K. Brennan and W. Dong, *J. Chem. Phys.*, 2002, **116**, 8948.
178. M. Lee, K.Y. Chan and Y.W. Tang, *Mol. Phys.*, 2002, **100**, 2201.
179. P. Gallo, R. Pellarin and M. Rovere, *Physica A*, 2002, **314**, 530.
180. D. Kranbuehl, R. Knowles, A. Hossain and A. Gilchriest, *J. Non-Cryst. Solids*, 2002, **307–310**, 495.
181. J.D. McCoy and J.G. Curro, *J. Chem. Phys.*, 2002, **116**, 9154.
182. E.A. Mason, A.P. Malinauskas and R.B. Evans(III), *J. Chem. Phys.*, 1967, **46**, 3199.
183. E.A. Mason and L.A. Viehland, *J. Chem. Phys.*, 1978, **68**, 3562.
184. E.A. Mason and A.P. Malinauskas, *Gas Transport in Porous Media: The Dusty Gas Model*, Elsevier, Amsterdam, 1983.
185. K.P. Travis and K.E. Gubbins, *J. Chem. Phys.*, 2000, **112**, 1984.
186. K.P. Travis and K.E. Gubbins, *Mol. Sim.*, 2000, **25**, 209.
187. K.P. Travis, *unpublished work*,
188. J. Kärger and D.M. Ruthven, *Diffusion in Zeolites and other microporous solids*, Wiley, New York, 1992.
189. K.P. Travis, *Mol. Phys.*, 2002, **100**, 2317.
190. M.G. Ahunbay, J.R. Elliot Jr. and O. Talu, *J. Phys. Chem. B*, 2002, **106**, 5163.
191. J.M.D. MacElroy, L.A. Pozhar and S.-H. Suh, *Colloids and Surfaces A*, 2001, **187–188**, 493.
192. J. Kärger, *J. Phys. Chem.*, 1991, **95**, 5558.
193. J. Kärger and H. Pfeifer, *Zeolites*, 1992, **12**, 872.
194. Z. Su, J.H. Cushman and J.E. Curry, *J. Chem. Phys.*, 2003, **118**, 1417.
195. P. Adhangale and D. Keffer, *Mol. Phys.*, 2002, **100**, 2727.
196. K.K. Mon and J.K. Percus, *J. Chem. Phys.*, 2002, **117**, 2289.
197. S. Vasenkov and J. Kärger, *Phys. Rev. E*, 2002, **66**, 052601.
198. A.I. Skoulidas and David S. Sholl, *J. Phys. Chem. B*, 2002, **106**, 5058.
199. C. Tunca and D.M. Ford, *J. Phys. Chem. B*, 2002, **106**, 10982.
200. S. Fritzsche and J. Kärger, *J. Phys. Chem. B*, 2003, **107**, 3515.
201. T. Düren, F.J. Keil and N.A. Seaton, *Mol. Phys.*, 2002, **100**, 3741.
202. Q. Zhang, J. Zheng, A. Shevade, L. Zhang, S. H. Gehrke, G. S. Heffelfinger and S. Jiang, *J. Chem. Phys.*, 2002, **117**, 808.
203. H. Dominguez and M. Rivera, *Mol. Phys.*, 2002, **100**, 3829.
204. E. Akhmatskaya, B.D. Todd, P.J. Daivis, D.J. Evans, K.E. Gubbins and L.A. Pozhar, *J. Chem. Phys.*, 1997, **106**, 4684.

205. V.P. Sokhan, D. Nicholson and N. Quirke, *J. Chem. Phys.*, 2002, **117**, 8531.
206. de Gennes, *Langmuir*, 2002, **18**, 3413.
207. H. Spikes and S. Granick, *Langmuir*, 2003, **19**, 5065.
208. Y.K. Tovbin, R.Y. Tugazakov and V.N. Komarov, *Colloids and Surfaces A*, 2002, **206**, 377.
209. J. Horbach and K. Binder, *J. Chem. Phys.*, 2002, **117**, 10796.
210. F. Porcheron and M. Schoen, *Phys. Rev. E*, 2002, **66**, 041205.
211. Y.W. Tang, I. Szalai and K-Y. Chan, *J. Phys. Chem. A*, 2001, **105**, 9616.
212. C.H. Turner, J.K. Brennan, J.K. Johnson and K.E. Gubbins, *J. Chem. Phys.*, 2002, **116**, 2138.
213. J.K. Johnson, A.Z. Panagiotopoulos and K.E. Gubbins, *Mol. Phys.*, 1994, **81**, 717.
214. K.P. Travis and D.J. Searles, *In preparation*,
215. J.K. Brennan, K.T. Thomson and K.E. Gubbins, *Langmuir*, 2002, **18**, 5438.
216. J.K. Brennan, T.J. Bandosz, K.T. Thomson and K.E. Gubbins, *Colloids and Surfaces A*, 2001, **187–188**, 539.
217. D.W. Oxtoby, *Annu. Rev. Mater. Res.*, 2002, **32**, 39.
218. M. Schmidt, *Phys. Rev. E*, 2002, **66**, 041108.
219. T.X. Nguyen, S.K. Bhatia and D. Nicholson, *J. Chem. Phys.*, 2002, **117**, 10827.
220. M.B. Sweatman, *Phys. Rev. E*, 2001, **65**, 011102.
221. P.I. Ravikovitch, A. Vishnyakov and A.V. Neimark, *Phys. Rev. E*, 2001, **64**, 011602.
222. S.-C. Kim, Z.T. Németh, M. Heni and H. Löwen, *Mol. Phys.*, 2001, **99**, 1875.
223. C.N. Patra and S.K. Ghosh, *J. Chem. Phys.*, 2002, **117**, 8933.
224. Y.-X. Yu and J. Wu, *J. Chem. Phys.*, 2002, **117**, 2368.
225. M.R. Kamat and D. Keffer, *Mol. Phys.*, 2002, **100**, 2689.
226. H.E. Stanley, S.V. Buldyrev, N. Giovambattista, E. La Nave, A. Scala, F. Sciortino and F.W. Starr, *Physica A*, 2002, **306**, 230.
227. G. Garberoglio and R. Vallauri, *Physica A*, 2002, **314**, 492.
228. S. Chowdhuri and A. Chandra, *Chem. Phys. Lett*, 2003, **373**, 79.
229. G.T. Evans, *J. Chem. Phys.*, 2001, **115**, 9905.
230. A. Verdaguer and J.A. Padro, *Mol. Phys.*, 2002, **100**, 3401.
231. M. Matsumoto, S. Salto and I. Ohmine, *Nature*, 2002, **416**, 409.
232. K. Honda, *J. Chem. Phys.*, 2002, **117**, 3558.
233. I. Nezbeda, *Mol. Phys.*, 2001, **99**, 1631.
234. J. Kolafa, I. Nezbeda and M. Lisal, *Mol. Phys.*, 2001, **99**, 1751.
235. M. Predota, P.T. Cummings and A.A. Chialvo, *Mol. Phys.*, 2002, **100**, 2703.
236. A.V. Gubskaya and P.G. Kusalik, *J. Chem. Phys.*, 2002, **117**, 5290.
237. www.ccp5.ac.uk,
238. W. Smith, C.W. Yong and P.M. Rodger, *Mol. Sim.*, 2002, **28**, 385.
239. N.S. Martys and J.G. Hagedorn, *Mater. Struct.*, 2002, **35**, 650.
240. M.M. Dupin, I. Halliday and C.M. Care, *J. Phys. A-Math. Gen.*, 2003, **36**, 8517.
241. P.J. Love, M. Nekovee, P.V. Coveney, J. Chin, N. Gonzalez-Segredo and J.M.R. Martin, *Com. Phys. Comm*, 2003, **153**, 340.
242. P.J. Hoogerbrugge and J.M.V.A. Koelman, *Euro. Phys. Lett.*, 1992, **19**, 155.
243. C.C. Liew and M. Mikami, *Chem. Phys. Lett*, 2003, **368**, 346.
244. I. Pagonabarraga and D. Frenkel, *J. Chem. Phys.*, 2001, **115**, 5015.
245. S.Y. Trofimov, E.L.F. Nies and M.A.J. Michels, *J. Chem. Phys.*, 2002, **117**, 9383.
246. P. Español and M. Revenga, *Phys. Rev. E*, 2003, **67**, 026705.
247. S. Succi, I.V. Karlin and H. Chen, *Rev. Mod. Phys.*, 2002, **74**, 1203.

248. I. Halliday, L.A. Hammond, C.M. Care, K. Good and A. Stevens, *Phys. Rev. E*, 2001, **64**, 011208.
249. M. Hirabayashi, Y. Chen and H. Ohashi, *J. Mag. Mag. Mat.*, 2002, **252**, 138.
250. I. Ispolatov and M. Grant, *Phys. Rev. E*, 2002, **65**, 056704.
251. J. Chin and P. V. Coveney, *Phys. Rev. E*, 2002, **66**, 016303.
252. V. Sofonea, W.-G. Früh and A. Cristea, *J. Mag. Mag. Mat.*, 2002, **252**, 144.
253. R.L. McGreevy, *J. Phys. Condens. Matter*, 2001, **13**, 877.
254. R.L. McGreevy and P. Zetterstrom, *Curr Opin in Sol State and Mats Sci.*, 2003, **7**, 41.
255. L. Pusztai, H. Dominguez and O.A. Pizio, *Physica A*, 2002, **316**, 65.
256. L. Pusztai, J-C. Soetens and P.A. Bopp, *Physica A*, 2003, **323**, 42.
257. G. Toth and A. Baranyai, *J. Chem. Phys.*, 2001, **114**, 2027.
258. G. Tóth, *J. Chem. Phys.*, 2001, **115**, 4770.
259. E. Bourasseau, P. Ungerer and A. Boutin, *J. Phys. Chem. B*, 2002, **106**, 5483.
260. J.R. Errington and A.Z. Panagiotopoulos, *J. Chem. Phys.*, 1999, **111**, 9731.
261. E.C. Mbamala and G. Pastore, *Physica A*, 2002, **313**, 312.
262. W.R. Smith and M. Lísal, *Phys. Rev. E*, 2002, **66**, 011104.
263. L.I. Kioupis and E.J. Maginn, *Fluid Phase Equilibria*, 2002, **200**, 75.
264. B. Chen and J.I. Siepmann, *J. Phys. Chem. B*, 2000, **104**, 8725.
265. B. Chen and J.I. Siepmann, *J. Phys. Chem. B*, 2001, **105**, 11275.
266. G.C. Boulougouris, I.G. Economou and D.N. Theodorou, *J. Chem. Phys.*, 2001, **115**, 8231.
267. Q. Yan, R. Faller and J.J. de Pablo, *J. Chem. Phys.*, 2002, **116**, 8745.
268. S.H. Patil, K.T. Tang and J.P. Toennies, *J. Chem. Phys.*, 2002, **116**, 8118.
269. K.T. Tang and J.P. Toennies, *J. Chem. Phys.*, 1984, **80**, 3726.
270. T. Masaki, H. Aoki, S. Munejiri, Y. Ishii and T. Itami, *J. Non-Cryst. Solids*, 2002, **312–314**, 191.
271. L. Reatto, D. Levesque and J.J. Weis, *Phys. Rev. A*, 1986, **33**, 3451.
272. S. Hess and M. Kroger, *Phys. Rev. E*, 2000, **61**, 4629.
273. S. Hess and M. Kroger, *Phys. Rev. E*, 2001, **64**, 011201.
274. J. Vrabec, J. Stoll and H. Hasse, *J. Phys. Chem. B*, 2001, **105**, 12126.
275. R. Chelli, R. Righini, S. Califano and P. Procacci, *J. Molec. Liq.*, 2002, **96–97**, 87.
276. G. Marcelli, B.D. Todd and R.J. Sadus, *J. Chem. Phys.*, 2001, **115**, 9410.
277. N.N. Medvedev, *Physica A*, 2002, **314**, 678.
278. A.A. Louis, P.G. Bolhuis, R. Finken, V. Krokoviack, E. J. Meijer and J. P. Hansen, *Physica A*, 2002, **306**, 251.
279. S.J. Stuart, J.M. Hicks and M.T. Mury, *Mol. Sim*, 2003, **29**, 177.
280. T.M. Galea and P. Attard, *Phys. Rev. E*, 2002, **66**, 041207.
281. T.M.A.O.M. Barenbrug, E.A.J.F. Peters and J.D. Schieber, *J. Chem. Phys.*, 2002, **117**, 9202.
282. T. ikeshoji, B. Hafskold and H. Furuholt, *Mol. Sim*, 2003, **29**, 101.
283. J. Lopez-Lemus and J. Alejandre, *Mol. Phys.*, 2002, **100**, 2983.
284. F. Siperstein, A.L. Myers and O. Talu, *Mol. Phys.*, 2002, **100**, 2025.

7
Numerical Methods in Chemistry

BY T.E. SIMOS

1 Introduction

In this paper we will present the recent developments on the numerical solution of the Schrödinger equation and related problems.

More specifically we will present the recent advance for the numerical solution of the radial and two-dimensional time-independent Schrödinger equation and related problems.

The radial Schrödinger equation has the form:

$$y''(r) = [l(l+1)/r^2 + V(r) - K^2]\, y(r). \qquad (1)$$

Models of this type, which represent a boundary value problem, occur frequently in theoretical physics and chemistry, (see for example ref. 1–4).

In the following we present some notations for (1):

- The function $W(r) = l(l+1)/r^2 + V(r)$ denotes *the effective potential*. This satisfies $W(r) \to 0$ as $r \to \infty$
- k^2 is a real number denoting *the energy*
- l is a given integer representing *angular momentum*
- V is a given function which denotes the potential.
- The boundary conditions are:

$$y(0) = 0 \qquad (2)$$

- and a second boundary condition, for large values of r, determined by physical considerations.

It is known from the literature that the last decades many numerical methods have been constructed for the approximate solution of the Schrödinger equation (see ref. 5–6). The aim and the scope of the above activity was the development of fast and reliable methods.

The developed methods can be divided into two main categories:

- Methods with constant coefficients
- Methods with coefficients dependent on the frequency of the problem.†

In the present review we shall present recent advances in the construction of numerical methods for the numerical solution of the Schrödinger equation and related problems. We will also present remarks for improvement of each method. Finally for the most of the case we will give software for the construction and a software for the application of the above methods to the Schrödinger equation.

In Section 2 we analyse a new category of methods named multiderivative methods. We present multiderivative methods with minimal phase-lag and constant coefficients as well as exponentially-fitted multiderivative methods with coefficients dependent on the frequency of the problem. For these methods numerical results on the radial Schrödinger equation are given and analysed.

In Section 3 we analyse a new category of methods for the numerical solution of the radial Schrödinger equation, the purely symplectic integrators. We describe the construction of a second order, a third order and a fifth order symplectic integrator. For these methods numerical results on the radial eigenvalue Schrödinger equation are given and analysed.

In Section 4 we analyse two categories of methods for the numerical solution of the two-dimensional time-independent Schrödinger equation. The first category of method are the asymptotically symplectic integrators. We describe the construction of a third order and a fifth order symplectic integrator. The second category of methods are the well know Numerov-type methods. For these methods we describe its computational implementation. For the methods of both categories numerical results on the two-dimensional eigenvalue Schrödinger equation are given and analysed.

In Appendix A the Maple programme for the analysis of the multiderivative methods described in section 2 is presented. In Appendix B we present a Matlab programme for the phase shift problem of the radial Schrödinger equation. With this programme we have produced the results of section 2. This is a dynamical programme for the determination of the phase shifts. This means that the user can compute any phase shift of any potential changing the routine of the potential. In Appendix C we present, for the first time in the literature, the order conditions for the construction of symplectic integrators up to fifth order. In the Appendices D, E and F we present the Mathematica Programme for the construction of second order, third order and fifth order symplectic integrators.

† In the case of the Schrödinger equation the frequency of the problem is equal to: $\sqrt{|l(l+1)/r^2 + V(r) - k^2|}$

2 Multiderivative Methods

2.1 Stability and Phase-lag Analysis of Multiderivative Methods. – For the investigattion of the periodic stability properties of the numerical schemes for problems of Schrödinger type, Lambert and Watson[7] have introduced the scalar test equation

$$y'' = -q^2 y \tag{3}$$

and the **interval of periodicity**, where q is a constant.

Based on ref. 7 when a symmetric two-step explicit multiderivative method is applied to the scalar test equation (3), we obtain the difference equation:

$$y_{n+1} - 2B(H)y_n + y_{n-1} = 0 \tag{4}$$

and the associate characteristic equation:

$$z^2 - 2B(H)z + 1 = 0 \tag{5}$$

where $H = qh$.

We have the following definitions, theorems and remarks.

Definition 1. (see ref. 7) A symmetric two-step method with the characteristic equation given by (5) is said to have an *interval of periodicity* $(0, H_0^2)$ if, for all $H \in (0, H_0^2)$, the roots $z_i, i = 1,2$ satisfy

$$z_1 = e^{i\theta(H)}, \text{ and } z_2 = e^{-i\theta(H)} \tag{6}$$

where $\theta(H)$ is a real function of H.

Theorem 1. A method which has a characteristic equation given by (5) has a non-empty interval of periodicity $(0, H_0^2)$, if for all $H^2 \in [0, H_0^2]$, $|B(H)| < 1$.

So we have that in order a symmetric two-step explicit multiderivative method to have a non-empty interval of periodicity the following conditions must hold:

$$1 \pm B(H) > 0 \tag{7}$$

for all $H^2 \in [0, H_0^2]$.

Theorem 2. For all H in the interval of periodicity, we can write:

$$cos[\theta(H)] = B(H), \tag{8}$$

where $H^2 \in [0, H_0^2]$.

Definition 2. For any symmetric two-step method with the characteristic equation given by (10) the phase-lag† is equal to (see ref. 8 and 9):

$$t = H - \theta(H) = H - \cos^{-1}(B(H)) = c\,H^{p+1} + O(H^{p+3}) \quad (9)$$

where c is the **phase-lag constant** and p is **phase-lag order**.

Based on the above Coleman[29] has found the following remark:

Remark 1.

$$t = c\,H^{p+1} + O(H^{p+3}) \Rightarrow \cos(H) - B(H) = \\ \cos(H) - \cos(H - t) = c\,H^{p+2} + O(H^{p+4}) \quad (10)$$

where t is the phase-lag of the method.

2.2 A New Family of Multiderivative Methods.

We introduce the following family of methods to integrate $y'' = f(x)y(x)$:

$$\bar{y}_{n+1} = 2y_n - y_{n-1} + a_0 h^2 y_n'' + a_1 h^4 y_n^{(4)}$$
$$y_{n+1} = 2y_n - y_{n-1} + h^2 \left[c_0 y_n'' + c_1 (\bar{y}_{n+1}'' + y_{n-1}'') \right] \quad (11)$$

$$+ h^4 \left[c_2 y_n^{(4)} + c_3 \left(\bar{y}_{n+1}^{(4)} + y_{n-1}^{(4)} \right) \right] \quad (12)$$

where $y''_{n \pm i} = f_{n \pm i} y_{n \pm i}$, $y_{n \pm i}^{(4)} = \left(f''_{n \pm i} + f_{n \pm i}^2 \right) y_{n \pm i} + 2 f'_{n \pm i} y'_{n \pm i}$ and $i = -1(1)1$. We note also that $\bar{y}''_{n+1} = f_{n+1} \bar{y}_{n+1}$ where $\bar{y}_{n+1}$ is calculated from the relation (11). It is mentioned that $y_{n \pm i} = y[x_0 + (n \pm i)h]$, $i = 0, 1$ and x_0 is the initial point of integration. It is noted that h is the integration step. It is easy to see that in order the above method (11)–(12) to be applicable, then approximate schemes for the first derivatives of y are needed.

In order the above method (11)–(12) to be of algebraic order six, then the parameters must be given by (for the analysis see ref. 10):

$$a_0 = 1, \quad a_1 = \frac{1}{12}, \quad c_0 = \frac{115}{126},$$

$$c_1 = \frac{11}{252}, \quad c_2 = \frac{313}{7560} \quad \text{and} \quad c_3 = -\frac{13}{15120} \quad (13)$$

Based on the above coefficients we can find that the local truncation error of the above scheme (11)–(12) is given by:

---

† Phase-lag physically means how well the numerical method approximates the solution of the scalar test equation $y'' = -q^2 y$. If we have a method of phase-lag order p this means that |Solution Approximate − Solution Analytical | = $O(h^p)$.

7: Numerical Methods in Chemistry

$$L.T.E(h) = -\frac{11}{90720} h^8 y_n^{(6)} \tag{14}$$

For the above method (11)–(12) we have

$$B(H) = 1 - \frac{1}{2} H^2 \left(1 - \frac{11}{252} H^2 + \frac{11}{3024} H^4\right)$$
$$+ \frac{1}{2} H^4 \left(\frac{5}{126} + \frac{13}{15120} H^2 - \frac{13}{181440} H^4\right) \tag{15}$$

From (15) we have that (7) is hold for every $H^2 \in (0, 6.88)$ i.e. larger than the corresponding interval of periodicity of Numerov's method (which is equal to $(0, 6)$).

2.3 A New Family of Multiderivative Methods with Minimal Phase-Lag.

We introduce the following new family of methods to integrate $y'' = f(x, y)$:

$$\hat{y}_{n+1} = 2y_n - y_{n-1} + h^2 y_n'' \tag{16}$$

$$y_n^{[i]} = y_n - b_i h^2 (\hat{y}_{n+1}'' - 2 y_n''^{[i-1]} + y_{n-1}''), \quad i = 1(1)3 \tag{17}$$

$$\bar{y}_{n+1} = 2 y_n - y_{n-1} + a_0 h^2 y_n''^{[3]} + a_1 h^4 y_n^{(4)[3]} \tag{18}$$

$$y_{n+1} = 2 y_n - y_{n-1} + h^2 \left[c_0 y''_n + c_1 (\bar{y}''_{n+1} + y''_{n-1})\right]$$
$$+ h^4 \left[c_2 y_n^{(4)} + c_3 \left(\bar{y}_{n+1}^{(4)} + y_{n-1}^{(4)}\right)\right] \tag{19}$$

where $y''_{n\pm i} = f_{n\pm i} y_{n\pm i}$, $y_{n\pm i}^{(4)} = \left(f''_{n\pm i} + f^2_{n\pm i}\right) y_{n\pm i} + 2 f'_{n\pm i} y'_{n\pm i}$ and $i = -1(1)1$ and b_c is a constant. We note also that $\hat{y}''_{n+1} = f_{n+1} \hat{y}_{n+1}$ where y_{n+1} is calculated from the relation (16) and $y_n''^{[i]} = f_n y_n^{[i]}$, $i = 1(1)3$ where $y_n^{[i]}$, $i = 1(1)3$ is calculated from the relation (17). We note also that $y_n^{[0]} = y_n$.

It is easy to see that in order the above method (16)–(19) to be applicable, then approximate schemes for the first derivatives of y are needed.

Expanding in Taylor series the terms $y_{n\pm j}$, $y''_{n\pm j}$ and $y_{n\pm j}^{(4)}$, $j = -1, 1$ and substituting the expansions into the new method (16)–(19) we can find that its local truncation error is given by:

$$L.T.E(h) = -\frac{11}{90720} h^8 \left[y_n^{(6)} - 360 b_3 y_n^{(4)}\right] \tag{20}$$

Applying the method (16)–(19) into the scalar test equation (3), we obtain the difference equation (4) and the associate characteristic equation (5), where:

$$B(H) = 1 - \frac{1}{2}H^2 + \frac{1}{24}H^4 - \frac{1}{720}H^6 - \left(\frac{11}{504}b_3 + \frac{13}{362880}\right)H^8$$
$$+ b_3\left(\frac{11}{252}b_2 + \frac{1}{720}\right)H^{10} + b_3\left(-\frac{11}{126}b_2b_1 - \frac{1}{360}b_2 + \frac{13}{362880}\right)H^{12}$$
$$+ b_2b_3\left(\frac{1}{180}b_1 - \frac{13}{181440}\right)H^{14} + \frac{13}{90720}b_3b_2b_1H^{16} \tag{21}$$

Based on Definition 2 and Remark 1 we have that:

$$\cos(H) - B(H) = H^8\left(\frac{11}{181440} + \frac{11}{504}b_3\right) - H^{10}\left(\frac{1}{3628800} + \frac{11}{252}b_2b_3 + \frac{1}{720}b_3\right)$$
$$+ H^{12}\left(\frac{1}{479001600} + \frac{11}{126}b_1b_2b_3 + \frac{1}{360}b_2b_3 - \frac{13}{362880}b_3\right)$$
$$- H^{14}\left(\frac{1}{87178291200} + \frac{1}{180}b_1b_2b_3 - \frac{13}{181440}b_2b_3\right) + \cdots \tag{22}$$

In order to have minimal phase-lag, the following system of equations must hold:

$$\frac{11}{181440} + \frac{11}{504}b_3 = 0$$
$$\frac{1}{3628800} + \frac{11}{252}b_2b_3 + \frac{1}{720}b_3 = 0$$
$$\frac{1}{479001600} + \frac{11}{126}b_1b_2b_3 + \frac{1}{360}b_2b_3 - \frac{13}{362880}b_3 = 0 \tag{23}$$

The solution of the above system of equations is given by:

$$b_3 = -\frac{1}{360}, \quad b_2 = -\frac{13}{440}, \quad b_1 = -\frac{296}{6435} \tag{24}$$

Substituting the above values of b_i, $i = 1(1)3$ into the above formula (22) we find that:

$$\cos(H) - B(H) = \frac{15889}{591951360000}H^{14} + \cdots \tag{25}$$

Based on the above analysis the methods presented in the following Table 1 can be obtained:

2.4 Computational Implementation. – In order to be applicable a multiderivative method we need approximate schemes for the first derivatives of y.

7: Numerical Methods in Chemistry

Table 1 *Properties of the Multiderivative Methods with Constant Coefficients*

Method	b_1	b_2	b_3	Interval of Periodicity	Phase-Lag Order
I	0	0	$-\dfrac{1}{360}$	(0, 7.338713058)	8
II	0	$-\dfrac{13}{440}$	$-\dfrac{1}{360}$	(0, 7.608050475)	10
III	$-\dfrac{296}{6435}$	$-\dfrac{13}{440}$	$-\dfrac{1}{360}$	(0, 7.850664770)	12

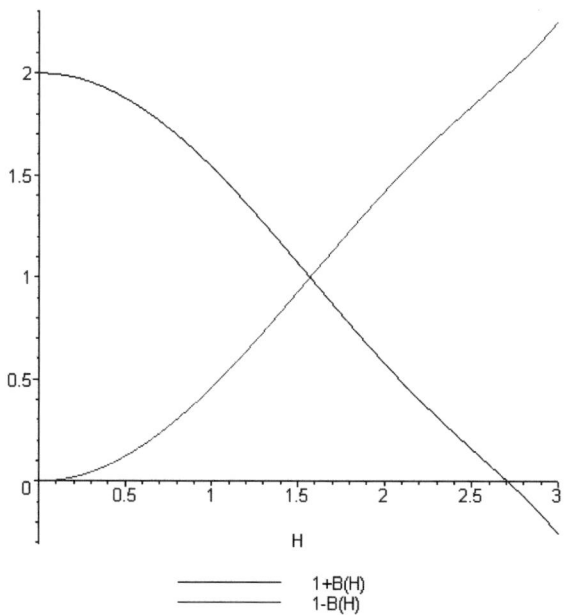

Figure 1 *Stability Functions for Case I*

For the new proposed methods we have the following formula:

$$y^{(4)}_{n\pm i} = \left(f''_{n\pm i} + f^2_{n\pm i}\right) y_{n\pm i} + 2 f'_{n\pm i} y'_{n\pm i} \quad \text{and} \quad i = -1(1)1. \tag{26}$$

The general formulae of the first derivatives on the points $x_i, i = n-1(1)n+1$ are given by:

$$h y'_{n+1} = a_{2,n+1} y_{n+1} + a_{1,n+1} y_n + a_{0,n+1} y_{n-1}$$
$$+ h^2 \left(b_{2,n+1} y''_{n+1} + b_{1,n+1} y''_n + b_{0,n+1} y''_{n-1}\right)$$
$$h y'_n = a_{2,n} y_{n+1} + a_{1,n} y_n + a_{0,n} y_{n-1}$$

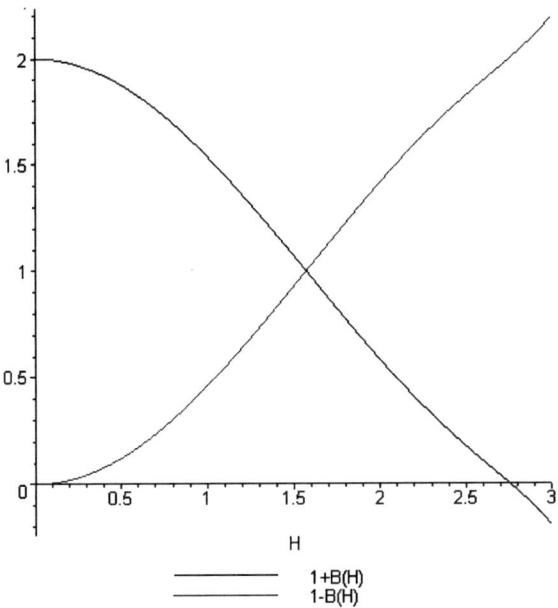

Figure 2 *Stability Functions for Case II*

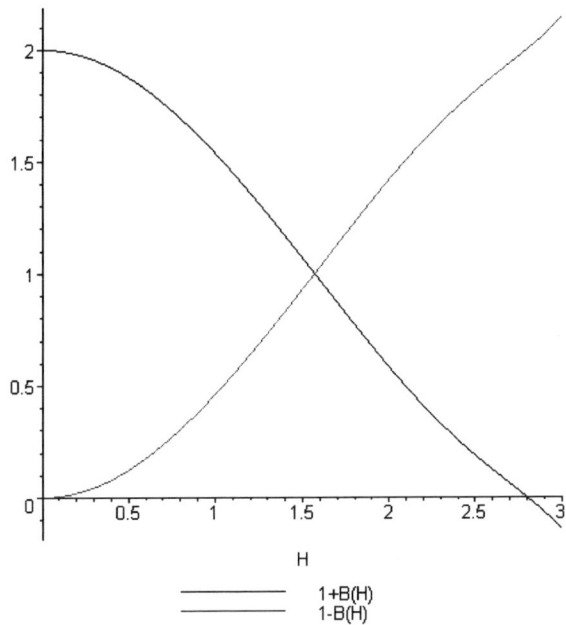

Figure 3 *Stability Functions for Case III*

$$+h^2 (b_{2,n} y''_{n+1} + b_{1,n} y''_n + b_{0,n} y''_{n-1})$$
$$h y'_{n-1} = a_{2,n-1} y_{n+1} + a_{1,n-1} y_n + a_{0,n-1} y_{n-1}$$
$$+h^2 (b_{2,n-1} y''_{n+1} + b_{1,n-1} y''_n + b_{0,n-1} y''_{n-1}) \tag{27}$$

In order the above methods to have maximal algebraic order and for the case: $b_{1,n+1} = b_{1,n} = b_{1,n-1} = 1$ the following values must hold (see for analysis ref. 10):

$$a_{2,n+1} = \frac{1}{10}, a_{1,n+1} = \frac{4}{5}, a_{0,n+1} = \frac{-9}{10}$$

$$b_{2,n+1} = \frac{11}{30}, b_{0,n+1} = \frac{1}{30}$$

$$a_{2,n} = \frac{-7}{10}, a_{1,n} = \frac{12}{5}, a_{0,n} = \frac{-17}{10}$$

$$b_{2,n} = \frac{1}{60}, b_{0,n} = \frac{11}{60}$$

$$a_{2,n-1} = \frac{-3}{2}, a_{1,n-1} = 4, a_{0,n-1} = \frac{-5}{2}$$

$$b_{2,n-1} = \frac{1}{6}, b_{0,n-1} \frac{-1}{6} \tag{28}$$

The local truncation error of the above formulae is given by:

$$L.T.E._{n+1} = L.T.E._{n-1} = -\frac{1}{45} h^5 y_n^{(5)} \text{ and}$$
$$L.T.E._n = \frac{7}{360} h^5 y_n^{(5)} \tag{29}$$

For the computational implementation of the multiderivative methods obtained here the following formula is also needed:

$$h y'_n = aa_{1,n} y_n + aa_{0,n} y_{n-1} + h^2 (bb_{1,n} y''_n + bb_{0,n} y''_{n-1}) \tag{30}$$

In order the above methods to have maximal algebraic order the following values must hold (see for analysis ref. 10):

$$bb_{0,n} = \frac{1}{6}, aa_{0,n} = -1, bb_{1,n} = \frac{1}{3}, aa_{1,n} = 1 \tag{31}$$

The local truncation error of the above formula is given by:

$$L.T.E._n = -\frac{1}{24} h^4 y_n^{(4)} \tag{32}$$

In Appendix A we present the construction of the multiderivative methods presented in this chapter.

2.5 Numerical Illustrations. – In this section we present some numerical results to illustrate the performance of our new methods. Consider the numerical integration of the Schrödinger equation (1) using the well-known Woods-Saxon potential (see refs. 1, 4–6, 8) which is given by

$$V(r) = V_w(r) = \frac{u_0}{(1+z)} - \frac{u_0 z}{[a(1+z)^2]}$$

with $z = \exp[(r - R_0)/a]$, $u_0 = -50$, $a = 0.6$ and $R_0 = 7.0$.

In the case of negative eigenenergies (i.e. when $E \in [-50,0]$) we have the well-known **bound-states problem** while in the case of positive eigenenergies (i.e. when $E \in [0,1000]$) we have the well-known **resonance problem** (see refs. 5, 6 and 15).

2.5.1 Resonance Problem. – In the asymptotic region the equation (1) effectively reduces to

$$y''(x) + (k^2 - \frac{l(l+1)}{x^2})y(x) = 0,$$

for x greater than some value X.

The above equation has linearly independent solutions $kxj_l(kx)$ and $kxn_l(kx)$, where $j_l(kx), n_l(kx)$ are the **spherical Bessel and Neumann functions** respectively. Thus the solution of equation (1) has the asymptotic form (when $x \to \infty$)

$$y(x) \approx Akxj_l(kx) - Bn_l(kx) \approx D[\sin(kx - \pi l/2) + \tan\delta_l \cos(kx - \pi l/2)]$$

where δ_l is the **phase shift** which may be calculated from the formula

$$\tan\delta_l = \frac{y(x_2)S(x_1) - y(x_1)S(x_2)}{y(x_1)C(x_2) - y(x_2)C(x_1)}$$

for x_1 and x_2 distinct points on the asymptotic region (for which we have that x_1 is the right hand end point of the interval of integration and $x_2 = x_1 - h$, h is the stepsize) with $S(x) = kxj_l(k_x)$ and $C(x) = kxn_l(k_x)$.

Since the problem is treated as an initial-value problem, one needs y_0 and y_1 before starting a two-step method. From the initial condition, $y_0 = 0$. The value y_1 is computed using the Runge-Kutta-Nyström 12(10) method of Dormand et al.[18–19] With these starting values we evaluate at x_1 of the asymptotic region the phase shift δ_l from the above relation.

As a test for the accuracy of our methods we consider the numerical integration of the Schrödinger equation (1) with $l = 0$ in the well-known case where the potential $V(r)$ is the Woods-Saxon one.

One can investigate the problem considered here, following two procedures. The first procedure consists of finding the **phase shift** $\delta(E) = \delta_l$ for $E \in [1,1000]$. The second procedure consists of finding those E, for $E \in [1,1000]$, at which δ equals $\pi/2$. In our case we follow the first procedure i.e. we try to find the phase shifts for given energies. The obtained phase shift is then compared to the analytic value of $\pi/2$.

The above problem is the so-called **resonance problem** when *the positive eigenenergies lie under the potential barrier*. We solve this problem, using the technique fully described in refs. 5-6, 10.

The boundary conditions for this problem are:

$$y(0) = 0,$$
$$y(x) \approx \cos[\sqrt{E}x] \text{ for large } x.$$

The domain of numerical integration is [0, 15].

For comparison purposes in our numerical illustration we use the well known Numerov's method (which is indicated as method [a]), the explicit Numerov-type method of Chawla[20] (which is indicated as method [b]), the multiderivative method with phase-lag of order 8 developed in this paper (which is indicated as method [c]) and the multiderivative method with phase-lag of order 10 developed in this paper (which is indicated as method [d]).

The numerical results obtained for the four methods, with stepsizes equal to $h = \frac{1}{2^n}$, were compared with the analytic solution of the Woods-Saxon potential resonance problem, rounded to six decimal places. In Figure 4 we show the errors $Err = log_{10} |E_{calculated} - E_{analytical}|$ of the highest eigenenergy $E_3 = 989.701916$ for several values of n.

In Appendix B we present a Matlab programme for the extraction of the results presented in Figure 4. This is a dynamical programme for the computation of the phase shifts. In order one to compute phase shifts for other potentials he must only changes the potential routines in the programme.

2.5.2 The Bound-States Problem. – For negative energies we solve the so-called bound-states problem, i.e. the equation (1) with $l = 0$ and boundary conditions given by

$$y(0) = 0,$$
$$y(x) \approx \exp(-\sqrt{-E}x) \text{ for large } x.$$

In order to solve this problem numerically we use a strategy which has been proposed by Cooley[21] and has been improved by Blatt.[22] This strategy involves integrating forward from the point $x = 0$, backward from the point $x_b = 15$ and matching up the solution at some internal point in the range of integration. As initial conditions for the backward integration we take:

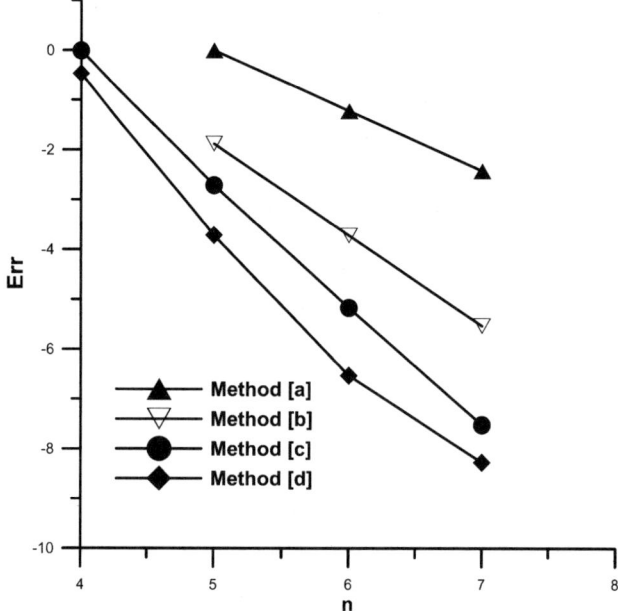

Figure 4 Error $Err = \log_{10} |E_{calculated} - E_{analytical}|$ for several values of n for the resonance $E_3 = 989.701916$. The nonexistence of a value for a method indicates that Err is positive.

$$y(x_b) = \exp(-\sqrt{-E}x_b) \text{ and } y(x_b - h) = \exp[-\sqrt{-E}(x_b - h)],$$

where h is the steplength of integration of the numerical method.

The true solutions to the Woods-Saxon bound-states problem were obtained correct to nine decimal places using the analytic solution and the numerical results obtained for the six methods mentioned above were compared to this true solution. The results are similar with those of resonance problem.

2.6 Remarks and Conclusions. – Based on the results presented in the Figure 4 we can concluded the following:

- The explicit Numerov-type method of Chawla[20] is more efficient than the method of Numerov
- The multiderivative method with phase-lag of order 8 is more efficient than the the explicit Numerov-type method of Chawla.[20]
- Finally the most efficient method is the multiderivative method with phase-lag of order 10.

From the above we can conclude that the class of the multiderivative methods is a new interesting family of methods for the numerical solution of the Schrödinger equation and related problems.

3 Symplectic Methods for the Numerical Solution of the Radial Shrödinger Equation

3.1 Introduction. – The one-dimensional time-independent Schrödinger equation has the form

$$-\frac{1}{2}\frac{d^2q}{dx^2} + V(x)q = Eq \tag{33}$$

where E is the energy eigenvalue, $V(x)$ is the potential and q is the wavefunction.

Liu et al.[11] has transformed (33) into Hamiltonian canonical equation using Legendre transformation. The Hamiltonian canonical equations are illustrated below

$$\begin{cases} \dot{p} = -\frac{\partial H}{\partial q} = -B(x)q \\ \dot{q} = \frac{\partial H}{\partial p} = p \end{cases} \tag{34}$$

where $B(x) = 2[E - V(x)]$ and H the hamiltonian function

$$H(q, p, x) = \frac{1}{2}p^2 + \frac{1}{2}B(x)q^2 \tag{35}$$

The symplectic integration of Hamiltonian dynamical systems is a recent research field. Third order integrators were constructed by Ruth,[12] fourth order integrators were obtained by Candy and Rozmus[13] and Forest and Ruth.[15] Yoshida[14] has constructed symplectic integrators of sixth and eighth order.

3.2 Symplectic Integrators – Basic Theory. – The characterization of a canonical transformation is done by using matrix algebra or by using differential forms (2-form).

Definition 3. (see ref. 16) A mapping is symplectic if

$$L^T J L = J \tag{36}$$

where L is the 2d-dimensional Jacobian matrix of the mapping and $J = \begin{pmatrix} 0_d & I_d \\ -I_d & 0_d \end{pmatrix}$, with I_d and 0_d denoting the unit and zero d-dimensional matrix

Proposition 1. (see ref. 16) A transformation $\begin{pmatrix} q \\ p \end{pmatrix} \rightarrow \begin{pmatrix} q^* \\ p^* \end{pmatrix}$ is symplectic (2-form) if and only if $\sum_{i=1}^{d} dq_i^* \wedge dp_i^* = \sum_{i=1}^{d} dq_i \wedge dp_i$, rewriting as $dq^* \wedge dp^* = dq \wedge dp$

In this chapter we will study symplectic integrators as proposed by Forest and Ruth,[15] Yoshida:[14]

$$\begin{cases} P_i = P_{i-1} - c_i h B^{n+\frac{1}{2}} Q_{i-1} \\ Q_i = Q_{i-1} + d_i h P_i \end{cases} \quad i = 1,...,k \tag{37}$$

where $Q_0 = q^n$, $P_0 = p^n$, $B^{n+\frac{1}{2}} = B(x_n + \frac{h}{2})$, c_i and d_i are free parameters and k is the number of stages.

The solution at the point x_{n+1} is given by:

$$Q_k = q^{n+1}, \quad P_k = p^{n+1} \tag{38}$$

The parameters c_i and d_i are determined by Yoshida[14]

$$\exp[h(A+B)] = \prod_{i=1}^{k} \exp(c_i h A) \exp(d_i h B) + O(h^{n+1}) \tag{39}$$

where k and n, are the number of stages and the order of method respectively.

3.3 Construction of Symplectic Integrators. – In order to determine the coefficients c_i, d_i, we follow the procedure described below:

1. We expand the left hand side of (39) in powers of h,

$$S(h) = \exp[h(A+B)] = 1 + h(A+B) + \tfrac{1}{2}h^2(A^2 + AB + BA + B^2) + ...$$

2. We expand the right hand side of (39)

$$\tilde{S}(h) = \prod_{i=1}^{k} \exp(c_i h A) \exp(d_i h B) = 1 + h\left(\sum_{i=1}^{k} c_i A + \sum_{i=1}^{k} d_i B\right) +$$
$$+ \frac{1}{2}h^2 \left[\left(\sum_{i=1}^{k} c_i\right)^2 A^2 + 2\sum_{i=1}^{k} d_i \sum_{j=1}^{i} c_j AB + 2\sum_{i=1}^{k} d_i \sum_{j=i+1}^{k} c_j BA + \left(\sum_{i=1}^{k} d_i\right)^2 B \right] + \cdots$$

3. We want the two expressions to agree up to h^n. The resulting equations for the coefficients c_i, d_i are given in the Appendix C where we present all the equations up to fifth order.

The above equations are depended linearly. The transformation into a linearly independent system of equations, leads to the following number of equations.

In the Appendix C we give also the exact linearly independent equations. Based on the above Table we conclude that a second order method needs at

7: Numerical Methods in Chemistry

Table 2 *The number of order condition equations for a Symplectic Integrator*

Algebraic Order for a Symplectic Integrator	Number of Equations
1	2
2	3
3	5
4	8
5	14

least two stages, a third order method needs at least three stages while a fifth order method needs seven stages.

3.4 Development of the New Methods. – *3.4.1 The second-order method.* In this chapter we construct a second-order method. From the previous chapter we conclude that such a method must have at least two-stages, of the form (37) i.e.

$$\begin{cases} P_1 = p^n - c_1 h B(x_n + \frac{h}{2}) q^n \\ Q_1 = q^n + d_1 h P_1 \\ p^{n+1} = P_1 - c_2 h B(x_n + \frac{h}{2}) Q_1 \\ q^{n+1} = Q_1 + d_2 h p^{n+1} \end{cases}$$

Following approach described in section 3.3, in order the method to be of second order the following system of equations must hold (see ref. 17 for more details)

$$\begin{cases} (a) & c_1 + c_2 = 1 \\ (b) & d_1 + d_2 = 1 \\ (c) & c_1 d_1 + (c_1 + c_2) d_2 = \frac{1}{2} \\ (d) & d_1 c_2 = \frac{1}{2} \end{cases} \quad (40)$$

For second order equations (a)–(c) should be satisfied, while (d) follows from the first three equations.

Yoshida[14] derived his second order method by letting $d_2 = 0$, then

$$c_1 = \frac{1}{2}, c_2 = \frac{1}{2}, d_1 = 1, d_2 = 0 \quad (41)$$

We follow a different approach here based on the minimization of the error. Since we have three equations, we let d_1 be a free parameter, then

$$c_1 = 1 - \frac{1}{2d_1}, c_2 = \frac{1}{2d_1}, d_2 = 1 - d_1 \quad (42)$$

We minimize the error–function in order to find the coefficients. Based on (42) the error function finally has the form (see for details ref. 17):

$$error(d_1) = \frac{\sqrt{6}}{12}\sqrt{\left(\frac{-3+4d_1}{2d_1}\right)^2 + (-2+3d_1)^2} \qquad (43)$$

Minimization of the error–function gives

$$d_1 = \frac{\sqrt{2}}{2} \quad (error(d_1) = 0.0350),$$

then

$$c_1 = 1 - \frac{\sqrt{2}}{2},\ c_2 = \frac{\sqrt{2}}{2},\ d_1 = \frac{\sqrt{2}}{2},\ d_2 = 1 - \frac{\sqrt{2}}{2} \qquad (44)$$

It is important to be noticed the symmetries, $c_1 = d_2$ and $c_2 = d_1$.

3.4.2 The Third-order Method. – In this chapter we construct a third–order method. From the previous chapter we conclude that such a method must have at least three–stages, of the form (37) i.e.

$$\begin{cases} P_1 = p^n - c_1 hB(x_n + \frac{h}{2})q^n \\ Q_1 = q^n + d_1 h P_1 \\ P_2 = P_1 - c_2 hB(x_n + \frac{h}{2})Q_1 \\ Q_2 = Q_1 + d_2 h P_2 \\ p^{n+1} = P_2 - c_3 hB(x_n + \frac{h}{2})Q_2 \\ q^{n+1} = Q_2 + d_3 h p^{n+1} \end{cases}$$

Following approach described in section 3.3, in order the method to be of third order the following system of equations must hold (see ref. 17 for more details):

$$\begin{cases} (a) & c_1 + c_2 + c_3 = 1 \\ (b) & d_1 + d_2 + d_3 = 1 \\ (c) & c_1 d_1 + (c_1 + c_2)d_2 + (c_1 + c_2 + c_3)d_3 = \frac{1}{2} \\ (d) & c_2 c_3 d_2 + c_1((c_2 + c_3)d_1 + c_3 d_2) = \frac{1}{6} \\ (e) & c_2 d_1 d_2 + d_3((d_1 + d_2)c_3 + c_2 d_1) = \frac{1}{6} \end{cases} \qquad (45)$$

Ruth[12] obtained his third order method (with $error = 0.0495164$), with coefficients given by:

$$c_1 = 1, c_2 = -\frac{2}{3}, c_3 = \frac{2}{3}, d_1 = -\frac{1}{24}, d_2 = \frac{3}{4}, d_3 = \frac{7}{24} \qquad (46)$$

We follow a different approach here based on the minimization of the error. Since we have five equations (and six parameters), we let d_2 be a free parameter. Based in the analysis presented in ref. 17 the Minimization of the error function gives:

$$d_2 = -\frac{58623767696137}{811561628596785} \text{ or } d_2 = \frac{5200281507982433}{4399167664813427}, \text{ (with } error(d_2)$$
$$= 0.0374228)$$

So the rest of the coefficients are given by (see ref. 17 for details):

$$c_1 = \frac{479561939695517}{1857710613287345} \quad c_2 = \frac{5200281507982433}{4399167664813427} \quad c_3 = -\frac{4618293127047827}{10490100451822575}$$

$$d_1 = \frac{108606835852797}{172086020422633} \quad d_2 = -\frac{58623767696137}{811561628596785} \quad d_3 = \frac{810034846678267}{1836329443349088}$$

or

$$c_1 = \frac{810034846678267}{1836329443349088} \quad c_2 = -\frac{58623767696137}{811561628596785} \quad c_3 = \frac{108606835852797}{172086020422633}$$

$$d_1 = -\frac{4618293127047827}{10490100451822575} \quad d_2 = \frac{5200281507982433}{4399167664813427} \quad d_3 = \frac{479561939695517}{1857710613287345}$$

3.4.3 The Fifth-order Method. – In this chapter we construct a fifth-order method. From the previous chapter we conclude that such a method must have at least seven–stages, of the form (37) i.e.

$$\begin{cases} P_1 = p^n - c_1 hB(x_n + \frac{h}{2})q^n \\ Q_1 = q^n + d_1 hP_1 \\ P_2 = P_1 - c_2 hB(x_n + \frac{h}{2})Q_1 \\ Q_2 = Q_1 + d_2 hP_2 \\ P_3 = P_2 - c_3 hB(x_n + \frac{h}{2})Q_2 \\ Q_3 = Q_2 + d_3 hP_3 \\ P_4 = P_3 - c_4 hB(x_n + \frac{h}{2})Q_3 \\ Q_4 = Q_3 + d_4 hP_4 \\ P_5 = P_4 - c_5 hB(x_n + \frac{h}{2})Q_4 \\ Q_5 = Q_4 + d_5 hP_5 \\ P_6 = P_5 - c_6 hB(x_n + \frac{h}{2})Q_5 \\ Q_6 = Q_5 + d_6 hP_6 \\ p^{n+1} = P_6 - c_7 hB(x_n + \frac{h}{2})Q_6 \\ q^{n+1} = Q_6 + d_7 hp^{n+1} \end{cases}$$

The linearly independent system of fifth-order equations are given in Appendix C

For fifth order equations we have fourteen equations and fourteen parameters. This set of equations can be solved numerically. Using the Newton method 46 solutions for fifth-order integrator have been obtained.

If we follow the procedure of the minimization of the sum of squares of order conditions, using the Levenberg Marquardt method, the same solutions have been obtained.

In Appendices D, E and F we present the Mathematica Programme for the construction of second order, third order and fifth order symplectic integrators.

For internal computations 40-digits of precision are used.

We have checked all the 46 produced solutions based on the error analysis and we have found that the most efficient one is the method given in Table 3.

3.5 Numerical Examples. – In order to apply the new proposed methods, the shooting technique is used. For comparison purposes with existing methods we use two potentials the harmonic oscillator and the hydrogen atom.

3.5.1 The Harmonic Oscillator. – Let the potential of the one-dimensional harmonic oscillator

$$V(x) = \frac{1}{2}x^2, \quad (-\infty < x < +\infty) \quad (47)$$

For this potential the exact eigenvalues are given by the formula

$$E_n = n + \frac{1}{2}, \quad (n = 0, 1, 2 ...) \quad (48)$$

Table 3 *Fifth Order Symplectic Integrator*

Fifth order symplectic integrator

$c_1 = 0.4515650720436605661536769078444402592772$
$c_2 = -0.0026255177260405503212166318348852182746$
$c_3 = -0.2887462490910128204969177168706319164955$
$c_4 = 0.4703720043422901320446596003861052932322$
$c_5 = 0.3704466763359327321311653913528413374090$
$c_6 = 0.1934796732533845648578957608942231680799$
$c_7 = -0.1944916591582146243692633117720558567511$
$d_1 = 1.9042327805084463874533312275884597491900$
$d_2 = -1.9395863664419246052720045464619766543740$
$d_3 = 0.3960766510231830289112974915012668544240$
$d_4 = 0.5133868104090695626740381054834970276760$
$d_5 = -2.9677394606045473652638582641642989867680$
$d_6 = 0.0041774095286693157391416751180472532000$
$d_7 = 3.0894521755771036757580543109350047566510$

7: Numerical Methods in Chemistry

In order to compute the eigenvalues, we take as boundary conditions

$$y(xmin) = 0, \quad y(xmax) = 0 \qquad (49)$$

where $xmin$ and $xmax$ are respectively the left and right boundaries. We define N as a positive integer and then the space $[xmin, xmax]$ is divided into N equal intervals. The length of each interval is equal to: $h = \frac{xmax - xmin}{N}$ and this denote that $x_n = xmin + nh, (n = 1, 2, \ldots, N-1)$. Then in order to calculate the eigenvalues, we use a symplectic scheme and the shooting method.

We have the following computations:

For comparison purposes we use two-block of methods:

First Block: Low order symplectic integrator and Runge-Kutta-Nyström methods

- The new two–stage second order
- The two–stage second order method obtained by Yoshida[14]
- The method Runge-Kutta-Nyström 6(4)6FD developed in ref 23.

Second Block: High order symplectic integrators

- The new three-stage third order symplectic integrator
- The new seven-stage fifth order symplectic integrator
- The Ruth's third order method developed in ref. 12, 15
- The Yoshida's fourth order symplectic integrator developed in ref. 14
- The Yoshida's sixth order symplectic integrator developed in ref. 14

In Figures 5–6 we present the

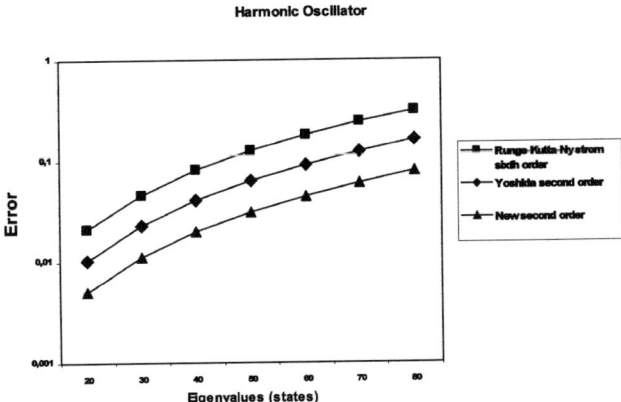

Figure 5 *Values of Error for the eigenvalues $E20, \ldots, E80$ of the Harmonic Oscillator. Methods used: i) Runge-Kutta-Nystrom[23] method of order of six, ii) Yoshida[14] symplectic-scheme method of two stage–second order, iii) New Symplectic-scheme of two stage–second*

Harmonic Oscillator

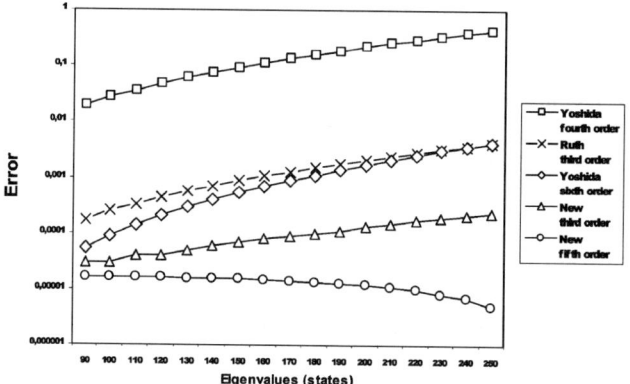

Figure 6 *Values of Error for the eigenvalues E90,..., E250 of the Harmonic Oscillator. Methods used: i) Yoshida[14] symplectic-scheme method of four stage–fourth order, ii) Ruth[12,15] symplectic-cheme method of three stage–third order, iii) Yoshida[14] symplectic-scheme method of eight stage–six order, iv) New Symplectic-scheme of three stage–third order, v) New Symplectic-scheme of seven stage–fifth order*

$$\text{Error} = -\log_{10}(E_{calculated} - E_{analytical})$$

for several eigenvalues.

3.5.2 The Hydrogen Atom. – For the hydrogen atom the radial wave function is determined by one–dimensional Shrödinger equation of the form:

$$y''(r) + (2E + \frac{2}{r} - \frac{l(l+1)}{r^2})y(r) = 0, \qquad 0 \leq r < +\infty \qquad (50)$$

where $l = 0, 1, 2, \ldots$.

In this paper we solve the eigenvalue problem for $l = 0$. The boundary conditions are

$$y(0) = 0, \quad y(+\infty) = 0,$$

and the exact eigenvalues are calculated by the formula

$$E_n = -\frac{1}{2n^2}, \qquad (n = 1, 2, 3 \ldots) \qquad (51)$$

For comparison purposes we use the following methods

- The new three–stage third order symplectic integrator
- The new seven-stage fifth order symplectic integrator
- The Ruth's third order method developed in ref. 12, 15

- The Yoshida's fourth order symplectic integrator developed in ref. 14
- The Yoshida's sixth order symplectic integrator developed in ref. 14

In Figures 7 we present the

$$\text{Error} = -\log_{10}(E_{\text{calculated}} - E_{\text{analytical}})$$

for several eigenvalues.

3.6 Remarks and Conclusions. – From the results presented in the Figures 5–7 we have the following remarks:

- The two–stage second order method obtained by Yoshida[14] is more efficient than the method Runge-Kutta-Nyström 6(4)6FD developed in ref. 23.
- The new two-stage second order symplectic integrator is more efficient than the two–stage second order method obtained by Yoshida.[14]
- The Ruth's third order method developed in ref. 12, 15 is more efficient than the Yoshida's fourth order symplectic integrator developed in ref. 14.
- The Ruth's third order method developed in ref. 12, 15 has similar behavior with the Yoshida's sixth order symplectic integrator developed in ref. 14.
- The new third order symplectic integrator is more efficient than the Ruth's third order method developed in ref. 12, 15 and the Yoshida's sixth order symplectic integrator developed in ref. 14.

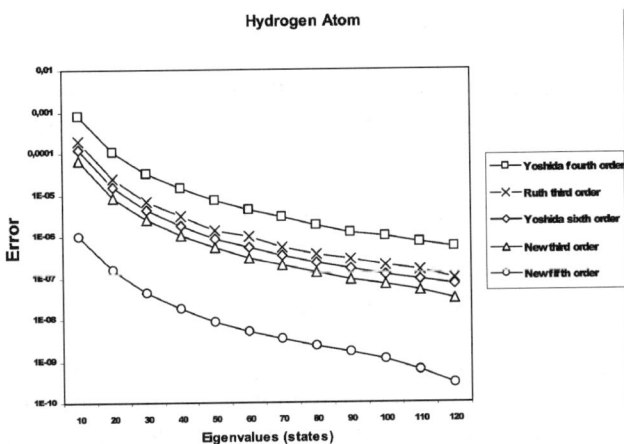

Figure 7 *Values of Error for the eigenvalues E10, E20, ..., E120 of the Hydrogen Atom. Methods used: i) Yoshida[14] symplectic-scheme method of four stage–fourth order, ii) Ruth[12,15] symplectic-scheme method of three stage–third order, iii) Yoshida[14] symplectic-scheme method of eight stage–six order, iv) New Symplectic-scheme of three stage–third order, v) New Symplectic-scheme of seven stage–fifth order*

- Finally the most efficient method is the new developed fifth order symplectic integrator.

Generally we can say that the symplectic integrators are a very interesting new insight for the numerical solution of the Schrödinger equation and related problems.

4 Numerical Solution of the Two-Dimensional Schrödinger Equation

4.1 Introduction. – In the literature the numerical solution of the two-dimensional Schrödinger equation was treated by means of discretization of both variables x and y (among others ref. 23–25).

In the new insights we use partial discretization only on the variable y. Then we have an ordinary differential equation problem.

In this chapter we apply to the two-dimensional problem:

- symplectic methods of higher order.
- Numerov's method
- and a modified Numerov method with minimal phase-lag (ref. 20,27)

Finally we apply:

- the five-point formula with full discretization.

The methods are tested on the two-dimensional harmonic oscillator and the two-dimensional Henon-Heils potential.

4.2 Partial Discretisation of the Two-dimensional Equation. – The two-dimensional time-independent Schrödinger equation can be written in the form

$$\frac{\partial^2 \psi}{\partial x^2} + \frac{\partial^2 \psi}{\partial y^2} + (2E - 2V(x,y))\psi(x,y) = 0,$$
$$\psi(x, \pm\infty) = 0, \quad -\infty < x < \infty,$$
$$\psi(\pm\infty, y) = 0, \quad -\infty < y < \infty \tag{52}$$

where E is the energy eigenvalue, $V(x,y)$ is the potential and $\psi(x,y)$ the wave function. The wave functions $\psi(x,y)$ asymptotically approaches infinity away from the origin. We consider $\psi(x,y)$ for y in the finite interval $[-R_y, R_y]$ and

$$\psi(x, -R_y) = 0 \quad \text{and} \quad \psi(x, R_y) = 0$$

the boundary conditions. We also consider partition of the interval $[-R_y, R_y]$

$$-R_y = y_{-N}, y_{-N+1}, \ldots, y_{-1}, y_0, y_1, \ldots, y_{N-1}, y_N = R_y$$

where $y_{j+1} - y_j = h_y = R_y/N$.

We approximate the partial derivative with respect to y with the difference quotient

$$\frac{\partial^2 \psi}{\partial y^2} = \frac{\psi(x,y_{j+1}) - 2\psi(x,y_j) + \psi(x,y_{j-1})}{h_y^2}$$

and substitute into the original equation

$$\frac{\partial^2 \psi}{\partial x^2} = -\frac{1}{h_y^2}\psi(x,y_{j+1}) - B(x,y_j)\psi(x,y_j) - \frac{1}{h_y^2}\psi(x,y_{j-1})$$

where

$$B(x,y_j) = 2\left(E - V(x,y_j) - \frac{1}{h_y^2}\right)$$

We also define the $2N-1$ length vector

$$\Psi(x) = \left(\psi(x,y_{-N+1}), \psi(x,y_{-N+2}), \ldots, \psi(x,y_0), \ldots, \psi(x,y_{N-2}), \psi(x,y_{N-1})\right)^T$$

then equation (1) can be written as

$$\frac{\partial^2 \Psi}{\partial x^2} = -S(x)\Psi(x) \tag{53}$$

where $S(x)$ is a $(2N-1) \times (2N-1)$ matrix

$$S(x) = \begin{pmatrix} B(x,y_{-N+1}) & 1/h_y^2 & & & \\ 1/h_y^2 & B(x,y_{-N+2}) & 1/h_y^2 & & \\ & \ddots & \ddots & \ddots & \\ & & 1/h_y^2 & B(x,y_{N-2}) & 1/h_y^2 \\ & & & 1/h_y^2 & B(x,y_{N-1}) \end{pmatrix}$$

The matrix $S(x)$ can be written in terms of three matrices the identity matrix I, the diagonal matrix V which contains the potential at the mesh points $y_{-N+1}, \ldots, y_{N-1}$ and the tridiagonal matrix M with diagonal elements -2 and off diagonal elements 1.

$$S(x) = 2EI - 2V(x) + \frac{1}{h_y^2}M$$

4.3 Application of Symplectic Methods. – We note here that if the matrix $S(x)$ is of order $k = 2N-1$, and the eigenvalue problem is of $l = k^2$ order, i.e. for a $N = 20$ points partition of the interval $[-R, R]$ we need to find the eigenvalues of a 1521 size matrix.

4.3.1 Third Order Asymptotically Symplectic Method. – The third order method of Ruth's[12,15] can be written as:

$$p_1 = \phi_n - c_1 h S \psi_n,$$
$$q_1 = \psi_n + d_1 h p_1,$$
$$p_2 = p_1 - c_2 h S q_1,$$
$$q_2 = q_1 + d_2 h p_2,$$
$$\phi_{n+1} = p_2 - c_3 h S q_2,$$
$$\psi_{n+1} = q_2 + d_3 h \phi_{n+1}$$

where

$$c_1 = \frac{7}{24}, \quad c_2 = \frac{3}{4}, \quad c_3 = -\frac{1}{24}, \quad d_1 = \frac{2}{3}, \quad d_2 = -\frac{2}{3}, \quad d_3 = 1$$

We write for each stage

$$\begin{pmatrix} 1 & 0 \\ -d_1 h & 1 \end{pmatrix} \begin{pmatrix} p_1 \\ q_1 \end{pmatrix} = \begin{pmatrix} 1 & -c_1 h S \\ 0 & 1 \end{pmatrix} \begin{pmatrix} \phi_n \\ \psi_n \end{pmatrix}$$

$$\begin{pmatrix} 1 & 0 \\ -d_2 h & 1 \end{pmatrix} \begin{pmatrix} p_2 \\ q_2 \end{pmatrix} = \begin{pmatrix} 1 & -c_2 h S \\ 0 & 1 \end{pmatrix} \begin{pmatrix} p_1 \\ q_1 \end{pmatrix}$$

$$\begin{pmatrix} 1 & 0 \\ -d_3 h & 1 \end{pmatrix} \begin{pmatrix} \phi_{n+1} \\ \psi_{n+1} \end{pmatrix} = \begin{pmatrix} 1 & -c_3 h S \\ 0 & 1 \end{pmatrix} \begin{pmatrix} p_2 \\ q_2 \end{pmatrix}$$

$$\begin{pmatrix} \phi_{n+1} \\ \psi_{n+1} \end{pmatrix} = M \begin{pmatrix} \phi_n \\ \psi_n \end{pmatrix}, \quad \text{where} \quad M = \begin{pmatrix} \alpha(S) & \beta(S) \\ \gamma(S) & \delta(S) \end{pmatrix}$$

where $\alpha(S)$, $\gamma(S)$ are second degree polynomials in S and $\beta(S)$ and $\delta(S)$ are third degree polynomials in S.

We write, now, the method as a two step method

$$\alpha_2 \psi^{n+1} - \alpha_1 \psi^n + \alpha_0 \psi^{n-1} = 0$$

where

$$\alpha_2 = \beta(S_{n-1/2})$$
$$\alpha_1 = \alpha(S_{n+1/2})\beta(S_{n-1/2}) + \alpha(S_{n-1/2})\beta(S_{n+1/2})$$
$$\alpha_0 = \beta(S_{n+1/2})$$

We note that the above is a generalized algebraic eigenvalue problem which involves powers of the eigenvalue E up to E^5 which is very expensive in terms of computation.

7: Numerical Methods in Chemistry

Now we explain what we mean with the term <u>asymptotically symplectic</u>. We subtract the terms corresponding to powers of h higher than h^3. So we obtain the following method

$$\alpha_0 = 1 - \frac{h^2}{6} S(x_{n+1/2})$$

$$\alpha_1 = 2 - \frac{2}{3} h^2 \left(S(x_{n-1/2}) + S(x_{n+1/2}) \right)$$

$$\alpha_2 = 1 - \frac{h^2}{6} S(x_{n-1/2}) \qquad (54)$$

For the Method (54) we have the condition:

$$M^T J M = J + O(h^5)$$

where

$$J = \begin{pmatrix} 0 & 1 \\ -1 & 0 \end{pmatrix}$$

Definition 4

A method is called **asymptotically symplectic method** if for the matrix form M we have:

$$M^T J M = J + O(h^n)$$

where J is given above and n is the order of powers of h from where we subtract the terms in the expansions of a_i.

We note here that these methods are almost symplectic for small values of h.

4.3.2 Fifth Order Asymptotically Symplectic Method. – Here we will develop a fifth order method

For this method we have

$$\alpha_0 = 1 - \frac{h^2}{3!} S(x_{n+1/2}) + \frac{1}{5!} h^4 S^2(x_{n+1/2})$$

$$\alpha_1 = 2 - \frac{2}{3} h^2 \left[S(x_{n-1/2}) + S(x_{n+1/2}) \right] - \frac{1}{20} h^4 \left[S^2(x_{n-1/2}) + S^2(x_{n+1/2}) \right]$$

$$\quad - \frac{1}{6} h^4 S(x_{n-1/2}) S(x_{n+1/2})$$

$$\alpha_2 = 1 - \frac{h^2}{3!} S(x_{n-1/2}) + \frac{1}{5!} h^4 S^2(x_{n-1/2}) \qquad (55)$$

For the Method (55) we have the condition:

$$M^T JM = J + O(h^7)$$

4.3.3 Computational Implementation. – In order to apply the above methods we consider x in the interval $[-R_x, R_x]$ with boundary conditions

$$\Psi(-R_x) = 0, \quad \Psi(R_x) = 0$$

we take a partition of the above interval of length N

$$-R_x = x_{-N}, x_{-N+1}, \ldots, x_{-1}, x_0, x_1, \ldots, x_{N-1}, x_N = R_x$$

then the step size is as before $x_{n+1} - x_n = h_x = R_x/N$.

We denote Ψ^i the k length vector $\Psi(x)$ evaluated at x_i

$$\Psi^i = \Psi(x_i), \text{ for } i = -N + 1, \ldots, N - 1$$

<u>Third Order Method</u>

Applying the third order method (54) to problem (53) we obtain:

$$\Psi^{n+1} - 2\Psi^n + \Psi^{n-1} = \frac{h_x^2}{3} E(\Psi^{n+1} - 8\Psi^n + \Psi^{n-1})$$

$$- \frac{h_x^2}{3}\left(V^{n-\frac{1}{2}}\Psi^{n+1} - 4(V^{n-\frac{1}{2}} + V^{n+\frac{1}{2}})\Psi^n + V^{n+\frac{1}{2}}\Psi^{n-1}\right)$$

$$+ \frac{1}{6}\frac{h_x^2}{h_y^2} M(\Psi^{n+1} - 8\Psi^n + \Psi^{n-1})$$

Let us consider the l length vector

$$\Psi = \left(\Psi^{-N+1}, \Psi^{-N+2}, \ldots, \Psi^0, \ldots, \Psi^{N-2}, \Psi^{N-1}\right)^T$$

Then the above discretization can be written as

$$(P + Eh_x^2 Q)\Psi = 0$$

where

$$P = A - \frac{1}{6}\frac{h_x^2}{h_y^2}CB + \frac{1}{3}h_x^2 V^a, \quad \text{and} \quad Q = -\frac{1}{3}B$$

The matrices A and B are block tridiagonal matrices of size $l \times l$, each block is a diagonal matrix of size $k \times k$. The diagonal blocks of A are $-2I$, while the diagonal blocks of B are $-8I$. The off diagonal blocks for both A and B are the identity matrix I. The block diagonal matrix C has diagonal blocks M. The above matrices are given by:

$$A = \begin{pmatrix} -2I_k & I_k & & \\ I_k & -2I_k & I_k & \\ & \ddots & \ddots & \ddots \\ & & I_k & -2I_k \end{pmatrix}, \quad B = \begin{pmatrix} -8I_k & I_k & & \\ I_k & -8I_k & I_k & \\ & \ddots & \ddots & \ddots \\ & & I_k & -8I_k \end{pmatrix}$$

$$M = \begin{pmatrix} -2 & 1 & & \\ 1 & -2 & 1 & \\ & \ddots & \ddots & \ddots \\ & & 1 & -2 \end{pmatrix}, \quad C = \begin{pmatrix} M & & & \\ & M & & \\ & & \ddots & \\ & & & M \end{pmatrix}$$

Finally, V^a is a block tridiagonal matrix.

$$V^a = \begin{pmatrix} -4(V_{-N+1/2}+V_{-N+3/2}) & V^{-N+1/2} & & & & \\ V_{-N+5/2} & -4(V_{-N+3/2}+V_{-N+5/2}) & V_{-N+3/2} & & & \\ & \ddots & \ddots & \ddots & & \\ & & V_{N-3/2} & -4(V_{N-3/2}+V_{N-5/2}) & V_{N-5/2} \\ & & & V_{N-1/2} & -4(V_{N-1/2}+V_{N-3/2}) \end{pmatrix}$$

where each V_j for $j = -N+1, \ldots, N-1$ is a $k \times k$ diagonal matrix, and V is a $l \times l$ diagonal matrix.

$$V_j = \begin{pmatrix} V(x_j, y_{-N+1}) & & & \\ & V(x_j, y_{-N+2}) & & \\ & & \ddots & \\ & & & V(x_j, y_{N-1}) \end{pmatrix}$$

Fifth Order Method

Applying the third order method (55) to problem (53) we obtain (for simplicity we set $h_x = h_y = h$):

$$\Psi^{n+1} - 2\Psi^n + \Psi^{n-1} = \frac{h^4}{30} E^2 (\Psi^{n+1} - 32\Psi^n + \Psi^{n-1})$$

$$+ \frac{h^2}{30} E M(\Psi^{n+1} - 32\Psi^n + \Psi^{n-1})$$

$$- \frac{h^4}{15} E \left(V_- \Psi^{n+1} - 16(V_- + V_+) \Psi^n + V_+ \Psi^{n-1} \right)$$

$$+ \frac{h^4}{30} \left(V_-^2 \Psi^{n+1} - \left(6V_-^2 + 6V_+^2 + 20V_- V_+ \right) \Psi^n + V_+^2 \Psi^{n-1} \right)$$

$$+\frac{1}{120}M^2\left(\Psi^{n+1}-32\Psi^n+\Psi^{n-1}\right)$$

$$-\frac{h^2}{60}\left(V_-\Psi^{n+1}-(26V_-+6V_+)\Psi^n+V_+\Psi^{n-1}\right)M$$

$$-\frac{h^2}{60}M\left(V_-\Psi^{n+1}-(6V_-+26V_+)\Psi^n+V_+\Psi^{n-1}\right)$$

where

$$V_-=V^{n-\frac{1}{2}} \quad \text{and} \quad V_+=V^{n+\frac{1}{2}}$$

Then the above discretization can be written as

$$(P+Eh^2Q-E^2h^4R)\,\Psi=0$$

where

$$P=A-\frac{1}{6}CB+\frac{1}{120}FD+h^2\left(\frac{1}{3}V^a-\frac{1}{60}V^cC-\frac{1}{60}CV^d\right)+h^4\frac{1}{30}V^e,$$

$$Q=-\frac{1}{3}B+\frac{1}{30}CD-h^2\frac{1}{15}V^b,$$

$$R=-\frac{1}{30}D$$

The matrix D is a block tridiagonal matrix of size $l\times l$, each block is a diagonal matrix of size $k\times k$. The diagonal blocks of D are $-32\,I$, while the off diagonal blocks are the identity matrix I. The block diagonal matrix F has diagonal blocks M^2. Finally, V^a, V^b, V^c, V^d and V^e are block tridiagonal matrices. The above matrices are given by:

$$D=\begin{pmatrix}-32I_k & I_k & & \\ I_k & -32I_k & I_k & \\ & \ddots & \ddots & \ddots \\ & & I_k & -32I_k\end{pmatrix}, \quad F=\begin{pmatrix}M^2 & & & \\ & M^2 & & \\ & & \ddots & \\ & & & M^2\end{pmatrix},$$

$$V_b=\begin{pmatrix}-16(V_-+V_+) & V_- & & \\ V_+ & -16(V_-+V_+) & V_- & \\ & \ddots & \ddots & \ddots \\ & & V_+ & -16(V_-+V_+)\end{pmatrix}$$

$$V_c=\begin{pmatrix}-26V_--6V_+ & V_- & & \\ V_+ & -26V_--6V_+ & V_- & \\ & \ddots & \ddots & \ddots \\ & & V_+ & -26V_--6V_+\end{pmatrix}$$

$$V_d = \begin{pmatrix} -6V_- - 26V_+ & V_- & & \\ V_+ & -6V_- - 26V_+ & V_- & \\ & \ddots & \ddots & \ddots \\ & & V_+ & -6V_- - 26V_+ \end{pmatrix}$$

$$V_e = \begin{pmatrix} -6V_-^2 - 6V_+^2 - 20V_-V_+ & V_-^2 & & \\ V_+^2 & -6V_-^2 - 6V_+^2 - 20V_-V_+ & V_-^2 & \\ & \ddots & \ddots & \ddots \\ & & V_+^2 & -6V_-^2 - 6V_+^2 - 20V_-V_+ \end{pmatrix}$$

4.4 Application of Numerov-type Methods. – Now we consider, again, x in the interval $[-R_x, R_x]$ with boundary conditions

$$\Psi(-R_x) = 0, \ \Psi(R_x) = 0$$

For convenience we consider $R_x = R_y = R$, we also take a partition of the above interval of length N

$$-R_x = x_{-N}, x_{-N+1}, \ldots, x_{-1}, x_0, x_1, \ldots, x_{N-1}, x_N = R_x$$

then the step size is as before $x_{n+1} - x_n = h_x = R/N$.

4.4.1 Numerov's Method. – The well known Numerov's method is

$$\psi_{n+1} - 2\psi_n + \psi_{n-1} = \frac{h^2}{12}(f_{n+1} + 10f_n + f_{n-1})$$

where f is the right hand side function in our case $f(x, \Psi) = -S(x)\Psi$.

We apply the above method to equation (53) and we have:

$$\Psi^{n+1} - 2\Psi^n + \Psi^{n-1} = \frac{h_x^2}{12}\left(-S(x_{n+1})\Psi^{n+1} - 10S(x_n)\Psi^n - S(x_{n-1})\Psi^{n-1}\right)$$

each Ψ^n for $n = -N+1, \ldots, 0, \ldots, N-1$ is the $k = (2N-1)$ length vector $\Psi(x)$ evaluated at x_n and $S(x_n)$ for $n = -N+1, \ldots, 0, \ldots, N-1$ is a $(2N-1) \times (2N-1)$ matrix.

Substitution of $S(x)$ ($S(x) = 2EI - 2V(x) + \frac{1}{h_y^2}M$) gives

$$\Psi^{n+1} - 2\Psi^n + \Psi^{n-1} = -\frac{h_x^2}{12}\left(2EI - 2V_{n+1} + \frac{1}{h_y^2}M\right)\Psi^{n+1}$$

$$+ \frac{10h_x^2}{12}\left(2EI - 2V_n + \frac{1}{h_y^2}M\right)\Psi^n - \frac{h_x^2}{12}\left(2EI - 2V_{n-1} + \frac{1}{h_y^2}M\right)\Psi^{n-1}$$

or

$$\Psi^{n+1} - 2\Psi^n + \Psi^{n-1} = -\frac{h_x^2}{6} E\left(\Psi^{n+1} + 10\Psi^n + \Psi^{n-1}\right)$$
$$+ \frac{h_x^2}{6}\left(V_{n+1}\Psi^{n+1} + 10V_n\Psi^n + V_{n-1}\Psi^{n-1}\right)$$
$$- \frac{1}{12}\frac{h_x^2}{h_y^2} M\left(\Psi^{n+1} + \Psi^n + \Psi^{n-1}\right) \quad (56)$$

Now let the $l = k^2 = (2N-1)^2$ length vector

$$\Psi = \left(\Psi^{-N+1}, \Psi^{-N+2}, \ldots, \Psi^0, \ldots, \Psi^{N-2}, \Psi^{N-1}\right)^T$$

From (56) the method can be written in matrix form as

$$A\Psi = -\frac{1}{6}h_x^2 EB\Psi + \frac{1}{6}h_x^2 BV\Psi - \frac{1}{12}\frac{h_x^2}{h_y^2}CB\Psi$$

or in the more general form as

$$(P + Eh_x^2 Q)\Psi = 0$$

where

$$P = A - \frac{1}{6}h_x^2 BV + \frac{1}{12}\frac{h_x^2}{h_y^2}CB, \quad \text{and} \quad Q = \frac{1}{6}B$$

4.4.2 Numerov's Method with Minimal Phase-lag. – We also applied a modification of Numerov's method with minimum phase-lag, developed by Chawla and Rao.[27] It is known that the classical Numerov method has phase lag $h^4/480$, the Chawla and Rao method has phase lag $h^6/12096$. The method is

$$\hat{y}_n = y_n - \alpha h^2(f_{n+1} - 2f_n + f_{n-1})$$
$$y_{n+1} - 2y_n + y_{n-1} = \frac{h^2}{12}(f_{n+1} + 10\hat{f}_n + f_{n-1})$$

with $\alpha = 1/200$.

We apply the method to equation (53)

$$\Psi^{n+1} - 2\Psi^n + \Psi^{n-1} = -\frac{h_x^2}{12}\left(S(x_{n+1})\Psi^{n+1} + 10S(x_n)\Psi^n + S(x_{n-1})\Psi^{n-1}\right)$$
$$- \alpha\frac{10h_x^4}{12}S(x_n)\left(S(x_{n+1})\Psi^{n+1} - 2S(x_n)\Psi^n + S(x_{n-1})\Psi^{n-1}\right)$$

Substitution of $S(x)$ ($S(x) = 2EI - 2V(x) + \frac{1}{h_y^2} M$) gives the following generalized eigenvalue problem

$$\Psi^{n+1} - 2\Psi^n + \Psi^{n-1} = -\alpha \frac{10 h_x^4}{3} E^2 (\Psi^{n+1} - 2\Psi^n + \Psi^{n-1})$$

$$-\frac{h_x^2}{6} E\left(\Psi^{n+1} + 10\Psi^n + \Psi^{n-1}\right) + \alpha \frac{10 h_x^4}{3} EV_n (\Psi^{n+1} - 2\Psi^n + \Psi^{n-1})$$

$$-\alpha \frac{10}{3} \frac{h_x^4}{h_y^2} EM(\Psi^{n+1} - 2\Psi^n + \Psi^{n-1})$$

$$+\alpha \frac{10 h_x^4}{3} E(V_{n+1} \Psi^{n+1} - 2V_n \Psi^n + V_{n-1} \Psi^{n-1})$$

$$+\frac{h_x^2}{6} \left(V_{n+1} \Psi^{n+1} + 10 V_n \Psi^n + V_{n-1} \Psi^{n-1}\right) - \frac{1}{12} \frac{h_x^2}{h_y^2} M\left(\Psi^{n+1} + \Psi^n + \Psi^{n-1}\right)$$

$$-\alpha \frac{10 h_x^4}{3} V_n (V_{n+1} \Psi^{n+1} - 2V_n \Psi^n + V_{n-1} \Psi^{n-1}) - \alpha \frac{10 h_x^4}{12 h_y^4} M^2 (\Psi^{n+1} - 2\Psi^n + \Psi^{n-1})$$

$$+\alpha \frac{20}{12} \frac{h_x^4}{h_y^2} V_n M (\Psi^{n+1} - 2\Psi^n + \Psi^{n-1})$$

$$+\alpha \frac{20}{12} \frac{h_x^4}{h_y^2} M(V_{n+1} \Psi^{n+1} - 2V_n \Psi^n + V_{n-1} \Psi^{n-1})$$

Again, this can be written in matrix form

$$A\Psi = -\alpha \frac{10 h_x^4}{3} E^2 \Psi + h_x^2 \left(\frac{1}{6} B + \alpha \frac{10}{3} \frac{h_x^2}{h_y^2} CA - \alpha h_x^2 \frac{10}{3}(AV + VA)\right) E\Psi$$

$$+ \left(-\frac{1}{6} h_x^2 BV + \frac{1}{12} \frac{h_x^2}{h_y^2} CB + \alpha \frac{5}{6} \frac{h_x^4}{h_y^2} \left(\frac{h_x^2}{h_y^2} DA - 2VCA - 2CAV\right) + \alpha h_x^4 \frac{10}{3} VAV\right) \Psi,$$

or

$$(P + E h^2 Q - E^2 h^4 R) \Psi = 0$$

where

$$P = A - \frac{1}{6} h^2 BV + \frac{1}{12} CB + \alpha h^2 \frac{5}{6} (DA - 2VCA - 2CAV) + \alpha h^4 \frac{10}{3} VAV,$$

$$Q = \frac{1}{6} B + \alpha \frac{10}{3} CA - \alpha h^2 \frac{10}{3} (AV + VA), \quad \text{and} \quad R = -\alpha \frac{10}{3} A$$

and D is a block diagonal matrix with each block equal to M^2.

Matrices P, Q, R are real, symmetric and sparse, they are very large even for small N (e.g. $l = 1521$ for $N = 20$). In order to manage to work as we increase

N we treat them as sparse matrices in terms of storage and computational work.

We note here that the matrices A and B block tridiagonal with dimension $l \times l$ of the form:

$$A = \begin{pmatrix} -2I_k & I_k & & \\ I_k & -2I_k & I_k & \\ & \ddots & \ddots & \ddots \\ & & I_k & -2I_k \end{pmatrix}, \quad B = \begin{pmatrix} 10I_k & I_k & & \\ I_k & 10I_k & I_k & \\ & \ddots & \ddots & \ddots \\ & & I_k & 10I_k \end{pmatrix}$$

The matrix M is tridiagonal of dimension $k \times k$, and the matrices C and D are block diagonal $l \times l$ matrices of the form:

$$M = \begin{pmatrix} -2 & 1 & & \\ 1 & -2 & 1 & \\ & \ddots & \ddots & \ddots \\ & & 1 & -2 \end{pmatrix}, \quad C = \begin{pmatrix} M & & & \\ & M & & \\ & & \ddots & \\ & & & M \end{pmatrix}, \quad D = \begin{pmatrix} M^2 & & & \\ & M^2 & & \\ & & \ddots & \\ & & & M^2 \end{pmatrix}$$

We also note that each V_j for $j = -N+1, \ldots, N-1$ is a $k \times k$ diagonal matrix, and V is a $l \times l$ diagonal matrix.

$$V_j = \begin{pmatrix} V(x_j, y_{-N+1}) & & & \\ & V(x_j, y_{-N+2}) & & \\ & & \ddots & \\ & & & V(x_j, y_{N-1}) \end{pmatrix}, \quad V = \begin{pmatrix} V_{-N+1} & & & \\ & V_{-N+2} & & \\ & & \ddots & \\ & & & V_{N-1} \end{pmatrix}$$

4.4.3 Numerical Results. – We applied both numerical methods developed above to the calculation of the eigenvalues of the two-dimensional harmonic oscillator and the two-dimensional Henon-Heiles potential. Results are compared with those produced using the full discretisation technique.

4.4.3.1 Two-dimensional harmonic oscillator. – The potential of the two-dimensional harmonic oscillator is

$$V(x, y) = \frac{1}{2}(x^2 + y^2)$$

The exact eigenvalues are given by

$$E_n = n + 1, \quad n = n_x + n_y, \quad n_x, n_y = 0, 1, 2, \ldots$$

We have two cases of application:

7: Numerical Methods in Chemistry

- Symplectic Integrators
- Finite Difference Methods

In the first case we use the new developed second order asymptotically symplectic integrator and the fifth order asymptotically symplectic integrator.

In the second case we use for comparison purposes the methods:

1. five point formula (full discretisation)
2. Numerov's method described in paragraph 4.4.1
3. Numerov-type method with minimal phase-lag described in paragraph 4.4.2

In Figures 8 and 9 we present the error:

$$\text{Error} = \log_{10}(E_{\text{calculated}} - E_{\text{analytical}})$$

for the two cases mentioned above.

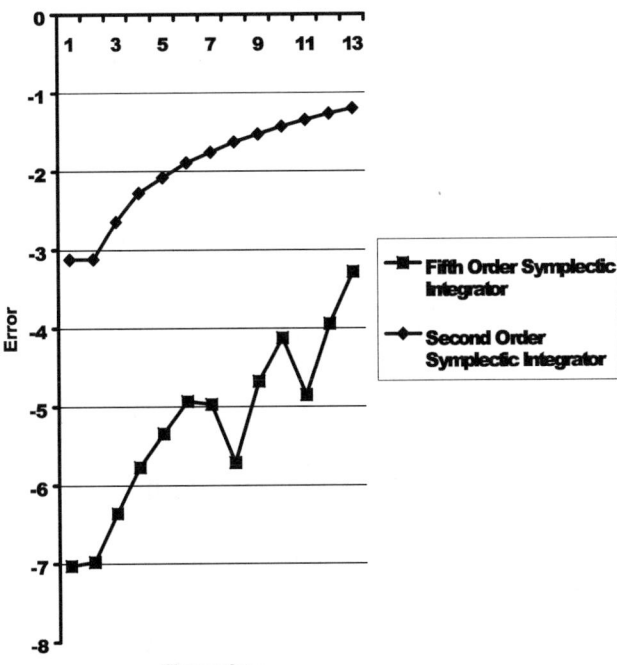

Figure 8 *Values of Error for several values of eigenenergy for the third order and fifth order symplectic integrator*

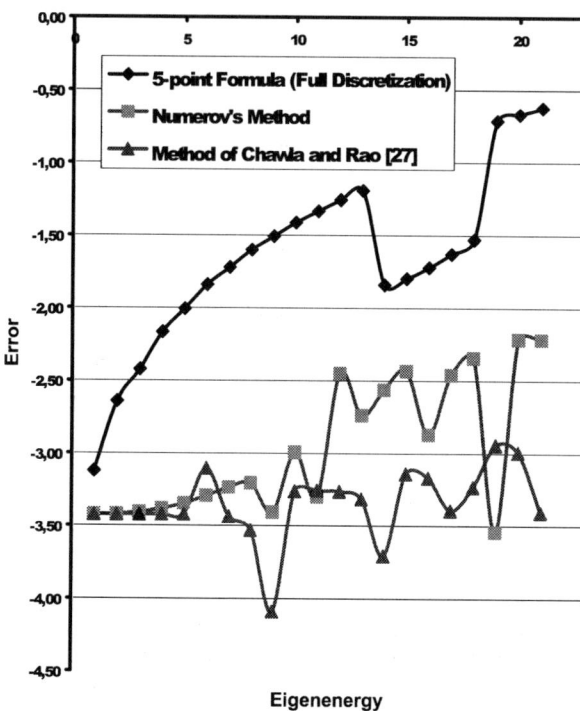

Figure 9 *Values of Error for several values of eigenenergy for the methods: (1) 5-points formula (full discretization), (2) Numerov's method and (3) Numerov-type method with minimal phase-lag*

4.4.3.2 Two-dimensional Henon-Heiles potential. – The Henon-Heiles potential is

$$V(x,y) = \frac{1}{2}(x^2 + y^2) + (0.0125)^{1/2}\left(x^2 y - \frac{y^3}{3}\right)$$

In the next table we give results using Numerov method for different values of N, as well as the eigenvalues given by Davis and Heller.[3]

The results are similar we those presented in the figures 8 ans 9.

4.5 Remarks and Conclusions. – From the results presented in Figures 8–9 we have the following remarks:

- The fifth order asymptotically symplectic integrator is much more efficient than the second order asymptotically symplectic integrator.
- The Numerov's method is more efficient than the five-point formula (full discretisation).

7: *Numerical Methods in Chemistry*

- The Numerov-type method with minimal phase-lag described in the paragraph 4.4.2 is much more efficient than the five-point formula and generally more efficient than the Numerov's method.
- The second order asymptotically symplectic integrator generally has the same behaviour with the five-point formula
- The fifth order asymptotically symplectic integrator is generally more efficient than the Numerov-type method with minimal phase-lag.

In the feature review we will present Matlab programmes for the two dimensional Schrödinger equation. In the feature review we will study the problem of the time dependent Schrödinger equation.

5 General Comments on the Bibliography of the Numerical Methods in Chemistry

Simos[28–29] has developed two two-stage exponentially-fitted and trigonometrically-fitted symmetric multistep methods for which the problems of numerical instabilities and resonances have been solved. We note here that the linear symmetric multistep methods are much simpler than the hybrid (Runge–Kutta type) ones. For long time integration of initial-value problems with oscillating solutions these methods are very important, because of the factors of simplicity and accuracy (especially orbital problems).

In ref. 30 Psihoyios and Simos have constructed a special form two-stage trigonometrically-fitted and exponentially-fitted symmetric multistep method and they also discussed in some detail its stability characteristics. This special form behaves better than the corresponding methods especially in orbital problems.

In ref. 31 Simos developed a family of trigonometrically-fitted symmetric ten-step methods for the efficient solution of the Schrödinger equation and related problems. He described the construction and the stability of the new methods. Numerical results obtained for the resonance problem of the radial Schrödinger equation are also presented. For comparison purposes well know methods from literature are used.

In ref. 32 Psihoyios ans Simos have studied exponentially-fitted and trigonometrically-fitted predictor-corrector methods of the class of explicit advanced step point (EAS) methods. The new methods have been applied to several well know problems with oscillating solutions and the results obtained show the efficiency of the newly introduced methods compared with the classical EAS methods, Adams-Bashforth predictor-corrector methods and the well known Runge-Kutta Dormand-Prince methods.

In ref. 33 Simos and Aguiar introduced a non-symmetric (dissipative) explicit two-step hybrid exponentially-fitted method of algebraic order eight. Dissipative methods are useful in the literature since they have much more free parameters for definition. As a result of the above is the production of "*economic*" methods, i.e. of methods which use fewer function evaluations per step than the corresponding symmetric methods. Another result of the above is the construction of "*accurate*" methods since one can optimize the algebraic error (because there are more free parameters). This method has the following characteristics:

- It is more accurate than the exponentially-fitted method of Raptis and Cash[47] (which is an exponentially-fitted method of the same kind) since it integrates more algebraic polynomials than the method of Raptis and Cash[47] and
- It is more efficient also than the classical dissipative two-step method since it integrates exponential and trigonometric functions.
- It is also simpler than the method of Cash and Raptis[47] and the exponentially-fitted method (Case I) of Thomas, Simos and Mitsou[48] since it is explicit (while the methods of ref. 47–48 are implicit and in order to calculate a new approximation of the solution an equation must be solved).

The method developed in ref. 33 has been applied to *the resonance problem* (which arises from the one-dimensional Schrödinger equation) with two different types of potential and to the coupled differential equations arising from the Schrödinger equation. For the coupled differential equations a variable-step method is produced based also on Raptis and Cash[47] method. The above applications indicate the efficiency of the new approach.

In ref. 34 Simos has developed a dissipative sixth order trigonometrically-fitted method. The method has the following characteristics:

1. It requires only four function evaluations per step
2. Is more accurate than the classical dissipative sixth algebraic order method of Papageorgiou et al.[49] for problems with solution with oscillating behaviour since it integrates trigonometric functions.
3. Is simpler than other sixth order symmetric multistep methods in the literature.

In ref. 36 Konguetsof and Simos developed an exponentially-fitted and trigonometrically method. More specifically they have considered the method:

$$y_{n+4} - y_{n+3} - y_{n-3} + y_{n-4} =$$
$$= h^2 \left[d_3 (f_{n+3} + f_{n-3}) + d_2 (f_{n+2} + f_{n-2}) + d_1 (f_{n+1} + f_{n-1}) + d_0 f_n \right]$$

and they required the above method to be exactly for any linear combination of the functions:

$$\{1, x, x^2, \cdots, x^7, \exp(\pm wx)\}$$

or

$$\{1, x, x^2, \cdots, x^7, \cos(\pm wx), \sin(\pm wx)\}$$

For the above method detailed stability analysis is given. Numerical results show the efficiency of the developed method.

In ref. 37 Simos and Aguiar have developed a symplectic integrator which has the following properties:

1. It is of second algebraic order
2. It integrates exactly exponential or trigonometric functions. So, is more accurate than the classical methods.†
3. It has the symplecticness property

The numerical results show that this new method has much better behaviour than the classical symplectic integrator developed by Calvo and Sanz-Serna.[50]

In ref. 38 Aguiar and Simos considered the following family of twelve step symmetric methods:

$$a_6(y_{n+6}+y_{n-6})+a_5(y_{n+5}+y_{n-5})+a_4(y_{n+4}+y_{n-4})$$
$$+a_3(y_{n+3}+y_{n-3})+a_2(y_{n+2}+y_{n-2})+a_1(y_{n+1}+y_{n-1})=$$
$$h^2\big[b_6(f_{n+6}+f_{n-6})+b_5(f_{n+5}+f_{n-5})+b_4(f_{n+4}+f_{n-4})$$
$$+b_3(f_{n+3}+f_{n-3})+b_2(f_{n+2}+f_{n-2})+b_1(f_{n+1}+f_{n-1})+b_0 f_n\big]$$

They considered two cases:

Case I $b_6 = 0$ (Explicit case)

In this case the method is constructed such that to integrate exactly any linear combination of the functions:

$$\{1, x, x^2, \cdots, x^{11}, \cos(\pm wx), \sin(\pm wx)\}$$

and the local truncation error of the method is given by:

$$LTE = C_{14} h^{14} (w^2 y_n^{(14)} + y_n^{(16)}) + O(h^{16})$$

Case II $b_6 \neq 0$ (Implicit case)

In this case the method is constructed such that to integrate exactly any linear combination of the functions:

$$\{1, x, x^2, \cdots, x^{13}, \cos(\pm wx), \sin(\pm wx)\}$$

† We call classical the corresponding method with constant coefficients.

and the local truncation error of the method is given by:

$$LTE = C_{16}h^{16}(w^2 y_n^{(14)} + y_n^{(16)}) + O(h^{18})$$

For comparison purposes in their numerical illustration they have used the following methods:

1. the well-known Numerov's method,
2. the exponentially fitted method of Raptis and Allison,[51]
3. the exponentially-fitted method of Ixaru and Rizea,[52]
4. the exponentially-fitted method of Raptis,[53]
5. the classical Cowell method of order 8 mentioned in Henrici,[54]
6. the Cowell method of fourth algebraic order which integrates exactly functions of the form $\{1, x, x^2, x^3, \cos(\pm wx), \sin(\pm wx)\}$, which has been developed by Stiefel and Bettis,[55]
7. the exponentially-fitted Cowell method of fourth algebraic order which integrates exactly functions of the form $\{1, x, \cos(\pm wx), \sin(\pm wx), x\cos(\pm wx), x\sin(\pm wx)\}$, which has been developed by Stiefel and Bettis,[55]
8. the exponentially-fitted Cowell method of sixth algebraic order which integrates exactly functions of the form $\{1, x, x^2, x^3, x^4, x^5, \cos(\pm wx), \sin(\pm wx)\}$, which has been developed by Stiefel and Bettis,[55]
9. the exponentially-fitted Cowell method of sixth algebraic order which integrates exactly functions of the form $\{1, x, x^2, x^3, \cos(\pm wx), \sin(\pm wx), x\cos(\pm wx), x\sin(\pm wx)\}$, which has been developed by Stiefel and Bettis,[55]
10. the exponentially-fitted Cowell method of sixth algebraic order which integrates exactly functions of the form $\{1, x, \cos(\pm wx), \sin(\pm wx), x\cos(\pm wx), x\sin(\pm wx), x^2\cos(\pm wx), x^2\sin(\pm wx)\}$, which has been developed by Stiefel and Bettis[55] and
11. the new exponentially-fitted symmetric twelve step method of algebraic order twelve.

The results show the efficiency of the new developed method.

We can produce more efficient methods with the following way: Consider the method:

$$y_{n+6} + y_{n-6} - 2(y_{n+5} + y_{n-5}) + 2(y_{n+4} + y_{n-4})$$
$$-2(y_{n+3} + y_{n-3}) + 2(y_{n+2} + y_{n-2}) - 2(y_{n+1} + y_{n-1}) + 2y_n =$$
$$h^2 \left[b_5(f_{n+5} + f_{n-5}) + b_4(f_{n+4} + f_{n-4}) + b_3(f_{n+3} + f_{n-3}) \right.$$
$$\left. + b_2(f_{n+2} + f_{n-2}) + b_1(f_{n+1} + f_{n-1}) + b_0 f_n \right]$$

We require the above method to be accurate for any linear combination of the functions which are shown in Table 4.

7: Numerical Methods in Chemistry

Table 4 *Exponential fitted Multistep Twelfth Algebraic Order Methods*

CASES	Set of functions for which the method is accurate
I	$\{1, x, x^2, \cdots, x^9, \cos(\pm wx), \sin(\pm wx), x\cos(\pm wx), x\sin(\pm wx)\}$
II	$\{1, x, x^2, \cdots, x^7, \cos(\pm wx), \sin(\pm wx),$ $x\cos(\pm wx), x\sin(\pm wx), x^2\cos(\pm wx), x^2\sin(\pm wx)\}$
III	$\{1, x, x^2, \cdots, x^5, \cos(\pm wx), \sin(\pm wx), x\cos(\pm wx), x\sin(\pm wx),$ $x^2\cos(\pm wx), x^2\sin(\pm wx), x^3\cos(\pm wx), x^3\sin(\pm wx)\}$
IV	$\{1, x, x^2, x^3, \cos(\pm wx), \sin(\pm wx), x\cos(\pm wx), x\sin(\pm wx),$ $x^2\cos(\pm wx), x^2\sin(\pm wx), x^3\cos(\pm wx), x^3\sin(\pm wx),$ $x^4\cos(\pm wx), x^4\sin(\pm wx)\}$

CASE I

In order the method to be accurate for the set of functions determined by Case I of Table 4 the following system of equations must hold:

$$12 = 4b_5 + 4b_4 + 4b_3 + 4b_2 + 4b_1 + 2b_0$$
$$852 = 600b_5 + 384b_4 + 216b_3 + 96b_2 + 24b_1$$
$$44532 = 37500b_5 + 15360b_4 + 4860b_3 + 960b_2 + 60b_1$$
$$2033652 = 1750000b_5 + 458752b_4 + 81648b_3$$
$$+ 7168b_2 + 112b_1$$

$$-4\cos(5w) + 4\cos(4w) - 4\cos(3w) + 4\cos(2w) + 2\cos(6w) - 4\cos(w) + 2 =$$
$$= -w^2(2b_5\cos(5w) + 2b_4\cos(4w) + 2b_3\cos(3w) + 2b_2\cos(2w) + 2b_1\cos(w) + b_0)$$

$$-4(\sin(w) - 2\sin(2w) - 3\sin(6w) - 4\sin(4w) + 5\sin(5w) + 3\sin(3w))$$
$$= -2w(b_1 w \sin(w) + 3b_3 w \sin(3w) - b_0 + 5b_5 w \sin(5w) + 4b_4 w \sin(4w)$$
$$+ 2b_2 w \sin(2w) - 2b_4 \cos(4w) - 2b_2 \cos(2w)$$
$$- 2b_5 \cos(5w) - 2b_3 \cos(3w) - 2b_1 \cos(w))$$

Solving the above system of equations we obtain the coefficients of the new proposed method.

The local truncation error of the new method is given by:

$$LTE = \frac{h^{14}}{435891456000}(-642378126767 y_n^{(14)} + 48933158186 w^2 y_n^{(12)} + 24466579093 w^4 y_n^{(10)})$$

CASE II

In order the method to be accurate for the set of functions determined by Case II of Table 4 the following system of equations must hold:

$$12 = 4b_5 + 4b_4 + 4b_3 + 4b_2 + 4b_1 + 2b_0$$

$$852 = 600b_5 + 384b_4 + 216b_3 + 96b_2 + 24b_1$$

$$44532 = 37500b_5 + 15360b_4 + 4860b_3 + 960b_2 + 60b_1$$

$$-4\cos(5w) + 4\cos(4w) - 4\cos(3w) + 4\cos(2w) + 2\cos(6w) - 4\cos(w) + 2 =$$
$$= -w^2(2b_5\cos(5w) + 2b_4\cos(4w) + 2b_3\cos(3w) + 2b_2\cos(2w) + 2b_1\cos(w) + b_0)$$

$$-4(\sin(w) - 2\sin(2w) - 3\sin(6w) - 4\sin(4w) + 5\sin(5w) + 3\sin(3w))$$
$$= -2w(b_1w\sin(w) + 3b_3w\sin(3w) - b_0 + 5b_5w\sin(5w)$$
$$+ 4b_4w\sin(4w) + 2b_2w\sin(2w) - 2b_4\cos(4w) - 2b_2\cos(2w)$$
$$- 2b_5\cos(5w) - 2b_3\cos(3w) - 2b_1\cos(w))$$

$$64\cos(4w) - 100\cos(5w) - 36\cos(3w) + 16\cos(2w) + 72\cos(6w) - 4\cos(w)$$
$$= 2b_0 - 32b_4w\sin(4w) - 16b_2w\sin(2w) - 24b_3w\sin(3w) - 40b_5w\sin(5w) - 8b_1w\sin(w)$$
$$- 2b_1w^2\cos(w) - 32b_4w^2\cos(4w) - 8b_3w^2\cos(3w) - 50b_5w^2\cos(5w) + 4b_2\cos(2w)$$
$$+ 4b_4\cos(4w) + 4b_3\cos(3w) + 4b_1\cos(w) + 4b_5\cos(5w) - 8b_2w^2\cos(2w)$$

Solving the above system of equations we obtain the coefficients of the new proposed method.

The local truncation error of the new method is given by:

$$LTE = \frac{h^{14}}{435891456000}(-642378126767\,y_n^{(14)} + 73399737279w^2 y_n^{(12)}$$
$$+ 73399737279w^4 y_n^{(10)} + 24466579093w^6 y_n^{(8)})$$

CASE III

In order the method to be accurate for the set of functions determined by Case III of Table 4 the following system of equations must hold:

$$12 = 4b_5 + 4b_4 + 4b_3 + 4b_2 + 4b_1 + 2b_0$$

$$852 = 600b_5 + 384b_4 + 216b_3 + 96b_2 + 24b_1$$

$$-4\cos(5w) + 4\cos(4w) - 4\cos(3w) + 4\cos(2w) + 2\cos(6w) - 4\cos(w) + 2 =$$
$$= -w^2(2b_5\cos(5w) + 2b_4\cos(4w) + 2b_3\cos(3w) + 2b_2\cos(2w) + 2b_1\cos(w) + b_0)$$

$$-4(\sin(w) - 2\sin(2w) - 3\sin(6w) - 4\sin(4w) + 5\sin(5w) + 3\sin(3w))$$
$$= -2w(b_1 w \sin(w) + 3b_3 w \sin(3w) - b_0 + 5b_5 w \sin(5w)$$
$$+ 4b_4 w \sin(4w) + 2b_2 w \sin(2w) - 2b_4 \cos(4w) - 2b_2 \cos(2w)$$
$$- 2b_5 \cos(5w) - 2b_3 \cos(3w) - 2b_1 \cos(w))$$

$$64\cos(4w) - 100\cos(5w) - 36\cos(3w) + 16\cos(2w) + 72\cos(6w) - 4\cos(w)$$
$$= 2b_0 - 32b_4 w \sin(4w) - 16b_2 w \sin(2w) - 24b_3 w \sin(3w) - 40b_5 w \sin(5w) - 8b_1 w \sin(w)$$
$$- 2b_1 w^2 \cos(w) - 32b_4 w^2 \cos(4w) - 8b_3 w^2 \cos(3w) - 50b_5 w^2 \cos(5w) + 4b_2 \cos(2w)$$
$$+ 4b_4 \cos(4w) + 4b_3 \cos(3w) + 4b_1 \cos(w) + 4b_5 \cos(5w) - 8b_2 w^2 \cos(2w)$$

$$432\sin(6w) - 4\sin(w) + 32\sin(2w) - 108\sin(3w) + 256\sin(4w) - 500\sin(5w)$$
$$= 24b_2 \sin(2w) + 48b_4 \sin(4w) + 12b_1 \sin(w) + 60\sin(5w)b_5 + 36b_3 \sin(3w)$$
$$+ 12b_1 w \cos(w) + 300b_5 w \cos(5w) - 128b_4 w^2 \sin(4w) - 250b_5 w^2 \sin(5w)$$
$$- 54b_3 w^2 \sin(3w) + 48b_2 w \cos(2w) + 108b_3 w \cos(3w) + 192b_4 w \cos(4w)$$
$$- 16b_2 w^2 \sin(2w) - 2b_1 w^2 \sin(w)$$

Solving the above system of equations we obtain the coefficients of the new proposed method.

The local truncation error of the new method is given by:

$$LTE = \frac{h^{14}}{435891456000}(-642378126767 y_n^{(14)} + 97866316372 w^2 y_n^{(12)}$$
$$+ 146799474558 w^4 y_n^{(10)} + 97866316372 w^6 y_n^{(8)} + 24466579093 w^8 y_n^{(6)})$$

CASE IV

In order the method to be accurate for the set of functions determined by Case IV of Table 4 the following system of equations must hold:

$$12 = 4b_5 + 4b_4 + 4b_3 + 4b_2 + 4b_1 + 2b_0$$

$$-4\cos(5w) + 4\cos(4w) - 4\cos(3w) + 4\cos(2w) + 2\cos(6w) - 4\cos(w) + 2 =$$
$$= -w^2(2b_5 \cos(5w) + 2b_4 \cos(4w) + 2b_3 \cos(3w) + 2b_2 \cos(2w) + 2b_1 \cos(w) + b_0)$$

$$-4(\sin(w) - 2\sin(2w) - 3\sin(6w) - 4\sin(4w) + 5\sin(5w) + 3\sin(3w))$$
$$= -2w(b_1 w \sin(w) + 3b_3 w \sin(3w) - b_0 + 5b_5 w \sin(5w) + 4b_4 w \sin(4w)$$
$$+ 2b_2 w \sin(2w) - 2b_4 \cos(4w) - 2b_2 \cos(2w)$$
$$- 2b_5 \cos(5w) - 2b_3 \cos(3w) - 2b_1 \cos(w))$$

$64\cos(4w) - 100\cos(5w) - 36\cos(3w) + 16\cos(2w) + 72\cos(6w) - 4\cos(w) = 2b_0$
$- 32b_4w\sin(4w) - 16b_2w\sin(2w) - 24b_3w\sin(3w) - 40b_5w\sin(5w) - 8b_1w\sin(w)$
$- 2b_1w^2\cos(w) - 32b_4w^2\cos(4w) - 8b_3w^2\cos(3w) - 50b_5w^2\cos(5w) + 4b_2\cos(2w)$
$+ 4b_4\cos(4w) + 4b_3\cos(3w) + 4b_1\cos(w) + 4b_5\cos(5w) - 8b_2w^2\cos(2w)$

$432\sin(6w) - 4\sin(w) + 32\sin(2w) - 108\sin(3w) + 256\sin(4w)$
$- 500\sin(5w) = 24b_2\sin(2w) + 48b_4\sin(4w) + 12b_1\sin(w) + 60\sin(5w)b_5$
$+ 36b_3\sin(3w) + 12b_1w\cos(w) + 300b_5w\cos(5w) - 128b_4w^2\sin(4w)$
$- 250b_5w^2\sin(5w) - 54b_3w^2\sin(3w) + 48b_2w\cos(2w) + 108b_3w\cos(3w)$
$+ 192b_4w\cos(4w) - 16b_2w^2\sin(2w) - 2b_1w^2\sin(w)$

$- 2500\cos(5w) - 324\cos(3w) + 64\cos(2w) + 1024\cos(4w) + 2592\cos(6w)$
$- 4\cos(w) = - 128b_2w\sin(2w) - 432b_3w\sin(3w) - 1024b_4w\sin(4w)$
$- 2000b_5w\sin(5w) - 16b_1w\sin(w) - 2b_1w^2\cos(w) - 32b_2w^2\cos(2w)$
$- 512b_4w^2\cos(4w) - 162b_3w^2\cos(3w) - 1250b_5w^2\cos(5w) + 216b_3\cos(3w)$
$+ 96b_2\cos(2w) + 600b_5\cos(5w) + 24b_1\cos(w) + 384b_4\cos(4w)$

Solving the above system of equations we obtain the coefficients of the new proposed method.

The local truncation error of the new method is given by:

$$LTE = \frac{h^{14}}{435891456000}(-642378126767 y_n^{(14)} + 122332895465 w^2 y_n^{(12)} + 244665790930 w^4 y_n^{(10)}$$
$$+ 244665790930 w^6 y_n^{(8)} + 122332895465 w^8 y_n^{(6)} + 24466579093 w^{10} y_n^{(4)})$$

We have applied the optimization procedure developed above (Cases I–IV) in the resonance problem of the Schrödinger equation (see section 2.5). For comparison purposes we have used, also, the method developed in ref. 38 which was the optimal methods over the eleven described above. In Figure 10 we present the Error for the computation of $E_3 = 989.701916$ for several values of n where:

$$\text{Error} = -\log_{10}(E_{\text{calculated}} - E_{\text{analytical}})$$

and n is an integer which is determined in the selection of stepsize $h = \frac{1}{2^n}$.

In ref. 56 Ixaru et al. have developed the exponential fitting extension of the classical variable step-size two-step BDF algorithm. They have investigated procedures for the determination of the optimal value of frequency and they have introduced algorithms for selection of the optimal stepsizes for the specific accuracy.

In ref. 57 Ixaru et al. have studied the linear multistep algorithms for first order Ordinary Differential Equations and they have investigated the problem

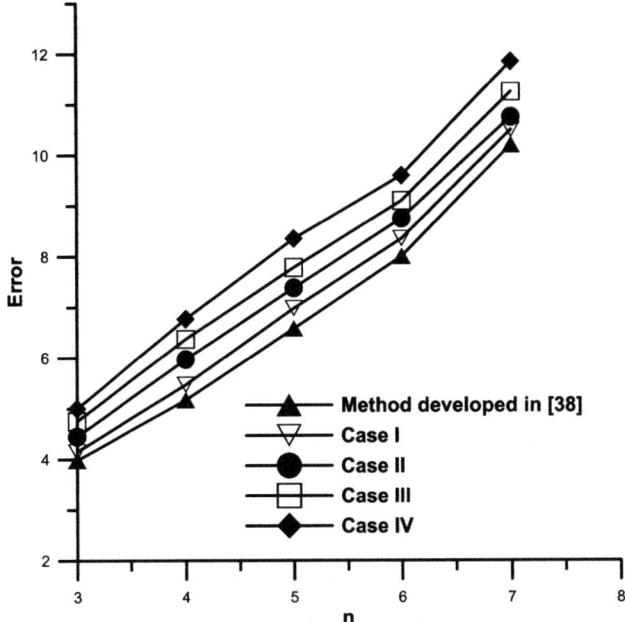

Figure 10 *Values of Error for several values of n for the resonance $E_3 = 989.701916$*

of the determination of the optimal frequency. As in other studies has been mentioned, the optimal frequency is dependent on the behaviour of the error.

We note here that the methods developed in ref. 56–57 and by Psihopyios and Simos[30,32] will be investigated further in the next review.

Appendix A

Analysis of Multiderivative Methods

> restart;
>
> 'Taylor Series Expansions of the Terms y_{n \pm i} and their derivatives';
>
> ynp1:=convert(taylor(y(x+h),h=0,13),polynom);
> ynm1:=convert(taylor(y(x−h),h=0,13),polynom);
> yn:=y(x);
> dnp1:=convert(taylor(diff(y(x+h),x$2),h=0,13),polynom);
> dnm1:=convert(taylor(diff(y(x−h),x$2),h=0,13),polynom);
> dn:='@@'(D,2)(y)(x);
> yn:=y(x);
> d1np1:=convert(taylor(diff(y(x+h),x$1),h=0,13),polynom);
> d1nm1:=convert(taylor(diff(y(x−h),x$1),h=0,13),polynom);

```
> d1n: = D(y)(x);
> d4n: = '@@'(D,4)(y)(x);
> d4np1: = convert(taylor(diff(y(x + h),x$4),h = 0,13),polynom);
> d4nm1: = convert(taylor(diff(y(x − h),x$4),h = 0,13),polynom);
>
> 'Error Analysis for the Layers and for the Multiderivative Method';
>
>
> a[0]: = 1;
> a[1]: = 1/12;
> c[0]: = 115/126;
> c[1]: = 11/252;
> c[2]: = 313/7560;
> c[3]: = −13/15120;
>
> 'Layers';
>
> err1: = simplify(ynp1 − 2*yn + ynm1 − h^2*dn);
> err2: = simplify(b[1]*h^2*(dnp1 − 2*dn + dnm1));
> err3: = simplify(b[2]*h^2*(dnp1 − 2*(dn + err2) + dnm1));
> err4: = simplify(b[3]*h^2*(dnp1 − 2*(dn + err3) + dnm1));
> err5: = simplify(ynp1 − 2*yn + ynm1 − a[0]*h^2*(dn + err4) − a[1]*h^4*
> (d4n + err4));
>
> 'Final Method';
>
> errfinal: = simplify(ynp1 − 2*yn + ynm1 −
> h^2*(c[0]*dn + c[1]*(dnp1 + err5 + dnm1)) −
> h^4*(c[2]*d4n + c[3]*(d4np1 + err5 + d4nm1)));
>
> 'Local Truncation Error of the Method';
>
> err: = h^8*coeff(errfinal,h,8);
>
> restart;
>
> 'Stability and Phase-Lag Analysis of the Methods';
>
> a[0]: = 1;
> a[1]: = 1/12;
> c[0]: = 115/126;
> c[1]: = 11/252;
> c[2]: = 313/7560;
> c[3]: = −13/15120;
>
```

7: Numerical Methods in Chemistry

```
> kpnp1: = 2*kn − knm1 − H^2*kn;
> kpn1: = kn + b[1]*H^2*(kpnp1 − 2*kn + knm1);
> kpn2: = kn + b[2]*H^2*(kpnp1 − 2*kpn1 + knm1);
> kpn3: = kn + b[3]*H^2*(kpnp1 − 2*kpn2 + knm1);
> kppnp1: = 2*kn − knm1 − a[0]*H^2*kpn3 + a[1]*H^4*kpn3;
>
> stab: = simplify(knp1 − 2*kn + knm1 + H^2*(c[0]*kn + c[1]*
> (kppnp1 + knm1)) − H^4*(c[2]*kn + c[3]*(kppnp1 + knm1)));
>
> BH: = simplify(coeff(stab,kn)/(−2));
>
> 'Phase-Lag Analysis';
>
> pl: = simplify(convert(taylor(cos(H),H = 0,20),polynom)-BH);
>
>
> eq1: = coeff(pl,H,8);
> eq2: = coeff(pl,H,10);
> eq3: = coeff(pl,H,12);
>
> solut: = solve({eq1,eq2,eq3},{b[1],b[2],b[3]});
>
> assign(solut);
>
> 'Final Phase-Lag';
> pl;
>
> 'Stability and Phase-Lag- Case I';
>
> b[3]: = −1/360;
> b[2]: = 0;
> b[1]: = b[2];
> pl;

> BH;
>
> plot([1 + BH,1 − BH],H = 0..3,axes = NORMAL,title = "Stability Functions
> for the Case I",legend = ["1 + B(H)", "1 − B(H)"]);
>
> BH: = 1 − 1/2*H^2 + 1/24*H^4 + 1/40320*H^8 − 1/720*H^6 − 1/259200*
> H^10 − 13/130636800*H^12;

> fsolve(1 + BH);
> 2.709005917*2.709005917;
>
> 'Stability and Phase-Lag- Case II';
```

```
>
> b[3]:=-1/360;
> b[2]:=-13/440;
> b[1]:=0;
> pl;
> BH;
>
> plot([1+BH,1-BH],H=0..3,axes=NORMAL,title="Stability Functions
> for the Case II",legend=["1+B(H)", "1-B(H)"]);
>
> BH:=1-1/2*H^2+1/24*H^4+1/40320*H^8-1/3628800*H^10-1/
> 720*H^6-2353/7185024000*H^12-169/28740096000*H^14;
> fsolve(1+BH);
> 2.758269471*2.758269471;
>
> 'Stability and Phase-Lag- Case III';
>
> b[3]:=-1/360;
> b[2]:=-13/440;
> b[1]:=-296/6435;
> pl;
> BH;
>
> plot([1+BH,1-BH],H=0..3,axes=NORMAL,title="Stability Functions
> for the Case III",legend=["1+B(H)", "1-B(H)"]);
>
> BH:=1-1/2*H^2+1/24*H^4+1/40320*H^8-1/3628800*H^10+1/
> 479001600*H^12-1/720*H^6-14149/526901760000*H^14-481/
> 889146720000*H^16;

> fsolve(1+BH);
> 2.801903776*2.801903776;
>
> 'Error Analysis of the Derivatives used for Computational Implementation';
>
> 'Computation of the error for y'[n+1]';
> a[2,n+1]:=1/10;
> a[1,n+1]:=4/5;
> a[0,n+1]:=-9/10;
> b[2,n+1]:=11/30;
> b[1,n+1]:=1;
> b[0,n+1]:=1/30;
>
> errynp1:=simplify(h*d1np1-(a[2,n+1]*ynp1+a[1,n+1]*yn+a[0,n+1]*
> ynm1+h^2*(b[2,n+1]*dnp1+b[1,n+1]*dn+b[0,n+1]*dnm1)));
>
```

```
>
> 'Computation of the error for y'[n]';
>  a[2,n]:=-7/10;
>  a[1,n]:=12/5;
>  a[0,n]:=-17/10;
>  b[2,n]:=1/60;
>  b[1,n]:=1;
>  b[0,n]:=11/60;
>
> erryn:=simplify(h*d1n-(a[2,n]*ynp1+a[1,n]*yn+a[0,n]*ynm1
>  +h^2*(b[2,n]*dnp1+b[1,n]*dn+b[0,n]*dnm1)));
>
>
> 'Computation of the error for y'[n-1]';
>  a[2,n-1]:=-3/2;
>  a[1,n-1]:=4;
>  a[0,n-1]:=-5/2;
>  b[2,n-1]:=1/6;
>  b[1,n-1]:=1;
>  b[0,n-1]:=-1/6;
>
> errynm1:=simplify(h*d1nm1-(a[2,n-1]*ynp1+a[1,n-1]*yn+a[0,n-1]*
>  ynm1+h^2*(b[2,n-1]*dnp1+b[1,n-1]*dn+b[0,n-1]*dnm1)));
>
> 'Computation of the error for y'[n] for the beginnig of the algorithm';
>
>  aa[1,n]:=-1;
>  aa[0,n]:=-1;
>  bb[1,n]:=1/3;
>  bb[0,n]:=1/6;
>  erryn:=simplify(h*d1n-(aa[1,n]*yn+aa[0,n]*ynm1+h^2*(bb[1,n]*dn
>  +bb[0,n]*dnm1)));
>
```

Appendix B

ROUTINES FOR THE NUMERICAL SOLUTION OF THE RADIAL SCHRÖDINGER EQUATION PHASE-SHIFT PROBLEM

Programme driverf.m

envir

Subroutine envir.m

```
function envir
clc;
clear all;
disp('****************************************************')
disp('                        Author                    *')
disp('                      Dr.T.E.SIMOS                *')
disp('                                                  *')
disp('   Active Member of the European Academy of Sciences an Arts  *')
disp('        Department of Computer Science and Technology,        *')
disp('    Faculty of Sciences and Technology,University of Peloponnese,  *')
disp('                   GR-22100 Tripolis, Greece      *')
disp('                                                  *')
disp('                    Email addresses:              *')
disp('            tsimos@mail.ariadne-t.gr, simos@uop.gr *')
disp('                                                  *')
disp('             Email address (for the journals):    *')
disp('                  simos-editor@uop.gr             *')
disp('                                                  *')
disp('                    Postal address:               *')
disp(' Amfithea-Paleon Faliron, 26 Menelaou Street, GR-175 64 Athens,
      Greece                                            *')
disp('                                                  *')
disp('****************************************************')

disp('SELECT ONE OF THE METHODS')
method = input('(1) Numerov \n (2) NumerovType \n (3) Multiderivative \n (4) Multiderivative with minimal phase-lag(8) \n (5) Multiderivative with minimal phase-lag(10) \n');
while method < 1 | method > 5
    method = input('SELECT ONE OF THE METHODS \n (1) Numerov \n (2) NumerovType \n (3) Multiderivative \n (4) Multiderivative with minimal phase-lag(8) \n (5) Multiderivative with minimal phase-lag(10) \n');
end

disp('INPUTS: INITIAL INTERVAL H0 and EIGENVALUE');
H0 = input('Initial Interval H0 =  );
while H0 < =0
    H0 = input('H0 must be possitive. Initial Interval H0 =  );
end

L = input('Select one of the Eigenvalues \n 1) 53.588872000 \n 2) 163.215341000 \n 3) 341.495874000 \n 4) 989.701916000 \n');
while L < 1 | L > 4
    L = input('Select one of the Eigenvalues \n 1) 53.588872000 \n 2) 163.215341000 \n 3) 341.495874000 \n 4) 989.701916000 \n');
end
fprintf('————————————————\n')
```

```
if method = = 1
  numerov_pl(H0,L)
elseif method = = 2
  numerovtype4_pl(H0,L)
-elseif method = = 3
  multiderivative1_pl(H0,L)
elseif method = = 4
  multiderivative2_pl(H0,L)
elseif method = = 5
  multiderivative3_pl(H0,L)
end

xx = input('NEW CALCULATION ? (1) YES, (2) N0 \n');
if xx = = 1
  driverf
end
```

Subroutine potent.m

```
function potent = potent(X)

U0 = -50.0;
A = 0.6;
X0 = 7.0;
U1 = U0/A;
S = exp((X - X0)/A);
potent = U0/(1 + S) - U1*S/(1 + S)^2;
%
```

Subroutine d1potent.m

```
function d1potent = d1potent(X)

U0 = -50;
A = 0.6;
X0 = 7;
U1 = U0/A;
S = exp((X - X0)/A);
d1potent = -U0*S*(A + A*S + 1 - S)/(A*A*(1 + S)^3);
%
```

Subroutine d2potent.m

```
function d2potent = d2potent(X)

U0 = -50;
A = 0.6;
```

```
X0 = 7;
U1 = U0/A;
S = exp((X − X0)/A);
S2 = exp(2*(X − X0)/A);
d2potent = U0*S*(−A + A*S2 − 1 + 4*S − S2)/(A*A*A*(1 + S)^4);
%————————————————————
```

Subroutine numerov_pl.m

```
function numerov_pl(H0,L)

%————————————————
SSSK(1) = 53.588872000;
SSSK(2) = 163.215341000;
SSSK(3) = 341.495874000;
SSSK(4) = 989.701916000;

SSK = SSSK(L);
fprintf('***** NUMEROV METHOD *****\n')
fprintf('EIGENVALUE = %12.12f\n',SSK)

SK = sqrt(SSK) ;
SNFE = 1/H0*15;
M0 = 0;
J = 0;
R0 = 0;
%————————————————
SJ = J*(J + 1);
R = R0;
K = 0;
H = H0/2;
H2 = H*H;
H3 = H2*H;
X = 0;
M = M0 + 1;
Y(M) = 0 ;
FB = 0;
F2B = 0.003175726349;
RH = R0 + 0.5*H;
UB = feval('potent',RH);
CB = SJ/RH^2 + UB;
FBB = H2*(SSK − CB);
M = M + 1 ;
R = R + H ;
Y(M) = H^(J + 1);
```

```
U(M) = feval('potent',R);
CM = SJ/R^2 + U(M);
FA = H2*(SSK - CM);

while M < 17
   M = M + 1;
   R = R + H;
   FAA = FBB;
   FC = FB;
   FB = FA;
   U(M) = feval('potent',R);
   CM = SJ/R^2 + U(M);
   RH = RH + H;
   UB = feval('potent',RH);
   CB = SJ/RH^2 + UB;
   FA = H2*(SSK - CM);
   FBB = H2*(SSK - CB);
   PAR1 = 2*Y(M - 1)*(FB*(341*FAA - 719*FBB + 4056) + 48.*(20*FAA
     + 73*FBB - 390)) + Y(M - 2)*(FC*(271*FAA - 253*FBB + 312) + 48*
     (58*FAA - 47*FBB + 390));
   Y(M) = PAR1/(FA*(41*FAA - 59*FBB - 312) - 48.*(6*FAA + 5*
     (FBB + 78)));
end

H = 2*H;
H2 = H*H;
FB = 4*FA;
FC = 4*FC;
KM1 = M;
KM2 = M - 2;
KM = KM1;

while abs(R - 15) > 10^(-8)
   KM = KM + 1;
   R = R + H;
   U(KM) = feval('potent',R);
   CM = SJ/R^2 + U(KM);
   FA = H2*(SSK - CM);
   FFA = -FA/H2;
   FFB = -FB/H2;
   FFC = -FC/H2;
   Y(KM) = ((2 + (10*H2*FFB)/12)*Y(KM1) - (1 - (FFC*H2)/12)*Y(KM2))/
(1 - (FFA*H2)/12);

   if (KM1 > = 17)
      KM1 = KM1 + 1;
```

```
    else
       KM1 = KM1 + 2;
    end
    if (KM2 > = 17)
       KM2 = KM2 + 1;
    else
       KM2 = KM2 + 2;
    end
    FC = FB;
    FB = FA;
end
%
%   CALCULATE PHASE SHIFT
%
KM1 = KM1 - 1;
KM2 = KM2 - 1;
NM = KM;
AR = R;
[SB1,SN1] = feval('sfrbsl',J,SK*R);        % CALL SFRBSL――
[SB2,SN2] = feval('sfrbsl',J,SK*(R - H));  % CALL SFRBSL――
TNETA = ((R - H)*Y(KM)*SB2 - R*Y(KM1)*SB1)/(R*Y(KM1)*SN1 -
(R - H)*Y(KM)*SN2) ;
ETA = atan(((-1)^J)*TNETA);

SER = abs(ETA) - 2*atan(1);
ERROR = abs(SER);
DLERR = log10(ERROR);

%
%   END OF INTEGRATION LOOP - - - NORMALIZE
%
% AN = Y(M)/(SK*R*(cos(ETA)*SB1 + (-1)^J*sin(ETA)*SN1));
fprintf('INITIAL INTERVAL = %6.6f\n',H0)
fprintf('PHASE SHIFT = %12.12f\n',ETA)
fprintf('THE ABS ERROR  = %12.12e\n',ERROR)
fprintf('NFE = %4.0f\n',SNFE)
fprintf('LERR = %12.12f\n',DLERR)
fprintf('―――――――――――――――――――\n');
%  ***********************************************************
```

Subroutine numerovtype4_pl

function numerovtype4_pl (H0,L)

%―――――――――――――
SSSK(1) = 53.588872000;

7: Numerical Methods in Chemistry

```
SSSK(2) = 163.215341000;
SSSK(3) = 341.495874000;
SSSK(4) = 989.701916000;

SSK = SSSK(L);
fprintf('***** NUMEROV-TYPE METHOD *****\n')
fprintf('EIGENVALUE = %12.12f\n',SSK)

SK = sqrt(SSK) ;
SNFE = 1/H0*15;
M0 = 0;
J = 0;
R0 = 0;
%————————————————
a = -1/300;
SJ = J*(J+1);
R = R0;
K = 0;
H = H0/2;
H2 = H*H;
H3 = H2*H;
X = 0;
M = M0+1;
Y(M) = 0;
FB = 0;
RH = R0+0.5*H;
UB = feval('potent',RH);
CB = SJ/RH^2+UB;
FBB = H2*(SSK-CB);
M = M+1;
R = R+H;
Y(M) = H^(J+1);
U(M) = feval('potent',R);
CM = SJ/R^2+U(M);
FA = H2*(SSK-CM);

while M < 17
    M = M+1;
    R = R+H;
    FAA = FBB;
    FC = FB;
    FB = FA;
    U(M) = feval('potent',R);
    CM = SJ/R^2+U(M);
    RH = RH+H;
    UB = feval('potent',RH);
```

```
    CB = SJ/RH^2 + UB;
    FA = H2*(SSK - CM);
    FBB = H2*(SSK - CB);
    PAR1 = 2*Y(M - 1)*(FB*(341*FAA - 719*FBB + 4056) + 48*(20*FAA +
    73*FBB - 390)) + Y(M - 2)*(FC*(271*FAA - 253*FBB + 312) + 48*(58*
    FAA - 47*FBB + 390));
    Y(M) = PAR1/(FA*(41*FAA - 59*FBB - 312) - 48.*(6*FAA + 5*(FB +
    78)));
end

H = 2*H;
H2 = H*H;
FB = 4*FA;
FC = 4*FC;
KM1 = M;
KM2 = M - 2;
KM = KM1;

while abs(R - 15) > 10^(-8)
    KM = KM + 1;
    R = R + H;
    U(KM) = feval('potent',R);
    CM = SJ/R^2 + U(KM);
    FA = H2*(SSK - CM);
    FFA = -FA/H2;
    FFB = -FB/H2;
    FFC = -FC/H2;

    YHP = 2*Y(KM1) - Y(KM2) + H2*FFB*Y(KM1);
    FHP = FFA*YHP;
    YHN = Y(KM1) - a*H2*(FHP - 2*FFB*Y(KM1) + FFC*Y(KM2));
    FHN = FFB*YHN;

    Y(KM) - 2*Y(KM1) - Y(KM2) + (1/12)*H2*(FHP + 10*FHN + FFC*
    Y(KM2));

    if (KM1 > = 17)
        KM1 = KM1 + 1;
    else
        KM1 = KM1 + 2;
    end
    if (KM2 > = 17)
        KM2 = KM2 + 1;
    else
        KM2 = KM2 + 2;
    end
```

7: Numerical Methods in Chemistry

```
  FC=FB;
  FB=FA;
end
%
%   CALCULATE PHASE SHIFT
%
KM1 = KM1 - 1;
KM2 = KM2 - 1;
NM = KM;
AR = R;
[SB1,SN1] = feval('sfrbsl',J,SK*R) ;        % CALL SFRBSL———
[SB2,SN2] = feval('sfrbsl',J,SK*(R - H)) ; % CALL SFRBSL———
TNETA = ((R - H)*Y(KM)*SB2 - R*Y(KM1)*SB1)/(R*Y(KM1)*SN1 -
(R - H)*Y(KM)*SN2) ;
ETA = atan(((-1)^J)*TNETA);

SER = abs(ETA) - 2*atan(1);
ERROR = abs(SER);
DLERR = log10(ERROR);

%
% END OF INTEGRATION LOOP - - - NORMALIZE
%
% AN = Y(M)/(SK*R*(cos(ETA)*SB1 + (-1)^J*sin(ETA)*SN1));

fprintf('INITIAL INTERVAL = %6.6f\n',H0)
fprintf('PHASE SHIFT = %12.12f\n',ETA)
fprintf('THE ABS ERROR  = %12.12e\n',ERROR)
fprintf('NFE = %4.0f\n',SNFE)
fprintf('LERR = %12.12f\n',DLERR)
fprintf('————————————————————\n');
%   ****************************************************************
```

Subroutine multiderivative1_p1

```
function multiderivative1_p1 (H0,L)

SSSK(1) = 53.588872000;
SSSK(2) = 163.215341000;
SSSK(3) = 341.495874000;
SSSK(4) = 989.701916000;

SSK = SSSK(L);
fprintf('***** MULTIDERIVATIVE METHOD *****\n')
fprintf('EIGENVALUE = %12.12f\n',SSK)
```

```
SK = sqrt(SSK) ;
SNFE = 1/H0*15;
M0 = 0;
J = 0;
R0 = 0;
%————————
A0 = 1;
A1 = 1/12;
C0 = 115/126;
C1 = 11/252;
C2 = 313/7560;
C3 = -13/15120;
%————————
BBNN = 1/3;
BBNNM1 = 1/6;
AANN = 1;
AANNM1 = -1;
%————————

ANP1NP1 = 1/10;
ANP1N = 4/5;
ANP1NM1 = -9/10;
BNP1NP1 = 11/30;
BNP1N = 1;
BNP1NM1 = 1/30;

ANNP1 = -7/10;
ANN = 12/5;
ANNM1 = -17/10;
BNNP1 = 1/60;
BNN = 1;
BNNM1 = 11/60;

ANM1NP1 = -3/2;
ANM1N = 4;
ANM1NM1 = -5/2;
BNM1NP1 = 1/6;
BNM1N = 1;
BNM1NM1 = -1/6;
%————————

SJ = J*(J + 1);
R = R0;
K = 0;
H = H0/2;
H2 = H*H;
```

7: Numerical Methods in Chemistry

```
H3 = H2*H;
H4 = H2*H2;
X = 0;
M = M0 + 1;
Y(M) = 0 ;
FB = 0;
F2B = 0.003175726349;
RH = R0 + 0.5*H;
UB = feval('potent',RH);
CB = SJ/RH^2 + UB;
FBB = H2*(SSK - CB);
F2BB = feval('d2potent',RH);
M = M + 1 ;
R = R + H ;
Y(M) = H^(J + 1);
U(M) = feval('potent',R);
CM = SJ/R^2 + U(M);
FA = H2*(SSK - CM);
F2A = feval('d2potent',R);

while M < 17
   M = M + 1;
   R = R + H;
   FAA = FBB;
   FC = FB;
   FB = FA;
   F2AA = F2BB;
   F2C = F2B;
   F2B = F2A;
   U(M) = feval('potent',R);
   CM = SJ/R^2 + U(M);
   FA = H2*(SSK - CM);
   RH = RH + H;
   UB = feval('potent',RH);
   CB = SJ/RH^2 + UB;
   FBB = H2*(SSK - CB);
   F2A = feval('d2potent',R);
   F2BB = feval('d2potent',RH);
   PAR1 = 2*Y(M - 1)*(FB*(341*FAA - 719*FBB + 4056) + 48.*(20*FAA
    + 73*FBB - 390)) + Y(M - 2)*(FC*(271*FAA - 253*FBB + 312) + 48*(58
    *FAA - 47*FBB + 390));
   Y(M) = PAR1/(FA*(41*FAA - 59*FBB - 312) - 48.*(6*FAA + 5*
    (FBB + 78)));
end

H = 2*H;
H2 = H*H;
```

```
H3 = H2*H;
H4 = H2*H2;
FB = 4*FA;
FC = 4*FC;
KM1 = M;
KM2 = M - 2;
KM = KM1;

while abs(R - 15) > 10^(-8)

  KM = KM + 1;
  R = R + H;
  U(KM) = feval('potent',R);

  CM = SJ/R^2 + U(KM);
  FA = H2*(SSK - CM);
  F2A = feval('d2potent',R);
  F2B = feval('d2potent',R - H);    % f(n)"
  F2C = feval('d2potent',R - 2*H);  % f(n - 1)"
  F1A = feval('d1potent',R);
  F1B = feval('d1potent',R - H);    % f(n)'
  F1C = feval('d1potent',R - 2*H);  % f(n - 1)'
  FFA = -FA/H2;
  FFB = -FB/H2;
  FFC = -FC/H2;

D1YNPR =((H2*BBNN*FFB*Y(KM1)+H2*BBNNM1*FFC*Y(KM2)+AANN*
Y(KM1)+AANNM1*Y(KM2)))/H;
D4YNPR = F2B*Y(KM1)+2*F1B*D1YNPR+FFB*FFB*Y(KM1);
  YPNP1 = 2* Y(KM1) - Y(KM2) + H2* A0* FFB* Y(KM1) + H4* A1*
D4YNPR; % yh(n+1)

D1YN =((H2*BNNP1*FFA*YPNP1+H2*BNN*FFB*Y(KM1)+H2*BNNM1*FFC*
Y(KM2)+ANNP1*YPNP1+ANN*Y(KM1)+ANNM1*Y(KM2)))/H;

D1YNP1 =((H2*BNP1NP1*FFA*YPNP1+H2*BNP1N*FFB*Y(KM1)+H2*BNP1NM1*
FFC*Y(KM2)+ANP1NP1*YPNP1+ANP1N*Y(KM1)+ANP1NM1*Y(KM2)))/
H;

D1YNM1 =((H2*BNM1NP1*FFA*YPNP1+H2*BNM1N*FFB*Y(KM1)+H2*BNM1NM1*
FFC*Y(KM2)+ANM1NP1*YPNP1+ANM1N*Y(KM1)+ANM1NM1*Y(KM2)))/
H;
  D4YN = F2B*Y(KM1)+2*F1B*D1YN+FFB*FFB*Y(KM1);
  D4YNP1 = F2A*YPNP1+2*F1A*D1YNP1+FFA*FFA*YPNP1;
  D4YNM1 = F2C*Y(KM2)+2*F1C*D1YNM1+FFC*FFC*Y(KM2);
```

7: Numerical Methods in Chemistry

```
  Y(KM) = 2*Y(KM1) – Y(KM2) + ...
    H2*(C0*FFB*Y(KM1) + C1*(FFA*YPNP1 + FFC*Y(KM2))) + ...
    H4*(C2*D4YN + C3*(D4YNP1 + D4YNM1));

  if (KM1 > = 17)
     KM1 = KM1 + 1;
  else
     KM1 = KM1 + 2;
  end
  if (KM2 > = 17)
     KM2 = KM2 + 1;
  else
     KM2 = KM2 + 2;
  end

  FC = FB;
  FB = FA;
end
%
%   CALCULATE PHASE SHIFT
%
KM1 = KM1 – 1;
KM2 = KM2 – 1;
NM = KM;
AR = R;
[SB1,SN1] = feval('sfrbsl',J,SK*R) ; % CALL SFRBSL‒‒‒‒
[SB2,SN2] = feval('sfrbsl',J,SK*(R – H)) ; % CALL SFRBSL‒‒‒‒
TNETA = ((R – H)*Y(KM)*SB2 – R*Y(KM1)*SB1)/(R*Y(KM1)*SN1 –
(R – H)*Y(KM)*SN2) ;
ETA = atan(((–1)^J)*TNETA);

SER = abs(ETA) – 2*atan(1);
ERROR = abs(SER);
DLERR = log10(ERROR);
%
%   END OF INTEGRATION LOOP - - - NORMALIZE
%
AN = Y(M)/(SK*R*(cos(ETA)*SB1 + (–1)^J*sin(ETA)*SN1));
fprintf('INITIAL INTERVAL = %6.6f\n',H0)
fprintf('PHASE SHIFT = %12.12f\n',ETA)
fprintf('THE ABS ERROR  = %12.12e\n',ERROR)
fprintf('NFE = %4.0f\n',SNFE)
fprintf('LERR = %12.12f\n',DLERR)
fprintf('———————————————\n');
%   *************************************************************
```

Subroutine multiderivative2_pl

function multiderivative2_pl(H0,L)

SSSK(1) = 53.588872000;
SSSK(2) = 163.215341000;
SSSK(3) = 341.495874000;
SSSK(4) = 989.701916000;

SSK = SSSK(L);
fprintf('***** MULTIDERIVATIVE METHOD WITH MINIMAL PHASE-LAG(8) *****\n')
fprintf('EIGENVALUE= %12.12f\n',SSK)

SK = sqrt(SSK) ;
SNFE = 1/H0*15;
M0 = 0;
J = 0;
R0 = 0;

%—————————
BC = −1/360;

A0 = 1;
A1 = 1/12;
C0 = 115/126;
C1 = 11/252;
C2 = 313/7560;
C3 = −13/15120;
%—————————
BBNN = 1/3;
BBNNM1 = 1/6;
AANN = 1;
AANNM1 = −1;
%—————————

ANP1NP1 = 1/10;
ANP1N = 4/5;
ANP1NM1 = −9/10;
BNP1NP1 = 11/30;
BNP1N = 1;
BNP1NM1 = 1/30;

ANNP1 = −7/10;
ANN = 12/5;
ANNM1 = −17/10;

7: Numerical Methods in Chemistry

```
BNNP1 = 1/60;
BNN = 1;
BNNM1 = 11/60;

ANM1NP1 = -3/2;
ANM1N = 4;
ANM1NM1 = -5/2;
BNM1NP1 = 1/6;
BNM1N = 1;
BNM1NM1 = -1/6;
%————————

SJ = J*(J+1);
R = R0;
K = 0;
H = H0/2;
H2 = H*H;
H3 = H2*H;
H4 = H2*H2;
X = 0;
M = M0+1;
Y(M) = 0 ;
FB = 0;
F2B = 0.003175726349;
RH = R0+0.5*H;
UB = feval('potent',RH);
CB = SJ/RH^2 + UB;
FBB = H2*(SSK – CB);
F2BB = feval('d2potent',RH);
M = M+1 ;
R = R+H ;
Y(M) = H^(J+1);
U(M) = feval('potent',R);
CM = SJ/R^2 + U(M);
FA = H2*(SSK – CM);
F2A = feval('d2potent',R);

while M < 17
   M = M+1;
   R = R+H;
   FAA = FBB;
   FC = FB;
   FB = FA;
   F2AA = F2BB;
   F2C = F2B;
   F2B = F2A;
```

```
U(M) = feval('potent',R);
CM = SJ/R^2 + U(M);
FA = H2*(SSK - CM);
RH = RH + H;
UB = feval('potent',RH);
CB = SJ/RH^2 + UB;
FBB = H2*(SSK - CB);
F2A = feval('d2potent',R);
F2BB = feval('d2potent',RH);
PAR1 = 2*Y(M - 1)*(FB*(341*FAA - 719*FBB + 4056) + 48.*(20*FAA +
73*FBB - 390)) + Y(M - 2)*(FC*(271*FAA - 253*FBB + 312) + 48*(58*
FAA - 47*FBB + 390));
Y(M) = PAR1/(FA*(41*FAA - 59*FBB - 312) - 48.*(6*FAA + 5*(FBB +
78)));
end

H = 2*H;
H2 = H*H;
H3 = H2*H;
H4 = H2*H2;
FB = 4*FA;
FC = 4*FC;
KM1 = M;
KM2 = M - 2;
KM = KM1;

while abs(R - 15) > 10^(-8)
  KM = KM + 1;
  R = R + H;
  U(KM) = feval('potent',R);
  CM = SJ/R^2 + U(KM);
  FA = H2*(SSK - CM);
  F2A = feval('d2potent',R);
  F2B = feval('d2potent',R - H);
  F2C = feval('d2potent',R - 2*H);
  F1A = feval('d1potent',R);
  F1B = feval('d1potent',R - H);
  F1C = feval('d1potent',R - 2*H);
  FFA = -FA/H2;
  FFB = -FB/H2;
  FFC = -FC/H2;

  YHAP = 2*Y(KM1) - Y(KM2) + H2*FFB*Y(KM1);
  YHN = Y(KM1) - BC*H2*(FFA*YHAP - 2*FFB*Y(KM1) + FFC*
  Y(KM2));
```

7: Numerical Methods in Chemistry

```
  DH1YNPR = ((H2*BBNN*FFB*Y(KM1) + H2*BBNNM1*FFC*Y(KM2)
+ AANN*Y(KM1) + AANNM1*Y(KM2)))/H;
  DH4YNPR = F2B*YHN + 2*F1B*DH1YNPR + FFB*FFB*YHN;
  YPNP1 = 2*Y(KM1) - Y(KM2) + H2*A0*FFB*YHN + H4*A1*
DH4YNPR;

D1YN = ((H2*BNNP1*FFA*YPNP1 + H2*BNN*FFB*Y(KM1) + H2*BNNM1*
FFC*Y(KM2) + ANNP1*YPNP1 + ANN*Y(KM1) + ANNM1*Y(KM2)))/H;

D1YNP1 = ((H2*BNP1NP1*FFA*YPNP1 + H2*BNP1N*FFB*Y(KM1) + H2*BNP1NM1*
FFC*Y(KM2) + ANP1NP1*YPNP1 + ANP1N*Y(KM1) + ANP1NM1*Y(KM2)))/
H;

D1YNM1 = ((H2*BNM1NP1*FFA*YPNP1 + H2*BNM1N*FFB*Y(KM1) + H2*BNM1NM1*
FFC*Y(KM2) + ANM1NP1*YPNP1 + ANM1N*Y(KM1) + ANM1NM1*Y(KM2)))/
H;
  D4YN = F2B*Y(KM1) + 2*F1B*D1YN + FFB*FFB*Y(KM1);
  D4YNP1 = F2A*YPNP1 + 2*F1A*D1YNP1 + FFA*FFA*YPNP1;
  D4YNM1 = F2C*Y(KM2) + 2*F1C*D1YNM1 + FFC*FFC*Y(KM2);

  Y(KM) = 2*Y(KM1) - Y(KM2) + ...
    H2*(C0*FFB*Y(KM1) + C1*(FFA*YPNP1 + FFC*Y(KM2))) + ...
    H4*(C2*D4YN + C3*(D4YNP1 + D4YNM1));

  if (KM1 > = 17)
    KM1 = KM1 + 1;
  else
    KM1 = KM1 + 2;
  end
  if (KM2 > = 17)
    KM2 = KM2 + 1;
  else
    KM2 = KM2 + 2;
  end
  FC = FB;
  FB = FA;
end
%
%   CALCULATE PHASE SHIFT
%
KM1 = KM1 - 1;
KM2 = KM2 - 1;
NM = KM;
AR = R;
[SB1,SN1] = feval('sfrbsl',J,SK*R) ;
[SB2,SN2] = feval('sfrbsl',J,SK*(R - H)) ;
```

```
TNETA = ((R – H)*Y(KM)*SB2 – R*Y(KM1)*SB1)/(R*Y(KM1)*SN1 – (R
– H)*Y(KM)*SN2) ;
ETA = atan(((–1)^J)*TNETA);

SER = abs(ETA) – 2*atan(1);
ERROR = abs(SER);
DLERR = log10(ERROR);
%
%   END OF INTEGRATION LOOP - - - NORMALIZE
%
AN = Y(M)/(SK*R*(cos(ETA)*SB1 + (–1)^J*sin(ETA)*SN1));
fprintf('INITIAL INTERVAL = %6.6f\n',H0)
fprintf('PHASE SHIFT = %12.12f\n',ETA)
Pl:fprintf('THE ABS ERROR  = %12.12e\n',ERROR)
fprintf('NFE = %4.0f\n',SNFE)
fprintf('LERR = %12.12f\n',DLERR)
fprintf('————————————————\n');
% ************************************************************
```

Subroutine multiderivative3_pl

```
function multiderivative3_pl(H0,L)

SSSK(1) = 53.588872000;
SSSK(2) = 163.215341000;
SSSK(3) = 341.495874000;
SSSK(4) = 989.701916000;

SSK = SSSK(L);
fprintf('***** MULTIDERIVATIVE METHOD WITH MINIMAL PHASE-
LAG(10) *****\n')
fprintf('EIGENVALUE = %12.12f\n',SSK)

SK = sqrt(SSK) ;
SNFE = 1/H0*15;
M0 = 0;
J = 0;
R0 = 0;

%————————————
BC = –1/360;
CC = –13/440;

A0 = 1;
A1 = 1/12;
C0 = 115/126;
```

```
C1 = 11/252;
C2 = 313/7560;
C3 = -13/15120;
%————————————
BBNN = 1/3;
BBNNM1 = 1/6;
AANN = 1;
AANNM1 = -1;
%————————————

ANP1NP1 = 1/10;
ANP1N = 4/5;
ANP1NM1 = -9/10;
BNP1NP1 = 11/30;
BNP1N = 1;
BNP1NM1 = 1/30;

ANNP1 = -7/10;
ANN = 12/5;
ANNM1 = -17/10;
BNNP1 = 1/60;
BNN = 1;
BNNM1 = 11/60;

ANM1NP1 = -3/2;
ANM1N = 4;
ANM1NM1 = -5/2;
BNM1NP1 = 1/6;
BNM1N = 1;
BNM1NM1 = -1/6;
%————————————

SJ = J*(J+1);
R = R0;
K = 0;
H = H0/2;
H2 = H*H;
H3 = H2*H;
H4 = H2*H2;
X = 0;
M = M0+1;
Y(M) = 0 ;
FB = 0;
F2B = 0.003175726349;
RH = R0+0.5*H;
UB = feval('potent',RH);
```

```
CB = SJ/RH^2 + UB;
FBB = H2*(SSK - CB);
F2BB = feval('d2potent',RH);
M = M + 1 ;
R = R + H ;
Y(M) = H^(J+1);
U(M) = feval('potent',R);
CM = SJ/R^2 + U(M);
FA = H2*(SSK - CM);
F2A = feval('d2potent',R);

while M < 17
   M = M + 1;
   R = R + H;
   FAA = FBB;
   FC = FB;
   FB = FA;
   F2AA = F2BB;
   F2C = F2B;
   F2B = F2A;
   U(M) = feval('potent',R);
   CM = SJ/R^2 + U(M);
   FA = H2*(SSK - CM);
   RH = RH + H;
   UB = feval('potent',RH);
   CB = SJ/RH^2 + UB;
   FBB = H2*(SSK - CB);
   F2A = feval('d2potent',R);
   F2BB = feval('d2potent',RH);
   PAR1 = 2*Y(M - 1)*(FB*(341*FAA - 719*FBB + 4056) + 48.*(20*FAA +
   73*FBB - 390)) + Y(M - 2)*(FC*(271*FAA - 253*FBB + 312) + 48*(58*
   FAA - 47*FBB + 390));
   Y(M) = PAR1/(FA*(41*FAA - 59*FBB - 312) - 48.*(6*FAA + 5*(FBB +
   78)));
end

H = 2*H;
H2 = H*H;
H3 = H2*H;
H4 = H2*H2;
FB = 4*FA;
FC = 4*FC;
KM1 = M;
KM2 = M - 2;
KM = KM1;
```

7: Numerical Methods in Chemistry 337

```
while abs(R – 15) > 10^(–8)
  KM = KM + 1;
  R = R + H;
  U(KM) = feval('potent',R);
  CM = SJ/R^2 + U(KM);
  FA = H2*(SSK – CM);
  F2A = feval('d2potent',R);
  F2B = feval('d2potent',R – H);
  F2C = feval('d2potent',R – 2*H);
  F1A = feval('d1potent',R);
  F1B = feval('d1potent',R – H);
  F1C = feval('d1potent',R – 2*H);
  FFA = –FA/H2;
  FFB = –FB/H2;
  FFC = –FC/H2;

  YHAP = 2*Y(KM1) – Y(KM2) + H2*FFB*Y(KM1);
  YHHN = Y(KM1) – CC*H2*(FFA*YHAP – 2*FFB*Y(KM1) + FFC*
  Y(KM2));
  YHN = Y(KM1) – BC*H2*(FFA*YHAP – 2*FFB*YHHN + FFC*
  Y(KM2));

DH1YNPR = ((H2*BBNN*FFB*Y(KM1) + H2*BBNNM1*FFC*Y(KM2)
 + AANN*Y(KM1) + AANNM1*Y(KM2)))/H;
  DH4YNPR = F2B*YHN + 2*F1B*DH1YNPR + FFB*FFB*YHN;
  YPNP1 = 2 * Y(KM1) – Y(KM2) + H2 * A0 * FFB * YHN + H4 * A1*
  DH4YNPR;

D1YN = ((H2*BNNP1*FFA*YPNP1 + H2*BNN*FFB*Y(KM1) + H2*BNNM1*
FFC*Y(KM2) + ANNP1*YPNP1 + ANN*Y(KM1) + ANNM1*Y(KM2)))/H;

D1YNP1 = ((H2*BNP1NP1*FFA*YPNP1 + H2*BNP1N*FFB*Y(KM1) + H2*
BNP1NM1*FFC*Y(KM2)+ANP1NP1*YPNP1 +ANP1N*Y(KM1)+ANP1NM1*Y(KM2)))
/H;

D1YNM1 = ((H2*BNM1NP1*FFA*YPNP1 + H2*BNM1N*FFB*Y(KM1) + H2*
BNM1NM1*FFC*Y(KM2)+ANM1NP1*YPNP1+ANM1N*Y(KM1)+ANM1NM1*Y(KM2)))
/H;
  D4YN = F2B*Y(KM1) + 2*F1B*D1YN + FFB*FFB*Y(KM1);
  D4YNP1 = F2A*YPNP1 + 2*F1A*D1YNP1 + FFA*FFA*YPNP1;
  D4YNM1 = F2C*Y(KM2) + 2*F1C*D1YNM1 + FFC*FFC*Y(KM2);

  Y(KM) = 2*Y(KM1) – Y(KM2) + ...
    H2*(C0*FFB*Y(KM1) + C1*(FFA*YPNP1 + FFC*Y(KM2))) + ...
    H4*(C2*D4YN + C3*(D4YNP1 + D4YNM1));
```

```
    if (KM1 > = 17)
       KM1 = KM1 + 1;
    else
       KM1 = KM1 + 2;
    end
    if (KM2 > = 17)
       KM2 = KM2 + 1;
    else
       KM2 = KM2 + 2;
    end
      FC = FB;
      FB = FA;
end
%
%   CALCULATE PHASE SHIFT
%
KM1 = KM1 - 1;
KM2 = KM2 - 1;
NM = KM;
AR = R;
[SB1,SN1] = feval('sfrbsl',J,SK*R) ;
[SB2,SN2] = feval('sfrbsl',J,SK*(R - H)) ;
TNETA = ((R - H)*Y(KM)*SB2 - R*Y(KM1)*SB1)/(R*Y(KM1)*SN1 - (R
 - H)*Y(KM)*SN2) ;
ETA = atan(((-1)^J)*TNETA);

SER = abs(ETA) - 2*atan(1);
ERROR = abs(SER);
DLERR = log10(ERROR);
%
%   END OF INTEGRATION LOOP - - - NORMALIZE
%
AN = Y(M)/(SK*R*(cos(ETA)*SB1 + (-1)^J*sin(ETA)*SN1));
fprintf('INITIAL INTERVAL = %6.6f\n',H0)
fprintf('PHASE SHIFT = %12.12f\n',ETA)
fprintf('THE ABS ERROR = %12.12e\n',ERROR)
fprintf('NFE = %4.0f\n',SNFE)
fprintf('LERR = %12.12f\n',DLERR)
fprintf('————————————\n');
% ****************************************************************
```

Subroutine sfrbsl

```
function [BP, BN] = sfrbsl (N,X)

if N - X < = 0
   BP = feval('strr',X,N);
```

7: Numerical Methods in Chemistry

```
  BN=feval('strr',X,-N-1);
elseif N-X >0
  BP=feval('skey',X,N) ;
  BN=feval('strr',X,-N-1) ;
end
% ************************************************************
```

Subroutine skey

```
function D= skey (X,N)

NW=N;
C=feval('sprs',X,NW+1);
B=feval('sprs',X,NW+2);
A=feval('sprs',X,NW+3);
t=NW;

for K=N:NW
   D=C-X*X/((t+t+3)*(t+t+5))*B;
   A=(1-X*X/((t+t+3)*(t+t+5)))*B-X*X/((t+t+5)*(t+t+7))*A;
   D=(A+D)/2;
   A=B;
   B=C;
   C=D;
   t=t-1;
end

S=0;
LN=N+1;
LNN=N+N+1;

for t=LN:LNN
   S=S+log(t);
end

D=D*exp(N*log(2*X)-S);
% *********************************************
```

Subroutine sprs

```
function D= sprs (X,N)

IS=0;
SUM=1;
TERM=1;
```

```
while abs(TERM/SUM) > 10^(-8)
  IS = IS + 1;
  TERM = TERM*X*X/(-2*IS*(2*(N + IS) + 1));
  SUM = SUM + TERM;
  D = SUM;
end
```

% **

Subroutine strr

```
function D = strr (X,N)

A = sin(X)/X;
B = cos(X)/X;
C = -(B + sin(X))/X ;

if N + 2 < 0
  LN = -2 - N;
  for I = 1:LN
    D = -(I + I + 1)*C/X - B;
    B = C ;
    C = D ;
  end
elseif N + 2 = = 0
  D = C;
elseif N + 2 > 0
  if N < 0
    D = B;
  elseif N = = 0
    D = A;
  elseif N > 0
    for I = 1:N
      D = (I + I - 1)*A/X - B ;
      B = A ;
      A = D;
    end
  end
end
```
% **

Appendix C

Order Conditions for the construction of Symplectic Integrators

7: Numerical Methods in Chemistry

$$e1 = -1 + \sum_{i=1}^{k} c_i$$

$$e2 = -1 + \sum_{i=1}^{k} d_i$$

$$e3 = -1 + \left(\sum_{i=1}^{k} c_i\right)^2$$

$$e4 = -1 + \left(\sum_{i=1}^{k} d_i\right)^2$$

$$e5 = -\frac{1}{2} + \sum_{i=1}^{k} d_i \left(\sum_{j=1}^{i} c_j\right)$$

$$e6 = -\frac{1}{2} + \sum_{i=1}^{k} d_i \left(\sum_{j=i+1}^{k} c_j\right)$$

$$e7 = -1 + \left(\sum_{i=1}^{k} c_i\right)^3$$

$$e8 = -1 + \left(\sum_{i=1}^{k} d_i\right)^3$$

$$e9 = -\frac{1}{6} + \frac{\sum_{i=1}^{k} d_i \left(\sum_{j=1}^{i} c_j\right)^2}{2}$$

$$e10 = -\frac{1}{6} + \frac{\sum_{i=1}^{k} c_i \left(\sum_{j=i}^{k} d_j\right)^2}{2}$$

$$e11 = -\frac{1}{6} + \frac{\sum_{i=1}^{k} d_i \left(\sum_{j=i+1}^{k} c_j\right)^2}{2}$$

$$e12 = -\frac{1}{6} + \frac{\sum_{i=2}^{k} c_i \left(\sum_{j=1}^{i-1} d_j\right)^2}{2}$$

$$e13 = -\frac{1}{6} + \sum_{i=1}^{k} c_i \left(\sum_{j=i}^{k} d_j \left(\sum_{l=j+1}^{k} c_l\right)\right)$$

$$e14 = -\frac{1}{6} + \sum_{j=2}^{k} d_j \left(\sum_{i=2}^{j} c_i \left(\sum_{j=1}^{i-1} d_j\right)\right)$$

$$e15 = -1 + \left(\sum_{i=1}^{k} c_i\right)^4$$

$$e16 = -1 + \left(\sum_{i=1}^{k} d_i\right)^4$$

$$e17 = -\frac{1}{24} + \frac{\sum_{i=1}^{k} d_i \left(\sum_{j=1}^{i} c_j\right)^3}{6}$$

$$e18 = -\frac{1}{24} + \frac{\sum_{i=1}^{k} c_i \left(\sum_{j=i}^{k} d_j \right)^3}{6}$$

$$e19 = -\frac{1}{24} + \frac{\sum_{i=1}^{k} d_i \left(\sum_{j=i+1}^{k} c_j \right)^3}{6}$$

$$e20 = -\frac{1}{24} + \frac{\sum_{i=2}^{k} c_i \left(\sum_{j=1}^{i-1} d_j \right)^3}{6}$$

$$e21 = -\frac{1}{24} + \frac{\sum_{m=1}^{k} c_m \left(\sum_{i=m+1}^{k} c_i \left(\sum_{j=i}^{k} d_j \left(\sum_{l=j+1}^{k} c_l \right) \right) \right)}{2} + \frac{\sum_{m=1}^{k} c_m \left(\sum_{i=m}^{k} c_i \left(\sum_{j=i}^{k} d_j \left(\sum_{l=j+1}^{k} c_l \right) \right) \right)}{2}$$

$$e22 = -\frac{1}{24} + \frac{\sum_{j=1}^{k-1} c_j \left(\sum_{i=j}^{k} d_i \left(\sum_{m=i+1}^{k} d_m \left(\sum_{l=m+1}^{k} c_l \right) \right) \right)}{2} + \frac{\sum_{j=1}^{k-1} c_j \left(\sum_{i=j}^{k} d_i \left(\sum_{m=i}^{k} d_m \left(\sum_{l=m+1}^{k} c_l \right) \right) \right)}{2}$$

$$e23 = -\frac{1}{24} + \frac{\sum_{j=1}^{k-1} c_j \left(\sum_{i=j}^{k} d_i \left(\sum_{j=i+1}^{k} c_j \right)^2 \right)}{2}$$

$$e24 = -\frac{1}{24} + \frac{\sum_{j=2}^{k} d_j \left(\sum_{i=2}^{j} c_i \left(\sum_{j=1}^{i-1} d_j \right)^2 \right)}{2}$$

$$e25 = -\frac{1}{24} + \frac{\sum_{m=1}^{k} d_m \left(\sum_{i=m+1}^{k} c_i \left(\sum_{j=i+1}^{k} c_j \left(\sum_{l=j}^{k} d_l \right) \right) \right)}{2} + \frac{\sum_{m=1}^{k} d_m \left(\sum_{i=m+1}^{k} c_i \left(\sum_{j=i}^{k} c_j \left(\sum_{l=j}^{k} d_l \right) \right) \right)}{2}$$

$$e26 = -\frac{1}{24} + \frac{\sum_{j=1}^{k} d_j \left(\sum_{i=j+1}^{k} c_i \left(\sum_{j=i}^{k} d_j \right)^2 \right)}{2}$$

$$e27 = -\frac{1}{24} + \sum_{m=1}^{k-1} c_m \left(\sum_{j=m}^{k-1} d_j \left(\sum_{i=j+1}^{k} c_i \left(\sum_{l=i}^{k} d_l \right) \right) \right)$$

$$e28 = -\frac{1}{24} + \sum_{m=1}^{k-2} d_m \left(\sum_{j=m+1}^{k-1} c_j \left(\sum_{i=j}^{k-1} d_i \left(\sum_{l=i+1}^{k} c_l \right) \right) \right)$$

$$e29 = -\frac{1}{24} + \frac{1}{4} \sum_{p=1}^{k} \sum_{j=p}^{k} \sum_{i=j}^{k} \sum_{m=i}^{k} If[(p=j,a=1,a=2), If(i=m,b=1,b=2), abc_p c_j d_i d_m]$$

$$e30 = -\frac{1}{24} + \frac{1}{4} \sum_{p=1}^{k-1} \sum_{j=p}^{k-1} \sum_{i=j+1}^{k} \sum_{m=i}^{k} If[(p=j,a=1,a=2), If(i=m,b=1,b=2), abd_p d_j c_i c_m]$$

$$e31 = -1 + \left(\sum_{i=1}^{k} c_i \right)^5$$

$$e32 = -1 + \left(\sum_{i=1}^{k} d_i \right)^5$$

$$e33 = -\frac{1}{120} + \frac{\sum_{i=1}^{k} d_i \left(\sum_{j=1}^{i} c_j \right)^4}{24}$$

7: Numerical Methods in Chemistry

$$e34 = -\frac{1}{120} + \frac{\sum_{i=1}^{k} c_i \left(\sum_{j=i}^{k} d_j\right)^4}{24}$$

$$e35 = -\frac{1}{120} + \frac{\sum_{i=1}^{k} d_i \left(\sum_{j=i+1}^{k} c_j\right)^4}{24}$$

$$e36 = -\frac{1}{120} + \frac{\sum_{i=2}^{k} c_i \left(\sum_{j=1}^{i-1} d_j\right)^4}{24}$$

$$e37 = -\frac{1}{120} + \frac{\sum_{i=1}^{k} d_i \left(\sum_{j=1}^{i} c_j\right)^3 \left(\sum_{m=i+1}^{k} d_m\right)}{12} + \frac{\sum_{i=1}^{k} d_i \left(\sum_{j=1}^{i} c_j\right)^3 \left(\sum_{m=i}^{k} d_m\right)}{12}$$

$$e38 = -\frac{1}{120} + \frac{\sum_{m=1}^{k} c_m \left(\sum_{i=m+1}^{k} c_i \left(\sum_{j=i}^{k} d_j\right)\right)^3}{12} + \frac{\sum_{m=1}^{k} c_m \left(\sum_{i=m}^{k} c_i \left(\sum_{j=i}^{k} d_j\right)\right)^3}{12}$$

$$e39 = -\frac{1}{120} + \frac{\sum_{i=2}^{k} c_i \left(\sum_{l=i+1}^{k} c_l \left(\sum_{j=1}^{i-1} d_j\right)\right)^3}{12} + \frac{\sum_{i=2}^{k} c_i \left(\sum_{l=i}^{k} c_l \left(\sum_{j=1}^{i-1} d_j\right)\right)^3}{12}$$

$$e40 = -\frac{1}{120} + \frac{\sum_{m=1}^{k} d_m \left(\sum_{i=m+1}^{k} d_i \left(\sum_{j=i+1}^{k} c_j\right)\right)^3}{12} + \frac{\sum_{m=1}^{k} d_m \left(\sum_{i=m}^{k} d_i \left(\sum_{j=i+1}^{k} c_j\right)\right)^3}{12}$$

$$e41 = -\frac{1}{120} + \frac{\sum_{m=2}^{k} c_m \left(\sum_{i=1}^{m-1} d_i \left(\sum_{j=1}^{i} c_j\right)\right)^3}{6}$$

$$e42 = -\frac{1}{120} + \frac{\sum_{m=1}^{k} c_m \left(\sum_{i=m+1}^{k} c_i \left(\sum_{j=m}^{i-1} d_j\right)\right)^3}{6}$$

$$e43 = -\frac{1}{120} + \frac{\sum_{m=1}^{k} c_m \left(\sum_{i=m}^{k} d_i \left(\sum_{j=i+1}^{k} c_j\right)\right)^3}{6}$$

$$e44 = -\frac{1}{120} + \frac{\sum_{i=2}^{k} c_i \left(\sum_{j=1}^{i-1} d_j\right)^3 \left(\sum_{m=i}^{k} d_m\right)}{6}$$

$$e45 = -\frac{1}{120} + \frac{\sum_{i=1}^{k} d_i \left(\sum_{m=1}^{i-1} d_m \left(\sum_{j=m+1}^{i} c_j\right)\right)^3}{6}$$

$$e46 = -\frac{1}{120} + \frac{\sum_{m=1}^{k-1} d_m \left(\sum_{i=m+1}^{k} c_i \left(\sum_{j=i}^{k} d_j\right)\right)^3}{6}$$

$$e47 = -\frac{1}{120} + \sum_{i=1}^{k-1}\sum_{j=i}^{k-1}\sum_{m=j}^{k-1}\sum_{n=m}^{k-1}\sum_{p=n+1}^{k} If[(i=j, a=\tfrac{1}{2}, a=1),$$
$$If(m=n, b=\tfrac{1}{2}, b=1), abc_i\, c_j\, d_m\, d_n\, c_p]$$

$$e48 = -\tfrac{1}{120} + \sum_{i=1}^{k-1}\sum_{j=i}^{k-1}\sum_{m=j}^{k-1}\sum_{n=m+1}^{k}\sum_{p=n}^{k} If[(i=j,a=\tfrac{1}{2},a=1), If(n=p,b=\tfrac{1}{2},b=1), abc_i c_j d_m c_n c_p]$$

$$e49 = -\tfrac{1}{120} + \sum_{i=1}^{k-1}\sum_{j=i}^{k-1}\sum_{m=j}^{k-1}\sum_{n=m+1}^{k}\sum_{p\neq n}^{k} If[(j=m,a=\tfrac{1}{2},a=1), If(n=p,b=\tfrac{1}{2},b=1), abc_i d_j d_m c_n c_p]$$

$$e50 = -\tfrac{1}{120} + \sum_{i=1}^{k-1}\sum_{j=i}^{k-1}\sum_{m=j+1}^{k}\sum_{n=m}^{k}\sum_{p=n}^{k} If[(i=j,a=\tfrac{1}{2},a=1), If(m=n,b=\tfrac{1}{2},b=1), abd_i d_j c_m c_n d_p]$$

$$e51 = -\tfrac{1}{120} + \sum_{i=1}^{k-1}\sum_{j=i}^{k-1}\sum_{m=j+1}^{k}\sum_{n=m}^{k}\sum_{p=n}^{k} If[(i=j,a=\tfrac{1}{2},a=1), If(n=p,b=\tfrac{1}{2},b=1), abd_i d_j c_m d_n d_p]$$

$$e52 = -\tfrac{1}{120} + \sum_{i=1}^{k-1}\sum_{j=i+1}^{k}\sum_{m=j}^{k}\sum_{n=m}^{k}\sum_{p=n}^{k} If[(j=m,a=\tfrac{1}{2},a=1), If(n=p,b=\tfrac{1}{2},b=1), abd_i c_j c_m d_n d_p]$$

$$e53 = -\tfrac{1}{120} + \frac{\sum_{j=1}^{k-1} c_j \left(\sum_{i=j+1}^{k-1} c_i \left(\sum_{m=i}^{k-1} d_m \left(\sum_{l=m+1}^{k} c_l \left(\sum_{p=l}^{k} d_p\right)\right)\right)\right)}{2} + \frac{\sum_{j=1}^{k-1} c_j \left(\sum_{i=j}^{k-1} c_i \left(\sum_{m=i}^{k-1} d_m \left(\sum_{l=m+1}^{k} c_l \left(\sum_{p=l}^{k} d_p\right)\right)\right)\right)}{2}$$

$$e54 = -\tfrac{1}{120} + \frac{\sum_{j=1}^{k-1} c_j \left(\sum_{i=j}^{k-1} d_i \left(\sum_{m=i+1}^{k-1} d_m \left(\sum_{l=m+1}^{k} c_l \left(\sum_{p=l}^{k} d_p\right)\right)\right)\right)}{2} + \frac{\sum_{j=1}^{k-1} c_j \left(\sum_{i=j}^{k-1} d_i \left(\sum_{m=i}^{k-1} d_m \left(\sum_{l=m+1}^{k} c_l \left(\sum_{p=l}^{k} d_p\right)\right)\right)\right)}{2}$$

$$e55 = -\tfrac{1}{120} + \frac{\sum_{j=1}^{k-1} c_j \left(\sum_{i=j}^{k-1} d_i \left(\sum_{m=i+1}^{k} c_m \left(\sum_{l=m+1}^{k} c_l \left(\sum_{p=l}^{k} d_p\right)\right)\right)\right)}{2} + \frac{\sum_{j=1}^{k-1} c_j \left(\sum_{i=j}^{k-1} d_i \left(\sum_{m=i+1}^{k} c_m \left(\sum_{l=m}^{k} c_l \left(\sum_{p=l}^{k} d_p\right)\right)\right)\right)}{2}$$

$$e56 = -\tfrac{1}{120} + \frac{\sum_{j=1}^{k-1} c_j \left(\sum_{i=j}^{k-1} d_i \left(\sum_{m=i+1}^{k} c_m \left(\sum_{l=m}^{k} d_l \left(\sum_{p=l+1}^{k} d_p\right)\right)\right)\right)}{2} + \frac{\sum_{j=1}^{k-1} c_j \left(\sum_{i=j}^{k-1} d_i \left(\sum_{m=i+1}^{k} c_m \left(\sum_{l=m}^{k} d_l \left(\sum_{p=l}^{k} d_p\right)\right)\right)\right)}{2}$$

$$e57 = -\tfrac{1}{120} + \frac{\sum_{j=1}^{k-1} d_j \left(\sum_{i=j+1}^{k-1} d_i \left(\sum_{m=i+1}^{k} c_m \left(\sum_{l=m}^{k} d_l \left(\sum_{p=l+1}^{k} c_p\right)\right)\right)\right)}{2} + \frac{\sum_{j=1}^{k-1} d_j \left(\sum_{i=j}^{k-1} d_i \left(\sum_{m=i+1}^{k} c_m \left(\sum_{l=m}^{k} d_l \left(\sum_{p=l+1}^{k} c_p\right)\right)\right)\right)}{2}$$

$$e58 = -\tfrac{1}{120} + \frac{\sum_{j=1}^{k-1} d_j \left(\sum_{i=j+1}^{k-1} c_i \left(\sum_{m=i+1}^{k} c_m \left(\sum_{l=m}^{k} d_l \left(\sum_{p=l+1}^{k} c_p\right)\right)\right)\right)}{2} + \frac{\sum_{j=1}^{k-1} d_j \left(\sum_{i=j+1}^{k-1} c_i \left(\sum_{m=i}^{k} c_m \left(\sum_{l=m}^{k} d_l \left(\sum_{p=l+1}^{k} c_p\right)\right)\right)\right)}{2}$$

$$e59 = -\tfrac{1}{120} + \frac{\sum_{j=1}^{k-1} d_j \left(\sum_{i=j+1}^{k-1} c_i \left(\sum_{m=i}^{k} d_m \left(\sum_{l=m+1}^{k} d_l \left(\sum_{p=l+1}^{k} c_p\right)\right)\right)\right)}{2} + \frac{\sum_{j=1}^{k-1} d_j \left(\sum_{i=j+1}^{k-1} c_i \left(\sum_{m=i}^{k} d_m \left(\sum_{l=m}^{k} d_l \left(\sum_{p=l}^{k} c_p\right)\right)\right)\right)}{2}$$

$$e60 = -\tfrac{1}{120} + \frac{\sum_{j=1}^{k-1} d_j \left(\sum_{i=j+1}^{k-1} c_i \left(\sum_{m=i}^{k} d_m \left(\sum_{l=m+1}^{k} c_l \left(\sum_{p=l+1}^{k} c_p\right)\right)\right)\right)}{2} + \frac{\sum_{j=1}^{k-1} d_j \left(\sum_{i=j+1}^{k-1} c_i \left(\sum_{m=i}^{k} d_m \left(\sum_{l=m+1}^{k} c_l \left(\sum_{p=l}^{k} c_p\right)\right)\right)\right)}{2}$$

$$e61 = -\tfrac{1}{120} + \sum_{j=1}^{k-1} c_j \left(\sum_{i=j}^{k-1} d_i \left(\sum_{m=i+1}^{k} c_m \left(\sum_{l=m}^{k} d_l \left(\sum_{p=l+1}^{k} c_p\right)\right)\right)\right)$$

$$e62 = -\tfrac{1}{120} + \sum_{j=1}^{k-1} d_j \left(\sum_{i=j+1}^{k-1} c_i \left(\sum_{m=i}^{k} d_m \left(\sum_{l=m+1}^{k} c_l \left(\sum_{p=l}^{k} d_p\right)\right)\right)\right)$$

Finally the number of linearly independent equations are 14: e1, e2, e5, e13, e14, e19, e20, e28, e35, e36, e37, e38, e61 and e62

7: Numerical Methods in Chemistry

Appendix D

Construction of Two-Step Second Order Symplectic Integrator

Mathematica Programme

Two step -- Second Order   (Direct Basis)

k = 2; (* Number of steps *)

Order Equations

$$g[1] = \sum_{i=1}^{k} c_i - 1; \quad g[2] = \sum_{i=1}^{k} d_i - 1; \quad g[3] = \sum_{i=1}^{k} d_i \sum_{j=1}^{i} c_j - \frac{1}{\text{Factorial}[2]};$$

$$g[4] = \sum_{i=1}^{k} d_i \sum_{j=i+1}^{k} c_j - \frac{1}{\text{Factorial}[2]};$$

Error -- Functions

$$g[5] = \frac{1}{\text{Factorial}[2]} \sum_{i=1}^{k} d_i \left(\sum_{j=1}^{i} c_j \right)^2 - \frac{1}{\text{Factorial}[3]};$$

$$g[6] = \frac{1}{\text{Factorial}[2]} \sum_{i=1}^{k} c_i \left(\sum_{j=i}^{k} d_j \right)^2 - \frac{1}{\text{Factorial}[3]};$$

$$g[7] = \frac{1}{\text{Factorial}[2]} \sum_{i=1}^{k} d_i \left(\sum_{j=i+1}^{k} c_j \right)^2 - \frac{1}{\text{Factorial}[3]};$$

$$g[8] = \frac{1}{\text{Factorial}[2]} \sum_{i=2}^{k} c_i \left(\sum_{j=1}^{i-1} d_j \right)^2 - \frac{1}{\text{Factorial}[3]};$$

$$g[9] = \sum_{i=1}^{k} c_i \sum_{j=1}^{k} d_j \sum_{l=j+1}^{k} c_l - \frac{1}{\text{Factorial}[3]};$$

$$g[10] = \sum_{j=2}^{k} d_j \sum_{i=2}^{j} c_i \sum_{j=1}^{i-1} d_j - \frac{1}{\text{Factorial}[3]};$$

Independent Equations

ek1 = {g[1] == 0, g[2] == 0, g[3] == 0}; S1 = Solve[ek1, {c_1, c_2, d_2}]

Error function

(* Least-Squares Method *)
Error = Simplify[Sqrt[$\sum_{i=5}^{10}$g[i]^2] /. S1[[1]]] /. $d_1 \to x$]

(* Minimize the Error function *)
FindMinimum[Error, {x, 0.5}]

$d_1 = 0.707107 = \frac{\sqrt{2}}{2}$

(* Coefficients of Two Step -- Second Order Method *)
COEFF = Flatten[{S1[[1]] /. $d_1 \to \frac{\sqrt{2}}{2}$, $d_1 \to \frac{\sqrt{2}}{2}$}]

CHECKING the Order Equations

Chop[Simplify[Table[g[i] /. COEFF, {i, 1, 4}]]]

Appendix E

Construction of Three-Step Third Order Symplectic Integrator – Mathematica Programme

Three Step – Third Order   (Direct Basis)

$k = 3;$ (* Number of steps *)

Order Equations

$$f[1] = \sum_{i=1}^{k} c_i - 1; \quad f[2] = \sum_{i=1}^{k} d_i - 1; \quad f[3] = \sum_{i=1}^{k} d_i \sum_{j=1}^{i} c_j - \frac{1}{\text{Factorial}[2]};$$

$$f[4] = \sum_{i=1}^{k} d_i \sum_{j=i+1}^{k} c_j - \frac{1}{\text{Factorial}[2]};$$

$$f[5] = \frac{1}{\text{Factorial}[2]} \sum_{i=1}^{k} d_i \left(\sum_{j=1}^{i} c_j \right)^2 - \frac{1}{\text{Factorial}[3]};$$

$$f[6] = \frac{1}{\text{Factorial}[2]} \sum_{i=1}^{k} c_i \left(\sum_{j=i}^{k} d_j \right)^2 - \frac{1}{\text{Factorial}[3]};$$

$$f[7] = \frac{1}{\text{Factorial}[2]} \sum_{i=1}^{k} d_i \left(\sum_{j=i+1}^{k} c_j \right)^2 - \frac{1}{\text{Factorial}[3]};$$

$$f[8] = \frac{1}{\text{Factorial}[2]} \sum_{i=2}^{k} c_i \left(\sum_{j=1}^{i-1} d_j \right)^2 - \frac{1}{\text{Factorial}[3]};$$

$$f[9] = \sum_{i=1}^{k} c_i \sum_{j=i}^{k} d_j \sum_{l=j+1}^{k} c_l - \frac{1}{\text{Factorial}[3]};$$

$$f[10] = \sum_{j=2}^{k} d_j \sum_{i=2}^{j} c_i \sum_{j=1}^{i-1} d_j - \frac{1}{\text{Factorial}[3]};$$

Error -- Functions

$$f[11] = \frac{1}{\text{Factorial}[3]} \sum_{i=1}^{k} d_i \left(\sum_{j=1}^{i} c_j \right)^3 - \frac{1}{\text{Factorial}[4]};$$

$$f[12] = \frac{1}{\text{Factorial}[3]} \sum_{i=1}^{k} c_i \left(\sum_{j=i}^{k} d_j \right)^3 - \frac{1}{\text{Factorial}[4]};$$

$$f[13] = \frac{1}{\text{Factorial}[3]} \sum_{i=1}^{k} d_i \left(\sum_{j=i+1}^{k} c_j \right)^3 - \frac{1}{\text{Factorial}[4]};$$

$$f[14] = \frac{1}{\text{Factorial}[3]} \sum_{i=2}^{k} c_i \left(\sum_{j=1}^{i-1} d_j \right)^3 - \frac{1}{\text{Factorial}[4]};$$

$f[15] =$

$$\frac{1}{\text{Factorial}[2]\,\text{Factorial}[2]} \text{Sum}[\text{If}[p == j, aa = 1, aa = 2]; \text{If}[i == m, bb = 1, bb = 2];$$
$$c_p\, c_j\, aa\, d_i\, d_m\, bb,\, \{p, 1, k\},\, \{j, p, k\},\, \{i, j, k\},\, \{m, i, k\}] - \frac{1}{\text{Factorial}[4]};$$

7: Numerical Methods in Chemistry

$$f[16] = \frac{1}{\text{Factorial}[2]\,\text{Factorial}[2]} \text{Sum}[\text{If}[p == j, aa = 1, aa = 2];$$
$$\text{If}[i == m, bb = 1, bb = 2]; d_p\, d_j\, aa\, c_i\, c_m\, bb,$$
$$\{p, 1, k-1\}, \{j, p, k-1\}, \{i, j+1, k\}, \{m, i, k\}] - \frac{1}{\text{Factorial}[4]};$$

$$f[17] = \frac{1}{\text{Factorial}[2]} \sum_{m=1}^{k} c_m \sum_{i=m}^{k} c_i \sum_{j=i}^{k} d_j \sum_{l=j+1}^{k} c_l +$$
$$\frac{1}{\text{Factorial}[2]} \sum_{m=1}^{k} c_m \sum_{i=m+1}^{k} c_i \sum_{j=i}^{k} d_j \sum_{l=j+1}^{k} c_l - \frac{1}{\text{Factorial}[4]};$$

$$f[18] = \frac{1}{\text{Factorial}[2]} \sum_{j=1}^{k-1} c_j \sum_{i=j}^{k} d_i \sum_{m=i}^{k} d_m \sum_{l=m+1}^{k} c_l +$$
$$\frac{1}{\text{Factorial}[2]} \sum_{j=1}^{k-1} c_j \sum_{i=j}^{k} d_i \sum_{m=i}^{k} d_m \sum_{l=m+1}^{k} c_l - \frac{1}{\text{Factorial}[4]};$$

$$f[19] = \frac{1}{\text{Factorial}[2]} \sum_{j=1}^{k-1} c_j \sum_{i=j}^{k} d_i \left(\sum_{j=i+1}^{k} c_j\right)^2 - \frac{1}{\text{Factorial}[4]};$$

$$f[20] = \frac{1}{\text{Factorial}[2]} \sum_{j=2}^{k} d_j \sum_{i=2}^{j} c_i \left(\sum_{j=1}^{i-1} d_j\right)^2 - \frac{1}{\text{Factorial}[4]};$$

$$f[21] = \frac{1}{\text{Factorial}[2]} \sum_{m=1}^{k} d_m \sum_{i=m+1}^{k} c_i \sum_{j=i}^{k} c_j \sum_{l=j}^{k} d_l +$$
$$\frac{1}{\text{Factorial}[2]} \sum_{m=1}^{k} d_m \sum_{i=m+1}^{k} c_i \sum_{j=i+1}^{k} c_j \sum_{l=j}^{k} d_l - \frac{1}{\text{Factorial}[4]};$$

$$f[22] = \frac{1}{\text{Factorial}[2]} \sum_{j=1}^{k} d_j \sum_{i=j+1}^{k} c_i \left(\sum_{j=i}^{k} d_j\right)^2 - \frac{1}{\text{Factorial}[4]};$$

$$f[23] = \sum_{m=1}^{k-1} c_m \sum_{j=m}^{k-1} d_j \sum_{i=j+1}^{k} c_i \sum_{l=i}^{k} d_l - \frac{1}{\text{Factorial}[4]};$$

$$f[24] = \sum_{m=1}^{k-2} d_m \sum_{j=m+1}^{k-1} c_j \sum_{i=j}^{k-1} d_i \sum_{l=i+1}^{k} c_l - \frac{1}{\text{Factorial}[4]};$$

Independent Equations

$\text{indeq} = \{f[1] == 0, f[2] == 0, f[3] == 0, f[9] == 0, f[10] == 0\};$

$S0 = \text{Solve}[\text{indeq}, \{c_1, c_2, c_3, d_1, d_3\}];$

ERROR -- Function

(* Least-Squares Method *)

$\text{Error31} = \text{Sqrt}\left[\sum_{i=11}^{24} f[i]^2 \,/.\, S0[[1]]\right] /.\, d_2 \to x;$

```
d20 = SetAccuracy[FindMinimum[Error31, {x, 1}, WorkingPrecision → 16,
   MaxIterations → 40], 18]

(* OR *)

d21 = SetAccuracy[FindMinimum[Error31, {x, -0.01}, WorkingPrecision → 16,
   MaxIterations → 40], 18]

(* The second solution S0[[2]] of independent equations,
 gives bigger error than above *)
```

$d_2 = 1.182105778737709834$ OR $d_2 = -0.072235766125881368$

(* Coefficients of Three Step--Third Order Method *)

```
CASE1 = Flatten[{S0[[1]] /. d₂ -> 1.18210577873770983359236197433`18.0727 ,
   d₂ -> 1.18210577873770983359236197433`18.0727 }]

(* OR *)

CASE2 =
 Chop[Flatten[{S0[[1]] /. d₂ -> -0.07223576612588136769410596116358616.8588 ,
   d₂ -> -0.07223576612588136769410596116358616.8588}]]
```

CHECKING the Order Equations

```
Chop[Table[f[i] /. CASE1, {i, 1, 10}]]

Chop[Table[f[i] /. CASE2, {i, 1, 10}]]
```

7: Numerical Methods in Chemistry

Appendix F

Construction of Seven-Step Fifth Order Symplectic Integrator

Mathematica Programme

Seven Step - Fifth Order   (Direct Basis)

$k = 7;$ (* Number of steps *)

Order Equations

$$f[1] = \sum_{i=1}^{k} c_i - 1; \quad f[2] = \sum_{i=1}^{k} d_i - 1; \quad f[3] = \sum_{i=1}^{k} d_i \sum_{j=1}^{i} c_j - \frac{1}{\text{Factorial}[2]};$$

$$f[4] = \sum_{i=1}^{k} d_i \sum_{j=i+1}^{k} c_j - \frac{1}{\text{Factorial}[2]}; \quad f[5] = \frac{1}{\text{Factorial}[2]} \sum_{i=1}^{k} d_i \left(\sum_{j=1}^{i} c_j \right)^2 - \frac{1}{\text{Factorial}[3]};$$

$$f[6] = \frac{1}{\text{Factorial}[2]} \sum_{i=1}^{k} c_i \left(\sum_{j=1}^{k} d_j \right)^2 - \frac{1}{\text{Factorial}[3]};$$

$$f[7] = \frac{1}{\text{Factorial}[2]} \sum_{i=1}^{k} d_i \left(\sum_{j=i+1}^{k} c_j \right)^2 - \frac{1}{\text{Factorial}[3]};$$

$$f[8] = \frac{1}{\text{Factorial}[2]} \sum_{i=2}^{k} c_i \left(\sum_{j=1}^{i-1} d_j \right)^2 - \frac{1}{\text{Factorial}[3]};$$

$$f[9] = \sum_{i=1}^{k} c_i \sum_{j=1}^{k} d_j \sum_{l=j+1}^{k} c_l - \frac{1}{\text{Factorial}[3]};$$

$$f[10] = \sum_{j=2}^{k} d_j \sum_{i=2}^{j} c_i \sum_{j=1}^{i-1} d_j - \frac{1}{\text{Factorial}[3]};$$

$$f[11] = \frac{1}{\text{Factorial}[3]} \sum_{i=1}^{k} d_i \left(\sum_{j=1}^{i} c_j \right)^3 - \frac{1}{\text{Factorial}[4]};$$

$$f[12] = \frac{1}{\text{Factorial}[3]} \sum_{i=1}^{k} c_i \left(\sum_{j=1}^{k} d_j \right)^3 - \frac{1}{\text{Factorial}[4]};$$

$$f[13] = \frac{1}{\text{Factorial}[3]} \sum_{i=1}^{k} d_i \left(\sum_{j=i+1}^{k} c_j \right)^3 - \frac{1}{\text{Factorial}[4]};$$

$$f[14] = \frac{1}{\text{Factorial}[3]} \sum_{i=2}^{k} c_i \left(\sum_{j=1}^{i-1} d_j \right)^3 - \frac{1}{\text{Factorial}[4]};$$

$f[15] =$
$$\frac{1}{\text{Factorial}[2]\,\text{Factorial}[2]} \text{Sum}[\text{If}[p == j, aa = 1, aa = 2]; \text{If}[i == m, bb = 1, bb = 2];$$
$$c_p\, c_j\, aa\, d_i\, d_m\, bb, \{p, 1, k\}, \{j, p, k\}, \{i, j, k\}, \{m, i, k\}] - \frac{1}{\text{Factorial}[4]};$$

$f[16] =$
$$\frac{1}{\text{Factorial}[2]\,\text{Factorial}[2]} \text{Sum}[\text{If}[p == j, aa = 1, aa = 2]; \text{If}[i == m, bb = 1, bb = 2];$$
$$d_p\, d_j\, aa\, c_i\, c_m\, bb, \{p, 1, k-1\}, \{j, p, k-1\}, \{i, j+1, k\}, \{m, i, k\}] - \frac{1}{\text{Factorial}[4]};$$

$$f[17] = \frac{1}{\text{Factorial}[2]} \sum_{m=1}^{k} c_m \sum_{i=m}^{k} c_i \sum_{j=i}^{k} d_j \sum_{l=j+1}^{k} c_l +$$

$$\frac{1}{\text{Factorial}[2]} \sum_{m=1}^{k} c_m \sum_{i=m+1}^{k} c_i \sum_{j=i}^{k} d_j \sum_{l=j+1}^{k} c_l - \frac{1}{\text{Factorial}[4]} ;$$

$$f[18] = \frac{1}{\text{Factorial}[2]} \sum_{j=1}^{k-1} c_j \sum_{i=j}^{k} d_i \sum_{m=i}^{k} d_m \sum_{l=m+1}^{k} c_l +$$

$$\frac{1}{\text{Factorial}[2]} \sum_{j=1}^{k-1} c_j \sum_{i=j}^{k} d_i \sum_{m=i+1}^{k} d_m \sum_{l=m+1}^{k} c_l - \frac{1}{\text{Factorial}[4]} ;$$

$$f[19] = \frac{1}{\text{Factorial}[2]} \sum_{j=1}^{k-1} c_j \sum_{i=j}^{k} d_i \left(\sum_{j=i+1}^{k} c_j \right)^2 - \frac{1}{\text{Factorial}[4]} ;$$

$$f[20] = \frac{1}{\text{Factorial}[2]} \sum_{j=2}^{k} d_j \sum_{i=2}^{j} c_i \left(\sum_{j=1}^{i-1} d_j \right)^2 - \frac{1}{\text{Factorial}[4]} ;$$

$$f[21] = \frac{1}{\text{Factorial}[2]} \sum_{m=1}^{k} d_m \sum_{i=m+1}^{k} c_i \sum_{j=i}^{k} c_j \sum_{l=j}^{k} d_l +$$

$$\frac{1}{\text{Factorial}[2]} \sum_{m=1}^{k} d_m \sum_{i=m+1}^{k} c_i \sum_{j=i+1}^{k} c_j \sum_{l=j}^{k} d_l - \frac{1}{\text{Factorial}[4]} ;$$

$$f[22] = \frac{1}{\text{Factorial}[2]} \sum_{j=1}^{k} d_j \sum_{i=j+1}^{k} c_i \left(\sum_{j=i}^{k} d_j \right)^2 - \frac{1}{\text{Factorial}[4]} ;$$

$$f[23] = \sum_{m=1}^{k-1} c_m \sum_{j=m}^{k-1} d_j \sum_{i=j+1}^{k} c_i \sum_{l=i}^{k} d_l - \frac{1}{\text{Factorial}[4]} ;$$

$$f[24] = \sum_{m=1}^{k-2} d_m \sum_{j=m+1}^{k-1} c_j \sum_{i=j}^{k-1} d_i \sum_{l=i+1}^{k} c_l - \frac{1}{\text{Factorial}[4]} ;$$

$$f[25] = \frac{1}{\text{Factorial}[4]} \sum_{i=1}^{k} d_i \left(\sum_{j=1}^{i} c_j \right)^4 - \frac{1}{\text{Factorial}[5]} ;$$

$$f[26] = \frac{1}{\text{Factorial}[4]} \sum_{i=1}^{k} c_i \left(\sum_{j=i}^{k} d_j \right)^4 - \frac{1}{\text{Factorial}[5]} ;$$

$$f[27] = \frac{1}{\text{Factorial}[4]} \sum_{i=1}^{k} d_i \left(\sum_{j=i+1}^{k} c_j \right)^4 - \frac{1}{\text{Factorial}[5]} ;$$

$$f[28] = \frac{1}{\text{Factorial}[4]} \sum_{i=2}^{k} c_i \left(\sum_{j=1}^{i-1} d_j \right)^4 - \frac{1}{\text{Factorial}[5]} ;$$

$$f[29] = eA3B2 = \frac{1}{\text{Factorial}[3]\,\text{Factorial}[2]} \sum_{i=1}^{k} d_i \left(\sum_{j=1}^{i} c_j \right)^3 \sum_{m=i}^{k} d_m +$$

$$\frac{1}{\text{Factorial}[3]\,\text{Factorial}[2]} \sum_{i=1}^{k} d_i \left(\sum_{j=1}^{i} c_j \right)^3 \sum_{m=i+1}^{k} d_m - \frac{1}{\text{Factorial}[5]} ;$$

$$f[30] = \frac{1}{\text{Factorial}[2]\,\text{Factorial}[3]} \sum_{m=1}^{k} c_m \sum_{i=m}^{k} c_i \left(\sum_{j=1}^{k} d_j\right)^3 +$$
$$\frac{1}{\text{Factorial}[2]\,\text{Factorial}[3]} \sum_{m=1}^{k} c_m \sum_{i=m+1}^{k} c_i \left(\sum_{j=1}^{k} d_j\right)^3 - \frac{1}{\text{Factorial}[5]};$$

$$f[31] = \frac{1}{\text{Factorial}[3]\,\text{Factorial}[2]} \sum_{i=2}^{k} c_i \sum_{l=i}^{k} c_l \left(\sum_{j=1}^{i-1} d_j\right)^3 +$$
$$\frac{1}{\text{Factorial}[3]\,\text{Factorial}[2]} \sum_{i=2}^{k} c_i \sum_{l=i+1}^{k} c_l \left(\sum_{j=1}^{i-1} d_j\right)^3 - \frac{1}{\text{Factorial}[5]};$$

$$f[32] = \frac{1}{\text{Factorial}[3]\,\text{Factorial}[2]} \sum_{m=1}^{k} d_m \sum_{i=m}^{k} d_i \left(\sum_{j=i+1}^{k} c_j\right)^3 +$$
$$\frac{1}{\text{Factorial}[3]\,\text{Factorial}[2]} \sum_{m=1}^{k} d_m \sum_{i=m+1}^{k} d_i \left(\sum_{j=i+1}^{k} c_j\right)^3 - \frac{1}{\text{Factorial}[5]};$$

$$f[33] = \frac{1}{\text{Factorial}[3]} \sum_{m=2}^{k} c_m \sum_{i=1}^{m-1} d_i \left(\sum_{j=1}^{i} c_j\right)^3 - \frac{1}{\text{Factorial}[5]};$$

$$f[34] = \frac{1}{\text{Factorial}[3]} \sum_{m=1}^{k} c_m \sum_{i=m+1}^{k} c_i \left(\sum_{j=m}^{i-1} d_j\right)^3 - \frac{1}{\text{Factorial}[5]};$$

$$f[35] = \frac{1}{\text{Factorial}[3]} \sum_{m=1}^{k} c_m \sum_{i=m}^{k} d_i \left(\sum_{j=i+1}^{k} c_j\right)^3 - \frac{1}{\text{Factorial}[5]};$$

$$f[36] = \frac{1}{\text{Factorial}[3]} \sum_{i=2}^{k} c_i \left(\sum_{j=1}^{i-1} d_j\right)^3 \sum_{m=i}^{k} d_m - \frac{1}{\text{Factorial}[5]};$$

$$f[37] = \frac{1}{\text{Factorial}[3]} \sum_{i=1}^{k} d_i \sum_{m=1}^{i-1} d_m \left(\sum_{j=m+1}^{i} c_j\right)^3 - \frac{1}{\text{Factorial}[5]};$$

$$f[38] = \frac{1}{\text{Factorial}[3]} \sum_{m=1}^{k-1} d_m \sum_{i=m+1}^{k} c_i \left(\sum_{j=1}^{k} d_j\right)^3 - \frac{1}{\text{Factorial}[5]};$$

```
f[39] = Sum[If[i == j, aa = 1/Factorial[2], aa = 1]; If[m == mm, bb = 1/Factorial[2], bb = 1];
    c_i c_j d_m d_mm c_ii aa bb, {i, 1, k - 1}, {j, i, k - 1}, {m, j, k - 1}, {mm, m, k - 1},
    {ii, mm + 1, k}] - 1/Factorial[5];


f[40] = Sum[If[i == j, aa = 1/Factorial[2], aa = 1];
    If[mm == ii, bb = 1/Factorial[2], bb = 1]; c_i c_j d_m c_mm c_ii aa bb,
    {i, 1, k - 1}, {j, i, k - 1}, {m, j, k - 1}, {mm, m + 1, k}, {ii, mm, k}] - 1/Factorial[5];
```

$f[41] = \text{Sum}\left[\text{If}\left[j == m, aa = \frac{1}{\text{Factorial}[2]}, aa = 1\right]; \text{If}\left[mm == ii, bb = \frac{1}{\text{Factorial}[2]}, bb = 1\right];\right.$
$c_i d_j d_m c_{mm} c_{ii} aa bb,$
$\left.\{i, 1, k-1\}, \{j, i, k-1\}, \{m, j, k-1\}, \{mm, m+1, k\}, \{ii, mm, k\}\right] - \frac{1}{\text{Factorial}[5]};$

$f[42] = \text{Sum}\left[\text{If}\left[i == j, aa = \frac{1}{\text{Factorial}[2]}, aa = 1\right];\right.$
$\text{If}\left[m == mm, bb = \frac{1}{\text{Factorial}[2]}, bb = 1\right]; d_i d_j c_m c_{mm} d_{ii} aa bb,$
$\left.\{i, 1, k-1\}, \{j, i, k-1\}, \{m, j+1, k\}, \{mm, m, k\}, \{ii, mm, k\}\right] -$
$\frac{1}{\text{Factorial}[5]};$

$f[43] = \text{Sum}\left[\text{If}\left[i == j, aa = \frac{1}{\text{Factorial}[2]}, aa = 1\right]; \text{If}\left[mm == ii, bb = \frac{1}{\text{Factorial}[2]}, bb = 1\right];\right.$
$\left.d_i d_j c_m d_{mm} d_{ii} aa bb, \{i, 1, k-1\}, \{j, i, k-1\}, \{m, j+1, k\}, \{mm, m, k\}, \{ii, mm, k\}\right] -$
$\frac{1}{\text{Factorial}[5]};$

$f[44] = \text{Sum}\left[\text{If}\left[j == m, aa = \frac{1}{\text{Factorial}[2]}, aa = 1\right]; \text{If}\left[mm == ii, bb = \frac{1}{\text{Factorial}[2]}, bb = 1\right];\right.$
$\left.d_i c_j c_m d_{mm} d_{ii} aa bb, \{i, 1, k-1\}, \{j, i+1, k\}, \{m, j, k\}, \{mm, m, k\}, \{ii, mm, k\}\right] -$
$\frac{1}{\text{Factorial}[5]};$

$f[45] = \frac{1}{\text{Factorial}[2]} \sum_{j=1}^{k-1} c_j \sum_{i=j}^{k-1} c_i \sum_{m=i}^{k-1} d_m \sum_{l=m+1}^{k} c_1 \sum_{p=1}^{k} d_p +$
$\frac{1}{\text{Factorial}[2]} \sum_{j=1}^{k-1} c_j \sum_{i=j+1}^{k-1} c_i \sum_{m=i}^{k-1} d_m \sum_{l=m+1}^{k} c_1 \sum_{p=1}^{k} d_p - \frac{1}{\text{Factorial}[5]};$

$f[46] = \frac{1}{\text{Factorial}[2]} \sum_{j=1}^{k-1} c_j \sum_{i=j}^{k-1} d_i \sum_{m=i}^{k-1} d_m \sum_{l=m+1}^{k} c_1 \sum_{p=1}^{k} d_p +$
$\frac{1}{\text{Factorial}[2]} \sum_{j=1}^{k-1} c_j \sum_{i=j}^{k-1} d_i \sum_{m=i+1}^{k-1} d_m \sum_{l=m+1}^{k} c_1 \sum_{p=1}^{k} d_p - \frac{1}{\text{Factorial}[5]};$

$f[47] = \frac{1}{\text{Factorial}[2]} \sum_{j=1}^{k-1} c_j \sum_{i=j}^{k-1} d_i \sum_{m=i+1}^{k} c_m \sum_{l=m}^{k} c_1 \sum_{p=1}^{k} d_p +$
$\frac{1}{\text{Factorial}[2]} \sum_{j=1}^{k-1} c_j \sum_{i=j}^{k-1} d_i \sum_{m=i+1}^{k} c_m \sum_{l=m+1}^{k} c_1 \sum_{p=1}^{k} d_p - \frac{1}{\text{Factorial}[5]};$

$f[48] = \frac{1}{\text{Factorial}[2]} \sum_{j=1}^{k-1} c_j \sum_{i=j}^{k-1} d_i \sum_{m=i+1}^{k} c_m \sum_{l=m}^{k} d_1 \sum_{p=1}^{k} d_p +$
$\frac{1}{\text{Factorial}[2]} \sum_{j=1}^{k-1} c_j \sum_{i=j}^{k-1} d_i \sum_{m=i+1}^{k} c_m \sum_{l=m}^{k} d_1 \sum_{p=l+1}^{k} d_p - \frac{1}{\text{Factorial}[5]};$

7: Numerical Methods in Chemistry

$$f[49] = \frac{1}{\text{Factorial}[2]} \sum_{j=1}^{k-1} d_j \sum_{i=j}^{k-1} d_i \sum_{m=i+1}^{k} c_m \sum_{l=m}^{k} d_1 \sum_{p=l+1}^{k} c_p +$$

$$\frac{1}{\text{Factorial}[2]} \sum_{j=1}^{k-1} d_j \sum_{i=j+1}^{k-1} d_i \sum_{m=i+1}^{k} c_m \sum_{l=m}^{k} d_1 \sum_{p=l+1}^{k} c_p - \frac{1}{\text{Factorial}[5]};$$

$$f[50] = \frac{1}{\text{Factorial}[2]} \sum_{j=1}^{k-1} d_j \sum_{i=j+1}^{k-1} c_i \sum_{m=i}^{k} c_m \sum_{l=m}^{k} d_1 \sum_{p=l+1}^{k} c_p +$$

$$\frac{1}{\text{Factorial}[2]} \sum_{j=1}^{k-1} d_j \sum_{i=j+1}^{k-1} c_i \sum_{m=i+1}^{k} c_m \sum_{l=m}^{k} d_1 \sum_{p=l+1}^{k} c_p - \frac{1}{\text{Factorial}[5]};$$

$$f[51] = \frac{1}{\text{Factorial}[2]} \sum_{j=1}^{k-1} d_j \sum_{i=j+1}^{k-1} c_i \sum_{m=i}^{k} d_m \sum_{l=m}^{k} d_1 \sum_{p=l+1}^{k} c_p +$$

$$\frac{1}{\text{Factorial}[2]} \sum_{j=1}^{k-1} d_j \sum_{i=j+1}^{k-1} c_i \sum_{m=i}^{k} d_m \sum_{l=m+1}^{k} d_1 \sum_{p=l+1}^{k} c_p - \frac{1}{\text{Factorial}[5]};$$

$$f[52] = \frac{1}{\text{Factorial}[2]} \sum_{j=1}^{k-1} d_j \sum_{i=j+1}^{k-1} c_i \sum_{m=i}^{k} d_m \sum_{l=m+1}^{k} c_1 \sum_{p=l}^{k} c_p +$$

$$\frac{1}{\text{Factorial}[2]} \sum_{j=1}^{k-1} d_j \sum_{i=j+1}^{k-1} c_i \sum_{m=i}^{k} d_m \sum_{l=m+1}^{k} c_1 \sum_{p=l+1}^{k} c_p - \frac{1}{\text{Factorial}[5]};$$

$$f[53] = \sum_{j=1}^{k-1} c_j \sum_{i=j}^{k-1} d_i \sum_{m=i+1}^{k} c_m \sum_{l=m}^{k-1} d_1 \sum_{p=l+1}^{k} c_p - \frac{1}{\text{Factorial}[5]};$$

$$f[54] = \sum_{j=1}^{k-1} d_j \sum_{i=j+1}^{k-1} c_i \sum_{m=i}^{k} d_m \sum_{l=m+1}^{k} c_1 \sum_{p=l}^{k} d_p - \frac{1}{\text{Factorial}[5]};$$

Independent Equations

```
sym = {d₁ → d1, d₂ → d2, d₃ → d3, d₄ → d4, d₅ → d5, d₆ → d6, d₇ → d7,
       c₁ → c1, c₂ → c2,
       c₃ → c3, c₄ → c4, c₅ → c5, c₆ → c6, c₇ → c7};

ieq1 =
  Flatten[
    {f[1]² + f[2]² + f[3]² + f[9]² + f[10]² + f[13]² + f[14]² + f[24]² + f[25]² + f[26]² +
     f[29]² + f[30]² + f[53]² + f[54]²} /. sym];
```

(* Levenberg Marquardt Method *)

(* 46 CASES *)

1 CASE

```
s01 = SetAccuracy[FindMinimum[ieq1[[1]],
    {c1, 1.5225441490806472356407814586668913810617606310158748},
    {d1, -0.0276076577102055608043214464048438206623778968802`34.161},
    {d2, 1.1806851009657162510944225701818271853926639065215`35.9986},
    {d3, -1.22457162739038799237864284533111110579089608626299455},
    {c2, -1.1500107005529907056029514933`18.0607},
    {c3, -0.0098760842661124864649835330965295`15.9946},
    {c4, 1.1980080119142311190216787509`18.0785},
    {c5, -1.3948478584648866807071954099`18.1445},
    {c6, 0.4701662641796056640863810116570685`17.6723},
    {c7, 0.36401621810964757131046098947990775`17.5611},
    {d4, 0.02291200056153355757437850570568155`16.3601},
    {d5, 0.4344300275783983811273003539099595`17.6379},
    {d6, 0.4825530233435921800833057204727085`17.6835},
    {d7, 0.13159913265135619875323413907608555`17.1193},
    MaxIterations -> 4000, Method -> LevenbergMarquardt,
    AccuracyGoal → 40, WorkingPrecision → 100], 40]

(* Checking Order-Equations *)
Table[{i, f[i]} /. sym /. s01[[2]], {i, 1, 54}]
```

2 CASE

```
s02 = SetAccuracy[FindMinimum[ieq1[[1]],
    {c1, 0.47501834514453949720351208570106713494289203770372945`38.8986},
    {d1, -0.40202099502883859942041233324125017291469047232925`35.6779},
    {d2, 0.3458217808647417833780552420386768069307650284295`35.624},
    {d3, 0.40096296748537135014791802587765775357750422749976915`37.2687},
    {c2, 0.0218565947411023496493864826106801046989858150482178`38.3396},
    {c3, -0.33494829803581194216377525663119740784168243408203125`39.525},
    {c4, 0.51263817465269123641036230765166692435741424560546875`39.7098},
    {c5, -0.01197870102055153309972445110133776324801146984100345`38.0784},
    {c6, -0.0321200042630466336968098066224337080193569374004472755`38.5068},
    {c7, 0.36953388878114484805692541158350650221093902587890625`39.5677},
    {d4, 0.98092653187935097136573858733754605054855346679687585`39.9916},
    {d5, -1.3620648986697867499628955556545406579971313476562585`40.1342},
    {d6, 0.92380502900080929684634156728861853480339050292968755`39.9656},
    {d7, 0.11256958446834731957242325961487949825823307037335325`39.0514},
    Method -> LevenbergMarquardt, MaxIterations -> 4000, AccuracyGoal → 40,
    WorkingPrecision → 40], 40]

(* Checking Order-Equations *)
Table[{i, f[i]} /. sym /. s02[[2]], {i, 1, 54}]
```

7: *Numerical Methods in Chemistry*   355

3 CASE

```
s03 = SetAccuracy[FindMinimum[ieq1[[1]],
    {c1, 0.7725669421529699366845236412632597952534226497001855`39.1098},
    {d1, -0.9770079649704520702216439037182820995967410985289321`34.2527},
    {d2, 2.5893194922152082428171201965499341454007603265472616`34.7097},
    {d3, -2.5891381814263986320415806706036137052857066162762107`35.803},
    {c2, -0.115030140392799837267823193087679`17.0608},
    {c3, -0.005272900515674524418718860374611`15.722},
    {c4, 0.118145106915978265194056007203471`17.0724},
    {c5, -0.614357026549179341401440979097797`17.7884},
    {c6, 0.474756419912400529081963895805529`17.6765},
    {c7, 0.369191598476344584600639109339681`17.5673},
    {d4, 0.940985394548075615439586272259476`17.9736},
    {d5, 0.378820564878818710763397348273429`17.5784},
    {d6, 0.541852660039297107097411299037049`17.7339},
    {d7, 0.115168034716447931220528744233889`17.0613},
    Method -> LevenbergMarquardt, MaxIterations -> 4000, AccuracyGoal -> 100,
    WorkingPrecision -> 100], 40]

(* Checking Order-Equations *)
Table[{i, f[i]} /. sym /. s03[[2]], {i, 1, 54}]
```

4 CASE

```
s04 = SetAccuracy[FindMinimum[ieq1[[1]],
    {c1, 0.2360281420950061070226173893990360362892149425249543`38.5948},
    {d1, 1.1230974244936490505021222815889528599102771823213349`35.8812},
    {d2, 0.0075409197289692030198310677337411873156955186173`33.9924},
    {d3, -1.2512191996732658721903564509371846336632424685034328`36.2468},
    {c2, -0.419927529628676965423039746383438`17.6232},
    {c3, 0.398023740650237545235512470753747`17.5999},
    {c4, -0.057257198615491756987783134036363`16.7578},
    {c5, 1.344468239324921654542777105`18.1286},
    {c6, -0.855606343080717368643206555134384`17.9323},
    {c7, 0.354270949254762490543413377963589`17.5493},
    {d4, 0.539977640011261472707815300964285`17.7324},
    {d5, -0.002111601986253623317035810202924`15.3246},
    {d6, 0.451712671004243948580381129431771`17.6549},
    {d7, 0.131002146421484499949627888781833`17.1173},
    Method -> LevenbergMarquardt, MaxIterations -> 4000, AccuracyGoal -> 100,
    WorkingPrecision -> 100], 40]

(* Checking Order-Equations *)
Table[{i, f[i]} /. sym /. s04[[2]], {i, 1, 54}]
```

5 CASE

```
s05 = SetAccuracy[FindMinimum[ieq1[[1]],
    {c1, 0.6282136999127506169810707855956037158061272805571447`39.02},
    {d1, -0.5986463167353405834658618080788419026659010921679991`35.108},
    {d2, 1.7635439024081566300154749773079218164643095670127946`35.582},
    {d3, -1.7887567142583139691279326555137241349922032241310615`37.5493},
    {c2, -0.01257735030793877273325165333517`16.0996},
    {c3, -0.2397603213075872874426863745611163`17.3798},
    {c4, 0.17566193873980409656532231110759`17.2447},
    {c5, 0.05352032582205892841109573510038`16.7285},
    {c6, -0.349987652308169860937425710289972`17.5441},
    {c7, 0.744929359449076011756574189348612`17.8721},
    {d4, 2.02744328916753957514629291847`18.3069},
    {d5, -1.858675755417187325235772732`18.2692},
    {d6, 1.3235745058146826114153782328`18.1217},
    {d7, 0.1315170890203837597245023007417`17.119},
    Method -> LevenbergMarquardt, MaxIterations -> 4000, AccuracyGoal -> 100,
    WorkingPrecision -> 100], 40]

(* Checking Order-Equations *)
Table[{i, f[i]} /. sym /. s05[[2]], {i, 1, 54}]
```

6 CASE

```
s06 = SetAccuracy[FindMinimum[ieq1[[1]],
    {c1, 0.1856481785546217189682453093063419625324461402449637`38.4905},
    {d1, -0.7608818032181425576690006491714040511434666130928499`35.4545},
    {d2, 0.0232822433358203544784268165607621050416005339169312`34.1129},
    {d3, 1.1151868364180990072244010659121226269315739920147822`36.1047},
    {c2, 0.17771230550653213664524798787169`17.2497},
    {c3, -0.19054864684689526432848083459248`17.28},
    {c4, 0.47615620194531510511737337765225`17.6777},
    {c5, -0.03437628634325576304497928958880`16.5363},
    {c6, -0.00943559634302340750899862342748`15.9748},
    {c7, 0.39484384352675339080462890706257`17.5964},
    {d4, 0.77485883700971946996816086539183`17.8892},
    {d5, -1.47329253687270500527972671`18.1683},
    {d6, 1.21099168945504120920020341`18.0831},
    {d7, 0.10985473387218021301681147861018`17.0408},
    Method -> LevenbergMarquardt, MaxIterations -> 4000, AccuracyGoal -> 100,
    WorkingPrecision -> 60], 40]

(* Checking Order-Equations *)
Table[{i, f[i]} /. sym /. s06[[2]], {i, 1, 54}]
```

7: Numerical Methods in Chemistry

7 CASE

```
s07 = SetAccuracy[FindMinimum[ieq1[[1]],
    {c1, 0.157762897725930697917532845268493185145171597347747`38.4199},
    {d1, 0.384787429156706730527801924229129496190015613871902`36.0604},
    {d2, 0.537152555989684389515918151892791320029762528710427`36.4687},
    {d3, -2.75516429385844724941034954654155289949208685674067`37.2587},
    {c2, 0.467438995915168927591319039089291173788561271507678`39.6697},
    {c3, -0.27595733515581923937771238732735704328152559965587`39.4408},
    {c4, 0.361045461966483627494012220630290681520002028199`39.5576},
    {c5, -0.362532672640351934138235919380803829139116247420566`39.5593},
    {c6, 0.65302672638564088872094264533249980251067075307591`39.8149},
    {c7, -0.000784074197052968207858443612413970543763803054151`36.8944},
    {d4, 0.00246910992120186533425550860901781522105338274956`37.3925},
    {d5, 2.70669057175892270962720674626688773244314244945308`40},
    {d6, -1.78271827185360917346059742414586082937973577902984`40},
    {d7, 1.90678289888554072786576463968958736498784866098554`40},
    Method -> LevenbergMarquardt, MaxIterations -> 4000, AccuracyGoal -> 100,
    WorkingPrecision -> 60], 40]

(* Checking Order-Equations *)
Table[{i, f[i]} /. sym /. s07[[2]], {i, 1, 54}]
```

8 CASE

```
s08 = SetAccuracy[FindMinimum[ieq1[[1]],
    {c1, 1.160505518178975662642692253224809111419043387937046`39.2865},
    {d1, -0.022456133653776644572553453559932595831557433189993`35.1308},
    {d2, 0.98976140421570730084132288530135859582919536547445`36.9354},
    {d3, -1.35795784772609439390783429346202499026254689498887`37.4261},
    {c2, -0.70912454370979066676966709401557095067773796066488`39.8507},
    {c3, -0.189687270732951460221166003273924563048058586999315`39.278},
    {c4, 1.091740832974322464899301479921861169652108945557969`40},
    {c5, 0.00092831166587088293465838584827667644354090364736`36.9677},
    {c6, -1.16771206117785945166322938111507999875371968392793`40},
    {c7, 0.813349212801432568177410359409628554964822994449755`39.9103},
    {d4, 3.05110723845648465273840438869845079575002039631123`40},
    {d5, -3.04388358806618283991107164905055490712090205271937`40},
    {d6, 1.17503030484614689769758107069368137101747010296650`40},
    {d7, 0.208398621927715025041974015003349645340436088159`39.3189},
    Method -> LevenbergMarquardt, MaxIterations -> 4000, AccuracyGoal -> 100,
    WorkingPrecision -> 60], 40]

(* Checking Order-Equations *)
Table[{i, f[i]} /. sym /. s08[[2]], {i, 1, 54}]
```

9 CASE

```
s09 =
  SetAccuracy[FindMinimum[ieq1[[1]],
    {c1, 1.21391493938824496956796275507354388878299042578919943`39.306},
    {d1, 0.65188871003409818704778781191978966965088160134136.8365},
    {d2, -0.269008276449660237378475158967022584848629601943`6.6474},
    {d3, 0.46323898185710606433157702785337535938098563568879723`7.1343},
    {c2, -1.127331425930025332716021180910683185404909059733144740},
    {c3, 0.5825618771809948948893020249917254289042994585877885`39.7653},
    {c4, 0.3489088675941861532396537993899433353597726387354855`39.5427},
    {c5, -0.8871028359164190319738683258974139107824801852587075`39.948},
    {c6, 1.0730293658697156399841274529016531338764926046059540},
    {c7, -0.2039807881866972929911565255487686907361658827265762`39.3096},
    {d4, -0.8480519704790304958875114383839096128867284670525193`39.9284},
    {d5, 0.6688188258993956491237208277553986227550319008521496`39.8253},
    {d6, -0.6635604205399942061389624532361564310486893921687984`39.8219},
    {d7, 0.996674149678085038897235739988204650633381681532170739.9986},
    Method -> LevenbergMarquardt, MaxIterations -> 4000, AccuracyGoal -> 100,
    WorkingPrecision -> 60], 40]

(* Checking Order-Equations *)
Table[{i, f[i]} /. sym /. s09[[2]], {i, 1, 54}]
```

10 CASE

```
s10 = SetAccuracy[FindMinimum[ieq1[[1]],
  {c1, 0.6585180505914448657097531869607281`7.0404},
  {d1, 4.5607269695621617742428487971405735`17.8809},
  {c2, -0.000700151532114397017955709204453`14.8452},
  {c3, -0.5003886233362296875881725099874727`17.6993},
  {c4, 0.4925720688848692074834900722412347`17.6925},
  {c5, 0.3505889178392951333051996698353697`17.5448},
  {c6, -0.38769633044551848888303879901960717.5885},
  {c7, 0.3871060679982533669907240891741997`17.5878},
  {d2, -4.5905381768518438434512063394875111`8.6619},
  {d3, 0.3821655925422511868028152548248837`17.5823},
  {d4, 0.5475919167548284338664643655461867`17.7385},
  {d5, 3.1487090623595808125401163123743148`18.4981},
  {d6, 0.0041194176381848055662482366088145`15.6148},
  {d7, -3.0527747820051631677529258012724218`18.4847}, Method -> LevenbergMarquardt,
  MaxIterations -> 4000, AccuracyGoal -> 100, WorkingPrecision -> 100], 40]

(* Checking Order-Equations *)
Table[{i, f[i]} /. sym /. s10[[2]], {i, 1, 54}]
```

7: Numerical Methods in Chemistry 359

11 CASE

s11 =
 SetAccuracy[FindMinimum[ieq1[[1]],
  {c1, 0.4515650720436605615367690784440259277162663565550780`38.8766},
  {d1, 1.904232780508446387453331227588459749189981517608`35.0349},
  {d2, -1.939586366441924605272004546461976654374494247563`35.0445},
  {d3, 0.396076651023180289112974915012668544242974567581577`36.7875},
  {c2, -0.002625517726040550321216631834885218246405348766761`37.4192},
  {c3, -0.288746249091012820496917716870631916494703745953472`39.4605},
  {c4, 0.470372004342290132044659600386105293232475851825280`39.6724},
  {c5, 0.370446676335932732131165391352841937409097523478947`39.5687},
  {c6, 0.193479673253384564857895760894223168079286129168673`39.2866},
  {c7, -0.194491659158214624369263311772055856751377045408175`39.2889},
  {d4, 0.513386810409069562674038105483497027676454866781561`39.7104},
  {d5, -2.96773946060454736526385826416429898676763712032755 06`40},
  {d6, 0.00417740952866931573914167511804725320020175143408 87`37.6209},
  {d7, 3.089452175577103675758054310935004756651195775309342 1`40},
  Method -> LevenbergMarquardt, MaxIterations -> 4000, AccuracyGoal → 100,
  WorkingPrecision → 60], 40]

(* Checking Order-Equations *)
Table[{i, f[i]} /. sym /. s11[[2]], {i, 1, 54}]

12 CASE

s12 =
 SetAccuracy[FindMinimum[ieq1[[1]],
  {c1, 0.5198279300409124443524814688381800858875108521728 96`38.9377},
  {d1, -0.8747531757059798445626269775116185918285507300647 667`35.1604},
  {d2, 2.706672984826799656726163541782250596628445952809 2512`35.7551},
  {d3, -2.690592842236979455975391963897640764143797716629 0086`36.3193},
  {c2, 0.391127454478455426289053037158213301818835755808 8402`39.5923},
  {c3, 0.005680899644185065712316739150909581215656954509 1911`37.7544},
  {c4, -0.404738796316473750623431836433431969220344757643 8434`39.6072},
  {c5, -0.164675388520224123807252493051875127138838155917 3575`39.2166},
  {c6, -0.113672264159751693186596663880946113138065965011 4132`39.0557},
  {c7, 0.766450164832896631263429748218950240575245316081 6867`39.8845},
  {d4, 1.92395512249739801797744163079565160346582238486472 83`40},
  {d5, -1.36524465519793006927671182763421435489845568411 6165`40},
  {d6, 1.11609267720558103821500488175342760593853385601253 92`40},
  {d7, 0.18386988861111065689612071471214390483800193712342 17`39.2645},
  Method -> LevenbergMarquardt, MaxIterations -> 4000, AccuracyGoal → 100,
  WorkingPrecision → 60], 40]

(* Checking Order-Equations *)
Table[{i, f[i]} /. sym /. s12[[2]], {i, 1, 54}]

13 CASE

s13 =
 SetAccuracy[FindMinimum[ieq1[[1]],
  {c1, 0.4107924030304732009021600714841693814605746320779928`38.8355},
  {d1, -0.5221752158944872712061530374881420091805816719155`36.3309},
  {d2, 1.186836399295929973385029409832433995416595420055`36.7007},
  {d3, -1.050313642183783689772327931741505253000734959183`38.1788},
  {c2, -0.083974361757121582517396974652681043279854500298018`38.9241},
  {c3, 1.333397639082588717390737853644245726003991904259342`40},
  {c4, -0.316982037566279699301798270404815315958529333051285`39.501},
  {c5, 0.374983982862989001186733679741933355151262342338756`39.574},
  {c6, -0.753016552800424903519209812245563563971971464784334`39.8768},
  {c7, 0.034798927147775265858773452432711460594526419457545`38.5416},
  {d4, 0.135791197097509982220558638252277345290056258800350`39.1329},
  {d5, 0.815034886705722713617186649738855260752271982021439`39.9112},
  {d6, -1.515571300718636336540797918350016114751065419642566`40},
  {d7, 1.950397675977446282965041897560967754734583898644632`40},
  Method -> LevenbergMarquardt, MaxIterations -> 4000, AccuracyGoal -> 100,
  WorkingPrecision -> 60], 40]

(* Checking Order-Equations *)
Table[{i, f[i]} /. sym /. s13[[2]], {i, 1, 54}]

14 CASE

s14 =
 SetAccuracy[FindMinimum[ieq1[[1]],
  {c1, 7.92671488865472218228098402283296754624646475193110`40},
  {d1, -0.000098458576731711244047978849669170276261249858`33.3762},
  {d2, 0.342996415697657093624632977133759613773541464744263`37.1962},
  {d3, 0.501083510387527963678830280348361358035074491819`37.4171},
  {c2, -7.82820900895099041479104156336639101180838642029377`40},
  {c3, 0.553674984678513674554533803446455684249412119715946`39.7433},
  {c4, 0.379166810743287038912425052690522037383321727120050`39.5788},
  {c5, -1.55574633564104984061976406634280140948830351789703`40},
  {c6, 6.62905635095816257700622094315655773521959784486512`40},
  {c7, -5.10465769044264521734335819241731058180210650544142`40},
  {d4, -0.739315233906230912855085534538963116740515669448650`39.8688},
  {d5, -0.003772637348990384315214766640691003818620267628890`37.5766},
  {d6, 0.000313927107832139637548837316179708874312805288658`36.4968},
  {d7, 0.898792476638935811473336185231022610152468425084160`39.9537},
  Method -> LevenbergMarquardt, MaxIterations -> 4000, AccuracyGoal -> 100,
  WorkingPrecision -> 60], 40]

(* Checking Order-Equations *)
Table[{i, f[i]} /. sym /. s14[[2]], {i, 1, 54}]

7: Numerical Methods in Chemistry   361

15 CASE

```
s15 =
 SetAccuracy[FindMinimum[ieq1[[1]],
   {c1, 0.1401281121758607547564899537726051124509528772002407`38.3684},
   {d1, 1.0195086305634474456773120285613122599921139667555403`35.6356},
   {d2, -0.6491252848234459252421810833433973994469667498516693`35.442},
   {d3, 1.0247076334964868305142013233859395367118768825458058`37.8917},
   {c2, -0.0072918399179441305758532506792176822518556350328431`37.8628},
   {c3, 0.4795823275954200956728875053197298671108865407617624`39.6809},
   {c4, -0.1336602855589300061521496959021161170924438874206797`39.126},
   {c5, -0.0006138082123165036106844608855473377969970235608335`36.788},
   {c6, 0.0846465964218295438949082500412977721182643598677447`38.9276},
   {c7, 0.4372088974960802460144016983332483854611927681846092`39.6407},
   {d4, -2.7182291572325748018708677513423400835929565945865459`40},
   {d5, 2.8296635817349436324891890739726954207658076039303024`40},
   {d6, -0.6192181082230426876614178826459456792592152756545547`39.7918},
   {d7, 0.1126927044841855060937642914117359448293401668611136`39.0519},
   Method -> LevenbergMarquardt, MaxIterations -> 4000, AccuracyGoal -> 100,
   WorkingPrecision -> 60], 40]

(* Checking Order-Equations *)
Table[{i, f[i]} /. sym /. s15[[2]], {i, 1, 54}]
```

16 CASE

```
s16 =
 SetAccuracy[FindMinimum[ieq1[[1]],
   {c1, 0.1185725214765229371220595312279486532215452102622776`38.2958},
   {d1, 0.3409834692494895556096646337810510684463906842269`36.0454},
   {d2, -0.0675889611521251018389440973867856490890162391319`35.6363},
   {d3, 0.5652364829065519006768349019339952587508871146634203`36.5744},
   {c2, 0.7561404701413999145488475016196445490028681061914`39.8786},
   {c3, -0.2253241467252794646350141847731890931723094847348468`39.3528},
   {c4, 0.3948170446081320441949787583013306319774203376490543`39.5964},
   {c5, -2.4635061319824821041384511693926699689542625185232582`40},
   {c6, 2.8784560119831227621870931322726424858597180349835606`40},
   {c7, -0.4591557695014160892795135692557072579349796858281875`39.662},
   {d4, -0.4102966281594581738733878955108777382442334707983334`39.6131},
   {d5, -0.0004393091620731343547917244581657957238709222310357`36.6428},
   {d6, 0.0159795522326675699248884586693490787220378709530393`38.2036},
   {d7, 0.5561253940849473838557357229714337771378049622741598`39.7452},
   Method -> LevenbergMarquardt, MaxIterations -> 4000, AccuracyGoal -> 100,
   WorkingPrecision -> 60], 40]

(* Checking Order-Equations *)
Table[{i, f[i]} /. sym /. s16[[2]], {i, 1, 54}]
```

17 CASE

```
s17 =
 SetAccuracy[FindMinimum[ieq1[[1]],
   {c1, 7.0887027930685437620129087174755666171717310927718654`40},
   {d1, 5.1193846118481367101215173875233592127904327`32.6724*^-6},
   {d2, 0.7872443814666632521747944752441166202962321545327`38.1103},
   {d3, -0.1249490806707022132537231885910501772795010753258005`37.4555},
   {c2, -6.7659259876020226752970006286254452395568328782215535`40},
   {c3, 0.8946519271125878675205540435472110510290929420508944`39.9517},
   {c4, -1.1960257877396021718883316004376101490552032629519957`40},
   {c5, 0.5203462283193135503908836783794890479437039704457532`39.7163},
   {c6, -0.5563780628630533799819979902342577478440472505920544`39.7454},
   {c7, 1.0146288897042330472429837798950464203115553864970906`40},
   {d4, -0.5308756841777124491826397443067635494768049183332453`39.725},
   {d5, -0.0801493285532888892355577830107450034483705419938722`38.9039},
   {d6, 0.4888145582828314714300491454780515500583808452299756`39.6891},
   {d7, 0.4599100342675969799303669736690030364908507454596735`39.6627},
   Method -> LevenbergMarquardt, MaxIterations -> 4000, AccuracyGoal -> 100,
   WorkingPrecision -> 60], 40]

(* Checking Order-Equations *)
Table[{i, f[i]} /. sym /. s17[[2]], {i, 1, 54}]
```

18 CASE

```
s18 =
 SetAccuracy[FindMinimum[ieq1[[1]],
   {c1, 0.1561972283647089531906583301704042054174680789849596`38.4155},
   {d1, 1.1235327055249202628390771246244326500740860478257`34.183},
   {d2, -2.6007402034131716209439868827230533872108959421477`34.8482},
   {d3, 2.6163097418066945147975561547256653903878559296105185`34.8515},
   {c2, -0.2694780316925820086119116004336312942019322508351584`39.4305},
   {c3, -0.0003071475231560441337774154277473975846208000182949`36.4873},
   {c4, 0.2478341947526400532256970480135731757173410755488067`39.3942},
   {c5, 0.8467823287913191445762790240128124112746695504066296`39.9278},
   {c6, -0.3035744438144149949165988997392712286976625154107247`39.4823},
   {c7, 0.3225458711214848966696535134038601280747368613237821`39.5086},
   {d4, -0.7176020855822318177159005350970551668501734449777946`39.8559},
   {d5, -0.0223379173989857731928667609846488918951021163172527`38.349},
   {d6, 0.4821441688286020658300970206028090717003111187411308`39.6832},
   {d7, 0.1186935902341723683860238788518503337939184072664521`39.0744},
   Method -> LevenbergMarquardt, MaxIterations -> 4000, AccuracyGoal -> 100,
   WorkingPrecision -> 60], 40]

(* Checking Order-Equations *)
Table[{i, f[i]} /. sym /. s18[[2]], {i, 1, 54}]
```

7: Numerical Methods in Chemistry

19 CASE

```
s19 =
 SetAccuracy[FindMinimum[ieq1[[1]],
    {c1, 0.13598682086023694477874560513405548574273592321605133`38.3553},
    {d1, 0.48194589189704243129263351017221169659764771053232`36.6685},
    {d2, -0.054403221054522651320567854364071246625794677505`35.7394},
    {d3, -0.068994793152332334848416507238308307862094623574392732`37.2518},
    {c2, -0.242736558973466884880817053652262195853600184490077632`39.3851},
    {c3, 2.889299049943851755697588253329192318721043449720799852`40},
    {c4, 0.022374359584696801032247208570560154312521284296441532`38.3498},
    {c5, -1.810638380744174969716690855834057657049050487432`34.2578*^-6},
    {c6, -2.10925576723923862869552844892911527745857363233569712`40},
    {c7, 0.30433390646230075624273415227569448010413916115577082`39.4834},
    {d4, 6.463841034581074010315920991021729593230027537743016332`40},
    {d5, -6.39737508770863403896824311308776415053317890267407922`40},
    {d6, 0.4626302510697121320972420087723659924001113694710696`39.6652},
    {d7, 0.1123559243676604514314309647238364227932815860073726`39.0506},
    Method -> LevenbergMarquardt, MaxIterations -> 4000, AccuracyGoal → 100,
    WorkingPrecision → 60], 40]

(* Checking Order-Equations *)
Table[{i, f[i]} /. sym /. s19[[2]], {i, 1, 54}]
```

20 CASE

```
s20 =
 SetAccuracy[FindMinimum[ieq1[[1]],
    {c1, 0.5588383604393864558615809025738642066621092984362765`38.9691},
    {d1, 2.3138571029874738893322985682115248146274148807292`32.1537},
    {d2, -2.7518392141778237671392525559198745427118765787852`32.2291},
    {d3, 0.39419882347674469996606773910748857561627844486861462`34.7301},
    {c2, -0.00051268425256126272877473463417215950302791251428072`36.7098},
    {c3, 0.01676859118233903606824808647996173956135392257799432`38.2245},
    {c4, -0.41359262999449504345524388922223439484201527507390222`39.6166},
    {c5, 0.5717963925992204010691189069103769287577425555951934`39.7572},
    {c6, -0.074328370443702792291661757740074307576333583654938132`38.8712},
    {c7, 0.34103034046981320547673248563227798694017099463365712`39.5328},
    {d4, 0.39515176517090370991600740566907815150657460174966792`39.5968},
    {d5, -0.20445345247249242571540259954650116192242871009883552`39.3106},
    {d6, 0.7340356319438480526015288964523985061870048806566004`39.8657},
    {d7, 0.11904934307134584103875254602588565669703248087978282`39.0757},
    Method -> LevenbergMarquardt, MaxIterations -> 4000, AccuracyGoal → 100,
    WorkingPrecision → 60], 40]

(* Checking Order-Equations *)
Table[{i, f[i]} /. sym /. s20[[2]], {i, 1, 54}]
```

21 CASE

```
s21 =
 SetAccuracy[FindMinimum[ieq1[[1]],
   {c1, 1.0528965897292023349852870810404036295227286963590465`39.2442},
   {d1, 0.8184949967434246102133722642411841952986026446414`36.4341},
   {d2, -0.8017261954151725104477938295805446897000750744425`36.4462},
   {d3, 0.7013098779873064247792187290204580618916987120311`37.7046},
   {c2, 0.0647250641408246840433406244023472833569562619019958`38.8111},
   {c3, -1.0467221400979323876308912993709311528046894007922293`40},
   {c4, 1.4303771081953293328860411518440346197293135951146367`40},
   {c5, 0.2362906890725930685799462613581147568028471940540151`39.3734},
   {c6, -1.8686076395382426048469872884924069843012439693846888`40},
   {c7, 1.1310403284982255719832634692184378476940876227472241`40},
   {d4, -0.2123782487539694790545856536793352466606784909787963`39.3271},
   {d5, 0.1013344133048720968662955210659890675265292337810174`39.0058},
   {d6, -0.2071645105181385817667082343065895534339534740163217`39.3163},
   {d7, 0.6001296666516774394102012032388381650778764489934861`39.7782},
   Method -> LevenbergMarquardt, MaxIterations -> 4000, AccuracyGoal -> 100,
   WorkingPrecision -> 60], 40]

(* Checking Order-Equations *)
Table[{i, f[i]} /. sym /. s21[[2]], {i, 1, 54}]
```

22 CASE

```
s22 =
 SetAccuracy[FindMinimum[ieq1[[1]],
   {c1, 0.1777292121959306045303056796812520206114236984594222`38.4716},
   {d1, 0.4562797565486187944846669551402461354149248574291471`36.176},
   {d2, 0.6057100941878303087882115570400315971768290754225296`36.612},
   {d3, -3.0640147019218270549021710263550483406686179191270105`37.2935},
   {c2, 0.5332267797923824087807713232299372353075686391410409`39.7269},
   {c3, -0.4306460201494765431566837255866990593092815712319309`39.6341},
   {c4, -0.0003457846453632553205822176699479162412763299410631`36.5388},
   {c5, 0.7441253805050216235093993457056432843497064658203868`39.8716},
   {c6, -0.7353170527903293902842844824470205931135922877649711`39.8665},
   {c7, 0.7112274850918455014107407708683502839545138551711511`39.852},
   {d4, 2.998510262545449806691961825484687424179376646536957`40},
   {d5, 0.5428596794747629686693691660139395822502513451886779`39.7347},
   {d6, 0.0125429908218237345866697115960653196377897209383432`38.0984},
   {d7, -0.5518880816566585583187081889199217179905537263886444`39.7419},
   Method -> LevenbergMarquardt, MaxIterations -> 4000, AccuracyGoal -> 100,
   WorkingPrecision -> 60], 40]

(* Checking Order-Equations *)
Table[{i, f[i]} /. sym /. s22[[2]], {i, 1, 54}]
```

7: Numerical Methods in Chemistry   365

23 CASE

s23 =
SetAccuracy[FindMinimum[ieq1[[1]],
  {c1, 0.3193664305308008428365849542564388842762757430565692`38.7261},
  {d1, 0.7842664923090667113608312281820877845186526484``36.7659},
  {d2, -0.1297313663061407662351740949260233613518977649290`36.5365},
  {d3, -0.5280703174095050404482546353473857467180441880959380`36.7389},
  {c2, 0.8952474113144769284343128633238535498513488074008983`39.9519},
  {c3, -1.2093101888917931521764517180309957034075388799505661`40},
  {c4, 0.5937950696255136147760121728983311990837029215796558`39.7736},
  {c5, 3.2747860288643686243914966737008371694311791048017992`40},
  {c6, -3.9097113590985364193798555921342420798128024561049891`40},
  {c7, 1.0358266076551695611179006459857769805778347592166327`40},
  {d4, -0.0686301399495523079147853676807676250696194205797594`38.8365},
  {d5, 0.0000678699493969236206910449816655959882281525341896`35.8317},
  {d6, 0.4743441848872657787564414335414528617246025135363611`39.6761},
  {d7, 0.4677532765194687008602503912489705273080780591135307`39.67}},
  Method -> LevenbergMarquardt, MaxIterations -> 4000, AccuracyGoal -> 100,
  WorkingPrecision -> 60], 40]

(* Checking Order-Equations *)
Table[{i, f[i]} /. sym /. s23[[2]], {i, 1, 54}]

+ 23 cases c_i < - > d_{8-i}

symsym = Flatten[Table[{c_i → d_{8-i}, d_i → c_{8-i}}, {i, 1, 7}]] /. sym

24 case < - (1 case) . . . . .

s24 = {c1 → 1.5225441490806472356407814586668913810617606310158748`40.1826,
  d1 → -0.0276076577102055608043214464048438206623778968076116`38.441,
  d2 → 1.1806851009657162510944225701818271853926639076053859`40.0721,
  d3 → -1.2245716273903879923786428453311111057908960873416552`40.088,
  c2 → -1.1500107005532435941190273029527185981897806736966554`40.0607,
  c3 → -0.0098760842661144384985297687650653460972512482084547`37.9946,
  c4 → 1.1980080119145096617108669102631742637556593323324128`40.0785,
  c5 → -1.3948478584650480494198918216818614613624129871749329`40.1445,
  c6 → 0.4701662641796147899294412383666039478215412619649482`39.6723,
  c7 → 0.3640162181096343947563592861029758130104836837668074`39.5611,
  d4 → 0.0229120005615084430720558295724715773366673854382655`38.3601,
  d5 → 0.4344300275784419746616623108218542208914957542564479`39.6379,
  d6 → 0.4825530233435728689255191056362255910925243451642721`39.6835,
  d7 → 0.1315991326513540154293044755235763517399225916848643`39.1193} /. symsym

(* Checking Order-Equations *)
Table[{i, f[i]} /. sym /. s24, {i, 1, 54}]

Appendix G

Maple programme for the Cases I and II presented in the section 5

```
> 'Case I';
>
> restart;
>
> y[n+6]:=(x+6*h)^n;
> y[n-6]:=(x-6*h)^n;
> y[n+5]:=(x+5*h)^n;
> y[n-5]:=(x-5*h)^n;
> y[n+4]:=(x+4*h)^n;
> y[n-4]:=(x-4*h)^n;
> y[n+3]:=(x+3*h)^n;
> y[n-3]:=(x-3*h)^n;
> y[n+2]:=(x+2*h)^n;
> y[n-2]:=(x-2*h)^n;
> y[n+1]:=(x+h)^n;
> y[n-1]:=(x-h)^n;
>
> f[n+6]:=diff(y[n+6],x$2);
> f[n-6]:=diff(y[n-6],x$2);
> f[n+5]:=diff(y[n+5],x$2);
> f[n-5]:=diff(y[n-5],x$2);
> f[n+4]:=diff(y[n+4],x$2);
> f[n-4]:=diff(y[n-4],x$2);
> f[n+3]:=diff(y[n+3],x$2);
> f[n-3]:=diff(y[n-3],x$2);
> f[n+2]:=diff(y[n+2],x$2);
> f[n-2]:=diff(y[n-2],x$2);
> f[n+1]:=diff(y[n+1],x$2);
> f[n-1]:=diff(y[n-1],x$2);;
> y[n]:=(x)^n;
> f[n]:=diff(y[n],x$2);
>
> errornew:=simplify(y[n+6]+y[n-6]-2*(y[n+5]+y[n-5])
+2*(y[n+4]+y[n-4])-2*(y[n+3]+y[n-3])+2*(y[n+2]+y[n-2])
-2*(y[n+1]+y[n-1])+2*y[n]=h^2*(b[5]*(f[n+5]+f[n-5])
+b[4]*(f[n+4]+f[n-4])+b[3]*(f[n+3]+f[n-3])+b[2]*(f[n+2]+f[n-2])
+b[1]*(f[n+1]+f[n-1])+b[0]*f[n]));
>
> n:=0;
> eq1:=simplify(errornew);
> n:=2;
> eq1:=simplify(errornew);
> eq1:=simplify(eq1/h^2);
```

7: Numerical Methods in Chemistry

```
>
> n:=4;
> eq2:=simplify(errornew);
> eq2:=simplify(eq2–6*x^2*h^2*eq1);
> eq2:=simplify(eq2/h^4);
>
> n:=6;
> eq3:=simplify(errornew);
> eq3:=simplify(eq3–15*x^2*h^4*eq2-15*x^4*h^2*eq1);
> eq3:=simplify(eq3/h^6);
> n:=8;
> eq4:=simplify(errornew);
> eq4:=simplify(eq4–28*x^2*h^6*eq3–70*x^4*h^4*eq2–28*x^6*h^2*eq1);
> eq4:=simplify(eq4/h^8);
>
> y[n+6]:=exp(I*w*(x+6*h));
> y[n–6]:=exp(I*w*(x–6*h));
> y[n+5]:=exp(I*w*(x+5*h));
> y[n–5]:=exp(I*w*(x–5*h));
> y[n+4]:=exp(I*w*(x+4*h));
> y[n–4]:=exp(I*w*(x–4*h));
> y[n+3]:=exp(I*w*(x+3*h));
> y[n–3]:=exp(I*w*(x–3*h));
> y[n+2]:=exp(I*w*(x+2*h));
> y[n–2]:=exp(I*w*(x–2*h));
> y[n+1]:=exp(I*w*(x+h));
> y[n–1]:=exp(I*w*(x–h));
> f[n–6]:=diff(y[n–6],x$2);
> f[n+6]:=diff(y[n+6],x$2);
> f[n–5]:=diff(y[n–5],x$2);
> f[n+5]:=diff(y[n+5],x$2);
> f[n–4]:=diff(y[n–4],x$2);
> f[n+4]:=diff(y[n+4],x$2);
> f[n–3]:=diff(y[n–3],x$2);
> f[n+3]:=diff(y[n+3],x$2);
> f[n–2]:=diff(y[n–2],x$2);
> f[n+2]:=diff(y[n+2],x$2);
> f[n–1]:=diff(y[n–1],x$2);
> f[n+1]:=diff(y[n+1],x$2);
>
> y[n]:=exp(I*w*(x));
> f[n]:=diff(y[n],x$2);;
> errornew:=simplify(y[n+6]+y[n–6]–2*(y[n+5]+y[n–
5])+2*(y[n+4]+y[n–4])–2*(y[n+3]+y[n–3])+2*(y[n+2]+y[n–2])–
2*(y[n+1]+y[n–1])+2*y[n]=h^2*(b[5]*(f[n+5]+f[n–
5])+b[4]*(f[n+4]+f[n–4])+b[3]*(f[n+3]+f[n–3])+b[2]*(f[n+2]+f[n–
2])+b[1]*(f[n+1]+f[n–1])+b[0]*f[n]));
```

```
> errornew: = simplify(errornew/exp(w*x*I));
> eq5: = errornew;
>
> y[n+6]: = (x+6*h)*exp(I*w*(x+6*h));
> y[n-6]: = (x-6*h)*exp(I*w*(x-6*h));
> y[n+5]: = (x+5*h)*exp(I*w*(x+5*h));
> y[n-5]: = (x-5*h)*exp(I*w*(x-5*h));
> y[n+4]: = (x+4*h)*exp(I*w*(x+4*h));
> y[n-4]: = (x-4*h)*exp(I*w*(x-4*h));
> y[n+3]: = (x+3*h)*exp(I*w*(x+3*h));
> y[n-3]: = (x-3*h)*exp(I*w*(x-3*h));
> y[n+2]: = (x+2*h)*exp(I*w*(x+2*h));
> y[n-2]: = (x-2*h)*exp(I*w*(x-2*h));
> y[n+1]: = (x+h)*exp(I*w*(x+h));
> y[n-1]: = (x-h)*exp(I*w*(x-h));
> f[n-6]: = diff(y[n-6],x$2);
> f[n+6]: = diff(y[n+6],x$2);
> f[n-5]: = diff(y[n-5],x$2);
> f[n+5]: = diff(y[n+5],x$2);
> f[n-4]: = diff(y[n-4],x$2);
> f[n+4]: = diff(y[n+4],x$2);
> f[n-3]: = diff(y[n-3],x$2);
> f[n+3]: = diff(y[n+3],x$2);
> f[n-2]: = diff(y[n-2],x$2);
> f[n+2]: = diff(y[n+2],x$2);
> f[n-1]: = diff(y[n-1],x$2);
> f[n+1]: = diff(y[n+1],x$2);
>
> y[n]: = x*exp(I*w*(x));
> f[n]: = diff(y[n],x$2);;
> errornew: = simplify(y[n+6]+y[n-6]-2*(y[n+5]+y[n-5])
   +2*(y[n+4]+y[n-4])-2*(y[n+3]+y[n-3])+2*(y[n+2]+y[n-2])
   -2*(y[n+1]+y[n-1])+2*y[n] = h^2*(b[5]*(f[n+5]+f[n-5])
   +b[4]*(f[n+4]+f[n-4])+b[3]*(f[n+3]+f[n-3])+b[2]*(f[n+2]+f[n-2])
   +b[1]*(f[n+1]+f[n-1])+b[0]*f[n]));
> errornew: = simplify(errornew/exp(w*x*I));
> lr: = simplify(evalc(Re(2*x-2*exp(-3*I*w*h)*x-2*exp(w*h*I)*h-
   6*exp(-6*I*w*h)*h+2*exp(-I*w*h)*h-2*exp(w*h*I)*x+
   6*exp(-3*I*w*h)*h-6*exp(3*I*w*h)*h-2*exp(-I*w*h)*x+
   exp(-6*I*w*h)*x+2*exp(-4*I*w*h)*x+2*exp(4*I*w*h)*x-
   2*exp(5*I*w*h)*x+2*exp(2*I*w*h)*x+8*exp(4*I*w*h)*h-
   2*exp(-5*I*w*h)*x+10*exp(-5*I*w*h)*h+exp(6*I*w*h)*x-
   8*exp(-4*I*w*h)*h-2*exp(3*I*w*h)*x+4*exp(2*I*w*h)*h+
   2*exp(-2*I*w*h)*x-4*exp(-2*I*w*h)*h+6*exp(6*I*w*h)*h-
   10*exp(5*I*w*h)*h)));
> li: = simplify(evalc(Im(2*x-2*exp(-3*I*w*h)*x-2*exp(w*h*I)*h-
   6*exp(-6*I*w*h)*h+2*exp(-I*w*h)*h-2*exp(w*h*I)*x+
```

7: Numerical Methods in Chemistry

```
    6*exp(-3*I*w*h)*h - 6*exp(3*I*w*h)*h - 2*exp(-I*w*h)*x +
exp(-6*I*w*h)*x + 2*exp(-4*I*w*h)*x + 2*exp(4*I*w*h)*x -
2*exp(5*I*w*h)*x + 2*exp(2*I*w*h)*x + 8*exp(4*I*w*h)*h -
2*exp(-5*I*w*h)*x + 10*exp(-5*I*w*h)*h + exp(6*I*w*h)*x -
8*exp(-4*I*w*h)*h - 2*exp(3*I*w*h)*x + 4*exp(2*I*w*h)*h +
2*exp(-2*I*w*h)*x - 4*exp(-2*I*w*h)*h + 6*exp(6*I*w*h)*h -
10*exp(5*I*w*h)*h)));
>   rr:=simplify(evalc(Re(-h^2*w*(b[0]*x*w + b[1]*w*exp(w*h*I)*x +
b[1]*w*exp(w*h*I)*h + b[4]*w*exp(-4*I*w*h)*x - 4*b[4]*w*
exp(-4*I*w*h)*h + b[3]*w*exp(3*I*w*h)*x +
3*b[3]*w*exp(3*I*w*h)*h + b[3]*w*exp(-3*I*w*h)*x -
b[1]*w*exp(-I*w*h)*h + b[2]*w*exp(-2*I*w*h)*x -
2*b[2]*w*exp(-2*I*w*h)*h + b[4]*w*exp(4*I*w*h)*x +
4*b[4]*w*exp(4*I*w*h)*h + b[2]*w*exp(2*I*w*h)*x +
2*b[2]*w*exp(2*I*w*h)*h + b[1]*w*exp(-I*w*h)*x +
b[5]*w*exp(5*I*w*h)*x + 5*b[5]*w*exp(5*I*w*h)*h +
b[5]*w*exp(-5*I*w*h)*x - 5*b[5]*w*exp(-5*I*w*h)*h -
3*b[3]*w*exp(- 3*I*w*h)*h - 2*I*b[0] - 2*I*b[1]*exp(w*h*I) -
2*I*b[5]*exp(-5*I*w*h) - 2*I*b[2]*exp(2*I*w*h) -
2*I*b[4]*exp(-4*I*w*h) - 2*I*b[3]*exp(3*I*w*h) -
2*I*b[3]*exp(-3*I*w*h) - 2*I*b[2]*exp(-2*I*w*h) -
2*I*b[5]*exp(5*I*w*h) - 2*I*b[4]*exp(4*I*w*h) -
2*I*b[1]*exp(-I*w*h)))));[:
>   ri:=simplify(evalc(Im(-h^2*w*(b[0]*x*w + b[1]*w*exp(w*h*I)*x +
b[1]*w*exp(w*h*I)*h + b[4]*w*exp(-4*I*w*h)*x -
4*b[4]*w*exp(- 4*I*w*h)*h + b[3]*w*exp(3*I*w*h)*x +
3*b[3]*w*exp(3*I*w*h)*h + b[3]*w*exp(-3*I*w*h)*x -
b[1]*w*exp(-I*w*h)*h + b[2]*w*exp(-2*I*w*h)*x -
2*b[2]*w*exp(- 2*I*w*h)*h + b[4]*w*exp(4*I*w*h)*x +
4*b[4]*w*exp(4*I*w*h)*h + b[2]*w*exp(2*I*w*h)*x +
2*b[2]*w*exp(2*I*w*h)*h + b[1]*w*exp(-I*w*h)*x +
b[5]*w*exp(5*I*w*h)*x + 5*b[5]*w*exp(5*I*w*h)*h +
b[5]*w*exp(-5*I*w*h)*x - 5*b[5]*w*exp(-5*I*w*h)*h -
3*b[3]*w*exp(-3*I*w*h)*h - 2*I*b[0] - 2*I*b[1]*exp(w*h*I) -
2*I*b[5]*exp(-5*I*w*h) - 2*I*b[2]*exp(2*I*w*h) -
2*I*b[4]*exp(-4*I*w*h) - 2*I*b[3]*exp(3*I*w*h) -
2*I*b[3]*exp(-3*I*w*h) - 2*I*b[2]*exp(-2*I*w*h) -
2*I*b[5]*exp(5*I*w*h) - 2*I*b[4]*exp(4*I*w*h) -
2*I*b[1]*exp(-I*w*h)))));
>   eq6:=lr=rr;
>   eq7:=li=ri;
>
>   'eq6 is equivalent with eq5';
>
>   solut:=solve({eq1,eq2,eq3,eq4},{b[0],b[1],b[2],b[3]});
>
```

```
> assign(solut);
> solut1:=solve({eq5,eq7},{b[4],b[5]});
> assign(solut1);
> b[4]:=combine(b[4]);
> b[5]:=combine(b[5]);
> b4t:=convert(taylor(b[4],w=0,50),polynom);
> b5t:=convert(taylor(b[5],w=0,22),polynom);
>
> 'Case II';
>
> restart;
>
> y[n+6]:=(x+6*h)^n;
> y[n-6]:=(x-6*h)^n;
> y[n+5]:=(x+5*h)^n;
> y[n-5]:=(x-5*h)^n;
> y[n+4]:=(x+4*h)^n;
> y[n-4]:=(x-4*h)^n;
> y[n+3]:=(x+3*h)^n;
> y[n-3]:=(x-3*h)^n;
> y[n+2]:=(x+2*h)^n;
> y[n-2]:=(x-2*h)^n;
> y[n+1]:=(x+h)^n;
> y[n-1]:=(x-h)^n;
>
> f[n+6]:=diff(y[n+6],x$2);
> f[n-6]:=diff(y[n-6],x$2);
> f[n+5]:=diff(y[n+5],x$2);
> f[n-5]:=diff(y[n-5],x$2);
> f[n+4]:=diff(y[n+4],x$2);
> f[n-4]:=diff(y[n-4],x$2);
> f[n+3]:=diff(y[n+3],x$2);
> f[n-3]:=diff(y[n-3],x$2);
> f[n+2]:=diff(y[n+2],x$2);
> f[n-2]:=diff(y[n-2],x$2);
> f[n+1]:=diff(y[n+1],x$2);
> f[n-1]:=diff(y[n-1],x$2);;
> y[n]:=(x)^n;
> f[n]:=diff(y[n],x$2);
> errornew:=simplify(y[n+6]+y[n-6]-2*(y[n+5]+y[n-5])
+2*(y[n+4]+y[n-4])-2*(y[n+3]+y[n-3])+2*(y[n+2]+y[n-2])
-2*(y[n+1]+y[n-1])+2*y[n]=h^2*(b[5]*(f[n+5]+f[n-5])
+b[4]*(f[n+4]+f[n-4])+b[3]*(f[n+3]+f[n-3])+b[2]*(f[n+2]+f[n-2])
+b[1]*(f[n+1]+f[n-1])+b[0]*f[n]));
> n:=0;
> eq1:=simplify(errornew);
```

7: Numerical Methods in Chemistry

```
> n:=2;
> eq1:=simplify(errornew);
> eq1:=simplify(eq1/h^2);
>
> n:=4;
> eq2:=simplify(errornew);
> eq2:=simplify(eq2-6*x^2*h^2*eq1);
> eq2:=simplify(eq2/h^4);
>
> n:=6;
> eq3:=simplify(errornew);
> eq3:=simplify(eq3-15*x^2*h^4*eq2-15*x^4*h^2*eq1);
> eq3:=simplify(eq3/h^6);
>
> y[n+6]:=exp(I*w*(x+6*h));
> y[n-6]:=exp(I*w*(x-6*h));
> y[n+5]:=exp(I*w*(x+5*h));
> y[n-5]:=exp(I*w*(x-5*h));
> y[n+4]:=exp(I*w*(x+4*h));
> y[n-4]:=exp(I*w*(x-4*h));
> y[n+3]:=exp(I*w*(x+3*h));
> y[n-3]:=exp(I*w*(x-3*h));
> y[n+2]:=exp(I*w*(x+2*h));
> y[n-2]:=exp(I*w*(x-2*h));
> y[n+1]:=exp(I*w*(x+h));
> y[n-1]:=exp(I*w*(x-h));
> f[n-6]:=diff(y[n-6],x$2);
> f[n+6]:=diff(y[n+6],x$2);
> f[n-5]:=diff(y[n-5],x$2);
> f[n+5]:=diff(y[n+5],x$2);
> f[n-4]:=diff(y[n-4],x$2);
> f[n+4]:=diff(y[n+4],x$2);
> f[n-3]:=diff(y[n-3],x$2);
> f[n+3]:=diff(y[n+3],x$2);
> f[n-2]:=diff(y[n-2],x$2);
> f[n+2]:=diff(y[n+2],x$2);
> f[n-1]:=diff(y[n-1],x$2);
> f[n+1]:=diff(y[n+1],x$2);
>
> y[n]:=exp(I*w*(x));
> f[n]:=diff(y[n],x$2);;
>  errornew:=simplify(y[n+6]+y[n-6]-2*(y[n+5]+y[n-5])
   +2*(y[n+4]+y[n-4])-2*(y[n+3]+y[n-3])+2*(y[n+2]+y[n-2])-
   2*(y[n+1]+y[n-1])+2*y[n]=h^2*(b[5]*(f[n+5]+f[n-5])
   +b[4]*(f[n+4]+f[n-4])+b[3]*(f[n+3]+f[n-3])+b[2]*(f[n+2]+f[n-2])
   +b[1]*(f[n+1]+f[n-1])+b[0]*f[n]));
```

```
>  errornew: = simplify(errornew/exp(w*x*I));
>  eq5: = errornew;
>
>
>  y[n+6]: = (x+6*h)*exp(I*w*(x+6*h));
>  y[n-6]: = (x-6*h)*exp(I*w*(x-6*h));
>  y[n+5]: = (x+5*h)*exp(I*w*(x+5*h));
>  y[n-5]: = (x-5*h)*exp(I*w*(x-5*h));
>  y[n+4]: = (x+4*h)*exp(I*w*(x+4*h));
>  y[n-4]: = (x-4*h)*exp(I*w*(x-4*h));
>  y[n+3]: = (x+3*h)*exp(I*w*(x+3*h));
>  y[n-3]: = (x-3*h)*exp(I*w*(x-3*h));
>  y[n+2]: = (x+2*h)*exp(I*w*(x+2*h));
>  y[n-2]: = (x-2*h)*exp(I*w*(x-2*h));
>  y[n+1]: = (x+h)*exp(I*w*(x+h));
>  y[n-1]: = (x-h)*exp(I*w*(x-h));
>  f[n-6]: = diff(y[n-6],x$2);
>  f[n+6]: = diff(y[n+6],x$2);
>  f[n-5]: = diff(y[n-5],x$2);
>  f[n+5]: = diff(y[n+5],x$2);
>  f[n-4]: = diff(y[n-4],x$2);
>  f[n+4]: = diff(y[n+4],x$2);
>  f[n-3]: = diff(y[n-3],x$2);
>  f[n+3]: = diff(y[n+3],x$2);
>  f[n-2]: = diff(y[n-2],x$2);
>  f[n+2]: = diff(y[n+2],x$2);
>  f[n-1]: = diff(y[n-1],x$2);
>  f[n+1]: = diff(y[n+1],x$2);
>
>  y[n]: = x*exp(I*w*(x));
>  f[n]: = diff(y[n],x$2);;
>  errornew: = simplify(y[n+6] + y[n-6] - 2*(y[n+5] + y[n-5])
   +2*(y[n+4] + y[n-4]) - 2*(y[n+3] + y[n-3]) + 2*(y[n+2] + y[n-2]) -
   2*(y[n+1] + y[n-1]) + 2*y[n] = h^2*(b[5]*(f[n+5] + f[n-5])
   + b[4]*(f[n+4] + f[n-4]) + b[3]*(f[n+3] + f[n-3]) + b[2]*(f[n+2] + f[n-2])
   + b[1]*(f[n+1] + f[n-1]) + b[0]*f[n]));
>  errornew: = simplify(errornew/exp(w*x*I));
>  lr: = simplify(evalc(Re(2*x - 2*exp(-3*I*w*h)*x - 2*exp(w*h*I)*h -
   6*exp(-6*I*w*h)*h + 2*exp(-I*w*h)*h - 2*exp(w*h*I)*x +
   6*exp(-3*I*w*h)*h - 6*exp(3*I*w*h)*h - 2*exp(-I*w*h)*x +
   exp(-6*I*w*h)*x + 2*exp(-4*I*w*h)*x + 2*exp(4*I*w*h)*x -
   2*exp(5*I*w*h)*x + 2*exp(2*I*w*h)*x + 8*exp(4*I*w*h)*h -
   2*exp(-5*I*w*h)*x + 10*exp(-5*I*w*h)*h + exp(6*I*w*h)*x -
   8*exp(-4*I*w*h)*h - 2*exp(3*I*w*h)*x + 4*exp(2*I*w*h)*h +
   2*exp(-2*I*w*h)*x - 4*exp(-2*I*w*h)*h + 6*exp(6*I*w*h)*h -
   10*exp(5*I*w*h)*h)));
```

7: Numerical Methods in Chemistry 373

```
>  li:=simplify(evalc(Im(2*x−2*exp(−3*I*w*h)*x−2*exp(w*h*I)*h−
6*exp(−6*I*w*h)*h+2*exp(−I*w*h)*h−2*exp(w*h*I)*x+
6*exp(−3*I*w*h)*h−6*exp(3*I*w*h)*h−2*exp(−I*w*h)*x+
exp(−6*I*w*h)*x+2*exp(−4*I*w*h)*x+2*exp(4*I*w*h)*x−
2*exp(5*I*w*h)*x+2*exp(2*I*w*h)*x+8*exp(4*I*w*h)*h−
2*exp(−5*I*w*h)*x+10*exp(−5*I*w*h)*h+exp(6*I*w*h)*x−
8*exp(−4*I*w*h)*h−2*exp(3*I*w*h)*x+4*exp(2*I*w*h)*h+
2*exp(−2*I*w*h)*x−4*exp(−2*I*w*h)*h+6*exp(6*I*w*h)*h−
10*exp(5*I*w*h)*h)));
>  rr:=simplify(evalc(Re(−h^2*w*(b[0]*x*w+b[1]*w*exp(w*h*I)*x+
b[1]*w*exp(w*h*I)*h+b[4]*w*exp(−4*I*w*h)*x−
4*b[4]*w*exp(−4*I*w*h)*h+b[3]*w*exp(3*I*w*h)*x+
3*b[3]*w*exp(3*I*w*h)*h+b[3]*w*exp(−3*I*w*h)*x−
b[1]*w*exp(−I*w*h)*h+b[2]*w*exp(−2*I*w*h)*x−
2*b[2]*w*exp(−2*I*w*h)*h+b[4]*w*exp(4*I*w*h)*x+
4*b[4]*w*exp(4*I*w*h)*h+b[2]*w*exp(2*I*w*h)*x+
2*b[2]*w*exp(2*I*w*h)*h+b[1]*w*exp(−I*w*h)*x+
b[5]*w*exp(5*I*w*h)*x+5*b[5]*w*exp(5*I*w*h)*h+
b[5]*w*exp(−5*I*w*h)*x−5*b[5]*w*exp(−5*I*w*h)*h−
3*b[3]*w*exp(−3*I*w*h)*h−2*I*b[0]−2*I*b[1]*exp(w*h*I)−
2*I*b[5]*exp(−5*I*w*h)−2*I*b[2]*exp(2*I*w*h)−
2*I*b[4]*exp(−4*I*w*h)−2*I*b[3]*exp(3*I*w*h)−
2*I*b[3]*exp(−3*I*w*h)−2*I*b[2]*exp(−2*I*w*h)−
2*I*b[5]*exp(5*I*w*h)−2*I*b[4]*exp(4*I*w*h)−
2*I*b[1]*exp(−I*w*h)))));
>  ri:=simplify(evalc(Im(−
h^2*w*(b[0]*x*w+b[1]*w*exp(w*h*I)*x+b[1]*w*exp(w*h*I)*h+
b[4]*w*exp(−4*I*w*h)*x−4*b[4]*w*exp(−4*I*w*h)*h+
b[3]*w*exp(3*I*w*h)*x+3*b[3]*w*exp(3*I*w*h)*h+
b[3]*w*exp(−3*I*w*h)*x−b[1]*w*exp(−I*w*h)*h+
b[2]*w*exp(−2*I*w*h)*x−2*b[2]*w*exp(−2*I*w*h)*h+
b[4]*w*exp(4*I*w*h)*x+4*b[4]*w*exp(4*I*w*h)*h+
b[2]*w*exp(2*I*w*h)*x+2*b[2]*w*exp(2*I*w*h)*h+
b[1]*w*exp(−I*w*h)*x+b[5]*w*exp(5*I*w*h)*x+
5*b[5]*w*exp(5*I*w*h)*h+b[5]*w*exp(−5*I*w*h)*x−
5*b[5]*w*exp(−5*I*w*h)*h−3*b[3]*w*exp(−3*I*w*h)*h−
2*I*b[0]−2*I*b[1]*exp(w*h*I)−2*I*b[5]*exp(−5*I*w*h)−
2*I*b[2]*exp(2*I*w*h)−2*I*b[4]*exp(−4*I*w*h)−
2*I*b[3]*exp(3*I*w*h)−2*I*b[3]*exp(−3*I*w*h)−
2*I*b[2]*exp(−2*I*w*h)−2*I*b[5]*exp(5*I*w*h)−
2*I*b[4]*exp(4*I*w*h)−2*I*b[1]*exp(−I*w*h)))));
>  eq6:=lr=rr;
>  eq7:=li=ri;
>
>  'eq6 is equivalent with eq5';
>
```

> y[n+6]:=(x+6*h)^2*exp(I*w*(x+6*h));
> y[n-6]:=(x-6*h)^2*exp(I*w*(x-6*h));
> y[n+5]:=(x+5*h)^2*exp(I*w*(x+5*h));
> y[n-5]:=(x-5*h)^2*exp(I*w*(x-5*h));
> y[n+4]:=(x+4*h)^2*exp(I*w*(x+4*h));
> y[n-4]:=(x-4*h)^2*exp(I*w*(x-4*h));
> y[n+3]:=(x+3*h)^2*exp(I*w*(x+3*h));
> y[n-3]:=(x-3*h)^2*exp(I*w*(x-3*h));
> y[n+2]:=(x+2*h)^2*exp(I*w*(x+2*h));
> y[n-2]:=(x-2*h)^2*exp(I*w*(x-2*h));
> y[n+1]:=(x+h)^2*exp(I*w*(x+h));
> y[n-1]:=(x-h)^2*exp(I*w*(x-h));
> f[n-6]:=diff(y[n-6],x$2);
> f[n+6]:=diff(y[n+6],x$2);
> f[n-5]:=diff(y[n-5],x$2);
> f[n+5]:=diff(y[n+5],x$2);
> f[n-4]:=diff(y[n-4],x$2);
> f[n+4]:=diff(y[n+4],x$2);
> f[n-3]:=diff(y[n-3],x$2);
> f[n+3]:=diff(y[n+3],x$2);
> f[n-2]:=diff(y[n-2],x$2);
> f[n+2]:=diff(y[n+2],x$2);
> f[n-1]:=diff(y[n-1],x$2);
> f[n+1]:=diff(y[n+1],x$2);
>
> y[n]:=x^2*exp(I*w*(x));
> f[n]:=diff(y[n],x$2);;
> errornew:=simplify(y[n+6]+y[n-6]-2*(y[n+5]+y[n-5])+
2*(y[n+4]+y[n-4])-2*(y[n+3]+y[n-3])+2*(y[n+2]+y[n-2])-
2*(y[n+1]+y[n-1])+2*y[n]=h^2*(b[5]*(f[n+5]+f[n-5])+
b[4]*(f[n+4]+f[n-4])+b[3]*(f[n+3]+f[n-3])+b[2]*(f[n+2]+f[n-2])+
b[1]*(f[n+1]+f[n-1])+b[0]*f[n]));
> errornew:=simplify(errornew/exp(w*x*I));
> lr:=simplify(evalc(Re(2*x^2+exp(-6*I*w*h)*x^2+
2*exp(-2*I*w*h)*x^2-2*exp(-I*w*h)*h^2-
20*exp(5*I*w*h)*x*h+12*exp(6*I*w*h)*x*h-2*exp(-I*w*h)*x^2-
2*exp(3*I*w*h)*x^2-2*exp(-3*I*w*h)*x^2-
18*exp(-3*I*w*h)*h^2+32*exp(-4*I*w*h)*h^2+2*exp(-4*I*w*h)*x^2-
50*exp(5*I*w*h)*h^2+exp(6*I*w*h)*x^2+36*exp(6*I*w*h)*h^2+
32*exp(4*I*w*h)*h^2-50*exp(-5*I*w*h)*h^2+2*exp(4*I*w*h)*x^2-
18*exp(3*I*w*h)*h^2+2*exp(2*I*w*h)*x^2+8*exp(2*I*w*h)*h^2-
2*exp(5*I*w*h)*x^2-2*exp(-5*I*w*h)*x^2-2*exp(w*h*I)*h^2-
2*exp(w*h*I)*x^2+8*exp(-2*I*w*h)*h^2-
12*exp(-6*I*w*h)*x*h+4*exp(-I*w*h)*x*h-
8*exp(-2*I*w*h)*x*h+8*exp(2*I*w*h)*x*h+20*exp(-5*I*w*h)*x*h-
4*exp(w*h*I)*x*h+12*exp(-3*I*w*h)*x*h-16*exp(-4*I*w*h)*x*h-

7: Numerical Methods in Chemistry

```
  12*exp(3*I*w*h)*x*h+16*exp(4*I*w*h)*x*h+36*exp(-6*I*w*h)*h^2)));
> li:=simplify(evalc(Im(2*x^2+exp(-6*I*w*h)*x^2+
  2*exp(-2*I*w*h)*x^2-2*exp(-I*w*h)*h^2-
  20*exp(5*I*w*h)*x*h+12*exp(6*I*w*h)*x*h-2*exp(-I*w*h)*x^2-
  2*exp(3*I*w*h)*x^2-2*exp(-3*I*w*h)*x^2-
  18*exp(-3*I*w*h)*h^2+32*exp(-4*I*w*h)*h^2+2*exp(-4*I*w*h)*x^2
  50*exp(5*I*w*h)*h^2+exp(6*I*w*h)*x^2+36*exp(6*I*w*h)*h^2+
  32*exp(4*I*w*h)*h^2-50*exp(-5*I*w*h)*h^2+2*exp(4*I*w*h)*x^2-
  18*exp(3*I*w*h)*h^2+2*exp(2*I*w*h)*x^2+8*exp(2*I*w*h)*h^2-
  2*exp(5*I*w*h)*x^2-2*exp(-5*I*w*h)*x^2-2*exp(w*h*I)*h^2-
  2*exp(w*h*I)*x^2+8*exp(-2*I*w*h)*h^2-12*exp(-6*I*w*h)*x*h+
  4*exp(-I*w*h)*x*h-8*exp(-2*I*w*h)*x*h+8*exp(2*I*w*h)*x*h+
  20*exp(-5*I*w*h)*x*h-4*exp(w*h*I)*x*h+12*exp(-3*I*w*h)*x*h-
  16*exp(-4*I*w*h)*x*h-12*exp(3*I*w*h)*x*h+16*exp(4*I*w*h)*x*h+
  36*exp(-6*I*w*h)*h^2)));
> rr:=simplify(evalc(Re(h^2*(2*b[0]-b[4]*w^2*exp(-4*I*w*h)*x^2-
  b[0]*x^2*w^2-b[4]*w^2*exp(4*I*w*h)*x^2-
  b[2]*w^2*exp(-2*I*w*h)*x^2-9*b[3]*w^2*exp(3*I*w*h)*h^2-
  b[3]*w^2*exp(-3*I*w*h)*x^2-9*b[3]*w^2*exp(-3*I*w*h)*h^2-
  4*b[2]*w^2*exp(2*I*w*h)*h^2-b[5]*w^2*exp(5*I*w*h)*x^2-
  b[3]*w^2*exp(3*I*w*h)*x^2-b[1]*w^2*exp(-I*w*h)*x^2-
  b[1]*w^2*exp(-I*w*h)*h^2-25*b[5]*w^2*exp(5*I*w*h)*h^2-
  b[2]*w^2*exp(2*I*w*h)*x^2-16*b[4]*w^2*exp(-4*I*w*h)*h^2-
  16*b[4]*w^2*exp(4*I*w*h)*h^2-
  2*b[1]*w^2*exp(w*h*I)*x*h+2*b[1]*exp(w*h*I)+2*b[3]*exp(3*I*w*h)+
  2*b[4]*exp(-4*I*w*h)+2*b[5]*exp(-5*I*w*h)+2*b[4]*exp(4*I*w*h)+
  2*b[3]*exp(-3*I*w*h)+2*b[1]*exp(-I*w*h)+2*b[2]*exp(2*I*w*h)+
  2*b[5]*exp(5*I*w*h)-b[1]*w^2*exp(w*h*I)*x^2-
  b[1]*w^2*exp(w*h*I)*h^2-4*b[2]*w^2*exp(-2*I*w*h)*h^2-
  b[5]*w^2*exp(-5*I*w*h)*x^2-25*b[5]*w^2*exp(-5*I*w*h)*h^2+
  4*I*b[2]*w*exp(2*I*w*h)*x+16*I*b[4]*w*exp(4*I*w*h)*h-
  20*I*b[5]*w*exp(-5*I*w*h)*h-12*I*b[3]*w*exp(-3*I*w*h)*h+
  4*I*b[2]*w*exp(-2*I*w*h)*x+4*I*b[3]*w*exp(3*I*w*h)*x-
  8*I*b[2]*w*exp(-2*I*w*h)*h+4*I*b[1]*w*exp(-I*w*h)*x+
  8*I*b[2]*w*exp(2*I*w*h)*h+4*I*b[3]*w*exp(-3*I*w*h)*x+
  12*I*b[3]*w*exp(3*I*w*h)*h-16*I*b[4]*w*exp(-4*I*w*h)*h+
  4*I*b[4]*w*exp(4*I*w*h)*x+4*I*b[5]*w*exp(-5*I*w*h)*x+
  4*I*b[4]*w*exp(-4*I*w*h)*x+20*I*b[5]*w*exp(5*I*w*h)*h+
  4*I*b[5]*w*exp(5*I*w*h)*x-4*I*b[1]*w*exp(-I*w*h)*h+
  4*I*b[1]*w*exp(w*h*I)*h+4*I*b[1]*w*exp(w*h*I)*x+
  4*I*b[0]*x*w-8*b[4]*w^2*exp(4*I*w*h)*x*h+
  6*b[3]*w^2*exp(-3*I*w*h)*x*h+8*b[4]*w^2*exp(-4*I*w*h)*x*h+
  10*b[5]*w^2*exp(-5*I*w*h)*x*h-10*b[5]*w^2*exp(5*I*w*h)*x*h+
  4*b[2]*w^2*exp(-2*I*w*h)*x*h+2*b[1]*w^2*exp(-I*w*h)*x*h-
  4*b[2]*w^2*exp(2*I*w*h)*x*h-6*b[3]*w^2*exp(3*I*w*h)*x*h+
  2*b[2]*exp(-2*I*w*h))))));
```

```
> ri:=simplify(evalc(Im(h^2*(2*b[0]-b[4]*w^2*exp(-4*I*w*h)*x^2-
b[0]*x^2*w^2-b[4]*w^2*exp(4*I*w*h)*x^2-
b[2]*w^2*exp(-2*I*w*h)*x^2-9*b[3]*w^2*exp(3*I*w*h)*h^2-
b[3]*w^2*exp(-3*I*w*h)*x^2-9*b[3]*w^2*exp(-3*I*w*h)*h^2-
4*b[2]*w^2*exp(2*I*w*h)*h^2-b[5]*w^2*exp(5*I*w*h)*x^2-
b[3]*w^2*exp(3*I*w*h)*x^2-b[1]*w^2*exp(-I*w*h)*x^2-
b[1]*w^2*exp(-I*w*h)*h^2-25*b[5]*w^2*exp(5*I*w*h)*h^2-
b[2]*w^2*exp(2*I*w*h)*x^2-16*b[4]*w^2*exp(-4*I*w*h)*h^2-
16*b[4]*w^2*exp(4*I*w*h)*h^2-2*b[1]*w^2*exp(w*h*I)*x*h+
2*b[1]*exp(w*h*I)+2*b[3]*exp(3*I*w*h)+2*b[4]*exp(-4*I*w*h)+
2*b[5]*exp(-5*I*w*h)+2*b[4]*exp(4*I*w*h)+2*b[3]*exp(-3*I*w*h)+
2*b[1]*exp(-I*w*h)+2*b[2]*exp(2*I*w*h)+2*b[5]*exp(5*I*w*h)-
b[1]*w^2*exp(w*h*I)*x^2-b[1]*w^2*exp(w*h*I)*h^2-
4*b[2]*w^2*exp(-2*I*w*h)*h^2-b[5]*w^2*exp(-5*I*w*h)*x^2-
25*b[5]*w^2*exp(-5*I*w*h)*h^2+4*I*b[2]*w*exp(2*I*w*h)*x+
16*I*b[4]*w*exp(4*I*w*h)*h-20*I*b[5]*w*exp(-5*I*w*h)*h-
12*I*b[3]*w*exp(-3*I*w*h)*h+4*I*b[2]*w*exp(-2*I*w*h)*x+
4*I*b[3]*w*exp(3*I*w*h)*x-8*I*b[2]*w*exp(-2*I*w*h)*h+
4*I*b[1]*w*exp(-I*w*h)*x+8*I*b[2]*w*exp(2*I*w*h)*h+
4*I*b[3]*w*exp(-3*I*w*h)*x+12*I*b[3]*w*exp(3*I*w*h)*h-
16*I*b[4]*w*exp(-4*I*w*h)*h+4*I*b[4]*w*exp(4*I*w*h)*x+
4*I*b[5]*w*exp(-5*I*w*h)*x+4*I*b[4]*w*exp(-4*I*w*h)*x+
20*I*b[5]*w*exp(5*I*w*h)*h+4*I*b[5]*w*exp(5*I*w*h)*x-
4*I*b[1]*w*exp(-I*w*h)*h+4*I*b[1]*w*exp(w*h*I)*h+
4*I*b[1]*w*exp(w*h*I)*x+4*I*b[0]*x*w-8*b[4]*w^2*exp(4*I*w*h)*x*h+
6*b[3]*w^2*exp(-3*I*w*h)*x*h+8*b[4]*w^2*exp(-4*I*w*h)*x*h+
10*b[5]*w^2*exp(-5*I*w*h)*x*h-10*b[5]*w^2*exp(5*I*w*h)*x*h+
4*b[2]*w^2*exp(-2*I*w*h)*x*h+2*b[1]*w^2*exp(-I*w*h)*x*h-
4*b[2]*w^2*exp(2*I*w*h)*x*h-6*b[3]*w^2*exp(3*I*w*h)*x*h+
2*b[2]*exp(-2*I*w*h))))) ;
> eq8:=lr=rr;
> eq9:=li=ri;
>
> 'It is proved that eq9/x is equivalent with 2*eq7';
>
>
> solut:=solve({eq1,eq2,eq3},{b[0],b[1],b[2]});
>
> assign(solut);
>
> solut1:=solve({eq5,eq7,eq8},{b[3],b[4],b[5]});
> assign(solut1);
>
> b[3]:=combine(b[3]);
> b[4]:=combine(b[4]);
> b[5]:=combine(b[5]);
> b3t:=convert(taylor(b[3],w=0,58),polynom);
```

```
> b4t: = convert(taylor(b[4],w = 0,64),polynom);
> b5t: = convert(taylor(b[5],w = 0,15),polynom);
>
```

References

1. L.Gr. Ixaru and M. Micu, *Topics in Theoretical Physics*. Central Institute of Physics, Bucharest, 1978.
2. L.D. Landau and F.M. Lifshitz: *Quantum Mechanics*. Pergamon, New York, 1965.
3. I. Prigogine, Stuart Rice (Eds): Advances in Chemical Physics Vol. 93: New Methods in Computational Quantum Mechanics, John Wiley & Sons, 1997.
4. G. Herzberg, *Spectra of Diatomic Molecules*, Van Nostrand, Toronto, 1950.
5. T.E. Simos, Atomic Structure Computations in Chemical Modelling: Applications and Theory (Editor: A. Hinchliffe, UMIST), *The Royal Society of Chemistry* 38–142(2000).
6. T.E. Simos, Numerical methods for 1D, 2D and 3D differential equations arising in chemical problems, *Chemical Modelling: Application and Theory*, The Royal Society of Chemistry, 2, 170–270(2002).
7. J.D. Lambert and I.A. Watson, *J. Inst. Math. Appl.*, 1976, **18**, 189.
8. P. J. Van Der Houwen and B. P. Sommeijer, *SIAM J. Numer. Anal.*, 1987, **24**, 595.
9. J.P. Coleman, Numerical methods for $y^n = f(x,y)$ in *Proc. of the First Intern. Colloq. on Numerical Analysis* (Bulgaria 1992), Edit. D. Bainov and V. Civachev, 27–38(1992).
10. T.E. Simos, Multiderivative methods for the numerical solution of the Schrödinger equation, *MATCH Commun. Math. Comput. Chem*, in press.
11. Xue-Shen Liu, Xiao-Yan Liu, Zhong-Yuan Zhou, Pei-Zhu Ding, Shou-Fu Pan, *Int. J. Quantum Chem.*, 2000, **79**, 343.
12. R.D. Ruth, *IEEE Trans. Nucl. Sci*, 1983, **NS-30**, 2669.
13. P.J. Candy and W. Rozmus, *J. Comp. Phys.*, 1991, **92**, 230.
14. H. Yoshida, *Physics Letters*, 1990, **150**, 262.
15. E. Forest and R.D. Ruth, *Physica D*, 1990, **43**, 105.
16. J.M. Sanz-Serna and M.P. Calvo, *Numerical Hamiltonian Problem*, Chapman and Hall London (1994).
17. K.Tselios and T.E. Simos, *J.Math.Chem.*, 2003, **34**, 83.
18. J.R. Dormand, M.E.A. El-Mikkawy and P.J. Prince, *IMA J. Numer. Anal*, 1987, **7**, 235.
19. J.R. Dormand, M.E El-Mikkawy and P.J. Prince, *IMA J. Numer. Anal*, 1987, **7**, 423.
20. M.M. Chawla and P.S. Rao, *J. Comput. Appl. Math.*, 1986, **15**, 329.
21. J.W. Cooley, *Math. Comp.*, 1961, **15**, 363.
22. J.M. Blatt, *J. Comput. Phys.*, 1967, **1**, 382.
23. J.R. Dormand and P.J. Prince, *Comput. Math. Applic.*, 1987, **13**, 937.
24. F.Y. Hajj, *Journal of Physics B: At. Mol. Phys.*, 1982, **15**, 683.
25. F.Y. Hajj, *Journal of Physics B: At. Mol. Phys.*, 1985, **18**, 1.
26. I.P Hamilton and J.C. Light, *Journal of Chemical Physics*, 1986, **84**, 306.
27. M.M. Chawla and P.S. Rao, *J. Comput. Appl. Math.*, 1984, **11**, 277.
28. T.E. Simos, *Phys. Lett. A*, 2003, **315**, 437.
29. T.E. Simos, *New Astron.*, 2003, **8**, 391.

30. G. Psihoyios and T.E. Simos, *New Astron.*, 2003, **8**, 679.
31. T.E. Simos, *J. Math. Chem.*, 2003, **34**, 39.
32. G. Psihoyios and T.E. Simos, *Int. J. Mod. Phys. C*, 2003, **14**, 175.
33. T.E. Simos and J. Vigo-Aguiar, *Comput. Phys. Commun.*, 2003, **152**, 274.
34. T.E. Simos, *Int. J. Mod. Phys. C*, 2002, **13**, 1333.
35. T.E. Simos, G. Avdelas and J. Vigo-Aguiar (Editors), Numerical methods in physics, chemistry, and engineering – Special Issue – Preface, *Comput. Math Appl.*, **45 (1–3)**(2003).
36. A. Konguetsof, T.E. Simos, *Comput. Math Appl.*, 2003, **45 (1–3)**, 547.
37. T.E. Simos and J. Vigo-Aguiar, *Phys. Rev. E*, **67**, art. no. 016701 Part 2 (2003).
38. J. Vigo-Aguiar and T.E. Simos, *J. Math. Chem.*, 2002, **32**, 257.
39. G. Avdelas, E. Kefalidis and T.E. Simos, *J. Math. Chem.*, 2002, **31**, 371.
40. C. Tsitouras and T.E. Simos, *J. Comput. Appl. Math.*, 2002, **147**, 397.
41. G. Avdelas, A. Konguetsof and T.E. Simos, *Comput. Phys. Commun.*, 2002, **148**, 59.
42. C. Tsitouras and T.E. Simos, *Appl. Math. Comput.*, 2002, **131**, 201.
43. T.E. Simos, J. Vigo-Aguiar, *J. Math. Chem.*, 2002, **31**, 135.
44. Z. Kalogiratou and T.E. Simos, *J. Math. Chem.*, 2002, **31**, 211.
45. T.E. Simos and P.S. Williams, *MATCH-Commun. Math. Co.*, 2002, **45**, 123.
46. T.E. Simos, *Appl. Math. Lett.*, 2002, **15**, 217.
47. A.D. Raptis and J.R. Cash, *Comput. Phys. Commun.*, 1987, **44**, 95.
48. R.M. Thomas, T.E. Simos and G.V. Mitsou, *J. Comput. Appl. Math.*, 1996, **67**, 255.
49. G. Papageorgiou, Ch. Tsitouras and I.Th. Famelis, *Int. J. Mod. Phys. C*, 2001, **12**, 657.
50. M.P. Calvo and J.M. Sanz-Serna, *SIAM J. Sci. Comput.*, 1993, **14**, 1237.
51. A.D. Raptis and A.C. Allison, *Comput. Phys. Commun.*, 1978, **14**, 1.
52. L.Gr. Ixaru and M. Rizea, *Comput. Phys. Commun.*,1980, **19**, 23.
53. A.D. Raptis, *Computing*, 1982, **28**, 373.
54. P. Henrici, *Discrete Variable Methods in Ordinary Differential Equations* (Wiley, New York, 1962).
55. E. Stiefel and D.G. Bettis, *Numer. Math.*, 1969, **13**, 154.
56. L. Gr. Ixaru, G. Vanden Berghe and H. De Meyer, *Comput. Phys. Commun.*, 2003, **150**, 116.
57. L. Gr. Ixaru, G. Vanden Berghe and H. De Meyer, *J. Compu. Appl. Math.*, 2002, **140**, 423.

8
Many-body Perturbation Theory and Its Application to the Molecular Structure Problem

BY S. WILSON

1 Introduction

This report covers developments in the theory and application of many-body perturbation theory to the molecular structure problem in the period June 2001 to May 2003. It thus continues the two earlier reviews[1,2] entitled *Many-body Perturbation Theory and its Application to the Molecular Electronic Structure Problem* published in this series, the first covering the published literature up to June 1999 and the second dealing with the period June 1999 to May 2001. The first review covered the many developments which had been made since my 1981 review[3] entitled *Many-body Perturbation Theory of Molecules* appeared in a previous Specialist Periodical Reports series. In my first report to this series, I described not only developments in the non-relativistic theory, which moved, for example, from the study of the triple excitation component of the correlation energy in my 1981 review to the fully diagrammatic analysis of the fifth order terms for closed-shell systems, but also the development and practical realization of the fully relativistic many-body perturbation theory, which did not exist in 1981 and which is now recognized as an essential ingredient of any *ab initio* treatment of molecules containing heavy elements. I described the systematic development of the underlying approximation which is ubiquitous in molecular studies — the use of finite basis sets — the algebraic approximation, which is continuing to drive down the basis set truncation error in molecular electronic structure calculations. I described the development of *cc*MPBT (*concurrent computation* Many-Body Perturbation Theory) and of algorithms and computer programs capable of exploiting the power of contemporary computing machines. My second report to this series provided a snapshot of both theoretical developments and application areas at the turn of the century with particular emphasis on application using finite- and low-order theory and especially 'MP2' methods. Second order Møller-Plesset theory continues to be the most widely used *ab initio* technique for the description of electron correlation effects in molecular systems.

The present report considers the many body perturbation theory and its application to the *molecular structure* problem in its entirety and considers both the motion of the electrons together with that of the nuclei. There have been important developments over the past two years in attempts to simultaneously describe both the nuclear and the electronic motions thereby avoiding the introduction of the Born-Oppenheimer approximation. These developments are summarized in section 2. There has been significant progress in the construction of low-order approximants which enhance the approximations supported by finite order many-body perturbation theory. A summary of this progress is given in section 3. The development of low-order scaling methods rest on the use of local correlation methods. The refinement of these methods is continuing to extend the size of molecular systems which can be attacked in low-order perturbation theory studies of correlation effects. Recent progress is reviewed in section 4. Second order many-body perturbation theory in its Møller-Plesset form continues to be the most widely used *ab initio* technique for the description of electron correlation effects in molecules. In section 5 a brief synopsis of applications published during the reporting period is given. Section 6 contains a summary of this report and considers the prospects for the years ahead.

2 Diagrammatic Many-body Perturbation Theory of Molecular Structure Including Nuclear and Electronic Motion

The vast majority of theoretical molecular structure studies attack the problem in two distinct stages. First the electrons are assumed to move in the field of fixed nuclei, that is, the Born-Oppenheimer approximation[4-6] is made. Solution of the electronic Schrödinger equation yields a potential energy curve or surface which then defines an effective potential in which nuclear motion takes place. The second stage therefore involves the solution of the nuclear Schrödinger equation for the motion of the nuclei in the effective potential generated by the electrons. The study of nuclear motion necessitates the determination of a potential energy hypersurface which in turn requires, in principle, the solution of the electronic Schrödinger equation for all possible nuclear configurations. This problem becomes increasingly intractable as molecular species containing larger numbers of atoms are considered. (For a recent review of the approximate separation of electronic and nuclear motion in the molecular structure problem within the framework of non-relativistic quantum mechanics see the recent work of Sutcliffe.[7-11])

In a recent review, Woolley and Sutcliffe[12] repeat a comment made by Löwdin[13] in 1990

> "One of the most urgent problems of modern quantum chemistry is to treat the motions of the atomic nuclei and the electrons on a more or less equivalent basis."

In 1969, Thomas published two papers[14,15] in which a molecular structure theory was developed without invoking the Born-Oppenheimer approximation. In these publications and two further papers published in 1970[16,17] Thomas studied methane, ammonia, water and hydrogen fluoride adding the kinetic energy operators of the protons to the electronic hamiltonian and using Slater-type orbitals centred on the heavier nuclei for the protonic wave functions. Over the years, a number of authors[18-26] have attempted the development of a non-Born-Oppenheimer theory of molecular structure, but problems of accuracy and/or feasibility remain for applications to arbitrary molecular systems.

Within the reporting period, Nakai[27] has presented a non-Born-Oppenheimer theory of molecular structure in which molecular orbitals (MO) are used to describe the motion of individual electrons and nuclear orbitals (NO) are introduced each of which describes the motion of single nuclei. Nakai presents an *ab initio* Hartree-Fock theory, which is designated "NO+MO/HF theory", which builds on the earlier work of Tachikawa *et al.*[28] In subsequent work published in 2003, Nakai and Sodeyama[29] apply many-body perturbation theory to the problem of simultaneously describing both the nuclear and electronic components of a molecular system. Their approach will be considered in some detail in the remainder of this section. In section 2.1 we define the total molecular hamiltonian operator describing both nuclear and electronic motion. The Hartree-Fock theory for nuclei and electrons is presented in section 2.2 and a many-body perturbation theory which uses this as a reference is developed in section 2.3. The diagrammatic perturbation theory of nuclei and electrons is reviewed in section 2.4. Section 2.5 considers the prospects for this area of research.

2.1 The Total Molecular Hamiltonian Operator. – The total molecular hamiltonian operator for a system containing N nuclei and n electrons may be written

$$\mathcal{H} = T + V \tag{1}$$

where the kinetic energy operator, T, is a sum of a two one-body terms, a nuclear term and an electronic term

$$T = T_n + T_e. \tag{2}$$

The nuclear kinetic energy operator, T_n, is a sum of one-particle operators, that is

$$T_n = -\sum_P^N \frac{1}{2m_P} \nabla_P^2$$
$$= \sum_P^N t_P \tag{3}$$

where

$$t_P = -\frac{1}{2m_P}\nabla_P^2 \qquad (4)$$

in which m_P is the mass of the nucleus labelled P. Similarly, the electronic kinetic energy operator, T_e, is a sum of one-electron operators, that is

$$T_e = -\sum_p^n \frac{1}{2}\nabla_p^2$$

$$= \sum_p^n t_p \qquad (5)$$

where

$$t_p = -\frac{1}{2}\nabla_p^2. \qquad (6)$$

The potential energy term, V, is a sum of three two-body terms, the first corresponding to nucleus-nucleus interactions, the second to nucleus-electron interactions and the third to electron-electron interactions.

$$V = V_{nn} + V_{ne} + V_{ee} \qquad (7)$$

The nucleus-nucleus interaction term has the form

$$V_{nn} = \sum_{P>Q}^N v_{nn}(P,Q)$$

$$= \sum_{P>Q}^N \frac{Z_P Z_Q}{r_{PQ}} \qquad (8)$$

where Z_P is the charge associated with nucleus P and r_{PQ} is the distance between nucleus P and nucleus Q. The nucleus-electron interaction term takes the form

$$V_{ne} = \sum_p^n \sum_P^N v_{en}(p,P)$$

$$= \sum_p^n \sum_P^N \frac{Z_P}{r_{pP}} \qquad (9)$$

where r_{pP} is the distance between nucleus P and electron p. The electron-electron interaction term takes the form

$$V_{ee} = \sum_{p>q}^n v_{ee}(p,p)$$

$$= \sum_{p>q}^n \frac{1}{r_{pq}} \qquad (10)$$

where r_{pq} is the distance between the electron labelled p and that labelled q.

2.2 The Hartree-Fock Theory of Nuclei and Electrons.

In order to develop a theory for the motion of both the nuclei and the electrons in a molecule, we write the total hamiltonian operator, $\mathcal{H}$, as a sum of an unperturbed or zero order hamiltonian, $\mathcal{H}_0$, and a perturbation, $\mathcal{H}_1$, that is

$$\mathcal{H} = \mathcal{H}_0 + \lambda H_1. \tag{11}$$

Here λ is a perturbation parameter which is introduced so as to define the order of different terms in the perturbation series but which is set equal to 1 in order to recover the physical situation.

The unperturbed hamiltonian operator is based on an independent particle model, that is, a model in which each particle, nucleus or electron, experiences an averaged interaction with the other particles in the system. The unperturbed hamiltonian operator is a sum of a kinetic energy term and an effective potential energy term

$$\mathcal{H}_0 = T + U \tag{12}$$

The kinetic energy component is the sum of one-particle terms defined in the previous section. The effective potential is a sum of a nuclear and an electronic component.

$$U = U_n + U_e \tag{13}$$

It is also a sum of one-particle terms.

The total wave function for a system of nuclei and electrons can be written as a product of a nuclear component

$$\Phi_n = \|\varphi_P \varphi_Q \ldots\| \tag{14}$$

in which φ_P is a single nucleus state function, or nuclear orbital (NO), and an electronic component

$$\Phi_e = \|\varphi_p \varphi_q \ldots\| \tag{15}$$

in which φ_p is a single electron state function, or electronic orbitals – more usually called a molecular orbital.

The single nucleus state function or nuclear orbital is an eigenfunction of a Hartree-Fock eigenvalue equation for the nuclear motion

$$\mathcal{F}_n \varphi_P = \varepsilon_P \varphi_P \tag{16}$$

in which the Fock operator has the form

$$\mathcal{F}_n = t_n + \sum_P^N (J_P \mp K_P) + \sum_p^n J_p$$

$$= t_n + u_n \tag{17}$$

where the nuclear Fock potential is

$$u_n = \sum_P^N (J_P \mp K_P) + \sum_p^n J_p \qquad (18)$$

J and K denote the Coulomb and exchange operators, respectively. In equation (16), the effective field of the nuclear orbital is due to the motion of the electrons and the remaining nuclei.

The Hartree-Fock equations for the electrons have the form

$$\mathcal{F}_e \varphi_p = \varepsilon_p \varphi_p \qquad (19)$$

where the Fock operator is given by

$$\begin{aligned}\mathcal{F}_e &= t_e + \sum_p^n (J_p - K_p) + \sum_P^N J_P \\ &= t_e + u_e \end{aligned} \qquad (20)$$

The effective potential for the electrons is

$$u_e = \sum_p^n (J_p - K_p) + \sum_P^N J_P \qquad (21)$$

which includes a mean-field coupling between the electronic and the nuclear motion. In equation (19) the effective field of the electronic (molecular) orbital is due to the motion of the nuclei and the other electrons in the system.

2.3 The Many-body Perturbation Theory of Nuclei and Electrons. – The unperturbed or zero order hamiltonian can be re-written in the form

$$\begin{aligned}\mathcal{H}_0 &= T + U \\ &= T_n + T_e + U_n + U_e \\ &= (T_n + U_n) + (T_e + U_e) \\ &= \mathcal{H}_{n0} + \mathcal{H}_{e0} \end{aligned} \qquad (22)$$

where $\mathcal{H}_{n0}$ is the unperturbed hamiltonian describing the motion of the nuclei

$$\mathcal{H}_{n0} = T_n + U_n \qquad (23)$$

and $\mathcal{H}_{e0}$ is the unperturbed hamiltonian for the motion of the electrons

$$\mathcal{H}_{e0} = T_e + U_e. \qquad (24)$$

The perturbing operator is the difference between the full hamiltonian and the zero order hamiltonian

$$\mathcal{H}_1 = \mathcal{H} - \mathcal{H}_0 \qquad (25)$$

so that
$$\mathcal{H}_1 = (T+V) - (T+U) \tag{26}$$
and thus
$$\begin{aligned}\mathcal{H}_1 &= V - U \\ &= V_{nn} + V_{ne} + V_{ee} - U_n - U_e\end{aligned} \tag{27}$$

Recall that the total molecular hamiltonian is written
$$\mathcal{H}(\lambda) = \mathcal{H}_0 + \lambda \mathcal{H}_1 \tag{28}$$
where the unperturbed hamiltonian, $\mathcal{H}(0) = \mathcal{H}_0$, has eigenvalues E_m and eigenfunctions $|\Phi_m\rangle$
$$\mathcal{H}_0 |\Phi_m\rangle = E_m |\Phi_m\rangle \tag{29}$$
The Schrödinger equation for the perturbed system can then be written[30]
$$\mathcal{H}(\lambda) |\Psi(\lambda)\rangle = \varepsilon(\lambda) |\Psi(\lambda)\rangle \tag{30}$$
where it is assumed that the exact eigenvalue is an analytic function of the perturbation parameter λ and can be expanded in a power series
$$\begin{aligned}\varepsilon &= \varepsilon(\lambda) \\ &= \sum_{k=0}^{\infty} E^{(k)} \lambda^k\end{aligned} \tag{31}$$
and similarly that the exact eigenfunction is an analytic function of λ and can also be written as a power series
$$\begin{aligned}|\Psi\rangle &= |\Psi(\lambda)\rangle \\ &= \sum_{k=0}^{\infty} |\chi^{(k)}\rangle \lambda^k\end{aligned} \tag{32}$$
Obviously, the constant term in the power series expansion for $\mathcal{E}(\lambda)$ is
$$E^{(0)} = E_0 \tag{33}$$
and the corresponding term in the power series for the exact wave function is
$$|\chi^{(0)}\rangle = |\Phi_0\rangle. \tag{34}$$
We write
$$\mathcal{E} = E + \Delta E \tag{35}$$

so that the "level shift" is

$$\Delta E = \mathcal{E} - E \tag{36}$$

In order to develop the Rayleigh-Schrödinger perturbation expansion for the energy and the wave function, we define the resolvent

$$\mathcal{R}_0 = \frac{\mathcal{Q}}{E_0 - \mathcal{H}_0} \tag{37}$$

in which $\mathcal{Q}$ is the projection operator

$$\mathcal{Q} = \sum_{m \neq 0} |\Phi_m\rangle\langle\Phi_m|$$
$$= 1 - |\Phi_0\rangle\langle\Phi_0|$$
$$1 - \mathcal{P} \tag{38}$$

and $\mathcal{P}$ is its orthogonal complement.

The Rayleigh-Schrödinger perturbation expansion for the energy has the form

$$\Delta E = \sum_{n=1}^{\infty} \langle \Phi_0 | \mathcal{H}_1 [\mathcal{R}_0(\mathcal{H}_1 - \Delta E)]^{n-1} | \Phi_0 \rangle$$
$$= \sum_{n=1}^{\infty} \langle \Phi_0 | \mathcal{H}_1 [\mathcal{R}_0 \mathcal{H}_1]^{n-1} | \Phi_0 \rangle + \Omega \tag{39}$$

where Ω represents the "renormalization terms".

The many-body perturbation expansion for the energy takes the form

$$\Delta E = \sum_{n=1}^{\infty} \langle \Phi_0 | \mathcal{H}_1 [\mathcal{R}_0 \mathcal{H}_1]^{n-1} | \Phi_0 \rangle_{\text{linked}} \tag{40}$$

where the subscript "linked" indicates that only terms corresponding to linked diagrams are included in the expansion.

A diagrammatic many-body perturbation theory describing both nuclei and electrons requires a second quantized formulation. (For a recent review of second quantization see the articles by Pickup[31-34] and by Karwowski.[34]) The unperturbed hamiltonian can be written in second quantized form as follows

$$\mathcal{H}_0 = \sum_{P}^{N} \varepsilon_P a_P^+ a_P + \sum_{p}^{n} \varepsilon_p a_p^+ a_p \tag{41}$$

where the first term on the right-hand side is associated with the nuclei and the second with the electrons. The ε_P are single particle energies for the nuclei. The ε_p are single particle energies for the electrons. The perturbing operator can be written as a sum of a one-particle and a two-particle part, that is

$$\mathcal{H}_1 = \mathcal{H}_1^{(1)} + \mathcal{H}_1^{(2)} \tag{42}$$

8: Perturbation Theory and Application to the Molecular Structure Problem

The one-particle component has the form

$$\mathcal{H}_1^{(1)} = -\sum_{P,Q}^{N} \langle P|u_n|Q\rangle a_P^+ a_Q + \sum_{p,q}^{n} \langle p|u_e|q\rangle a_p^+ a_q \tag{43}$$

where u_n is the Fock operator associated with the motion of the nuclei and u_e is the corresponding operator for the electrons. The two-particle component is

$$\begin{aligned}\mathcal{H}_1^{(2)} = &\frac{1}{4} \sum_{P,Q,R,S}^{N} \langle PQ|G|RS\rangle a_P^+ a_Q^+ a_S a_R \\ &+ \sum_{P,Q}^{N} \sum_{p,q}^{n} \langle Pp|\gamma|Qq\rangle a_P^+ a_p^+ a_Q a_q \\ &+ \frac{1}{4} \sum_{p,q,r,s}^{n} \langle pq|g|rs\rangle a_p^+ a_q^+ a_s a_r \end{aligned} \tag{44}$$

where the first term on the right-hand side describes antisymmetrized interactions between the nuclei, the second term describes nucleus-electron interactions, and the third term describes antisymmetrized interactions between the electrons.

The operators $\{a_P^+, a_Q^+, \ldots\}$ and $\{a_P, a_Q, \ldots\}$ are, respectively, the creation and annihilation operators for nuclei. These operators satisfy the following relations:

$$[a_P^+, a_Q]_\pm \equiv a_P^+ a_Q \pm a_Q a_P^+ = \delta_{PQ}$$

$$[a_P^+, a_Q^+]_\pm \equiv a_P^+ a_Q^+ \pm a_Q^+ a_P^+ = 0$$

$$[a_P, a_Q]_\pm \equiv a_P a_Q \pm a_Q a_P = 0$$

where the $+$ sign corresponds to nuclei that are fermions and the $-$ sign to those which are bosons. The creation and annihilation operators between different particles, μ and ν, say, satisfy the commutation relations

$$[a_\mu^+, a_\nu]_- \equiv a_\mu^+ a_\nu - a\nu^+ a_\mu = 0 \tag{45}$$

$$[a_\mu^+, a_\nu^+]_- \equiv a_\mu^+ a_\nu^+ - a\nu^+ a_\mu^+ = 0 \tag{46}$$

$$[a_\mu, a_\nu]_- \equiv a_\mu a_\nu - a_\nu a_\mu = 0 \tag{47}$$

The creation and annihilation operators for the electrons are $\{a_p^+, a_q^+, \ldots\}$ and $\{a_p, a_q, \ldots\}$, respectively. These operators satisfy the anticommutation relations

$$[a_p^+, a_q]_+ \equiv a_p^+ a_q + a_q a_p^+ = \delta_{pq} \tag{48}$$

$$[a_p^+, a_q^+]_+ \equiv a_p^+ a_q^+ + a_q^+ a_p^+ = 0 \tag{49}$$

$$[a_p, a_q]_+ \equiv a_p a_q + a_q a_p = 0 \tag{50}$$

For ground states and low-lying excited states it is convenient to adopt a particle-hole formalism. We use the indices

$$\{I, J, K, L, \ldots i, j, k, l, \ldots\} \tag{51}$$

for occupied single particle state functions and the indices

$$\{A, B, C, D, \ldots, a, b, c, d, \ldots\} \tag{52}$$

for unoccupied single particle state functions. The indices

$$\{P, Q, R, S, \ldots p, q, r, s, \ldots\} \tag{53}$$

are employed for arbitrary single particle state functions. The normal product of a second quantized operator is written

$$N[\ldots] \tag{54}$$

where the ellipsis denotes an arbitrary product of creation and annihilation operators and involves moving all annihilation operators to the right using the anticommutation and commutation relations given above.

The exact hamiltonian can be written in normal product form as

$$\mathcal{H}^N = \mathcal{H} - \langle \Phi_0 | \mathcal{H} | \Phi_0 \rangle \tag{55}$$

The unperturbed hamiltonian can also be written in normal product form as

$$\begin{aligned}\mathcal{H}_0^N &= \mathcal{H}_0 - \langle \Phi_0 | \mathcal{H}_0 | \Phi_0 \rangle \\ &= \sum_P^N \varepsilon_P N[a_P^+ a_P] + \sum_P^n \varepsilon_P N[a_P^+ a_P]\end{aligned} \tag{56}$$

The perturbing hamiltonian is written in normal product form using the relation

$$\begin{aligned}\mathcal{H}_1^N &= \mathcal{H}^N - \mathcal{H}_0^N \\ &= \mathcal{H}_1^{(1)N} + \mathcal{H}_1^{(2)N}\end{aligned} \tag{57}$$

where $\mathcal{H}_1^{(1)N}$ and $\mathcal{H}_1^{(2)N}$ are the normal product forms of the operators $\mathcal{H}_1^{(1)}$ and $\mathcal{H}_1^{(2)}$. Explicitly, the one-particle perturbation operator, $\mathcal{H}_1^{(1)N}$, can be written as

$$\begin{aligned}\mathcal{H}_1^{(1)N} = &\sum_{P,Q}^N \left[\left[\sum_R^N \langle PR|G|QR \rangle \right] - \langle P|u_n|Q \rangle \right] N[a_P^+ a_Q] \\ &+ \sum_{p,q}^n \left[\left[\sum_r^n \langle pr|g|qr \rangle \right] - \langle p|u_e|q \rangle \right] N[a_p^+ a_q]\end{aligned} \tag{58}$$

whilst the two-particle operator, $\mathcal{H}_1^{(2)N}$, has the form

$$\mathcal{H}_1^{(2)N} = V_{nn}^N + V_{ne}^N + V_{ee}^N$$

$$= \sum_{P,Q,R,S}^{N} \langle PQ|G|RS \rangle N\left[a_P^+ a_Q^+ a_S a_R\right] +$$

$$+ \sum_{P,Q}^{N} \sum_{p,q}^{n} \langle Pp|\gamma|Qq \rangle N\left[a_P^+ a_p^+ a_q a_Q\right] +$$

$$+ \sum_{p,q,r,s}^{n} \langle pq|g|rs \rangle N\left[a_p^+ a_q^+ a_s a_r\right] \quad (59)$$

Using the normal product unperturbed hamiltonian, the zero-order Schrödinger equation becomes

$$\mathcal{H}_0^N |\Phi_m\rangle = \Delta E_m^0 |\Phi_m\rangle \quad (60)$$

whilst the perturbed Schrödinger equation is

$$\mathcal{H}^N |\Psi\rangle = \Delta E |\Psi\rangle \quad (61)$$

where

$$\Delta E_m^0 = E_m^0 - E_0 \quad (62)$$

and

$$\Delta E = E - E_0 = E^{(1)} + E_{\text{correlation}}$$

2.4 The Diagrammatic Perturbation Theory of Nuclei and Electrons. – Diagrammatic methods are well established in handling the electron correlation problem which arising in the description of molecular structure within the Born-Oppenheimer approximation. In fact for the relativistic electronic structure problem which involves an infinite number of bodies, the use of diagrams becomes almost indispensable; cutting through complicated algebra to expose the essential physics of the various interactions taking place. In Lecture Notes for the 1980 Coulson Summer School in Theoretical Chemistry, the present author[35] wrote

> "It should perhaps be stated at this point that the use of diagrams in the manybody perturbation theory is not obligatory. The whole of the theoretical apparatus can be set up in entirely algebraic terms. However, the diagrams are both more physical and easier to handle than the algebraic expressions and it is well worth the effort required to familiarize oneself with the diagrammatic rules and conventions."

This point has been made by many authors. In his lecture notes for the 1972 Oxford Summer School, P.W. Atkins writes[36]

"In the early books on quantum theory the pages were covered with integral signs and expression such as

$$\int_{-\infty}^{\infty} dx \psi_n^*(x) H \psi_m(x). \tag{1.1}$$

But these soon gave way to the symbol

$$\langle n | H | m \rangle \tag{1.2}$$

which, as well as being more compact, enables the structure of an equation to be seen more clearly. In recent years, a new change has occurred, and instead of equations containing cumbersome integrals as in (1.1), or Dirac brackets as in (1.2), we now see the same thing written as

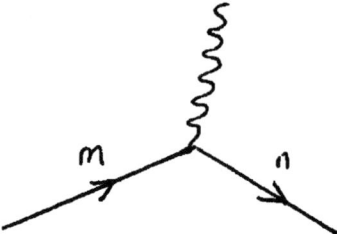

This diagram contains all the information contained in (1.1) and (1.2), but conveys it with remarkable clarity. We see that a system in a state m is deflected into a state n by the action of the operator H.

The diagrams are interpreted in terms of the particle-hole formalism. The Fermi level is defined such that all single particle states lying below it are occupied and all above it are unoccupied. In the particle-hole picture, the reference state is taken to be a vacuum state, containing no holes below the Fermi level and no particles above it. Excitation leads to the creation of particle-hole pairs, with particles above the Fermi level and holes below it.

The diagrammatic method can be extend to system containing both nuclei and electrons by defining nuclear and electronic vertices. A nuclear vertex is represented by an open dot o whereas an electronic vertex is represented a filled dot •. When it is necessary to describe the nuclear vertex associated with a particular nuclear species corresponding to a specific element or mass number then the details are written close to the relevant open dot. The basic components of the diagrams which are used to describe processes in the

o — — — x	one-nucleus operator
• — — — x	one-electron operator
o — — — o	nucleus-nucleus interaction operator
o — — — •	nucleus-electron interaction operator
• — — — •	electron-electron interaction operator

Figure 1 *Basic components of the diagrams for a system of nuclei and electrons*

particle-hole formalism are summarized in Figure 1. All changes in the state of a many-body system are caused by an interaction which is described by an operator. This operator may be a one-particle operator or a two-particle operator. A one-particle operator is represented by a horizontal dashed interaction line terminated by a cross. For the one-nucleus operator the other end of the horizontal dashed line is terminated by an open dot ○, that is

$$\circ\;-\;-\;-\;-\;\text{x}$$

whereas for the one-electron operator a filled dot • is used, that is

$$\bullet\;-\;-\;-\;-\;\text{x}.$$

Three types of two-particle interactions can occur. Each is represented by a horizontal dashed interaction line. For the nucleus-nucleus interaction this line is terminated by open dots, ○, at both end, that is

$$\circ\;-\;-\;-\;-\;\circ$$

The nucleus-electron interaction is represented by a horizontal dashed line terminated by an open dot, ○, at one end and a filled dot, •, at the other

$$\circ\;-\;-\;-\;-\;\bullet$$

The electron-electron interaction is represented by a horizontal dashed line terminated by filled dots, •, at both ends, that is

$$\bullet\;-\;-\;-\;-\;\bullet$$

In the convention which we shall follow here the two-electron interaction includes permutation of the two electrons. The two-nucleus interaction does not include permutation of the nuclei. The nucleus creation operators, $a_P^+, a_Q^+, \ldots$, are represented by arrows leaving the nucleus vertices. Similarly, the electron creation operators, $a_p^+, a_q^+, \ldots$, are represented by arrows leaving the electron vertices. The nucleus annihilation operators, $a_P, a_Q, \ldots$, are represented by arrows directed into the nucleus vertices. The electron annihilation operators, $a_p, a_q, \ldots$, are similarly represented by arrows directed into the electron vertices. Upward arrows represent "particle" lines whereas "downward" directed arrows "hole" lines. It should be noted that "particle" and "hole" lines may not connect a vertex corresponding to a nucleus to one associated with an electron and *vice versa*. Equally, "particle" and "hole" lines may not connect vertices corresponding to different elements or difference mass numbers.

2.4.1 Types of interaction. – The types of interactions which can arise in the diagrammatic perturbation theory of nuclei and electrons can be classified according to the number and type of particles involved. First we subdivide the interactions into one-particle and two-particle types.

One-particle interactions Obviously, the one-particle interactions can be subdivided into those involving a nucleus and those involving an electron. We consider each type in turn.

change in level of excitation

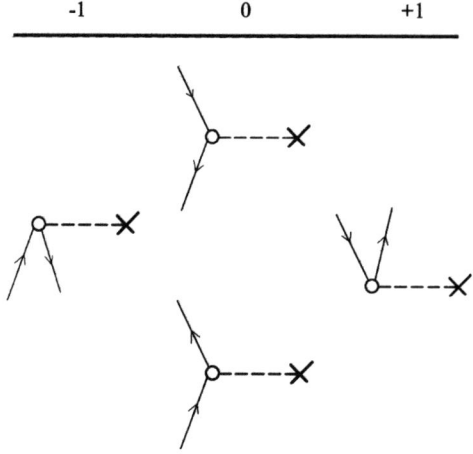

Figure 2 *Classification of one-nucleus interactions that can arise in energy diagrams according to the level of excitation involved*

One-nucleus interactions The one-nucleus interactions that can arise in energy diagrams are classified in Figure 2 according to the level of excitation involved. There is a total of four diagrams of this type. Two of these diagrams do not involve any change in the level of excitation. They involve the interaction of the one-nucleus operator with either a hole line or a particle line associate with a nuclear orbital. One diagram in Figure 2 involves a single de-excitation; that is, the destruction of a nuclear particle-hole pair. The remaining diagram in Figure 2 involves a single excitation; that is, the creation of a nuclear particle-hole pair.

One-electron interactions The one-electron interactions that can arise in energy diagrams are classified in Figure 3 according to the level of excitation involved. There are four diagrams of this type. Two of these diagrams do not involve any change in the level of excitation. They involve the interaction of the one-electron operator with either a hole line or a particle line associate with an electronic orbital. One diagram in Figure 3 involves a single de-excitation; that is, the destruction of a electronic particle-hole pair. The remaining diagram in Figure 3 involves a single excitation; that is, the creation of an electronic particle-hole pair.

Two-particle interactions The two-particle interactions can be subdivided into those between nuclei, those between nuclei and electrons and those between electrons.

Nucleus-nucleus interactions Nucleus-nucleus interactions that can arise in energy diagrams are classified in Figure 4 according to the level of excitation

8: Perturbation Theory and Application to the Molecular Structure Problem 393

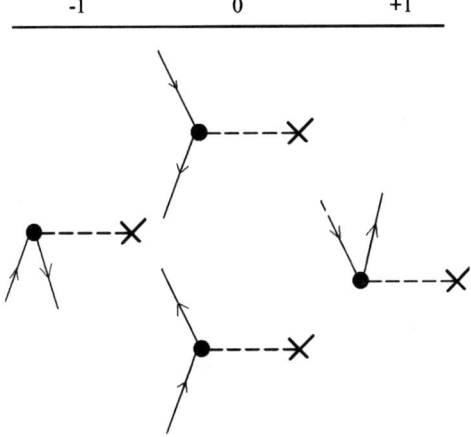

Figure 3 *Classification of one-electron interactions that can arise in energy diagrams according to the level of excitation involved*

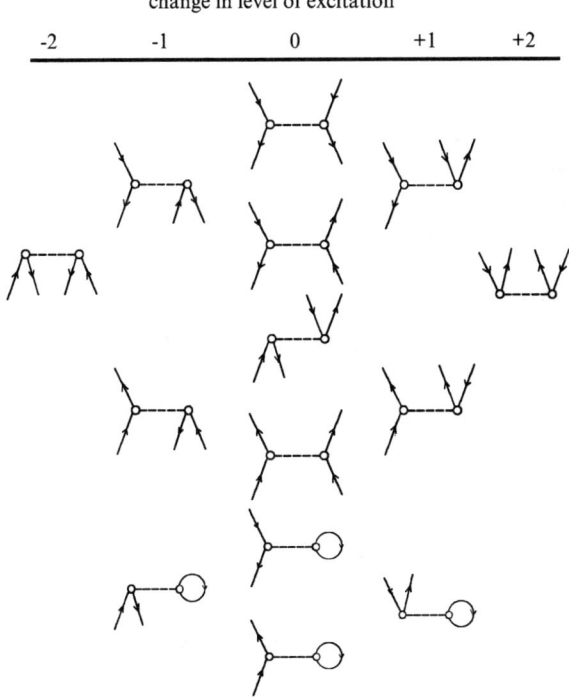

Figure 4 *Classification of two-nucleus interactions that can arise in energy diagrams according to the level of excitation involved*

involved. If the nuclei are identical then the matrix elements include a permutation operator which interchanges the co-ordinates of the two nuclei. There is a total of fourteen types of interaction between nuclei.

Four of these involve a "bubble" or self-energy. They can be classified in the same way as the one-nucleus interaction; so one bubble diagram involves a single excitation, two bubble diagrams involve no change in the level of excitation and the remaining bubble diagram involves a single de-excitation.

Of the remaining ten nucleus-nucleus interaction diagrams, one involves a double de-excitation; that is, the destruction of two nuclear particle-hole pairs, two diagrams involve a single de-excitation; that is, the destruction of one nuclear particle-hole pair, four diagrams involve no change in the level of excitation, two diagrams involve a single excitation; that is, the creation of a nuclear particle-hole pair, and, finally, one diagram involve a double excitation; that is, the creation of two nuclear particle-hole pairs.

Nucleus-electron interactions Nucleus-electron interactions that can arise in energy diagrams are classified in Figure 5 according to the level of excitation involved. There is a total of twenty-four types of interaction between nuclei.

Eight of these involve a "bubble" or self-energy. They can be classified in the same way as the one-particle interaction; so two bubble diagram involves a single excitation, four bubble diagrams involve no change in the level of excitation and the remaining two bubble diagrams involves a single de-excitation.

Of the remaining sixteen nucleus-nucleus interaction diagrams, one involves a double de-excitation; that is, the destruction of one nuclear particle-hole pair and one electron particle-hole pair, four diagrams involve a single de-excitation; that is, the destruction of either a nuclear particle-hole pair or an electron particle-hole pair, six diagrams involve no change in the level of excitation, four diagrams involve a single excitation; that is, the creation of a nuclear particle-hole pair or electron particle-hole pair, and, finally, one diagram involve a double excitation; that is, the creation of a nuclear particle-hole pair and an electron particle-hole pair.

Electron-electron interactions Electron-electron interactions that can arise in energy diagrams are classified in Figure 6 according to the level of excitation involved. Since the electrons are identical the matrix elements include a permutation operator which interchanges the co-ordinates of the two elecetrons. There is a total of fourteen types of interaction between electrons.

Four of these involve a "bubble" or self-energy. They can be classified in the same way as the one-electron interaction; so one bubble diagram involves a single excitation, two bubble diagrams involve no change in the level of excitation and the remaining bubble diagram involves a single de-excitation.

Of the remaining ten electron-electron interaction diagrams, one involves a double de-excitation; that is, the destruction of two electronic particle-hole pairs, two diagrams involve a single de-excitation; that is, the destruction of one electronic particle-hole pair, four diagrams involve no change in the level of excitation, two diagrams involve a single excitation; that is, the creation

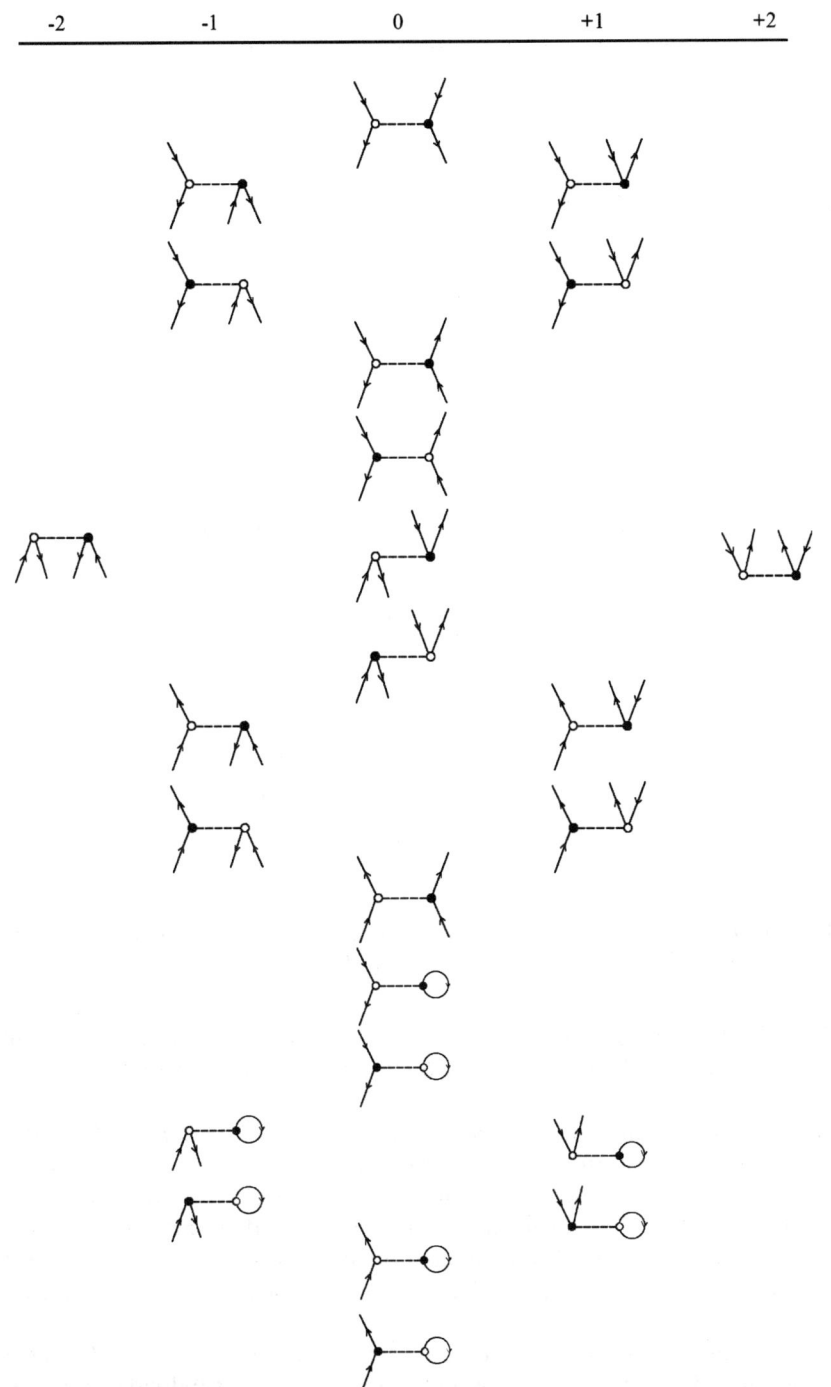

Figure 5 *Classification of nucleus-electron interactions that can arise in energy diagrams according to the level of excitation involved*

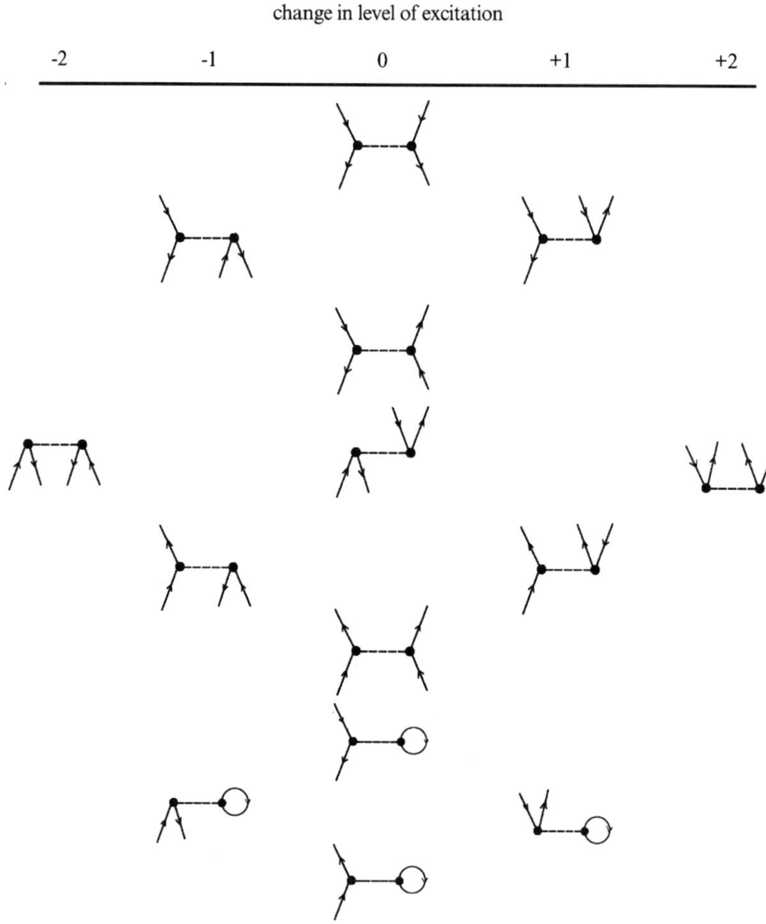

Figure 6 *Classification of two-electron interactions that can arise in energy diagrams according to the level of excitation involved*

of an electronic particle-hole pair, and, finally, one diagram involve a double excitation; that is, the creation of two electronic particle-hole pairs.

2.4.2 First-order Diagrammatic Perturbation Theory of Nuclei and Electrons. – There are five first-order energy terms in the diagrammatic perturbation theory expansion for the motion of nuclei and electrons. The diagrams are shown in Figure 7. Two of the diagrams in Figure 7 describe interactions with the mean field potential. The remaining three first order diagrams describe the interaction of the nuclei, of the electrons, and of the nuclei with the electrons.

2.4.3 Second-order Diagrammatic Perturbation Theory of Nuclei and Electrons. – The second order energy diagrams can be usefully subdivided into those involving only one-particle perturbations, those involving both one- and

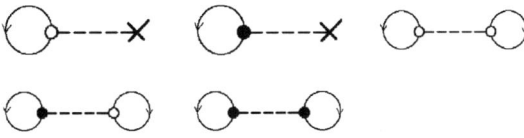

Figure 7 *First order diagrams in the perturbation theory of nuclei and electrons*

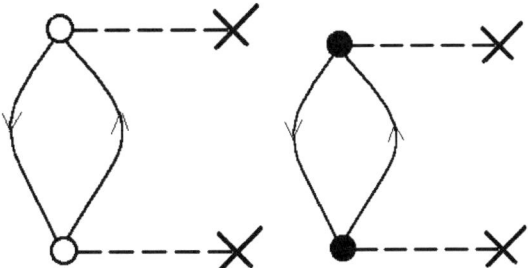

Figure 8 *Second order diagrams involving a one-particle interaction in the perturbation theory of nuclei and electrons*

two-particle perturbations and those involving two particle perturbations. We consider each of these classes of diagrams in turn.

Components involving a one-particle perturbation There are only two second order diagrams involving the one-particle mean field potential. They are displayed in Figure 8.

Components involving one- and two-particle perturbations The second order energy diagrams involving both one- and two-particle perturbations are collected in Figure 9.

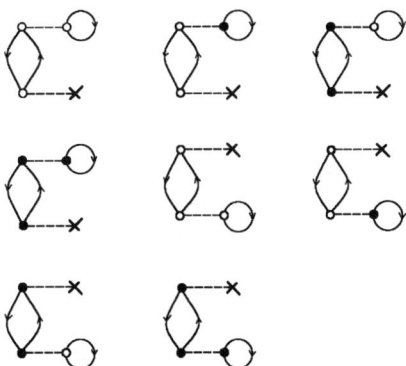

Figure 9 *Second order diagrams involving a one- and two-particle interactions in the perturbation theory of nuclei and electrons*

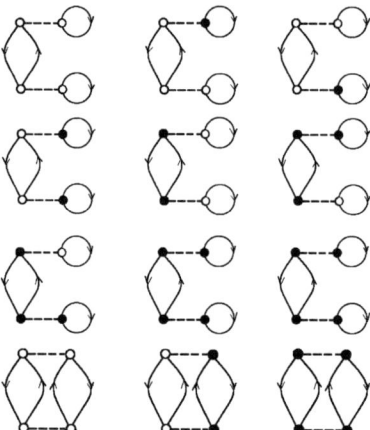

Figure 10 *Second order diagrams involving two-particle interactions in the perturbation theory of nuclei and electrons*

Components involving two-particle perturbations The second order energy diagrams involving two-particle perturbations are collected in Figure 10.

2.4.4 Third-order Diagrammatic Perturbation Theory of Nuclei and Electrons. – Some of examples of third order energy diagrams in the perturbation theory of nuclear and electronic motion are displayed in Figure 11.

2.4.5 Fourth-order Diagrammatic Perturbation Theory of Nuclei and Electrons. – Some of examples of fourth order energy diagrams in the perturbation theory of nuclear and electronic motion are displayed in Figure 12.

2.5 Prospects. – We began this section by recalling Löwdin's statement,[13] made in 1990, that there is an urgent need to describe the motions of electrons and nuclei in a more or less equivalent manner. We have demonstrated in this

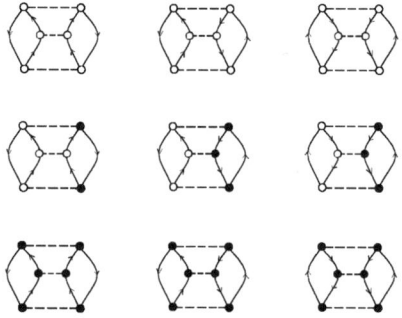

Figure 11 *Some examples of third order energy diagrams in the perturbation theory of nuclei and electrons*

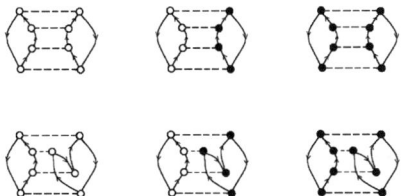

Figure 12 *Some examples of fourth order energy diagrams in the perturbation theory of nuclei and electrons*

section that significant progress towards this end has been achieved during the reporting period.

We emphasize again that the ability to perform practical calculations which go beyond the Born-Oppenheimer approximation is important not only for accurate studies of small molecular species but also in studies of larger molecules, including biomolecules, for which the number of nuclear configurations which have to be considered in the usual Born-Oppenheimer-based approach can become very large indeed. The work of Nakai and Sodeyama,[29] in particular, has established a firm foundation for the approach in which electrons and nuclei are treated on a more or less equivalent footing and further progress can be expected.

3 Diagrammatic Many-body Perturbation Theory of Molecular Electronic Structure: Low Order Approximants

Over the past forty years, summation approximants have been shown to offer significant advantages in low order perturbation studies. There have been some useful developments in the period under review, in particular, the continued studies of quadratic Padé approximants by Goodson.[37] In this section, we consider briefly summation approximants and their application in perturbation theory, Feenberg scaling, Padé approximants and quadratic Padé approximants, as well as the use of scaling and Padé approximants in conjunction with systematic sequences of basis sets.

3.1 Summation Approximants. – Perturbation theory affords the most systematic approach to the many-body problem. All terms are included in the expansion through a given order. No selection of configurations is required as in the method of configuration interaction. No summation of certain types of terms to infinite order whilst neglecting terms of lower order is undertaken as in the coupled cluster expansion. The perturbation expansion for the energy

$$\mathcal{E} = \mathcal{E}(\lambda) = \sum_{k=0}^{\infty} E^{(k)} \lambda^k \tag{63}$$

can be written

$$\mathcal{E}(\lambda) = \sum_{k=0}^{m} E^{(k)}\lambda^k + \sum_{k=m+1}^{\infty} E^{(k)}\lambda^k. \tag{64}$$

For the perturbed problem ($\lambda = 1$) the energy may be written

$$\mathcal{E}(1) = E[m] + R[m] \tag{65}$$

where

$$E[m] = \sum_{k=0}^{m} E^{(k)} \tag{66}$$

and

$$R[m] = \sum_{k=m+1}^{\infty} E^{(k)}. \tag{67}$$

The utility of the finite order power series, $E[m]$, rests on the smallness of the remainder term, $R[m]$. However, the range of problems to which perturbation theory can be applied can be significantly extended by employing functional forms which are more general than the Taylor expansion. The most well known of these approximants are those introduced by Padé[38] in 1892. Other approximants have been investigated during the period under review. In particular, quadratic Padé approximants have been shown to be useful in a series of papers by Goodson.[37,39–41]

Goodson[37] has recently described the rationale for the use of summation approximants in perturbation theory, in particular Møller-Plesset perturbation theory, as follows

> "A perturbation series should not be taken as an end in itself but rather as the starting point for a functional analysis. The [Møller-Plesset] series is the asymptotic series (a Taylor expansion) of a function $E(z)$, which is the analytic continuation of the energy eigenvalue in terms of the perturbation parameter z. The full configuration interaction energy, that is, the variational solution to the Schrödinger equation within a finite basis set expansion of the wave function, in principle corresponds to $E(1)$, while $E(0)$ is the sum of Hartree-Fock orbital energies."

He continues[37]

> "The [Møller-Plesset] series coefficients are proportional to the derivatives of $E(z)$ at $z = 0$, according to Taylor's theorem. These derivatives contain information about the functional form of $E(z)$ that can be used to construct summation approximants that model the function over a larger range of z than does the conventional power series expansion. The result is an extrapolation from $z = 0$ to $z = 1$ that is more accurate than the usual method for evaluating the [Møller-Plesset] series."

8: Perturbation Theory and Application to the Molecular Structure Problem

The energy coefficients in the Taylor expansion for the energy depend on the choice of zero-order hamiltonian. Changing the zero-order hamiltonian by multiplying by a scalar modifies the energy coefficients in the series and hence the magnitude of the remainder term R. We discuss this "Feenberg scaling" in section 3.2. As a digression, we recall the use of scaled many-body perturbation theory in conjunction with systematically extended basis sets in section 3.3. The use of Padé approximants in Rayleigh-Schrödinger and many-body perturbation theory is described in section 3.4. In section 3.5, quadratic Padé approximants and their use in Rayleigh-Schrödinger and many-body perturbation theory is surveyed. Prospects for this area of research are considered in section 3.6.

3.2 Feenberg Scaling. – Following Feenberg,[42,43] we can scale the zero-order Hamiltonian operator by multiplying it by an arbitrary scalar, μ, say, the zero-order hamiltonian is then

$$\mathcal{H}_0^\mu = \mu \mathcal{H}_0$$

and the perturbation is then

$$\mathcal{H}_1^\mu = \mathcal{H}_1 + (1-\mu)\mathcal{H}_0 \tag{68}$$

so that the full Hamiltonian is recovered when these two modified operators are added. It can be easily shown that the scaled second order energy, E_2^μ, is then given by

$$E_2^\mu = \frac{1}{\mu} E_2 \tag{69}$$

where E_2 is the second order energy corresponding to the zero-order hamiltonian before scaling. Now, if we know the exact correlation energy, $E_{exact}^{correlation}$, we can always put

$$\mu = \frac{E_2}{E_{exact}^{correlation}} \tag{70}$$

so that the second order energy given by the scaled zero-order hamiltonian is equal to the exact correlation energy. Although $E_{exact}^{correlation}$ is not known this simple modification shows that there is *some* choice of zero-order Hamiltonian which yields the exact correlation in second order. In such a case the remainder term is zero, R.

The energy coefficients in the modified many-body perturbation expansion, E_0^μ, E_1^μ, E_2^μ, ..., E_p^μ, ... may be written in terms of the original energy coefficients as follows:-

$$E_0^\mu = \mu E_0 \tag{71}$$

$$E_1^\mu = E_1 + (1-\mu)E_0 \tag{72}$$

$$= E_1 + (E_0 - E_0^\mu) \tag{73}$$

$$E_p^\mu = \frac{1}{\mu^{p-1}} \sum_{q=2}^{q=p} \binom{p-2}{q-2} (\mu-1)^{p-q} E_q \tag{74}$$

Explicitly, the first few orders take the form

$$E_2^\mu = \mu^{-1} E_2 \tag{75}$$

$$E_3^\mu = \mu^{-2} E_3 + \mu^{-2}(\mu-1) E_2 \tag{76}$$

$$E_4^\mu = \mu^{-3} E_4 + 2\mu^{-3}(\mu-1) E_3 + \mu^{-3}(\mu-1)^2 E_2 \tag{77}$$

Now, the total correlation energy may be written in the form

$$E_{\text{correlation}} = E_2^\mu + R_2^\mu \tag{78}$$

where μ is chosen so that the magnitude of remainder term is reduced

$$|R_2^\mu| < |R_2| \tag{79}$$

In the past, a number of prescriptions for the determination of μ have been investigated.[44-47] The most popular, which was discussed most recently by Schmidt et al.,[48] is to set the modified third order energy coefficient to zero

$$E_3^\mu = \mu^{-2} E_3 + \mu^{-2}(\mu-1) E_2 = 0 \tag{80}$$

an equation which can be solved for μ to give

$$\mu = \frac{E_2 - E_3}{E_2} \tag{81}$$

which, in turn, gives a scaled second order energy coefficient that is entirely equivalent to the [2/1] Padé approximant to the original perturbation expansion.

It does not seem to be widely recognized that the setting of scaled energy coefficients to zero in fourth and higher order does not lead to a unique value of μ. Thus, putting the modified fourth order coefficient to zero

$$E_2^\mu = 0 \tag{82}$$

gives two values of μ

$$\mu_\pm = \frac{1}{2E_2}\left(-2E_3 + 2E_2 \pm 2\sqrt{(E_3^2 - E_2 E_4)}\right) \tag{83}$$

The alternative approach of putting the sum of the third and fourth order terms in the modified series to zero

$$E_3^\mu + E_4^\mu = 0 \tag{84}$$

also leads to two solutions

$$u_\pm = \frac{1}{4E_2}\left(-3E_3 + 3E_2 \pm \sqrt{(9E_3^2 - 2E_3 E_2 + E_2^2 - 8E_2 E_4)}\right) \tag{85}$$

Putting higher order coefficients, or combinations of such coefficients, to zero always leads to a multivalued solution.

A procedure for the determination of a unique value of μ from higher order expansions was given by the present author.[49] The scaled second order energy coefficient may be set equal to the sum of the coefficients through third order

$$E_2^\mu = E_2 + E_3 \tag{86}$$

that is

$$\mu^{-1} E_2 = E_2 + E_3 \tag{87}$$

which can be solved for μ to give

$$\mu_{[3/0]} = \frac{E_2}{E_2 + E_3} \tag{88}$$

An alternative representation of the sum of the correlation energy coefficient thorough third order is given by the [2/1] Padé approximant

$$E[2/1] = \frac{E_2}{1 - \dfrac{E_3}{E_2}} \tag{89}$$

which has certain invariance properties which make its use attractive. Setting the modified second order energy equal to the [2/1] Padé approximant

$$\mu^{-1} E_2 = E[2/1] \tag{90}$$

gives

$$\mu^{-1} E_2 = \frac{E_2}{1 - \dfrac{E_3}{E_2}} \tag{91}$$

which can be solved for μ

$$\mu_{[2/1]} = \frac{E_2 - E_3}{E_2} \tag{92}$$

so that the scaled second order energy is itself the [2/1/] Padé approximant.

In fourth order, the scaled second order energy coefficient may be set equal to the sum of the second, third and fourth order terms in the original series

$$\mu^{-1} E_2 = E_2 + E_3 + E_4 \tag{93}$$

This equation has a unique solution

$$\mu_{[4/0]} = \frac{E_2}{E_2 + E_3 + E_4} \tag{94}$$

Alternatively, the fourth order Feenberg invariant[43,50,51]

$$\mathcal{E}_4 = E[2/1] + \left(\frac{E[2/1]}{E_2}\right)^3 \left(E_4 - \frac{E_3^2}{E_2}\right) \tag{95}$$

may be used. Setting the modified second order energy equal to this invariant gives the equation

$$\mu^{-1} E_2 = \mathcal{E}_4 \tag{96}$$

which has the unique solution

$$\mu = \frac{(E_3 - E_2)^3}{E_2^2 (2E_3 - E_2 - E_4)} \tag{97}$$

3.3 Digression: Scaled Many-Body Perturbation Theory and Systematically Extended Basis Sets.

In general, the scaled second order energy coefficient may be set equal to some higher order approximation to the total correlation energy, $\mathcal{E}$

$$E_2^\mu = \mathcal{E} \tag{98}$$

This provides a unique prescription for the determination of μ

$$\mu = \frac{E_2}{\mathcal{E}} \tag{99}$$

$\mathcal{E}$ may be an approximation to the total correlation energy, a finite order perturbation sum, a limited configuration interaction or coupled cluster energy.

Scaling parameters can transferred from a calculations carried out in a smaller basis set to one using a larger basis set.[49,52]

Consider a calculation performed using a basis set designated S_A. The relation between the scaled second order correlation component and the higher order approximation to the total correlation energy then takes the form

$$E_2^\mu [S_A] = \mathcal{E}[S_A] \tag{100}$$

giving

$$\mu[S_A] = \frac{E_2[S_A]}{\mathcal{E}[S_A]} \tag{101}$$

Now, consider a second calculation performed by using a basis set S_B. For this basis set, the scaled second order energy coefficient and the higher order approximation are related as follows:-

$$E_2^\mu[S_B] = \mathcal{E}[S_B] \tag{102}$$

The higher order approximation can be written

$$\mathcal{E}[S_B] = \frac{E_2[S_B]}{\mu[S_B]} \tag{103}$$

Assuming $\mu[S_B]$ may be replaced by $\mu[S_A]$ results in the estimate

$$\mathcal{E}[S_B] \sim \frac{E_2[S_B]}{\mu[S_A]} \tag{104}$$

3.4 Padé Approximants. – We have already mentioned the approximants introduced by Padé[38] in 1892. These are rational approximants which have the form

$$[L/M](\lambda) = \frac{P_L(\lambda)}{Q_M(\lambda)} \tag{105}$$

that is the ratio of a polynomial in λ of order L

$$P_L(\lambda) = p_0 + p_1\lambda + p_2\lambda^2 + \ldots + p_L\lambda^L \tag{106}$$

to a polynomial of order M

$$Q_M(\lambda) = 1 + q_1\lambda + q_2\lambda^2 + \ldots + q_L\lambda^M. \tag{107}$$

where, by convention, $q_0 = 1$. The usual Taylor series corresponds to the [N/0] Padé approximant. Other Padé approximants [L/M] of order N, i.e. for which $L + M = N$, can provide a convergent sequence of approximations when the usual power series, the [N/0] Padé approximants, diverge. Unlike the power series, some Padé approximants can handle a class of functions with various types of singularities still providing uniform convergence.

The [L/M] Padé approximant is defined by

$$P_L(\lambda) - Q_M(\lambda)E(\lambda) = \mathcal{O}(\lambda^{L+M+1}) \tag{108}$$

and

$$E_{[L/M]}(\lambda) = \frac{P_L(\lambda)}{Q_M(\lambda)} \tag{109}$$

where $\mathcal{O}(\lambda^{L+M+1})$ denotes terms of order $L+M+1$ and higher.

Consider again the scaling of the zero-order hamiltonian first proposed by Feenberg.[42,43] The total hamiltonian is invariant to the choice of the scaling parameter μ but the zero-order operator, $\mathcal{H}_0^\mu$, and the perturbation, $\mathcal{H}_1^\mu$, thus

$$\mathcal{H}(\lambda) = \mathcal{H}_0^\mu + \mathcal{H}_1 \mu \lambda \tag{110}$$

where

$$\mathcal{H}_1^\mu = \mathcal{H} - \mathcal{H}_0 \mu \tag{111}$$

The Padé approximants to the corresponding energy expansion has the form

$$\mathcal{E}^\mu[L/M](\lambda) = \frac{P_L\mu(\lambda)}{Q_M^\mu(\lambda)} \tag{112}$$

In general, the energy determined by setting the perturbation parameter λ to 1, $\mathcal{E}^\mu[L/M](1)$, depends on the choice of the scaling factor μ. However, it can be shown that, uniquely among all Padé approximants of order $2N+1$ the $[N+1/N]$ approximant is invariant to the choice of μ.[53] There is a large body of evidence demonstrating that the use of $[N+1/N]$ Padé approximants leads to improved results.[54]

In 1981, Cohen and Feldman[55] considered a rational fraction representation of the energy in developing a modified Rayleigh-Schrödinger perturbation series. Specifically, they wrote the perturbed Schrödinger equation in the form

$$Q_M(\lambda)(\mathcal{H}_0 + \mathcal{H}_1 \lambda)\Psi = P_L(\lambda)\Psi, \tag{113}$$

then made a Taylor series expansion for the wavefunction and equated powers of λ. In this way a Padé approximant to the energy was obtained directly rather than via the usual Taylor series expansion.

3.5 Quadratic Padé Approximants. – Quadratic Padé approximants were introduced by Padé in his original work.[38] They have a functional form which allows them to describe explicitly square-root branch-point singularities (see, for example, the text by Carrier, Krook and Pearson.[56] It is only fairly recently that the mathematical properties of the quadratic Padé approximants have been studied.[57-61] Applications of the quadratic Padé approximants have been realized in a various areas of atomic and molecular physics.[62-67,39,40,68,69]

Goodson has shown[40] that the accuracy of fourth order many-body perturbation theory calculations (designated "*MP4*") can be significantly increased by using the Feenberg scaling procedure to repartition the hamiltonian and summing the scaled perturbation series by using quadratic Padé approximants. This approach models the known mathematical structure of the perturbation theory explicitly. Goodson[40] shows that the accuracy achieved by this approach

"is typically somewhat higher than that of CCSD(T)"†

Consider again the usual Taylor expansion for the energy used in Rayleigh-Schrödinger perturbation theory

$$E(\lambda) \sim \sum_{p=0}^{\infty} \lambda^p E_p. \tag{114}$$

Practical applications of perturbation theory rest on a finite order expansion of the form

$$E^{[K]}(\lambda) = \sum_{p=0}^{K} \lambda^p E_p, \tag{115}$$

which is truncated at order K. If we define three polynomials

$$P_L(\lambda) = \sum_{l=0}^{L} \lambda^l p_l, \tag{116}$$

$$Q_M(\lambda) = 1 + \sum_{m=1}^{M} \lambda^m q_m, \tag{117}$$

and

$$R_N(\lambda) = \sum_{n=0}^{N} \lambda^n r_n \tag{118}$$

which satisfy the equation

$$Q_M(\lambda) E^2(\lambda) - P_L(\lambda) E(\lambda) + R_N(\lambda) \sim \mathcal{O}(\lambda^{L+M+N}) \tag{119}$$

then we can solve the equation

$$Q_M(\lambda) E^2(\lambda) - P_L(\lambda) E(\lambda) + R_N(\lambda) = 0 \tag{120}$$

for $E(\lambda)$ to obtain the summation approximant

$$S_{[L/M,N]}(\lambda) = \frac{1}{2} \left[\frac{P_L(\lambda)}{Q_M(\lambda)} \pm \frac{1}{Q_M(\lambda)} \sqrt{(P_L(\lambda))^2 - 4 Q_M(\lambda) R_N(\lambda)} \right] \tag{121}$$

which is termed a quadratic Padé approximant. The coefficients p_l, q_m and r_n are determined from a set of $L+M+N$ simultaneous linear equations obtained

† Coupled Cluster — Single and Doubles with perturbative Triples.

by collecting terms in (121) according to powers of λ. (121) is a multivalued approximant having two branches, which are connected by square root branch points at the roots of the discriminant polynomial

$$D_{[L/M,N]}(\lambda) = (P_L(\lambda))^2 - 4Q_M(\lambda)R_N(\lambda). \tag{122}$$

In 2000, Goodson[40] showed that the accuracy of fourth order many-body perturbation theory calculations with a Møller-Plesset zero order hamiltonian can be significantly improved by employing Feenberg scaling of the type discussed in section 3.2 and then summing the resulting series with quadratic Padé approximants. The Feenberg scaling was carried out using the repartitioned hamiltonian

$$\mathcal{H}^{(\mu)}(\lambda) = \mu\mathcal{H}_0 + (\mathcal{H}_1 - (1-\mu)\mathcal{H}_0)\lambda, \tag{123}$$

and the formulae

$$E_0^{(\mu)} = \mu E_0, \tag{124}$$

$$E_1^{(\mu)} = E_1 + (1-\mu)E_0, \tag{125}$$

and

$$E_p^{(\mu)} = \frac{1}{\mu^{p-1}} \sum_{k=2}^{p} \binom{p-2}{k-2}(\mu-1)^{p-k} E_k, \quad p \geq 2 \tag{126}$$

relating the scaled perturbation series with the original. The quadratic Padé approximants was determined according to the scheme described above.

3.6 Prospects. – Summation approximants improve the reliability of low order perturbation theory calculations with little additional computational effort. Given the tractability of low order perturbation studies there will undoubtedly be an effort to develop new approximants in the years ahead. The conventional "linear" Padé approximant is a function which is written as the ratio of the polynomials $P_L(\lambda)$ and $Q_M(\lambda)$ of degrees L and M, respectively, defined by the linear equation

$$P_L(\lambda) - Q_M(\lambda)E(\lambda) = \mathcal{O}(\lambda^{L+M+1}). \tag{127}$$

We have seen that the quadratic approximants are defined in terms of three polynomials $P_L(\lambda)$, $Q_M(\lambda)$ and $R_N(\lambda)$ defined by the equation

$$Q_M(\lambda)E^2(\lambda) - P_L(\lambda)E(\lambda) + R_N(\lambda) = 0 \tag{128}$$

Similar algebraic approximants of arbitrary degree can be defined. Indeed, this was done by Padé in his original work.[38] Writing the approximant of order t as $E[p_0, p_1, \ldots, p_t]$, we can introduce their definition as

$$\sum_{s=0}^{t} C_s(\lambda) E[p_0, p_1, \ldots, p_t](\lambda) = 0 \qquad (129)$$

The polynomials $C_s(\lambda)$ have degree p_s and satisfy the equations

$$\sum_{s=0}^{t} C_s(\lambda) E^s(\lambda) = \mathcal{O}(\lambda^\mu) \qquad (130)$$

where

$$u = t + \sum_{s=0}^{t} p_s \qquad (131)$$

Sergeev and Goodson[60] have discussed some of the mathematical properties of these higher order Padé approximants but they have not been applied to physical problems except in a few rare instances.

4 Diagrammatic Many-body Perturbation Theory of Molecular Electronic Structure for Larger Systems

4.1 Local Correlation Methods. – It is recognized that the steep scaling of algorithms for describing electron correlation in molecular systems is often an artifact of the orthogonal canonical basis, i.e. the solutions of the matrix Hartree-Fock equations, used to construct post-Hartree-Fock correlation theories. The steep scaling is not a consequence of the underlying physics. For example, dynamic correlation is a short-range effect decaying as r^{-6}. Schütz[70] has pointed out that

> "The delocalized character of [*Hartree-Fock*] canonical orbitals destroys the locality of correlation effects, leading to a quadratic scaling of the number of electron pairs and a cubic scaling of the number of orbital triples to be correlated, and an overall $\mathcal{O}(n^4)$ and $\mathcal{O}(n^6)$ increase in the number of pair and triple amplitudes, respectively."

The very steep scaling of the computational demands associated with conventional electron correlation studies arises from the fact that calculations are performed using a basis of canonical molecular orbitals which are, in general, delocalized over the entire molecular system. As Hampel and Werner[71] wrote in 1996

> "This not only prevents the omission of small correlation effects of distant electrons, but also leads to an unphysically steep increase in the number of virtual orbitals needed for the correlation of each particular electron pair."

They continue[71]

> "It is intuitively clear that a localized description of electron correlation is needed to avoid these problems."

Local correlation methods were first proposed in the mid-1960s by Sinanoğlu[72] and by Nesbet.[73] Since that time, many workers have suggested variants of the local correlation approach.[74-169] However, it is only in recent years that computational resources have emerged which allow applications to systems large enough to demonstrate the potential of local correlation methods.

4.2 Linear Scaling Correlation Methods. – The aim of local correlation methods is to realize algorithms which scale linearly with the size of the problem. The work of Pulay and Saebo[86,89,94,99,101,113,116,?,132,133,135,156,157,164] and of Werner et al.[132-135,140,141,146,153,160,169] is particularly noteworthy in this respect acheiving linear scaling for large molecules.

Much of this work is focused on the most widely used[1] *ab initio* correlation method, MP2, since this is the method which is least demanding of computational resources. The local MP2 algorithm described by Hetzer, Schütz, Stoll and Werner[146] is particularly efficient and these authors report that

"... the calculation of the MP2 energy is less expensive than the calculations of the Hartree-Fock energy for large systems."

Scuseria and Ayala[139] devised a linear scaling coupled cluster algorithm for double excitations, CCD, which, although based on an atomic orbital basis set, is equivalent, within the thresholds used, to the 'full' CCD method. Schütz, Werner and their co-workers have described local correlation methods for more complicated algorithms including coupled cluster 'singles and doubles' with and without perturbative estimates of the triple excitation component of the correlation energy. These methods, designated LCCSD (Local Coupled Cluster Singles and Doubles) and LCCSD(T) (Local Coupled Cluster Singles and Doubles with perturbative Triples) and described in References 71 and 160, and in References 141 and 70, respectively.

4.3 Local "MP2" Methods. – In the local MP2 method introduced by Pulay and Saebo[94] in 1986, the canonical occupied orbitals are replaced by orthogonal localized orbitals $\phi_i, \phi_j, \ldots$, which are represented by a coefficient matrix $\mathbf{L}$. A set of nonorthogonal and redundant virtual orbitals $\phi_a, \phi_b, \ldots$, are represented by the coefficient matrix

$$\mathbf{P} = \mathbf{I} - \mathbf{LL}^\dagger\mathbf{S}, \tag{132}$$

where $\mathbf{S}$ is the overlap matrix for the basis functions. Strong orthogonality implies that we have

$$\mathbf{L}^\dagger\mathbf{SP} = 0. \tag{133}$$

The first order wave function can be written

$$\Psi^{(1)} = \sum_{i,j} \sum_{a,b \in \mathcal{D}[i,j]} T_{ab}^{ij} \left| \Phi_{ij}^{ab} \right\rangle \tag{134}$$

in which

$$\left|\Phi_{ij}^{ab}\right\rangle = \frac{1}{2}E_{ai}E_{bj}|0\rangle \qquad (135)$$

where E_{ai} is a spin-adapted excitation operator, and the coefficients T_{ab}^{ij} are often termed amplitudes. The sum over i and j in (134) runs over all pairs of occupied orbitals. The sum over a and b in (134) runs over all pairs of virtual orbitals satisfying $a, b \in \mathcal{D}_{[i,j]}$, where $\mathcal{D}_{[i,j]}$ denotes the so-called pair domain which comprises orbitals localized in the same region of space as one or both of the occupied orbitals ϕ_i and ϕ_j.

The second order correlation energy can be written

$$E_2 = \sum_{i,j} \sum_{a,b \in \mathcal{D}_{[i,j]}} (2T_{ab}^{ij} - T_{ba}^{ij}) K_{ab}^{ij} \qquad (136)$$

where K_{ab}^{ij} is the integral

$$\begin{aligned} K_{ab}^{ij} &= (ai|bj) \\ &= \sum_p P_{pa} \sum_q P_{rb} \sum_r L_{qi} \sum_s L_{sj} (pq|rs) \end{aligned} \qquad (137)$$

in which the integral over basis functions is define as

$$(pq|rs) = \int d\mathbf{r}_1 \, d\mathbf{r}_2 \chi_p(\mathbf{r}_1) \chi_q(\mathbf{r}_1) \frac{1}{r_{12}} \chi_r(\mathbf{r}_2) \chi_s(\mathbf{r}_2). \qquad (138)$$

The four-index transformation in (137) is the most demanding part of a localized MP2 calculation if conventional methods are employed.

4.4 Splitting of the Coulomb Operator. – The Coulomb operator can be written as a sum of a long-range part, $L(r)$, and a short-range part, $S(r)$, as follows

$$\begin{aligned} \frac{1}{r} &= \frac{f(r)}{r} + \frac{1-f(r)}{r} \\ &= L(r) + S(r). \end{aligned} \qquad (139)$$

Here $f(r)$ is termed the separation function and is commonly taken to be the error function

$$f(r) = \text{erf}(\omega r) \qquad (140)$$

in which ω is a tunable decay parameter. This approach goes back to the work of Ewald in 1921[170] and has recently been re-examined by a number of authors[180-181]

The long-range function then takes the form

$$L(r) = \frac{\text{erf}(\omega r)}{r} \qquad (141)$$

which is slowly decaying but nonsingular, whilst the short-range function is

$$S(r) = \frac{1-\text{erf}(\omega r)}{r} \qquad (142)$$

and is rapidly decaying but singular.

Consider the exchange integral

$$K_{rs}^{ij} = (ai | bj)$$
$$= \int d\mathbf{r}_1 d\mathbf{r}_2 \phi_a(\mathbf{r}_1) \phi_i(\mathbf{r}_1) \frac{1}{r_{12}} \phi_b(\mathbf{r}_2) \phi_j(\mathbf{r}_2). \qquad (143)$$

Splitting the Coulomb operator into a long-range and a short-range part, this integral becomes

$$(ai | bj) = (ai | L(r) | bj) + (ai | S(r) | bj) \qquad (144)$$

where

$$(ai | L(r) | bj) = \int d\mathbf{r}_1 d\mathbf{r}_2 \phi_a(\mathbf{r}_1) \phi_i(\mathbf{r}_1) \frac{\text{erf}(\omega r)}{r} \phi_b(\mathbf{r}_2) \phi_j(\mathbf{r}_2) \qquad (145)$$

and

$$(ai | S(r) | bj) = \int d\mathbf{r}_1 d\mathbf{r}_2 \phi_a(\mathbf{r}_1) \phi_i(\mathbf{r}_1) \left[\frac{1-\text{erf}(\omega r)}{r} \right] \phi_b(\mathbf{r}_2) \phi_j(\mathbf{r}_2) \qquad (146)$$

Both of these partial integrals can be evaluated more efficiently than the original integral.

The short-range integral is obtained by a four-index transformation from the short-range integrals over the basis set. Negligible contributions are eliminated by screening the integral list. The Schwartz inequality

$$|(pq | S(r) | rs)| \leq |(pq | S(r) | pq)|^{\frac{1}{2}} |(rs | S(r) | rs)|^{\frac{1}{2}} \qquad (147)$$

leads to a computational scaling as $\mathcal{O}(N^2)$. $\mathcal{O}(N)$ scaling is achieved by taking account of the distances between p and r, p and s, etc. Hetzer et al.[146] use the condition

$$|(pq | S(r) | rs)| \lesssim \max(S_{pq}, S_{pr}, S_{qr}, S_{qs}) \qquad (148)$$

where

$$|(pq | S(r) | rs)| \lesssim \max(S_{pq}, S_{pr}, S_{qr}, S_{qs}) \qquad (149)$$

which, unlike the Schwarz inequality is not strict but "works well in practice"

4.5 Multipole Expansion of Long Range Integrals. – The integral K_{rs}^{ij} may be approximated as follows

8: Perturbation Theory and Application to the Molecular Structure Problem

$$(ai|bj) \approx \sum_{m,m'} Q_m^{ai} U_{mm'}^{R_{ij}} Q_{m'}^{bj} \tag{150}$$

where Q_m^{ai} is a multipole of the effective charge distribution

$$\rho_{ai} = \chi_a \chi_i \tag{151}$$

and $U_{mm'}^{R_{ij}}$ is the interaction coefficient depending only on the vector $\mathbf{R}_{ij}$ connecting the charge centroids of the occupied orbitals i and j. m and m' are compound indices which determine the type of multipole.

The distance between two electrons can be written

$$r_{12} = |\mathbf{R} - \mathbf{r}_1 + \mathbf{r}_2| \tag{152}$$

where the coordinates of electron 1 are

$$\mathbf{r}_1 = (x_1, y_1, z_1) \tag{153}$$

and those of electron 2 are

$$\mathbf{r}_2 = (x_2, y_2, z_2) \tag{154}$$

The vector

$$\mathbf{R} = (R_x, R_y, R_z) \tag{155}$$

connects the two centres. The long-range operator $L(r_{12})$ can be expanded as a polynomial as follows

$$L(r_{12}) = \sum_{l_x, l_y, l_z} D_{l_x l_y l_z}(\mathbf{R}) (x_2 - x_1)^{l_x} (y_2 - y_1)^{l_y} (z_2 - z_1)^{l_z} \tag{156}$$

in which the coefficients $D_{l_x l_y l_z}(\mathbf{R})$ are to be determined.

Rearranging the above expansion gives

$$L(r_{12}) = \sum_{\substack{m_x m_y m_z \\ n_x n_y n_z}} U_{(m_x m_y m_z)(n_x n_y n_z)}^{R} x_1^{m_x} y_1^{m_y} z_1^{m_z} x_2^{n_x} y_2^{n_y} z_2^{n_z}, \tag{157}$$

with

$$U_{(m_x m_y m_z)(n_x n_y n_z)}^{R} = (-1)^{m_x + m_y + m_z} \binom{m_x + n_x}{n_x} \binom{m_y + n_y}{n_y}$$
$$\times \binom{m_z + n_z}{n_z} D_{m_x + n_x, m_y + n_y, m_z + n_z}(\mathbf{R}) \tag{158}$$

Substituting this expansion for $L(r_{12})$ into the two-electron integral (145) yields the multipole expansion

$$(ai|L(r_{12})|bj) = \int d\mathbf{r}_1 \int d\mathbf{r}_2 \rho_{ai}(\mathbf{r}_1) L(\mathbf{r}_{12}) \rho_{bj}(\mathbf{r}_2)$$

$$= \sum_{m_x,m_y,m_z} \sum_{n_x,n_y,n_z} Q^{ai:R_{ai}}_{m_xm_ym_z} U^{R_{aibj}}_{(m_xm_ym_z)(n_xn_yn_z)} \times Q^{bj:R_{bj}}_{n_xn_yn_z} \quad (159)$$

in which

$$Q^{ai:R_{ai}}_{m_xm_ym_z} = \int d\mathbf{r} \rho_{ai}(\mathbf{r})(x - R_{ai,x})^{l_x}(y - R_{ai,y})^{l_y}(x - R_{ai,z})^{l_z} \quad (160)$$

is a "multipole moment".

4.6 Density Fitting Approximations. – The refinement of "local MP2" methods continues apace. In a recent paper, Werner et al.[169] describe a fast linear scaling second-order Møller-Plesset perturbation theory (MP2) using local and density fitting approximations. Density fitting approximations have a long history in ab initio quantum chemistry.[182–191]

In second order perturbation theory, two-electron integrals involving two occupied orbitals, ϕ_i and ϕ_j, and two virtual orbitals, ϕ_a and ϕ_b, are required. These integrals describe the electrostatic repulsion between two orbital product densities; that is,

$$K^{ij}_{ab} = (ai|bj)$$

$$= \int d\mathbf{r}_1 \int d\mathbf{r}_2 \frac{\rho_{ai}(\mathbf{r}_1)\rho_{bj}(\mathbf{r}_2)}{r_{12}} \quad (161)$$

where

$$\rho_{ai}(\mathbf{r}_1) = \phi_a(\mathbf{r}_1)\phi_i(\mathbf{r}_1)$$

and likewise for $\rho_{bj}(\mathbf{r}_2)$.

In the density fitting approximation, the one-electron densities are approximated as

$$\bar{\rho}_{ai}(\mathbf{r}) = \sum_{\mu}^{M} c^{ai}_{\mu} \chi\mu(\mathbf{r}) \quad (162)$$

where $\{\chi_\mu(\mathbf{r})\}$ is termed the fitting basis set. The expansion coefficients can be obtained by minimizing the positive definite functional[183–185]

$$\Delta_{ai} = \int d\mathbf{r}_1 \int d\mathbf{r}_2 \frac{[\rho_{ai}(\mathbf{r}_1) - \bar{\rho}_{ai}(\mathbf{r}_1)][\rho_{ai}(\mathbf{r}_2) - \bar{\rho}_{ai}(\mathbf{r}_2)]}{r_{12}} \quad (163)$$

It can be shown that this leads to

$$c^{ai}_\nu = \sum_\mu (ai|\mu)[J^{-1}]\mu\nu \quad (164)$$

and

$$\overline{K}_{ab}^{ij} = \sum_v c_v^{ai}(v|bj)$$
$$= \sum_{\mu,v}(ai|\mu)[\mathbf{J}^{-1}]\mu v\,(v|bj) \qquad (165)$$

where

$$J_{\mu v} = \int d\mathbf{r}_1 \int d\mathbf{r}_2 \frac{\chi_\mu(\mathbf{r}_1)\chi_v(\mathbf{r}_2)}{r_{12}} \qquad (166)$$

and

$$(ai|\mu) = \int d\mathbf{r}_1 \int d\mathbf{r}_2 \frac{\phi_a(\mathbf{r}_1)\phi_i(\mathbf{r}_1)\chi\mu(\mathbf{r}_2)}{r_{12}} \qquad (167)$$

In the work of Werner et al.,[169] density fitting approximations are employed to generate the two-electron integrals in method which they designate DF-LMP2. They point out that their method can be regarded as a local version of the well-known RI-MP2 method, which they proposed should be designated DF-MP2. Werner et al. demonstrate that

> "for large molecules DF-LMP2 is much faster (1–2 orders of magnitude) than either LMP2 or DF-MP2"

However, they also note that although

> "density fitting errors are found to be consistently small, ... the errors arising from local approximations are somewhat larger than expected from calculations on smaller systems"

In related work, Manby[192] has explored the use of density fitting in second-order linear-r_{12} Møller-Plesset perturbation theory. He uses density fitting to approximate all of the 4-index 2-electron integrals in the explicitly correlated MP2-R12 theory of Kutzelnigg and Klopper[193] and describes prototype calculations on the glycine molecule which demonstrate that

> "for large basis sets DF-MP2-R12 is faster than a standard MP2 calculation and takes only a small fraction of the time for the Hartree-Fock calculation".

In other related work, Schütz and Manby[195] use density fitting in a linear scaling formulation of local coupled cluster theory. They demonstrate the $\mathcal{O}(N)$ scaling of their "LDF-LCCSD" approach and conclude that

> "The approximate calculation of the 4-external integrals via density fitting in LDF-LCCSD is 10–100 times faster than the exact calculation via the $\mathcal{O}(N)$ 4-index transformation in LCCSD".

4.7 Prospects. – The increasing power of contemporary computing machines has opened up the possibility of applications of *ab initio* quantum chemical methods to larger molecular systems than has been possible previously. Indeed, "Moore's Law"† predicts a doubly of processor power every 18 months or so. However, these developments require the use of local correlation methods which lead to a linear scaling of algorithms. The systematic implementation of such algorithms is required if the errors associated with local approximations are to be controlled. Undoubtedly, future work will place local approximations on a firm foundation.

5 Diagrammatic Many-body Perturbation Theory of Molecular Electronic Structure: A Review of Applications

5.1 Incidence of the String "*MP2*" in Titles and/or Key-Words. – In my previous reports to this series,[1,2] the increasing use of many-body perturbation theory in molecular electronic structure studies was measured by interrogating the *Institute for Scientific Information* database[194]. In particular, I determined the number of incidences of the string "MP2" in titles and/or keywords. This acronym is frequently associated with the simplest form of many-body perturbation theory. This approach will undoubtedly miss many routine applications of second-order many-body perturbation theory but should serve to convey both the extent and breadth of contemporary application areas.

In my report for the period up to June 1999, I noted that the string "MP2" occurred in the titles and/or keywords of just 3 publications in 1989 but that this number had risen to 854 in 1998. In my report for the period June 1999 to May 2001, I was able to report 821 "hits" for 1999 and 883 for the year 2000. For this report, I found a total of 757 incidences for 2001 and 828 for 2002.

For my 2001 report, I analyzed the journals in which the 883 publications published in 2000 with the string "MP2" in the title and/or keywords. I have repeated this analysis for 2001 and 2002, to two complete years for which data became available in the present reporting period. The results are collected in table 1. Roughly 20% of the relevant publications appeared in the *Journal of Physical Chemistry A*, a figure which is in line with that obtained for the year 2000. Another ~10% of publications satisfying our search conditions appeared in J. Molec. Struct. *(THEOCHEM)*, a figure which is again in line with the year 2000 search. The numbers of qualifying publications appearing in others journals are summarized in table 1.

The wide range of journals in which these publications appear is indicative of the broad spectrum of application areas in which perturbative correlation treatments are being exploited. We can expect that there will be many more publications reporting work in which second-order many-body perturbation theory was exploited but which are not included in the above analysis because

---

† The exponential growth of processor speeds, memory sizes and communication bandwidths with a doubling period of about 18 months to two years was first formulated in the 1960s [G.E. Moore, *Cramming more components onto integrated circuits*, Electronics, 19 April 1960, 114–117.

8: Perturbation Theory and Application to the Molecular Structure Problem

Table 1 *Number of publications appearing in various journals in the years 2001 and 2002 for which the string "MP2" appears in the title and/or keywords*

Journal	2001		2002	
J. Phys. Chem. A	157	(21%)	158	(19%)
J. Molec. Struct. *(THEOCHEM)*	75	(10%)	74	(9%)
Chem. Phys. Lett.	54	(7%)	63	(8%)
J. Chem. Phys.	51	(7%)	40	(5%)
Phys. Chem. Chem. Phys.	38	(5%)	34	(4%)
J. Am. Chem. Soc	33	(4%)	32	(4%)
Int. J. Quant. Chem.	21	(3%)	33	(4%)
J. Comp. Chem.	14	(2%)	9	(1%)
J. Phys. Chem. B	8	(1%)	14	(2%)
Molec. Phys.	6	(1%)	11	(1%)
Theoret. Chem. Acc.	2	(0%)	9	(1%)
Total	757		828	

other details of a particular study were considered more important when assigning keywords.

5.2 Comparison with Other Methods. – In my report published in volume 1 of this series I compared the use of the second-order many-body perturbation theory with that of density functional theory and coupled cluster theory. In particular, figure 1 of that previous report shows how the use of "MP2" theory rose from 3 "hits" in 1989 to 854 in 1998, whilst for "DFT" the 1989 figure of 7 grew to 733 in 1998. For "CCSD" the numbers of "hits" in the database search stood at 244 in 1998. We provided an update to the indices for the string "MP2" in the 2002 report. In table 2, we present an update for "MP2", "DFT" and "CCSD" through to the end of the last complete year at the time of writing, 2002. The most striking observation about this table is the growth in the use of "DFT" which first exceeded "MP2" in 1999 and now stands at roughly a factor of two greater. Density functional theory forms the basis of the most widely used semi-empirical approach to the molecular electronic structure problem in contemporary work.

Table 2 *Incidence of the acronyms "MP2", "DFT" and "CCSD" in the title and/or keywords of publications over the period 1998 to 2002*

Year	"MP2"	"DFT"	"CCSD"
1998	854	738	244
1999	821	923	263
2000	883	1,221	283
2001	757	1,528	318
2002	828	1,723	303

Second order many-body perturbation theory is the most tractable of *ab initio* methods.

5.3 Synopsis of Applications of Second Order Many-body Perturbation Theory. – In this section, a brief synopsis of some of the many applications of second order many-body perturbation theory in its Møller-Plesset form published during the reporting period, June 2001 to May 2003, is given. In my 2001 report, I provided a more detailed review of applications of second-order theory giving a snapshot of contemporary work at the turn of the century. Because of space limitations the present report is more restricted. Interrogation of the ISI database to determine the number of incidences of the string "MP2" in the *title* during the period June 2001 to May 2003 resulted in a list of 56 publications.[196-251] The list of titles of these publications, given in reverse chronological order, serves to convey the variety of applications being reported.

1. A gradient-corrected density functional and MP2 study of phenol-ammonia and phenol-ammonia$^+$ hydrogen-bonded complexes[196]
2. VB-MP2: A hybrid method combining valence bond theory and many-body perturbation theory[197]
3. A DFT and MP2 theoretical study on the structure and spectroscopy of HO_3, HO_3^+, HO_3^-[198]
4. On the convergence of the (ΔE-CCSD(T)-ΔE-MP2) term for complexes with multiple H-bonds[199]
5. Classifications of families of homologous organic compounds based on energy by means of ab initio HF, MP2 and DFT optimizations[200]
6. Toward true DNA base-stacking energies: MP2, CCSD(T), and complete basis set calculations[201]
7. Comparative G2(MP2) molecular orbital study of $B_3H_7XH_3$ and H_3BXH_3 donor-acceptor complexes (X = N, P, and As)[202]
8. Internal rotation of amino and nitro groups in TATB†: MP2 versus DFT (B3LYP)[203]
9. A study of the isomers of C_{36} fullerene using single and multireference MP2 perturbation theory[204]
10. Norrish I vs II reactions of butanal: a combined CASSCF, DFT and MP2 study[205]
11. An efficient parallel algorithm for the calculation of canonical MP2 energies[206]
12. Ab initio/GIAO-MP2-calculated. structures and ^{11}B-^{13}C NMR chemical shift relationship in hypercoordinate onium-carbonium dications and isoelectronic onium-boronium cations[207]
13. A comparative DFT-MP2 study of the Creutz-Taube ion and related systems[208]
14. MP2 and QCISD study of hydrogen transfer reaction path of the reaction HNCO with carbon-hydrogen radicals CH_x (x = 1 similar to 3)[209]
15. Modeling intermolecular interactions in gas chromatography using MP2 molecular orbital calculations[210]
16. Timings of SCF, OFT, and MP2 calculations on a Beowulf cluster[211]
17. MP2 for periodic systems: Theory and applications[212]

† 1,3,5-triamino-2,4,6 trinitrobenzene

18. How good is Koopmans' approximation? G2(MP2) study of the vertical and adiabatic ionization potentials of some small molecules[213]
19. Comment on 'Efficient calculation of canonical MP2 energies' [P. Pulay, S. Saebo, K. Wolinski, Chem. Phys. Lett. 344 (2001) 543–552][214]
20. Reply to the comments on 'Efficient calculation of canonical MP2 energies' by A. Kohn and C. Hattig[215]
21. Planarization of 1,3,5,7-cyclooctatetraene as a result of a partial rehybridization at carbon atoms: an MP2/6-31G* and B3LYP/6-311G** study[216]
22. SCC-TB, DFT/B3LYP, MP2, AM1, PM3 and RHF study of ethylene oxide and propylene oxide structures, VA and VCD spectra[217]
23. Development of an improved Stillinger-Weber potential for tetrahedral carbon using *ab initio* (Hartree-Fock and MP2) methods[218]
24. The interaction between cytosine tautomers and water: an MP2 and coupled cluster electron correlation study[219]
25. Structures of small silver cluster cations (Ag_n^+), $n < 12$): ion mobility measurements versus density functional and MP2 calculations[220]
26. Local MP2 study of naphthalene, indole, and 2,3-benzofuran dimers[221]
27. Density function studies of peptides — Part I. Vibrational frequencies including isotopic effects and NMR chemical shifts of N-methylacetamide, a peptide model, from density function and MP2 calculations[222]
28. Conformational study of the alanine dipeptide at the MP2 and DFT levels[223]
29. An MP2 study of linear polarizabilities and second-order hyperpolarizabilities for centrosymmetric squaraines[224]
30. G3(MP2) calculations of enthalpies of hydrogenation, isomerization, and formation of [3]-radialene and related compounds[225]
31. Efficient use of the correlation consistent basis sets in resolution of the identity MP2 calculations[226]
32. The relationship between binding models of TMA† with furan and imidazole and the molecular electrostatic potentials: DFT and MP2 computational studies[227]
33. Postulation of the mechanism of the selective synthesis of isotactic poly(methyl methacrylate) catalysed by[Zr(Cp)(Ind)CMe$_2$(Me)(thf)](BPh$_4$): A Hartree-Fock, MP2 and density functional study[228]
34. Noncovalent interaction or chemical bonding between alkaline earth cations and benzene? A quantum chemistry study using MP2 and density-functional theory methods[229]
35. RI-MP2 calculations with extended basis sets — a promising tool for study of H-bonded and stacked DNA base pairs[230]
36. On the use of MP2 theory for electron correlation in atoms and molecules[231]
37. Ab initio MP2 calculations of the products of acetylene addition to $HgCl_2$[232]
38. Comparative G2(MP2) molecular orbital study of [H$_3$AlX(CH$_3$)$_2$]$^-$ (X=N, P, and As) and H$_3$AlY(CH$_3$)$_2$ (Y=O, S, and Se) donor-acceptor complexes[233]
39. Augmentation of the LANL2DZ basis set under B3LYP and MP2 models to improve calculated results for p block elements.[234]
40. Efficient calculation of canonical MP2 energies[235]
41. G2(MP2) investigation of alane-[X(CH$_3$)$_3$]$^-$ (X=C, Si, and Ge) and alane-Y(CH$_3$)$_3$ (Y=N, P, and As) interactions[236]
42. Theoretical insight into the interactions of TMA-benzene and TMA-pyrrole with B3LYP density-functional theory (DFT) and *ab initio* second order Moller-Plesset perturbation theory (MP2) calculations[237]

† tetramethyl ammonium

43. Harmonic frequency scaling factors for Hartree-Fock, S-VWN, B-LYP, B3-LYP, B3-PW91 and MP2 with the Sadlej pVTZ electric property basis set[238]
44. Influence of stacking interactions on NMR chemical shielding tensors in benzene and formamide homodimers as studied by HF, DFT and MP2 calculations[239]
45. Structure and conformational flexibility of uracil: A comprehensive study of performance of the MP2, B3LYP and SCC-DFTB methods[240]
46. Fast linear scaling second-order Moller-Plesset perturbation theory (MP2) using local and density fitting approximations[241]
47. X-ray, MP2 and DFT studies of the structure and vibrational spectra of trigonellinium chloride[242]
48. Influence of F and CF_3 groups at the C=C double bond on the structure and properties of olefinic-series compounds in the gas phase: An *ab initio* MP2 study[243]
49. Comparative G2(MP2) molecular orbital study of X_3AlYH_3 (X=F, Cl, Br; Y=N, P, As) donor-acceptor complexes[244]
50. *Ab initio* MP2 and DFT study of the thermal Syn elimination reaction in ethyl formate[245]
51. The tautomeric equilibria of cytosine studied by NQR spectroscopy and HF, MP2 and DFT calculations[246]
52. Application of MP2 results in comparative studies of semiempirical ground-state energies of large atoms[247]
53. Calculation of packing structure of methanol solid using *ab initio* lattice energy at the MP2 level[248]
54. Gas-phase molecular structure of nicotinamide studied by electron diffraction combined with MP2 calculations[249]
55. Oligothiophene radical cations: Polaron structure in hybrid DFT and MP2 calculations[250]
56. DFT and MP2 vibrational spectra and assignments for gauche N-methyleneformamide $CH_2=N-CHO$[251]

5.4 Prospects. – Applications of low order many-body perturbation theory in its "MP2" form have continued during the period under review. Applications have been reported in a wide range of application areas from studies of small silver cluster cations to the study of H-bonded and stacked DNA base pairs, from MP2 calculations on Beowulf clusters to MP2 techniques for periodic systems. There can be little doubt that applications of low order many-body perturbation theory will continue in the years ahead and this approach will remain the basis for the most widely used *ab initio* method for the study of molecular structure.

6 Summary and Prospects

This report has continued our biennial survey of *"Many-body Perturbation Theory and Its Application to the Molecular Structure Problem"*. We have identified three areas in which significant progress has been made during the reporting period June 2001 to May 2003. The first of these is the *"diagrammatic many-body perturbation theory of molecular structure including nuclear and electronic motion"* in which nuclear and electronic motion are considered on an equal footing. We have emphasize that the ability to perform practical calculations which go beyond the Born-Oppenheimer approximation is important not only for accurate studies of small molecular species but also in studies of larger molecules, including biomolecules, for which the number of nuclear configurations which have to be considered in the usual Born-Oppenheimer-based approach can become very large indeed. The second area in which significant progress has been made was in the development of *"low order approximants"* which can significantly improve the quality of finite order many-body perturbation theory calculations with little additional computational effort. Given the tractability of low order perturbation studies there will undoubtedly be an effort to develop new approximants and further analyse known approximants in the years ahead. In our third area *"diagrammatic many-body perturbation theory of molecular electronic structure for larger systems"* researchers are continuing to extend the size of molecular systems to which such method can be routinely applied. Section 5 contains a brief synopsis of some of the many applications of second order many-body perturbation theory in its Møller-Plesset form published during the reporting period, June 2001 to May 2003.

We close by considering the prospects for practical studies of electron correlation effects in molecular systems using many-body perturbation theory the coming years. There is a growing importance of applications to large molecules, for example, molecules of biological interest and molecular systems in nanoscience and nanotechnology. Often van der Waals interactions and hydrogen-bonded systems are important in studies of such systems. The description of such effects within density functional theory, the most popular semi-empirical method, remains problematic. The combination of local correlation methods giving linear scaling, with the simultaneous description of both nuclear and electronic motions, so as to avoid the need to explore the potential energy hypersurface, will, when used in conjunction with low order summation approximants, which combine higher accuracy with computational tractability, significantly extend the horizon of *ab initio* molecular structure studies facilitating more accurate applications to larger molecules.

References

1. S. Wilson, in Specialist Periodical Reports: Chemical Modelling — Applications and Theory, volume 1, Senior Reporter: A. Hinchliffe, Royal Society of Chemistry, London, 2000.
2. S. Wilson, in Specialist Periodical Reports: Chemical Modelling — Applications and Theory, volume 2, Senior Reporter: A. Hinchliffe, Royal Society of Chemistry, London, 2002.
3. S. Wilson, in Specialist Periodical Reports: Theoretical Chemistry, volume 4, Senior Reporter: C. Thomson, Royal Society of Chemistry, London, 1981.
4. M. Born and J.R. Oppenheimer, *Ann. der Phys.*, 1927, **84**, 457.
5. S.M. Blinder, *English translation of* "M. Born and J.R. Oppenheimer, 1927, Ann. der Phys. 84, 457" (with emendations by B.T. Sutcliffe and W. Geppert) in

Handbook of Molecular Physics and Quantum Chemistry, volume 1: Fundamentals, edited by S. Wilson, P.F. Bernath and R. McWeeny, John Wiley & Sons, Chichester, 2003.
6. M. Born and K. Huang, "Dynamical Theory of Crystal Lattices", Oxford University Press, Oxford, 1955.
7. B.T. Sutcliffe, "Coordinate systems and transformations", in Handbook of Molecular Physics and Quantum Chemistry, volume 1: Fundamentals, edited by S. Wilson, P.F. Bernath and R. McWeeny, chapter 31, John Wiley & Sons, Chichester, 2003.
8. B.T. Sutcliffe, "Molecular Hamiltonians", in Handbook of Molecular Physics and Quantum Chemistry, volume 1: Fundamentals, edited by S. Wilson, P.F. Bernath and R. McWeeny, chapter 32, John Wiley & Sons, Chichester, 2003.
9. B.T. Sutcliffe, "Potential energy curves and surfaces", in Handbook of Molecular Physics and Quantum Chemistry, volume 1: Fundamentals, edited by S. Wilson, P.F. Bernath and R. McWeeny, chapter 34, John Wiley & Sons, Chichester, 2003.
10. B.T. Sutcliffe, "Molecular structure and bonding", in Handbook of Molecular Physics and Quantum Chemistry, volume 1: Fundamentals, edited by S. Wilson, P.F. Bernath and R. McWeeny, chapter 35, John Wiley & Sons, Chichester, 2003.
11. B.T. Sutcliffe, "Breakdown of the Born-Oppenheimer approximation", in Handbook of Molecular Physics and Quantum Chemistry, volume 1: Fundamentals, edited by S. Wilson, P.F. Bernath and R. McWeeny, chapter 36, John Wiley & Sons, Chichester, 2003.
12. R.G. Woolley and B.T. Sutcliffe, in "Fundamental World of Quantum Chemistry. A tribute to the memory of P.-O. Löwdin" edited by E.J. Brandas and E.S. Kryachko, Kluwer Academic Publishers, Dordrecht, 2003.
13. P.-O. Löwdin, *J. Molec. Struct. Theochem*, 1991, **230**, 13.
14. I.L. Thomas, *Phys. Rev.*, 1969, **185**, 90.
15. I.L. Thomas, *Chem. Phys. Lett.*, 1969, **1**, 705.
16. I.L. Thomas, *Phys. Rev. A*, 1970, **2**, 1200.
17. I.L. Thomas, *Phys. Rev. A*, 1970, **3**, 1200.
18. W. Kolos and L. Wolniewicz, *Rev. Mod. Phys.*, 1963, **35**, 473.
19. D.M. Bishop, *Molec. Phys.*, 1974, **28**, 1397.
20. D.M. Bishop and L.M. Cheung, *Phys. Rev. A*, 1977, **16**, 640.
21. B.A. Pettite, *Chem. Phys. Lett.*, 1986, **130**, 399.
22. H.J. Monkhorst, *Phys. Rev. A*, 1987, **36**, 1544.
23. H. Nagao, K. Kodama, Y. Shigeta, H. Kawabe, K. Nishikawa, M. Makano and K. Yamaguchi, *Int. J. Quantum Chem.*, 1996, **60**, 45.
24. Y. Shigeta, Y. Ozaki, K. Kodama, H. Nagao, H. Kawabe and K. Nishikawa, *Int. J. Quantum Chem.* 1998, **69**, 629.
25. Y. Shigeta, H. Takahashi, S. Yamanaka, M. Mitani, H. Nagao, K. Yamaguchi, *Int. J. Quantum Chem.*, 1998, **70**, 659.
26. Y. Shigeta, H. Nagao, K. Nishikawa and K. Yamaguchi, *J. Chem. Phys.*, 1999, **111**, 6171.
27. H. Nakai, *Int. J. Quantum Chem.*, 2002, **86**, 511.
28. M. Tachikawa, K. Mori, H. Nakai and K. Iguchi, *Chem. Phys. Lett.*, 1998, **290**, 437.
29. H. Nakai and K. Sodeyama, *J. Chem. Phys.*, 2003, **118**, 1119.
30. S. Wilson, "Perturbation theory", in Handbook of Molecular Physics and Quantum Chemistry, volume 2: Molecular electronic structure, edited by S.

Wilson, P.F. Bernath and R. McWeeny, chapter 8, John Wiley & Sons, Chichester, 2003.
31. B.T. Pickup, "Classical field theory and second quantization", in Handbook of Molecular Physics and Quantum Chemistry, volume 1: Fundamentals, edited by S. Wilson, P.F. Bernath and R. McWeeny, chapter 26, John Wiley & Sons, Chichester, 2003.
32. B.T. Pickup, "The occupation number representation and the many-body problem", in Handbook of Molecular Physics and Quantum Chemistry, volume 1: Fundamentals, edited by S. Wilson, P.F. Bernath and R. McWeeny, chapter 27, John Wiley & Sons, Chichester, 2003.
33. B.T. Pickup, "Second quantization and Lie algebra", in Handbook of Molecular Physics and Quantum Chemistry, volume 1: Fundamentals, edited by S. Wilson, P.F. Bernath and R. McWeeny, chapter 28, John Wiley & Sons, Chichester, 2003.
34. J.A Karwowski, "Spectral density distribution moments", in Handbook of Molecular Physics and Quantum Chemistry, volume 1: Fundamentals, edited by S. Wilson, P.F. Bernath and R. McWeeny, chapter 29, John Wiley & Sons, Chichester, 2003.
35. S. Wilson, 1980, Lecture Notes, unpublished.
36. P.W. Atkins, 1972, Lecture Notes, unpublished.
37. D.Z. Goodson, *Int. J. Quantum Chem.*, 2003, **92**, 35.
38. H. Padé, *Ann. sci. Ecole norm. sup. Paris (Suppl.)*, 1892, **9**, 3.
39. D.Z. Goodson, *J. Chem. Phys.*, 2000, **112**, 4901.
40. D.Z. Goodson, *J. Chem. Phys.*, 2000, **113**, 6461.
41. D.Z. Goodson, *J. Chem. Phys.*, 2002, **116**, 6948.
42. E. Feenberg, *Phys. Rev.*, 1956, **103**, 1116.
43. E. Feenberg, *Ann. Phys. N.Y.*, 1958, **3**, 292.
44. A.T. Amos, *J. Chem. Phys.*, 1970, **52**, 603.
45. D.T. Tuan, *Chem. Phys.*, 1970, **7**, 115.
46. A.T. Amos, *Int. J. Quant. Chem.*, 1972, **6**, 125.
47. A.T. Amos, *J. Phys. B: At. Mol. Phys.*, 1978, **11**, 2053.
48. Ch. Schmidt, M. Warken and N.C. Handy, *Chem. Phys. Lett.*, 1993, **221**, 272.
49. S. Wilson, *J. Phys. B:At. Mol. Phys.*, 1979, **12**, L657.
50. S. Wilson, *Int. J. Quant. Chem*, 1980, **18**, 905.
51. S. Wilson and M.F. Guest, *J. Phys. B: At. Mol. Phys.*, 1981, **14**, 1709.
52. S. Wilson, *Theor. chim. Acta*, 1981, **59**, 71.
53. S. Wilson, D.M. Silver and R.A. Farrell, *Proc. Roy. Soc. London A*, 1977, **356**, 363.
54. S. Wilson, *J. Phys. B:At. Mol. Phys.*, 1979, **12**, 1623.
55. M. Cohen and Feldman, *J. Phys. B: At. Mol. Phys.*, 1981, **14**, 2535.
56. G.F. Carrier, M. Krook and C.E. Pearson, *Functions of a Complex Variable*, McGraw-Hill, New York, 1966.
57. R.E. Shafer, *SIAM (Soc. Ind. Appl. Math.) J. Math. Anal.*, 1975, **11**, 447.
58. G.A. Baker, Jr., and P. Graves-Morris, *Padé Approximants*, Cambridge University Press, Cambridge, 1996.
59. A.V. Sereyev, Zh. Vuchisl. Mat. Mat. Fiz., 1986, **26**, 348 [USSR Comput. Math. Math. Phys. **26**, 17.
60. A.V. Sergeev and D.Z. Goodson, *J. Phys. A: Math. Gen.*, 1998, **31**, 4301.
61. F.M. Fernández and C.G. Diaz, *Eur. Phys. J. D*, 2001, **15**, 41.
62. V.M. Vaĭnberg, V.D. Mur, V.S. Popov and A.V. Sergeev, *Pis'ma Zh. Eksp. Teor. Fiz.*, 1996, **44**, 9 [1986, JETP Lett. **44**, 9].

63. M. López-Cabrera, D.Z. Goodson, D.R. Herschbach and J.D. Morgan III, *Phys. Rev. Lett.*, 1992, **68**, 1992.
64. T.C. Germann and S. Kais, *J. Chem. Phys.*, 1993, **99**, 7739.
65. A.V. Sergeev, *J. Phys. A: Math. Gen.*, 1995, **28**, 4157.
66. D.Z. Goodson and A.V. Sergeev, *J. Chem. Phys.*, 1999, **110**, 8205.
67. F.M. Fernández and R.H. Tipping, *J. Molec. Struct. THEOCHEM* 1999, **488**, 157.
68. J.R. Walkup, M. Dunn and D.K. Watson, *J. Math. Phys.*, 2000, **41**, 218.
69. C.G. Diaz and F.M. Fernández, *J. Molec. Struct. THEOCHEM*, 2001, **541**, 29.
70. M. Schutz, *J. Chem. Phys.*, 2000, **113**, 9986.
71. C. Hampel and H.-J. Werner, *J. Chem. Phys.*, 1996, **104**, 6286.
72. O. Sinanoglu, *Adv. Chem. Phys.*, 1964, **6**, 315.
73. R. K. Nesbet, *Adv. Chem. Phys.*, 1965, **9**, 321.
74. R. Ahlrichs and W. Kutzelnigg, *J. Chem. Phys.*, 1968, **48**, 1819.
75. R. Ahlrichs and W. Kutzelnigg, *Theor. Chim. Acta.*, 1968, **10**, 377.
76. S. Diner, J. P. Malrieu and P. Calverie, *Theor. Chim. Acta*, 1969, **13**, 18.
77. J.M. Cullen and M.C. Zerner, *Int. J. Quantum Chem. Symp.*, 1975, **9**, 343.
78. G. Stollhoff and P. Fulde, *J. Chem. Phys.*, 1980, **73**, 4548.
79. B. Kirtman and C. deMeilo, *J. Chem. Phys.*, 1981, **75**, 4592.
80. J.M. Cullen and M.C. Zerner, *Theor. Chim. Acta*, 1982, **61**, 203.
81. J.M. Cullen and M.C. Zerner, *J. Chem. Phys.*, 1982, **77**, 4088.
82. B. Kirtman, *J. Chem. Phys.*, 1982, **86**, 1059.
83. P. Otto and J. Ladik, *Int. J. Quantum Chem.*, 1982, **22**, 169.
84. B. Kiel, G. Stollhoff, C. Weigel, P. Fulde and H. Stoll, *Z. Phys. B*, 1982, **46**, 1.
85. W.D. Laidig, G.D. Purvis III and R.J. Bartlett, *Int. J. Quantum Chem. Symp.*, 1982, **6**, 561.
86. P. Pulay, *Chem. Phys. Lett.*, 1983, **100**, 151.
87. W.D. Laidig, G.D. Purvis III and R.J. Bartlett, *Chem. Phys. Lett.*, 1983, **97**, 209.
88. E. Kapuy, Z. Csepes, and C. Kozmutza, *Int. J. Quantum Chem.*, 1983, **23**, 981.
89. S. Saebø and P. Pulay, *Chem. Phys. Lett.*, 1985, **113**, 13.
90. M. Takahashi and J. Paldus, *Phys. Rev. B*, 1985, **31**, 5121.
91. M. Takahashi and J. Paldus, *Int. J. Quantum Chem.*, 1985, **28**, 459.
92. W D. Laidig, G. D. Purvis III and R. J. Bartlett, *J. Phys. Chem.*, 1985, **89**, 2161.
93. W. Förner, J. Ladik, P. Otto and J. Čížek, *Chem. Phys.*, 1985, **97**, 251.
94. P. Pulay and S. Saebo, *Theor. Chim. Acta*, 1986, **69**, 357.
95. J. Pipek and J. Ladik, *Chem. Phys.* 1986, **102**, 445.
96. W. Meyer and L. Frommhold, *Phys. Rev. A* 1986, **33**, 3807.
97. B. Kirtman and C. E. Dykstra, *J. Chem. Phys.* 1986, **85**, 2791.
98. G. Stollhoff and P. Vasilopoulos, *J. Chem. Phys.*, 1986, **84**, 2744.
99. S. Saebø and P. Pulay, *J. Chem. Phys.*, 1987, **86**, 914.
100. E. Kapuy, F. Bartha, F. Bogar and C. Kozmutza, *Theor. Chim. Acta*, 1987, **72**, 337.
101. S. Saebø and P. Pulay, *J. Chem. Phys.*, 1988, **88**, 1884.
102. E. Kapuy, F. Bartha, C. Kozmutza, and F. Bogar, *J. Mol. Struct.*, 1988, **47**,59.
103. G. Stollhoff and K. P. Bohnen, *Phys. Rev. B*, 1988, 37,4678.
104. J. Pipek and P. G. Mezey, *J. Chem. Phys.*, 1989, **90**, 4916.
105. M. V. Ganduglia-Pirovano, G. Stollhoff, P. Fulde and K. P. Bohnen, *Phys. Rev. B*, 1989, **39**, 5156.
106. E. Kapuy, F. Bartha, F. Bogar, F. Csepes and C. Kozmutza, *Int. J. Quantum Chem.*, 1990, **38**, 139.

107. G. Konig and G. Stollhoff, *J. Phys. Chem.*, 1990, **91**, 2993.
108. G. Konig and G. Stollhoff, *Phys. Rev. Lett.*, 1990, **63**, 1239.
109. C. Kozmutza and E. Kapuy, *Int. J. Quantum. Chem.*, 1990, **38**, 665.
110. E. Kapuy, F. Bogar, F. Bartha and C. Kozmutza, *J. Mol. Struct.*, 1991, **79**, 61.
111. E. Kapuy and C. Kozmutza, *J. Chem. Phys.* 1991, **94**, 5565.
112. C. Kozmutza and E. Kapuy, *J. Comput. Chem.*, 1991, **12**, 953.
113. S. Saebø and P. Pulay, *Annu. Rev. Phys. Chem.*, 1993, **44**, 213.
114. R.J. Harrison and R.J. Shepard, *Annu. Rev. Phys. Chem.*, 1994, **45**, 623.
115. E. Kapuy and C. Kozmutza, *Chem. Phys. Lett.*, 1994, **226**, 484.
116. G. Rauhut, J.W. Boughton and P. Pulay, *J. Chem. Phys.*, 1995, **103**, 5662.
117. B. Kirtman, *Int. J. Quantum Chem.*, 1995, **55**, 103.
118. D.E. Bernholdt and R.J. Harrison, *J. Chem. Phys.*, 1995, **102**, 9582.
119. B. Kirtman, J.L. Toto, K.A. Robins and M. Hasan, *J. Chem. Phys.*, 1995, **102**, 5350.
120. D.B. Chesnut and E.F.C. Byrd, *J. Comput. Chem.*, 1996, **17**, 1431.
121. K. Raghavachari, J.B. Anderson, *J. Phys. Chem.*, 1996, **100**, 12960.
122. M. Head-Gordon, *J. Phys. Chem.*, 1996, **100**, 13213.
123. G. Stollhoff, *J. Chem. Phys.*, 1996, **105**, 227.
124. H. Stoll, *Ann. Phys.-Leip.*, 1996, **5**, 355.
125. J.D. Head and S.J. Silva, *J. Chem. Phys.* 1996, **104**, 3244.
126. C. Kozmutza, E. Kapuy, E.M. Evleth and J. Pipek, *Int. J. Quantum Chem.*, 1996, **57**, 775.
127. G. Rauhut and P. Pulay, *Chem. Phys. Lett.*, 1996, **248**, 223.
128. M.S. Lee and M. Head-Gordon, *J. Chem. Phys.*, 1997, **107**, 9085.
129. A.K. Wilson and J. Almlöf, *Theor. Chim. Acta*, 1997, **95**, 49.
130. G.A. DiLabio and J.S. Wright, *Chem. Phys. Lett.*, 1998, **297**, 181.
131. P.E. Maslen and M. Head-Gordon, *J. Chem. Phys.*, 1998, **109**, 7093.
132. A. ElAzhary, G. Rauhut, P. Pulay and H.-J. Werner, *J. Chem. Phys.*, 1998, **108**, 5185.
133. G. Rauhut, P. Pulay and H.-J. Werner, *J. Comput. Chem.*, 1998, **19**, 1241.
134. M. Schutz, , G. Rauhut, H.-J. Werner, *J. Phys. Chem. A*, 1998, **102**, 5997.
135. G. Hetzer, P. Pulay and H.-J. Werner, *Chem. Phys. Lett.*, 1998, **290**, 143.
136. P.E. Maslen and M. Head-Gordon, *Chem. Phys. Lett.*, 1998, **283**, 102.
137. G.A. DiLabio, *J. Phys. Chem. A*, 1999, **103**, 11414.
138. M. Nooijen, *J. Chem. Phys.*, 1999, **111**, 10815.
139. G.E. Scuseria and P.Y. Ayala, *J. Chem. Phys.*, 1999, **111**, 8330.
140. M. Schütz, G. Hetzer and H.-J. Werner, *J. Chem. Phys.*, 1999, **111**, 5691.
141. M. Schütz, R. Lindh and H.-J. Werner, *Molec. Phys.*, 1999, **96**, 719.
142. P.Y. Ayala, G.E. Scuseria and A. Savin, *Chem. Phys. Lett.*, 1999, **307**, 227.
143. G.E. Scuseria, *J. Phys. Chem. A*, 1999, **103**, 4782.
144. P.Y. Ayala and G.E. Scuseria, *J. Chem. Phys.*, 1999, **110**, 3660.
145. R.A. Friesner, R.B. Murphy, M.D. Beachy, M.N. Ringnalda, W.T. Pollard, B.D. Dunietz and Y. Cao, *J. Phys. Chem. A*, 1999, **103**, 1913.
146. G. Hetzer, M. Schütz, H. Stoll and H.-J. Werner, *J. Chem. Phys.*, 2000, **113**, 9443.
147. P. Constans, Ayala and G.E. Scuseria, *J. Chem. Phys.*, 2000, **113**, 10451.
148. P.Y. Ayala and G.E. Scuseria, *J. Comput. Chem.*, 2000, **21**, 1524.
149. J. Kong, C.A. White, A.I. Krylov, D. Sherrill, R.D. Adamson, T.R. Furlani, M.S. Lee, A.M. Lee, S.R. Gwaltney, T.R. Adams, C. Ochsenfeld, A.T.B. Gilbert, G.S. Kedziora, V.A. Rassolov, D.R. Maurice, N. Nair, Y.H. Shao, N.A. Besley, P.E. Maslen, J.P. Dombroski, H. Daschel, W.M. Zhang, P.P. Korambath, J. Baker,

E.F.C. Byrd, T. Van Voorhis, M. Oumi, S. Hirata, C.P. Hsu, N. Ishikawa, J. Florian, A. Warshel, B.G. Johnson, P.M.W. Gill, M. Head-Gordon, J.A. Pople, *J. Comput. Chem.*, 2000, **21**, 1532.
150. T.H. Dunning, Jr., *J. Phys. Chem. A*, 2000, **104**, 9062.
151. S.S. Iyengar, G.E. Scuseria, A. Savin, *Int. J. Quantum Chem.*, 2000, **79**, 222.
152. P.Y. Ayala and G.E. Scuseria, *Chem. Phys. Lett.*, 2000, **322**, 213.
153. M. Schutz, H.-J. Werner, *Chem. Phys. Lett.*, 2000, **318**, 370.
154. M.S. Lee, P.E. Maslen and M. Head-Gordon, *J. Chem. Phys.*, 2000, **112**, 3592.
155. P.Y. Ayala, K.N. Kudin, G.E. Scuseria, *J. Chem. Phys.*, 2001, **115**, 9698.
156. S. Saebø and P. Pulay, *J. Chem. Phys.*, 2001, **115**, 3975.
157. P. Pulay, S. Saebø and K. Wolinski, *Chem. Phys. Lett.* 2001, **344**, 543.
158. T. Head-Gordon and J.C. Wooley, *IBM Syst. J.*, 2001, **40**, 265.
159. R.A. Friesner and B.D. Dunietz, *Accounts Chem. Res.* 2001, **34**, 351.
160. M. Schütz and H.-J. Werner, *J. Chem. Phys.*, 2001, **114**, 661.
161. T.D. Crawford and R.A. King, *Chem. Phys. Lett.*, 2002, **366**, 611.
162. M. Schütz, *Phys. Chem. Chem. Phys.*, 2002, **4**, 3941.
163. G.J.O. Beran, S.R. Gwaltney and M. Head-Gordon, *J. Chem. Phys.*, 2002, **117**, 3040.
164. P. Pulay, J. Baker and K. Wolinski, *Chem. Phys. Lett.*, 2002, **358**, 354.
165. M. Schütz, *J. Chem. Phys.*, 2002, **116**, 8772.
166. S.H. Li, J. Ma and Y.S. Jiang, *J. Comput. Chem.*, 2002, **23**, 237.
167. M. Schütz and F.R. Manby, *Phys. Chem. Chem. Phys.*, 2003, **5**, 3349.
168. N. Flocke and R.J. Bartlett, *Chem. Phys. Lett.*, 2003, **367**, 80.
169. H.-J. Werner, F.R. Manby and P.J. Knowles, *J. Chem. Phys.*, 2003, **118**, 8149.
170. P.P Ewald, *Ann. Phys. (Paris)*, 1921, **64**, 253.
171. P.M.W. Gill, *Chem. Phys. Lett.*, 1997, **270**, 193.
172. J.P. Dombroski, S.W. Taylor, and P.M.W. Gill, *J. Phys. Chem.*, 1996, **100**, 6272.
173. R.D. Adamson, J.P. Dombroski, and P.M.W. Gill, *Chem. Phys. Lett.*, 1996, **254**, 329.
174. I. Panas, *Chem. Phys. Lett.*, 1995, **245**, 171.
175. H. Stoll and A. Savin, in *Density Functional Methods in Physics*, edited by R.M. Dreizler, p. 177, Plenum, New York, 1985.
176. T. Leininger, H. Stoll, H.-J. Werner and A. Savin, *Chem. Phys. Lett.*, 1997, **275**, 151.
177. I. Panas, *Mol. Phys.*, 1996, **89**, 239.
178. I. Panas and A. Snis, *Theor. Chim. Acta*, 1997, **97**, 232.
179. P.M.W. Gill, R.D. Adamson and J.A. Pople, *Mol. Phys.*, 1996, **88**, 1005.
180. P.M.W. Gill and R.D. Adamson, *Chem. Phys. Lett.*, 1997, **261**, 105.
181. A.M. Lee, S.W. Taylor, J.P. Dombroski, and P.M.W. Gill, *Phys. Rev. A*, 1997, **55**, 3233.
182. D.M. Schrader and S. Prager, *J. Chem. Phys.*, 1962, **37**, 3927.
183. J.L. Whitten, *J. Chem. Phys.*, 1973, **58**, 4496.
184. B.I. Dunlap, J.W.D. Connolly and J.R. Sabin, *Int. J. Quantum Chem. Symp.*, 1977, **11**, 81.
185. B.I. Dunlap, J.W.D. Connolly and J.R. Sabin, *J. Chem. Phys.*, 1979, **71**, 4993.
186. O. Vahtras, J. Almlof and M.W. Feyereisen, *Chem. Phys. Lett.*, 1993, **213**, 514.
187. M.W. Feyereisen, G. Fitzgerald and A. Komornicki, *Chem. Phys. Lett.*, 1993, **208**, 359.
188. A.P. Rendell and T.J. Lee, *J. Chem. Phys.*, 1994, **101**, 400.
189. S. Ten-no and S. Iwata, *Chem. Phys. Lett.*, 1995, **240**, 578.

190. B.I. Dunlap, *Phys. Chem. Chem. Phys.*, 2000, **2**, 2113.
191. F. Weigend, *Phys. Chem. Chem. Phys.*, 2002, **4**, 4285.
192. F.R. Manby, *J. Chem. Phys.*, 2003, **119**, 4607.
193. W. Klopper and W. Kutzelnigg, *Chem. Phys. Lett.*, 1987, **134**, 17; W. Kutzelnigg and W. Klopper, *J. Chem. Phys.*, 1991, **94**, 1985.
194. Institute for Scientific Information, Philadelphia, PA, USA.
195. M. Schütz and F.R. Manby, *Phys. Chem. Chem. Phys.*, 2003, **5**, 3349.
196. L. Pejov, *Chem. Phys.*, 2002, **285**, 177.
197. L.C.Song, L.Y. Lu, W. Wu, Z.X. Cao, Q.E. Zhang, *Chem. J. Chin. Univ.-Chin.*, 2002, **23**, 2133.
198. K.M. Pei, X.Y. Zhang, X.L. Kong, H.Y. Li, *Chin. J. Chem. Phys.*, 2002, **15**, 263.
199. P. Jurecka, P. Hobza, *Chem. Phys. Lett.*, 2002, **365**, 89.
200. A. Neugebauer and G. Hafelinger, *J. Phys. Org. Chem.*, 2002, **15**, 677.
201. P. Hobza and J. Sponer, *J. Am. Chem. Soc.*, 2002, **124**, 11802.
202. A. Es-sofi, C. Serrar, A. Ouassas, A. Jarid, A. Boutalib, I. Nebot-Gil and F. Tomas, *J. Phys. Chem. A*, 2002, **106**, 9065.
203. M.R. Manaa, R.H. Gee, and L.E. Fried, *J. Phys. Chem. A*, 2002, **106**, 8806.
204. S.A. Varganov, P.V. Avramov, S.G. Ovchinnikov, and M.S. Gordon, *Chem. Phys. Lett.*, 2002, **362**, 380.
205. X.B. Chen and W.H. Fang, *Chem. Phys. Lett.*, 2002, **361**, 473.
206. J. Baker, and P. Pulay, *J. Comput. Chem.*, 2002, **23**, 1150.
207. G. Rasul, G.K.S. Prakash, and G.A. Olah, *Proc. Natl. Acad. Sci. U. S. A.*, 2002, **99**, 9635.
208. J. Hardesty, S.K. Goh and D.S. Marynick, *Theochem-J. Mol. Struct.*, 2002, **588**, 223.
209. Y.Q. Ji, W.L. Feng, Z.F. Xu, M. Lei and M.R. Hao, *Acta Chim. Sin.*, 2002, **60**, 1167.
210. N.M. Trease, J.E. Struble and H.L.S. Holmes, *Abstr. Pap. Am. Chem. Soc.*, 2002, **223**, 101-ANYL.
211. M.B. Moore, D.H. Magers and M. Wiggins, *Abstr. Pap. Am. Chem. Soc.*, 2002, **223**, 183-COMP.
212. G.E. Scuseria, *Abstr. Pap. Am. Chem. Soc.*, 2002, **223**, 219-COMP.
213. Z.B. Maksic and R. Vianello, *J. Phys. Chem. A*, 2002, **106**, 6515.
214. A. Kohn and C. Hattig, *Chem. Phys. Lett.*, 2002, **358**, 350.
215. P. Pulay, J. Baker and K. Wolinski, *Chem. Phys. Lett.*, 2002, **358**, 354.
216. T.M. Krygowski, E. Pindelska, M.K. Cyranski and G. Hafelinger, G, *Chem. Phys. Lett.*, 2002, **359**, 158.
217. K. Frimand and K.J. Jalkanen, *Chem. Phys.*, 2002, **279**, 161.
218. A.S. Barnard and S.P. Russo, *Mol. Phys.*, 2002, **100**, 1517.
219. G. Fogarasi, and M.G. Szalay, *Chem. Phys. Lett.*, 2002, **356**, 383.
220. P. Weis, T. Bierweiler, S. Gilb and M.M. Kappes, *Chem. Phys. Lett.*, 2002, **355**, 355.
221. S. Fomine, M. Tlenkopatchev, S. Martinez and L. Fomina, *J. Phys. Chem. A*, 2002, **106**, 3941.
222. G. Cuevas, V. Renugopalakrishnan, G. Madrid and A.T. Hagler, *Phys. Chem. Chem. Phys.*, 2002, **4**, 1490.
223. R. Vargas, J. Garza, B.P. Hay and D.A. Dixon, *J. Phys. Chem. A*, 2002, **106**, 3213.
224. M.L. Yang, S.H. Li, J. Ma and Y.S. Jiang, *Chem. Phys. Lett.*, 2002, **354**, 316.
225. D.W. Rogers and F.J. McLafferty, *J. Phys. Chem. A*, 2002, **106**, 1054.
226. F. Weigend, A. Kohn and C. Hattig, *J. Chem. Phys*, 2002, **116**, 3175.

227. T. Liu, J.D. Gu, X.J. Tan, W.L. Zhu, X.M. Luo, H.L. Jiang, R.Y. Ji, K.X. Chen, I. Silman and J.L. Sussman, *J. Phys. Chem. A*, 2002, **106**, 157.
228. M. Holscher, H. Keul and H. Hocker, *Chem.-Eur. J.*, 2001, **7**, 5419.
229. X.J. Tan, W.L. Zhu, M. Cui, X.M. Luo, J.D. Gu, I. Silman, J.L. Sussman, H.L. Jiang, R.Y. Ji and K.X. Chen, *Chem. Phys. Lett.*, 2001, **349**, 113.
230. P. Jurecka, P. Nachtigall and P. Hobza, *Phys. Chem. Chem. Phys.*, 2001, **3**, 4578.
231. S. Wilson and I. Hubač, *Molec. Phys.*, 2001, **99**, 1813.
232. Y.A. Borisov and A.S. Peregudov, *Russ. Chem. Bull.*, 2001, **50**, 958.
233. A. Jarid, A. Boutalib, I. Nebot-Gil and F. Tomas, *Theochem-J. Mol. Struct.*, 2001, **572**, 161.
234. C.E. Check, T.M. Gilbert, L.S. Sunderlin, T.O. Faust and J.M. Bailey, *Abstr. Pap. Am. Chem. Soc.*, 2001, **222**, 203-COMP.
235. P. Pulay, S. Saebo and K. Wolinski, *Chem. Phys. Lett.*, 2001, **344**, 543.
236. A. Boutalib, A. Jarid, I. Nebot-Gil and F. Tomas, *J. Phys. Chem. A*, 2001, **105**, 6526.
237. T. Liu, J.D. Gu, X.J. Tan, W.L. Zhu, X.M. Luo, H.L. Jiang, R.Y. Ji, K.X. Chen, I. Silman and J.L. Sussman, *J. Phys. Chem. A*, 2001, **105**, 5431.
238. M.D. Halls, J. Velkovski and H.B. Schlegel, *Theor. Chem. Acc.*, 2001, **105**, 413.
239. J. Czernek, *J. Phys. Chem. A*, 2001, **107**, 3952.
240. O.V. Shishkin, L. Gorb, A.V. Luzanov, M. Elstner, S. Suhai and J. Leszczynski, *Theochem-J. Mol. Struct.*, 2001, **625**, 295.
241. H.-J.Werner, F.R. Manby and P.J. Knowles, *J. Chem. Phys.*, 2003, **118**, 8149.
242. M. Szafran, J. Koput, Z. Dega-Szafran, A. Katrusiak, M. Pankowski and K. Stobiecka, *Chem. Phys.*, 2003, **2**, 20189.
243. Y.A. Borisov, *J. Struct. Chem.*, 2002, **43**, 734.
244. A. Boutalib, *Theochem-J. Mol. Struct.*, 2003, **623**, 121.
245. J.M. Hermida-Ramon, J. Rodriguez-Otero and E.M. Cabaleiro-Lago, *J. Phys. Chem. A*, 2003, **107**, 1651.
246. J.N. Latosinska, M. Latosinska and J. Koput, *J. Mol. Struct.*, 2003, **648**, 9.
247. J.R. Flores, K. Jankowski and R. Slupski, *Collect. Czech. Chem. Commun.*, 2003, **68**, 240.
248. K. Nagayoshi, K. Kitaura, S. Koseki, S.Y. Re, K. Kobayashi, Y.K. Choe and S. Nagase, *Chem. Phys. Lett.*, 2003, **369**, 597.
249. T. Takeshima, H. Takeuchi, T. Egawa and S. Konaka, *J. Mol. Struct.*, 2003, **644**, 197.
250. V.M. Geskin, A. Dkhissi and J.L. Bredas, *Int. J. Quantum Chem.*, 2003, **91**, 350.
251. H.M. Badawi, *Theochem-J. Mol. Struct.*, 2002, **617**, 9.